Green Energy and Technology

Lingfeng Wang, Chanan Singh,
and Andrew Kusiak (Eds.)

Wind Power Systems

Applications of Computational Intelligence

With 256 Figures and 63 Tables

Springer

Editors

Dr. Lingfeng Wang
Department of Electrical Engineering
and Computer Science
University of Toledo
Toledo, OH 43606
USA
E-mail: l.f.wang@ieee.org

Dr. Andrew Kusiak
Mechanical and Industrial
Engineering Department
University of Iowa
3131 Seamans Center
Iowa City, IA 52242
USA
E-mail: andrew-kusiak@uiowa.edu

Dr. Chanan Singh
Electrical and Computer
Engineering Department
Texas A&M University
College Station, TX 77843-3128
USA
E-mail: singh@ece.tamu.edu

ISBN 978-3-642-13249-0 e-ISBN 978-3-642-13250-6

Springer Series in Green Energy and Technology ISSN 1865-3529

Library of Congress Control Number: 2010927161

Cover design: WMXDesign, Heidelberg

Printed on acid-free paper

9 8 7 6 5 4 3 2 1

springer.com

Preface

Renewable energy such as wind power has attracted much attention due to its several merits such as environmental friendliness and enhancement of nation's energy security. In recent years, large capacity of wind power is being integrated with conventional power grids. Therefore, it is necessary to address various challenging issues related to wind power systems, which are significantly different from the traditional generation systems. This book is intended as a resource for engineers, practitioners, and decision-makers interested in studying or using the power of computational intelligence based algorithms in handling various important problems in wind power systems at the levels of power generation, transmission, and distribution. This edited book includes the state-of-the-art studies on applications of computational intelligence, including evolutionary computation, neural networks, fuzzy logic, hybrid algorithms, multi-agent reinforcement learning, and several other approaches, to wind power systems. Chapters of original research on computational intelligence applications are included in various research areas including wind turbine control, wind turbine diagnosis, wind farm design, economic dispatch, conductor sizing, reliability analysis, power loss minimization, frequency regulation, and so forth.

In Chapter 1, P. Chen, P. Siano, Z. Chen, et al. present a hybrid optimization method that minimizes the annual system power losses. The method combines the Genetic Algorithm (GA), gradient-based constrained nonlinear optimization algorithm and sequential Monte Carlo simulation (MCS).

In Chapter 2, H. Falaghi and C. Singh propose a probabilistic approach for conductor sizing in electric power distribution systems accounting for wind power generators. The probabilistic evaluation of a solution related to the behavior of the wind power generators is embedded in a GA engine for the search of the optimal conductor planning solutions.

In Chapter 3, A. G. Gonzalez-Rodriguez, J. Serrano-Gonzalez, J. Castro-Mora, et al. discuss global optimization of wind farms based on GA. The proposed method combines a model of wind farm costs based on the life cycle of the facility and a method for searching optimal turbine location and wind farm configuration using GA.

In Chapter 4, M. Ramezani, H. Falaghi, and C. Singh propose three different methods for Capacity Benefit Margin (CBM) evaluation considering wind turbine generator which reflect different objectives. CBM determination is formulated as an optimization problem and Particle Swarm Optimization (PSO) method is used to solve the problem.

In Chapter 5, A. T. Al-Awami and M. A. El-Sharkawi present stochastic dispatch for a power system with both thermal and wind units. Multi-Objective Particle Swarm Optimization (MO-PSO) is employed to obtain the Pareto-front.

In Chapter 6, L. A. Osadciw, Y. Yan, X. Ye, et al. propose an inverse transformation based change detector, called Inverse Diagnostic Curve Detector (IDCD), to track the variation of power curve over time for diagnostics. IDCD is adaptable to different wind turbine types. The dynamic fitting is optimized by a PSO algorithm.

In Chapter 7, L. Yang, G. Y. Yang, Z. Xu, et al. propose the multi-objective optimal controller design of a Doubly Fed Induction Generator (DFIG) based wind turbine system using Differential Evolution (DE).

In Chapter 8, Y. Mishra, S. Mishra, F. Li, et al. discuss various operation modes of the DFIG-based wind farm system. The coordinated tuning of the damping controller to enhance the damping of the oscillatory modes using Bacteria Foraging (BF) technique is presented.

In Chapter 9, H. Bevrani and A. G. Tikdari propose an Artificial Neural Network (ANN) based power system emergency control scheme in the presence of high wind power penetration.

In Chapter 10, B. Singh, S. N. Singh, and E. Kyriakides present efficient control of power-electronic systems used in DFIG-based wind power generation. The conventional proportional-integral (PI) controller is replaced with a nonlinear Adaptive Neuro-Fuzzy Inference System (ANFIS) based controller.

In Chapter 11, H.-S. Ko, K. Y. Lee, and H.-C. Kim present the design of intelligent controllers for a wind-diesel power system equipped with a wind turbine driving an induction generator. The concepts of fuzzy-robust controller and fuzzy-neural hybrid controller are applied to design integrated non-linear controllers to provide control input for excitation system and governor system simultaneously.

In Chapter 12, V. Calderaro, C. Cecati, A. Piccolo, et al. design a sensorless peak power tracking control for maximum wind energy extraction and a voltage control allowing compensation of voltage variations at the wind turbine connection point are proposed. Both the controllers are based on fuzzy logic.

In Chapter 13, S. Mishra, Y. Mishra, F. Li, et al. propose the Tagaki-Sugino (TS) fuzzy controller for the DFIG-based wind generator. The conventional PI control loops for maintaining desired active power and DC capacitor voltage is compared with the TS fuzzy controllers.

In Chapter 14, P. J. Costa, A. S. Carvalho, A. J. Martins, et al. analyze wind power systems and propose a methodology to design fuzzy controllers in order to optimize turbine operation and farm operation.

In Chapter 15, H. Bevrani, F. Daneshfar, and R. Daneshmand present an overview of the key issues on frequency regulation concerning the integration of wind power units into the power systems. An intelligent agent based Load Frequency Control (LFC) using Multi-Agent Reinforcement Learning (MARL) is proposed.

We hope that this edited book has included a bunch of representative applications of computational intelligence techniques in the field of wind power systems. From the collection of these most recent studies, the readers are expected to find some chapters inspiring and useful to their own research. Undoubtedly, in the coming years, there will be increasing complexity and uncertainty in power systems due to the higher degree of wind power penetration. We believe that computational intelligence based methods will be more widely used and play a more important role in this research domain for dealing with various challenging and open-ended problems.

The editors would like to thank all the authors who have contributed their valuable work to this edited book. We are also grateful to all the reviewers who have generously devoted their time to reviewing the manuscripts in their tight schedules. Thanks also go to the Springer staffs who have interacted with us and provided continuous help throughout the production process of this book.

March 2010

L. Wang, Toledo, Ohio
C. Singh, College Station, Texas
A. Kusiak, Iowa City, Iowa

Contents

Optimal Allocation of Power-Electronic Interfaced Wind Turbines Using a Genetic Algorithm – Monte Carlo Hybrid Optimization Method

Peiyuan Chen, Pierluigi Siano, Zhe Chen, and Birgitte Bak-Jensen

Abstract. The increasing amount of wind power integrated to power systems presents a number of challenges to the system operation. One issue related to wind power integration concerns the location and capacities of the wind turbines (WTs) in the network. Although the location of wind turbines is mainly determined by the wind resource and geographic conditions, the location of wind turbines in a power system network may significantly affect the distribution of power flow, power losses, etc. Furthermore, modern WTs with power-electronic interface have the capability of controlling reactive power output, which can enhance the power system security and improve the system steady-state performance by reducing network losses. This chapter presents a hybrid optimization method that minimizes the annual system power losses. The optimization considers a 95%-probability of fulfilling the voltage and current limit requirements. The method combines the Genetic Algorithm (GA), gradient-based constrained nonlinear optimization algorithm and sequential Monte Carlo simulation (MCS). The GA searches for the optimal locations and capacities of WTs. The gradient-based optimization finds the optimal power factor setting of WTs. The sequential MCS takes into account the stochastic behaviour of wind power generation and load. The proposed hybrid optimization method is demonstrated on an 11 kV 69-bus distribution system.

Peiyuan Chen · Zhe Chen · Birgitte Bak-Jensen
Department of Energy Technology, Aalborg University, Pontoppidanstraede 101, Aalborg, Denmark
e-mail: pch@iet.aau.dk, zch@iet.aau.dk, bbj@iet.aau.dk

Pierluigi Siano
Department of Information & Electrical Engineering, University of Salerno, Fisciano (SA), Italy
e-mail: psiano@unisa.it

L. Wang et al. (Eds): *Wind Power Systems*, Green Energy and Technology, pp. 1–23.
springerlink.com

1 Introduction

Wind energy represents a renewable energy resource for electricity generation which contributes to the environment by reducing carbon-dioxide emission. Therefore, many European countries adopt policies to enhance wind energy utilization by means of incentives and financial options.

However, the connection of large amounts of wind turbines (WTs) to distribution systems presents a number of technical challenges to Distribution Network Operators (DNOs) (Masters 2002; Harrison et al. 2008). These challenges, such as steady-state voltage variation, power losses, voltage stability and reliability, are partly due to the mismatch between the location of energy resources and the local network capability of accommodating the new generation. Particularly, the location of WTs is determined by the local wind resources and geographical conditions. However, the capacity of the existing network where the WTs will be connected may not be sufficient to deliver the generated wind power. As a result, network reinforcement is required, which calls for a high capital investment.

System losses, being a major concern for DNOs, may be reduced or increased with the connection of WTs, depending on the locations and capacities of the connected WTs. System losses can be minimized by regulating WTs' power factors or reactive power outputs. This could benefit DNOs by reducing system operation costs without extra investment. Furthermore, DNOs may charge wind power producers for kWh energy flow through their networks by evaluating total network investment and system losses for a time span of 20 years (WTs' life time). Therefore, a reduction in system losses also benefits wind power producers by reducing their connection fee per kWh.

On the basis of the foregoing issues, DNOs would like to explore means to find the optimal locations and capacities of new WTs that can be accommodated within the existing networks, subject to constraints imposed by statutory regulations, equipment specifications and other operational or planning limits. In order to do this, DNOs require a reliable and repeatable method to identify the optimal capacity and location of new Distributed Generation (DG) so that the system power losses are minimized (Harrison et al. 2007a, 2007b, 2008).

A number of techniques were previously proposed to seek the optimal capacities and locations of DG. A common practice is to formulate an optimal power flow problem and to solve for the capacity and location of DG. The objective functions of the optimization formulation include minimizing the system power losses (Rau and Wan 1994; Nara et al. 2001; Kim et al. 2002; Celli et al. 2005), minimizing the investment and operating costs (El-Khaltam et al. 2004), maximizing the net revenue received by DNOs (Harrison 2007b; Piccolo and Siano 2009), maximizing the capacity (Keane and O'Malley 2005; Harrison et al. 2008;) or produced energy from DG (Keane and O'Malley 2007), etc. The constraints of the optimization normally consist of bus voltage limits and line thermal limits. Some also takes into account the short-circuit levels (Vovos et al. 2005) and reliability constraints (Greatbanks et al. 2003). As the DG location is a discrete variable, genetic algorithm (GA) can be used to find the optimal location (Kim et al. 2002;

Celli et al. 2005). Other algorithms such as Tabu search (Nara et al. 2001) can also be used.

Nevertheless, one major limitation in these methods is that the stochastic behaviour of load and wind power generation (WPG) has not been taken into account appropriately. In (Ochoa 2008), typical seasonal load profile is used to account for the seasonal and diurnal variation of load. A one-week time series of WPG has been used to account for its time-varying behavior. However, these are not sufficient to consider the stochastic behavior of load and WPG properly, as a typical daily or weekly curve cannot represent their stochastic behavior throughout a year. More sophisticated stochastic models of load and WPG needs to be adopted.

This chapter proposes a hybrid optimization method that aims at minimizing the total system losses while taking into account the stochastic behaviour of WPG and load during different seasons. The optimization algorithm considers the probability of fulfilling the main constraints, including voltage and current limits. The hybrid optimization method combines the Genetic Algorithm (GA), gradient-based constrained nonlinear optimization and the sequential Monte Carlo simulation (MCS) method. The GA is suitable for finding the optimal capacity and location of WTs as both control variables are integer values. The gradient-based constrained nonlinear optimization is adopted for the optimal power factor setting of WTs as the algorithm usually provides the fastest solution. The sequential MCS method can facilitate the use of time series models and is ideally suited to the analysis of stochastic generation such as wind power (Ubeda and Allan 1992).

The remainder of the chapter is organized as follows. First, a stochastic wind power model is introduced to simulate WPG in a sequential MCS. The cross-correlation of WPG between two wind farms is also considered. Then, the sequential MCS based optimization algorithm is presented for the optimal power factor setting of WTs. Third, the GA for the optimal allocation of WTs is discussed. Following this, the hybrid optimization method is proposed to combine the GA and the sequential MCS based optimization algorithm. Finally, the proposed hybrid optimization method is demonstrated on a 69-bus 11-kV radial distribution system.

2 Stochastic Wind Power Model

This section first introduces a stochastic model of wind power, referred to as the LARIMA model. Then, the cross-correlation modeling of multiple wind farms is presented. Finally, hourly measurements from Nysted offshore wind farm are used to illustrate the stochastic wind power model.

2.1 Single Wind Power Model

WPG shows a strong correlation in time. Such a temporal correlation or autocorrelation requires that realizations of WPG in a time sequence are not independent of each other. In other words, WPG cannot be simply sampled independently from a probability distribution. The following demonstrates in detail the statistical properties of WPG, followed by the stochastic modeling of WPG (Chen et al. 2009b).

The LARIMA(0,1,1) model developed in (Chen et al. 2009b) for stochastic WPG is briefly summarized as follows. The model was developed on the basis of one-year wind power data measured from the Nysted offshore wind farm with a rated capacity of 165.6 MW. The model is called 'LARIMA' because a limiter is added to a standard autoregressive integrated moving-average (ARIMA) model (Box et al. 1994). In a LARIMA(p, d, q) model, p represents the order of an autoregressive (AR) process, q represents the order of a moving average (MA) process, and d represents the degree of differencing operation. In this case, $p = 0$, $d = 1$, $q = 1$. The model is described by the block diagram shown in Fig. 1. The block diagram consists of a first-order MA model, i.e. the MA(1) model, an integration process, a limiter and a square transformation. The MA(1) model together with the integration process is also referred to as the ARIMA(0,1,1) model. B is a back shifter, similar to a unit delay.

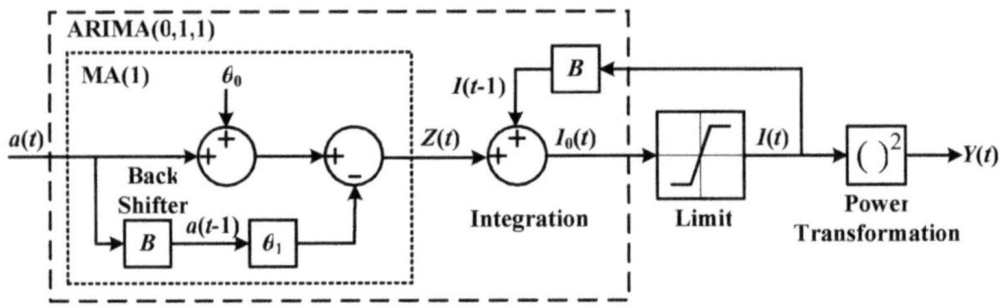

Fig. 1. Block diagram of the LARIMA(0,1,1) model of wind power generation

The LARIMA model can also be described mathematically by Eqs.(1)-(2):

$$Z(t) = \theta_0 + (1-\theta_1 B)\,a(t) \tag{1}$$

$$Z(t) = I_0(t) - I(t-1) \tag{2}$$

$$I(t) = \begin{cases} \mathrm{I}_{\max}, & I_0(t) > \mathrm{I}_{\max} \\ I_0(t), & \mathrm{I}_{\min} \le I_0(t) \le \mathrm{I}_{\max} \\ \mathrm{I}_{\min}, & I_0(t) < \mathrm{I}_{\min} \end{cases} \tag{3}$$

$$Y(t) = I^2(t) \tag{4}$$

where $\mathrm{I}_{\max}$ and $\mathrm{I}_{\min}$ denote the upper and lower bounds of the square-root of the wind farm power output, respectively. For the considered wind farm, the power output is bounded within [0, 165.6] MW, yielding $\mathrm{I}_{\max} = \sqrt{165.6} = 12.87$ and $\mathrm{I}_{\min} = 0$.

Equation (1) is the MA(1) model; θ_0 is the mean of the MA(1) model; θ_1 determines the temporal correlation (or autocorrelation) of the MA(1) model; $a(t)$ is the Gaussian white noise with variance σ_a. Equation (2) corresponds to one-degree of differencing ($d = 1$), accounting for the time-varying mean of the WPG. Equations (1) and (2) form the ARIMA(0,1,1) model as depicted in Fig. 1. Equation (3) represents the upper and lower limits of the WPG, accounting for the physical limitations of the wind farm. Equation (4) gives the final wind power time series $Y(t)$; the square-transformation accounts for the time-varying variance of the WPG. In summary, the model takes explicitly into account the temporal correlation, the random variation, the physical limitation, and the time-varying mean and variance of the WPG.

The detailed model identification and validation are presented in (Chen et al. 2009b). In brief, the model has in total three parameters (θ_0, θ_1 and σ_a). The model is validated against measurements in terms of temporal correlation and probability distribution in (Chen et al. 2009b). As demonstrated in (Chen et al. 2009b), the model requires much fewer parameters than a discrete Markov model; whereas it shows better performance than the discrete Markov model or an ARMA based model in terms of both the temporal correlation and probability distribution. Therefore, the LARIMA model will be used as the base model for the following correlation modeling.

2.2 Cross-Correlation Model of Wind Power

In the case of several WTs or wind farms in a power system, WPG from these WTs or wind farms may have strong correlations with each other, depending on their geographical locations. In contrast with the autocorrelation of WPG from a single WT or wind farm, this type of correlation is referred to as the cross-correlation of WPG among multiple WTs or wind farms.

In order to account for the cross-correlation of multiple WPGs, the LARIMA model is extended to a vector-LARIMA model. The vector-LARIMA model is established on the theory of vector time series (Wei, 1990), which describes relationships among several time series variables. The following illustration is based on a bivariate time series model. However, a multivariate time series model can be readily derived. Extending the MA(1) model in Fig. 1 to a vector time series model gives the vector-MA(1) model (Wei, 1990)

$$\mathbf{Z}(t) = \boldsymbol{\theta}_0 + (\mathbf{U} - \boldsymbol{\theta}_1 B)\mathbf{a}(t) \tag{5}$$

In the bivariate case,

$$\mathbf{Z}(t) = \begin{bmatrix} Z_1(t) \\ Z_2(t) \end{bmatrix}, \boldsymbol{\theta}_0 = \begin{bmatrix} \theta_{0,1} \\ \theta_{0,2} \end{bmatrix}, \mathbf{U} = \begin{bmatrix} 1 & 0 \\ 0 & 1 \end{bmatrix}, \boldsymbol{\theta}_1 = \begin{bmatrix} \theta_{11} & \theta_{12} \\ \theta_{21} & \theta_{22} \end{bmatrix}, \mathbf{a}(t) = \begin{bmatrix} a_1(t) \\ a_2(t) \end{bmatrix}$$

where $\boldsymbol{\theta}_0$ is the mean of the bivariate-MA(1) model. θ_{11} determines the weight of random noise a_1 at time $t-1$ retained at time t, and contributes to the autocorrelation of Z_1; θ_{22} is interpreted in the same way, but contributes to the autocorrelation of $Z_2(t)$. θ_{12} determines the weight of a_2 at time $t-1$ retained in Z_1 at time t, and contributes to the cross-correlation between $Z_2(t-1)$ and $Z_1(t)$ at time-lag one; θ_{21} determines the weight of a_1 at time $t-1$ retained in Z_2 at time t, and contributes to the cross-correlation between $Z_1(t-1)$ and $Z_2(t)$ at time-lag one. The white noise $a_1(t)$ and $a_2(t)$ have zero means and a covariance matrix $\mathbf{\Sigma}$.

By also extending Eqs. (2)-(4) to bivariate time series models, a complete bivariate-LARIMA(0,1,1) model is developed. Figure 2 shows the block diagram of the bivariate-LARIMA(0,1,1) model. The model can generate two correlated wind power time series $Y_1(t)$ and $Y_2(t)$. The parameter estimation of the bivariate-LARIMA model given the measured data is discussed in detail in (Chen et al. 2009c).

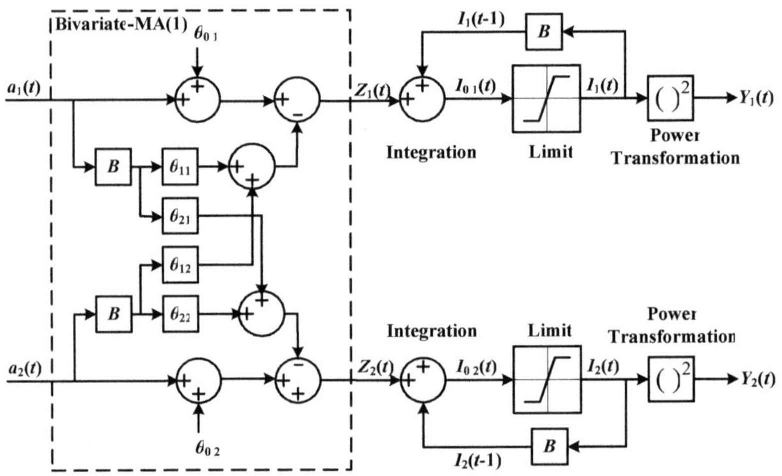

Fig. 2. Block diagram of the bivariate-LARIMA(0,1,1) model of wind power generation

2.3 *Bivariate-LARIMA Model for Wind Power Simulation*

The following presents a numerical example of the bivariate-LARIMA model based on the wind power data measured from the Nysted offshore wind farm. According to the determined bivariate-LARIMA model, two correlated wind power time series are simulated.

Applying Bivariate-LARIMA Model to Wind Power

Historical wind power data from Part A and Part B of the Nysted offshore wind farm are used for parameter estimations. Part A of the wind farm has the same capacity as Part B, i.e. 82.8 MW. The data are measured on an hourly basis, from January 1 to December 31 in 2005.

In order to account for the seasonal variation, the wind power data are grouped into summer and winter period. For each group of data ($y_1(t)$ and $y_2(t)$), the square-root and one-degree differencing transformation are applied to obtain two new time series $z_1(t)$ and $z_2(t)$.

$$\begin{cases} z_1(t)=(1-B)\sqrt{y_1(t)} \\ z_2(t)=(1-B)\sqrt{y_2(t)} \end{cases}, \qquad \text{for } t = 1,\ldots,\text{N}. \tag{6}$$

The sample covariance matrices at time-lag zero, $\underline{\boldsymbol{\Gamma}}(0)$, and that at time-lag one, $\underline{\boldsymbol{\Gamma}}(1)$, for summer and winter period are:

$$\underline{\boldsymbol{\Gamma}}_{\text{sm}}(0)=\begin{bmatrix}0.70 & 0.63\\ 0.63 & 0.70\end{bmatrix}, \underline{\boldsymbol{\Gamma}}_{\text{wt}}(0)=\begin{bmatrix}0.58 & 0.54\\ 0.54 & 0.60\end{bmatrix} \tag{7}$$

$$\underline{\boldsymbol{\Gamma}}_{\text{sm}}(1)=\begin{bmatrix}0.11 & 0.16\\ 0.10 & 0.11\end{bmatrix}, \underline{\boldsymbol{\Gamma}}_{\text{wt}}(1)=\begin{bmatrix}0.11 & 0.14\\ 0.11 & 0.11\end{bmatrix} \tag{8}$$

where subscripts ‘sm’ denotes summer, and ‘wt’ denotes winter. In order to understand how strong the autocorrelations and cross-correlations are, the corresponding correlation matrices of Eqs. (7) and (8) need to be calculated (Wei, 1990):

$$\underline{\mathbf{P}}_{\text{sm}}(0)=\underline{\mathbf{P}}_{\text{wt}}(0)=\begin{bmatrix}1 & 0.91\\ 0.91 & 1\end{bmatrix} \tag{9}$$

$$\underline{\mathbf{P}}_{\text{sm}}(1)=\begin{bmatrix}0.16 & 0.22\\ 0.15 & 0.16\end{bmatrix}, \underline{\mathbf{P}}_{\text{wt}}(1)=\begin{bmatrix}0.19 & 0.24\\ 0.19 & 0.19\end{bmatrix} \tag{10}$$

As shown in Eq. (9), the correlation matrix at time-lag zero is the same for summer and winter period. Whereas the correlation matrices at time-lag one, as shown in (10), differ slightly for summer and winter period. The off-diagonal elements of Eq. (9) are the cross-correlation coefficients at time-lag zero (0.91), which are very strong. The diagonal elements of Eq. (10) are the autocorrelation coefficients at time-lag one (0.16 and 0.19), which are rather weak. The off-diagonal elements are the cross-correlation coefficients at time-lag one, which are also very weak. However, these do not necessarily mean that the autocorrelation and cross-correlation at time-lag one of the wind power data $y_1(t)$ and $y_2(t)$ are weak. This is because of the square-root and differencing transformation between $y_i(t)$ and $z_i(t)$ as shown in Eq. (6).

Based on the estimated covariance matrices as in Eqs. (7) and (8), $\boldsymbol{\theta}_1$ and $\boldsymbol{\Sigma}$ for both summer and winter period can be estimated (Chen et al. 2009c):

$$\underline{\boldsymbol{\theta}}_{1,\text{sm}}=\begin{bmatrix}-0.18 & 0.01\\ -0.46 & 0.25\end{bmatrix}, \underline{\boldsymbol{\theta}}_{1,\text{wt}}=\begin{bmatrix}-0.11 & -0.10\\ -0.42 & 0.18\end{bmatrix} \tag{11}$$

$$\underline{\boldsymbol{\Sigma}}_{\text{sm}}=\begin{bmatrix}0.68 & 0.61\\ 0.61 & 0.65\end{bmatrix}, \underline{\boldsymbol{\Sigma}}_{\text{wt}}=\begin{bmatrix}0.56 & 0.51\\ 0.51 & 0.56\end{bmatrix} \tag{12}$$

Negative elements of $\boldsymbol{\theta}_1$ show a positive correlation and positive elements show a negative correlation. This is because of the minus sign in Eq. (5). Recall that $\boldsymbol{\theta}_1$ determines the weight of random noise at time $t-1$ retained at time t. A small value of $\boldsymbol{\theta}_1$ indicates a weak autocorrelation and cross-correlation of $\mathbf{Z}(t)$ at time-lag one. This is in accordance with the small values of the correlation matrices at time-lag one in Eq. (10). $\boldsymbol{\Sigma}$ is the covariance matrix of the random noise. The corresponding cross-correlation coefficients are also equal to 0.91, both for the summer and winter season. This indicates a strong cross-correlation at time-lag zero, which is in accordance with the large values of correlation matrices at time-lag zero in Eq. (9).

Finally, according to Fig. 2, $\theta_{0,i}$ is adjusted until the sample mean of the simulated time series $Y_i(t)$ coincides with that of the measured time series $y_i(t)$:

$$\underline{\boldsymbol{\theta}}_{0,\mathrm{sm}} = \begin{bmatrix} -0.03 \\ -0.04 \end{bmatrix}, \underline{\boldsymbol{\theta}}_{0,\mathrm{wt}} = \begin{bmatrix} 0.04 \\ 0.05 \end{bmatrix} \tag{13}$$

Wind Power Time Series Simulation

Bivariate wind power time series, $Y_1(t)$ and $Y_2(t)$, are simulated according to Fig. 2. The time-domain plot (only for 2 weeks) and the scatter plot of the two time series are shown in Fig. 3. $Y_1(t)$ is referred to as Wind power A and $Y_2(t)$ is referred to as Wind power B.

In the actual situation, wind may pass through Part A and Part B of the wind farm at the same time, which results in similar WPG from the two parts of the wind farm; whereas wind may pass from Part A to Part B (or from Part B to Part A) of the wind farm, which results in different WPG from the two parts of the wind farm. These two consequences are also observed in the simulated time series in Fig. 3 (a), where Wind power A and Wind power B have identical values during certain periods and discrepancy during other periods. The time-domain plot also shows that Wind power A fluctuates in a very similar way as Wind power B due to their strong cross-correlation. The strong correlation is also observed in the scatter plot in Fig. 3 (b), whose shape tends to follow a straight line. In addition, the two ends of the scatter plot are more condensed than the middle part. This is caused by the upper and lower limits of WPG due to cut-in and rated wind speed.

The sample probability distribution of the sum of Wind power A and Wind power B is shown in Fig. 4 (a). Although not shown here, the sample probability distribution of the model fits well with that of the measurements. For comparison, two wind power time series are simulated independently by using the LARIMA model in Fig. 1 for Part A and Part B of the wind farm without taking into account their cross-correlations. The sample probability distribution of the sum of the two uncorrelated time series is shown in Fig. 4 (b). Evidently, the probability distribution is very different from the one shown in Fig. 4 (a). This indicates the importance of correlation modeling when simulating WPG.

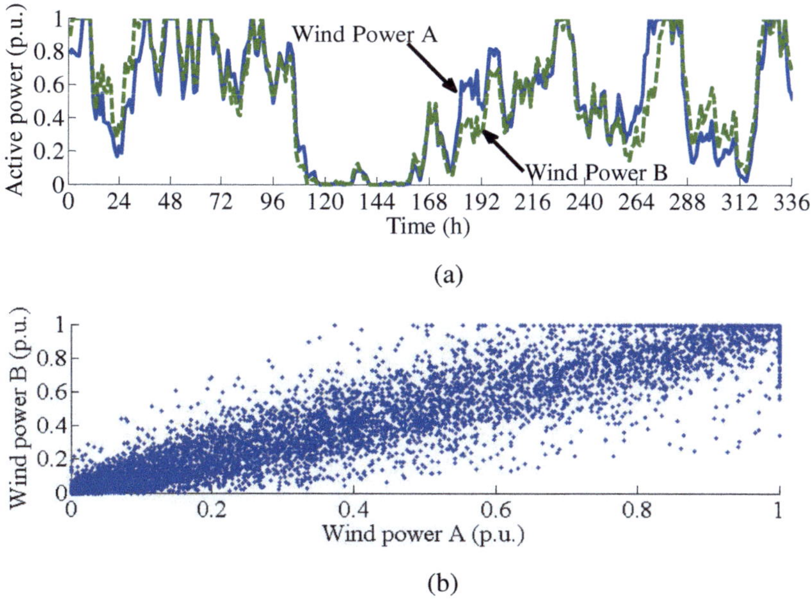

Fig. 3. Wind power A and Wind power B: (a) time-domain plot, and (b) scatter plot

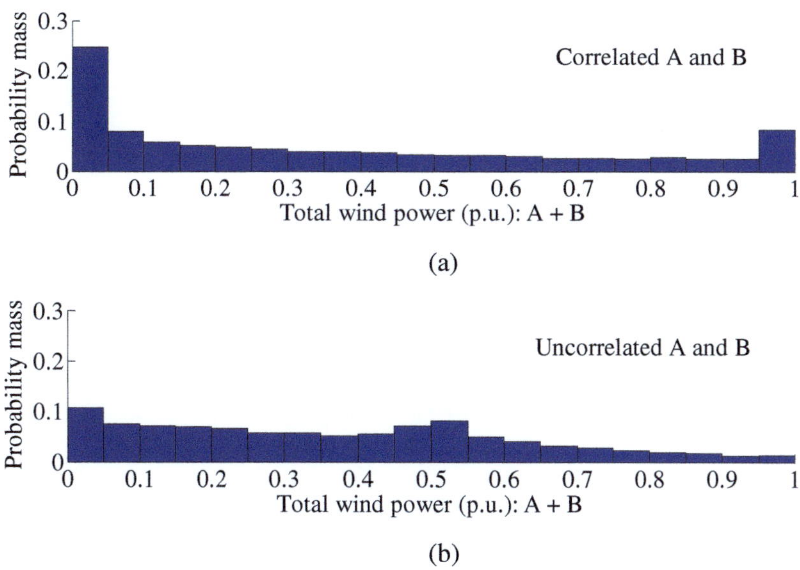

Fig. 4. Effect of cross-correlation on probability distribution of total wind power generation: (a) strong correlation with Eqs. (9) and (10), (b) no correlation

3 Optimization Approaches

During the planning stage of a modern distribution system, utilities are interested in knowing the optimal locations and capacity of WTs in the network so that the total system power losses can be minimized during system operation (Harrison et al. 2007a, 2007b, 2008). In addition, the utilities would like to know if the system power losses can be further reduced by controlling the power factor of WTs. However, the utilities usually confront a dilemma that how the stochastic behavior of wind power can be taken into account in a realistic way. The following demonstrates one solution to the issues addressed above.

The proposed solution combines standard optimization techniques with sequential Monte Carlo simulation (MCS), which is widely accepted as an effective approach to the analysis of stochastic generation (Ubeda and Allan, 1992). The hybrid optimization method is graphically illustrated in Fig. 5. The method consists of four parts: 1) load flow calculation for the evaluation of system steady-state performance, 2) sequential MCS for the probabilistic assessment of load flow results, 3) constrained nonlinear optimization for the optimal power factor setting of WTs, and 4) GA for the optimal allocation of WTs. The following describes the implementation of the hybrid optimization method in detail.

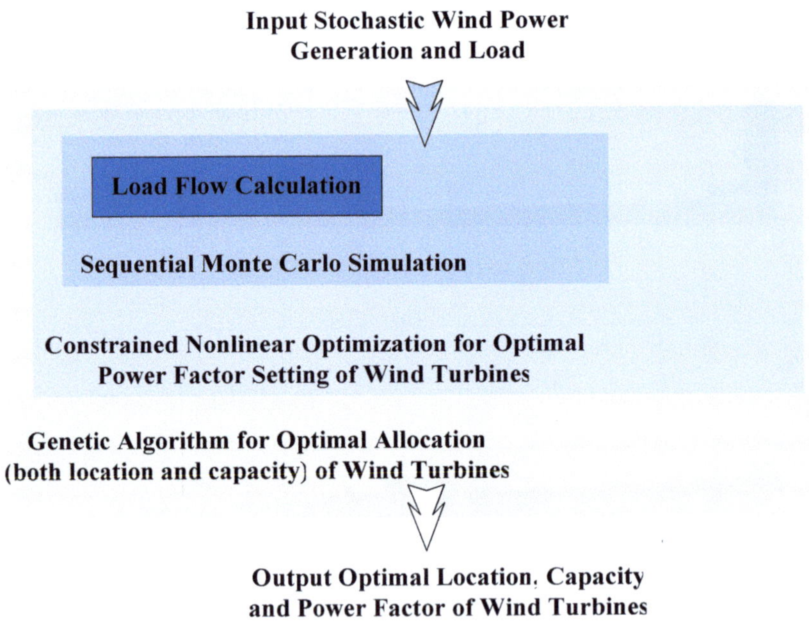

Fig. 5. Scheme of hybrid optimization method

3.1 Optimal Power Factor Setting of Wind Turbines

The constrained nonlinear optimization algorithm aims to minimize total system power losses by controlling the power factor of WTs. The optimization considers the voltage and current limits that are fulfilled at a 95%-probability. As shown in Fig. 5, the optimization requires inputs of total system power losses, bus voltages and line currents from the sequential MCS. The optimization provides outputs of minimum power losses to GA as well as corresponding optimal power factor of WTs. The algorithm for the constrained nonlinear optimization is based on the gradient and Hessian information of the Lagrangian. Mathematically, the objective function of the optimization is to

$$\text{minimize } \overline{P}_{\text{loss}} = \frac{1}{N}\sum_{i=1}^{N} P_{\text{loss}}(i), \tag{14}$$

where N is the length of a MCS, e.g. 8760 for a evaluation over a year; $P_{loss}(i)$ are the total system power losses at i^{th} hour; $\overline{P}_{loss}$ are the average system power losses over the study period. The total system losses are calculated by the sequential MCS shown in Fig. 6. The algorithm performs N consecutive load flow calculations in chronological order. The algorithm requires inputs of power factor of WTs, wind power time series and load time series. The algorithm provides outputs of average system power losses $\overline{P}_{loss}$ and time series of bus voltages and line currents over the studied period. The optimization is subject to the following constraints:

$$\tan \boldsymbol{\varphi} - \tan(\varphi_{\max}) \leq 0, \tag{15}$$

$$-\tan \boldsymbol{\varphi} + \tan(\varphi_{\min}) \leq 0, \tag{16}$$

$$\mathbf{F}_{\mathbf{V}}^{-1}(0.975) - V_{\max} \leq 0, \tag{17}$$

$$-\mathbf{F}_{\mathbf{V}}^{-1}(0.025) + V_{\min} \leq 0, \tag{18}$$

$$\mathbf{F}_{\mathbf{I}}^{-1}(0.95) - \mathbf{I}_{\max} \leq 0, \tag{19}$$

where $\boldsymbol{\varphi}$ is a vector of power-factor angles of WTs; φ_{max} and φ_{min} are the maximum and minimum power-factor angles, respectively. $\mathbf{P}_{WT}$ is time series of Wind power A and Wind power B shown in Fig. 3. $\mathbf{P}_{load}$ and $\mathbf{Q}_{load}$ are active and reactive load time series, respectively. V_{max} is the upper voltage limit and V_{min} is the lower voltage limit. $\mathbf{I}_{max}$ is a vector that contains current limits of all lines/transformers. The following explains Eqs. (17)-(19) in detail.

The inequality constraint of Eq. (17) requires that voltage is below V_{max} for minimum 97.5% of the time; and Eqs. (18) requires that voltage is below V_{min} for maximum 2.5% of the time. As a result, the voltage is within [V_{min}, V_{max}] for minimum 95% of the time. This is similar to the requirement set by the EN50160 standard, where 95% of the 10-min mean rms values of the supply voltage shall be within the limits (EN50160, 1999). According to the requirement, 5% voltage violation is tolerated.

$\mathbf{F}^{-1}{}_{V}(0.975)$ in Eq. (17) is a vector that contains 97.5%-quantile voltages of all buses. $\mathbf{F}^{-1}{}_{V}(0.025)$ in Eq. (18) is a vector that contains 2.5%-quantile voltages of all buses. Both 97.5%-quantile voltages and 2.5%-quantile voltages can be calculated from the voltage time series obtained from the MCS in Fig. 6. A

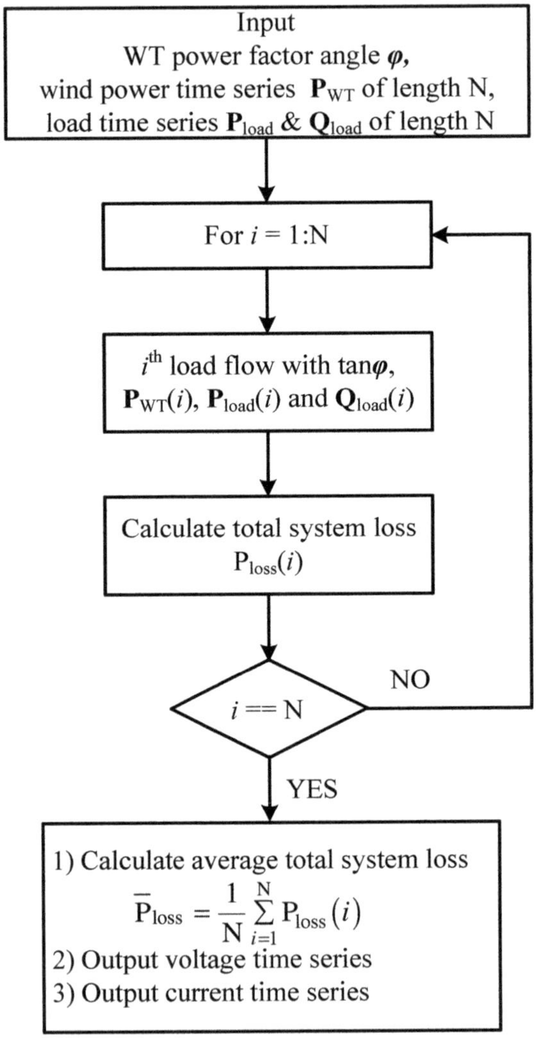

Fig. 6. Flow chart of sequential Monte Carlo simulation based load flow algorithm

97.5%-quantile voltage means that voltage values at a bus are below the quantile voltage for 97.5% of the time. Similar interpretation applies to a 2.5%-quanitle voltage. The two quantiles are also illustrated in a cumulative distribution function shown in the lower plot of Fig. 7. The upper plot of Fig. 7 illustrates the voltage constraints Eqs. (17) and (18) in a probability density function.

The inequality constraint of Eq. (19) requires that current of each line/transformer branch should be below I_{max} for minimum 95% of the time. As a result, overcurrent does not occur for more than 5% of the time. $\mathbf{F}^{-1}{}_{I}(0.95)$ is a vector that contains 95%-qauntile current of all branches and is calculated from the current time series obtained from the MCS in Fig. 6. A 95%-quantile current means that current values of a branch are below the quantile current for 95% of the time.

In summary, the proposed optimization algorithm, formulated as Eqs. (14)-(19), searches for optimal power-factor values of WTs in order to minimize the total system losses over a studied period. The algorithm requires that both the statutory voltage requirements and the current requirement should be fulfilled at a 95%-probability.

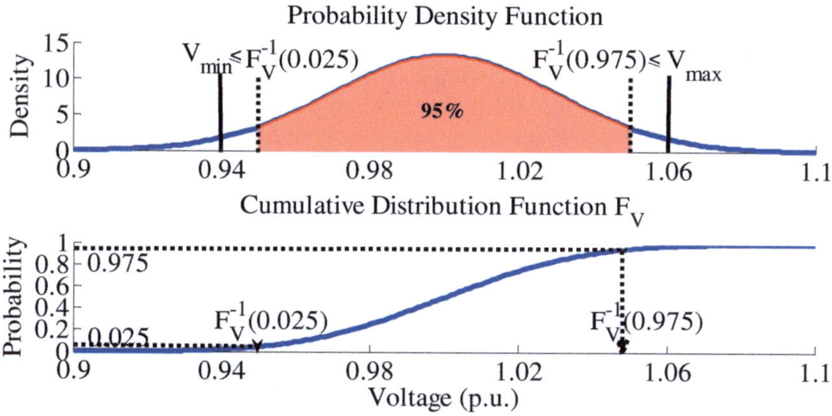

Fig. 7. Interpretation of $F^{-1}{}_{V}(0.975)$ and $F^{-1}{}_{V}(0.025)$

3.2 *Genetic Algorithm for Optimal Allocation of Wind Turbines*

The GA is used in order to select the types and number of WTs to be allocated at each candidate bus. The GA randomly generates the initial population of solutions (individuals) by defining a set of vectors. Each vector, or called a chromosome, has a size $N_e = N_C \times N_T$, where N_C is the number of candidate locations and N_T is the number of defined WT types. This is demonstrated in Fig. 8.

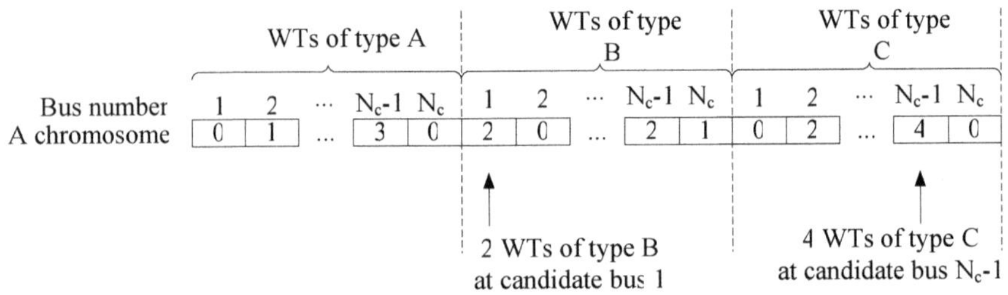

Fig. 8. Schematic of the GA chromosome

As shown in Fig. 8, a chromosome consists of a vector of integers, each of which represents the number of WTs of a given type to be allocated at a candidate bus. For instance, WTs of type A is associated with the first part of the vector with the size of N_C, which is the number of the candidate locations. Each element of this vector is an integer representing the number of WTs of type A connected to the corresponding bus. As such, the locations and types of WTs are expressed as a string of integers.

At each generation of the GA, a new set of improved individuals is created by selecting individuals according to their fitness; the selection mechanism used here is the normalized geometric ranking scheme. After the new population is selected, genetic operators are applied to selected individuals for a discrete number of times. These genetic operators are simple crossover and binary mutation. A simple crossover randomly selects a cut-point dividing each parent into two segments. Then, two segments from different parents are combined to form a new child (individual). A binary mutation changes each of the bits of the parent based on the probability of mutation. An elitism mechanism is also adopted to ensure the best member of the population is not lost. The iteration process continues until one of the stopping criteria is reached.

3.3 Genetic Algorithm - Monte Carlo Hybrid Optimization Method

For each chromosome of the GA, the constrained nonlinear optimization algorithm nested in the GA algorithm computes the fitness function used by the GA and the optimal power factor setting of WTs.

The constrained nonlinear optimization algorithm is based on a sequential MCS, which performs a number of load flow calculations in the chronological order of a year. The sequential MCS generates time series of system power losses, bus voltages and line currents. The system power losses are exported to the constrained nonlinear optimization algorithm as its objective function, with the bus voltages and line currents as its nonlinear constraints. The constrained nonlinear optimization provides outputs of minimum power losses to the GA given a specified number and location of WTs. Consequently, this hybrid method will deliver the best locations as well as the best WT types in the end. The flow chart of the foregoing hybrid optimization method is shown in Fig. 9.

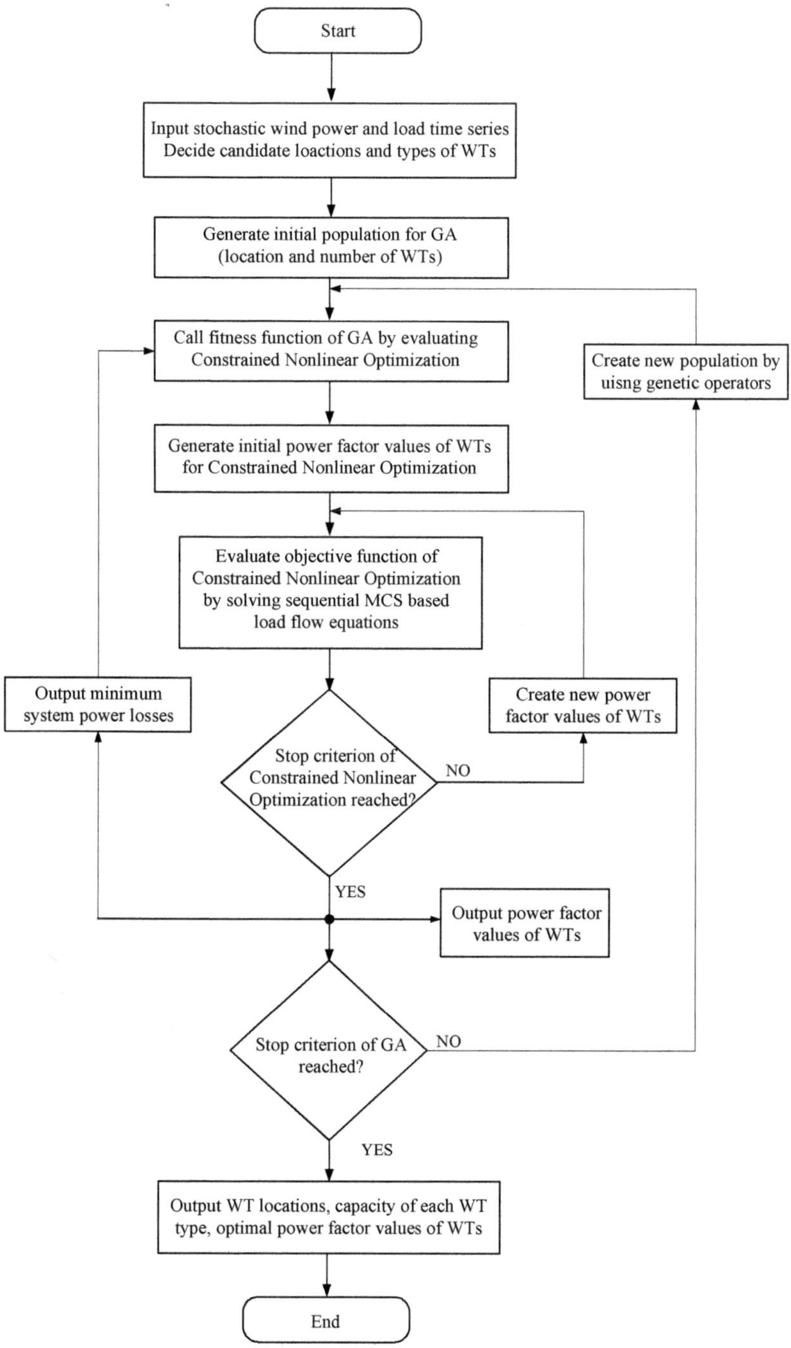

Fig. 9. Flow chart of the Hybrid optimization method

4 Case Study

4.1 System Description

This section describes the distribution system and data that are used to demonstrate the hybrid optimization approach proposed in the previous section for the optimal allocation (both sitting and sizing) and power factor setting of WTs.

69-Bus Distribution System

The 69-bus radial distribution system (Das, 2006) is used as a case network for the simulation studies. The network configuration is shown in Fig. 10 and the network data are provided in (Das, 2006). A 12 MVA 33/11 kV substation transformer is included in the network to connect the four main distribution feeders to the slack bus (bus 1). The upper two feeders are located in area A, and the lower two are located in area B. The 11-kV side of the transformer is denoted as bus 70. The voltage at the 11-kV side is controlled at 1.0167 p.u. by a tap regulator. There are in total 13 tap positions, with maximum six steps above and below the reference position. One tap step adjusts voltage by 0.0167 p.u. The voltage limits of all buses are set to ±6% of the nominal value (11 kV), i.e. $V_{max} = 1.06$ p.u. and $V_{min} = 0.94$ p.u. The current limit of all lines is 157A. In this case, the average active power losses of the network without the connection of WTs are 25 kW.

Wind Power and Load Time Series

As shown in Fig. 10, the possible locations of WTs are bus 7, 15, 22 and 29 in area A and bus 38, 43, 50, 56 and 64 in area B. Time series of WPG are simulated from the bivariate-LARIMA model. In particular, WPG from all the four WTs in area A follows the Wind Power A shown in Fig. 3. This assumes a perfect positive linear cross-correlation among the four WTs. A similar assumption is applied to the WTs in area B, where the WPG follows the Wind power B instead. However, the WPG in area A is not linearly correlated with that in area B, but follows the cross-correlation matrices defined in (9) and (10).

In contrast, loads are connected to all the buses from bus 2 to bus 69. The peak load data at each bus are given in (Das, 2006). The total peak load of the network is $(2.90 + j1.99)$ MVA. The active load power in area A and in area B follows the Load A and the Load B shown in Fig. 11 (a), respectively. The Load A and Load B are simulated according to the AR(12) based model developed in (Chen et al. 2009a). Load has a strong diurnal and weekly periodicity as shown in Fig. 11 (a). Similar to the WPG in area A, the loads in area A are also assumed to have a perfect positive linear cross-correlation with each other. The same assumption is

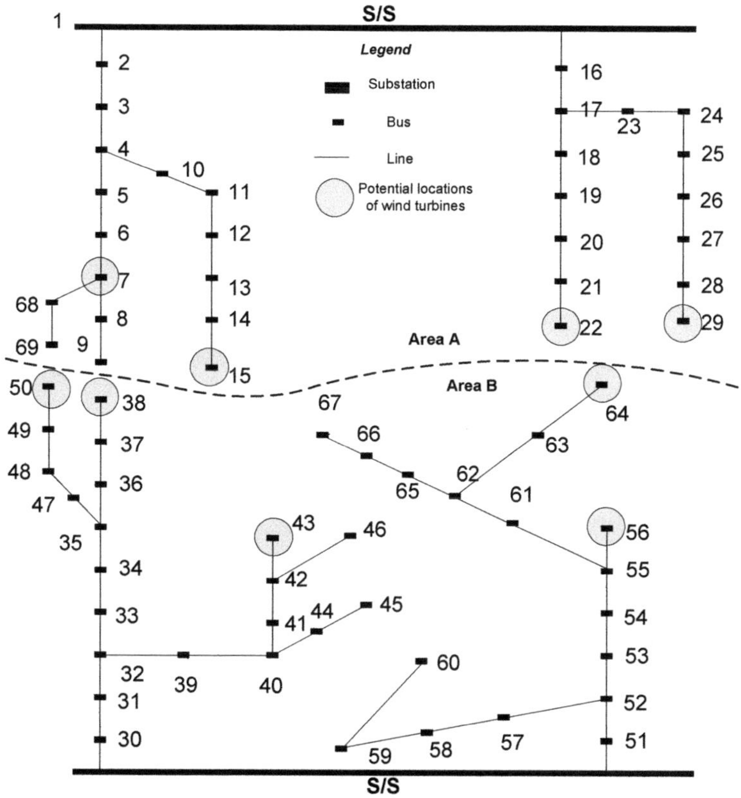

Fig. 10. The 69-bus 11-kV distribution system connected with WTs

applied to the loads in area B. However, the Load A and Load B do not have a perfect linear cross-correlation. Instead, a strong cross-correlation between the Load A and the Load B is present as shown in the scatter plot in Fig. 11 (b). The power factors of the loads are assumed time-invariant as provided in (Das, 2006).

In addition, wind power and load are assumed uncorrelated within summer or winter season. However, wind power and load within a whole year may still be correlated. This is reasonable as the mean values of wind power and load are both lower in summer than in winter. The correlation between wind power and load may be affected due to other factors such as temperature. However, it is rather weak (Papaefthymiou and Kurowicka, 2009) and is not taken into account here.

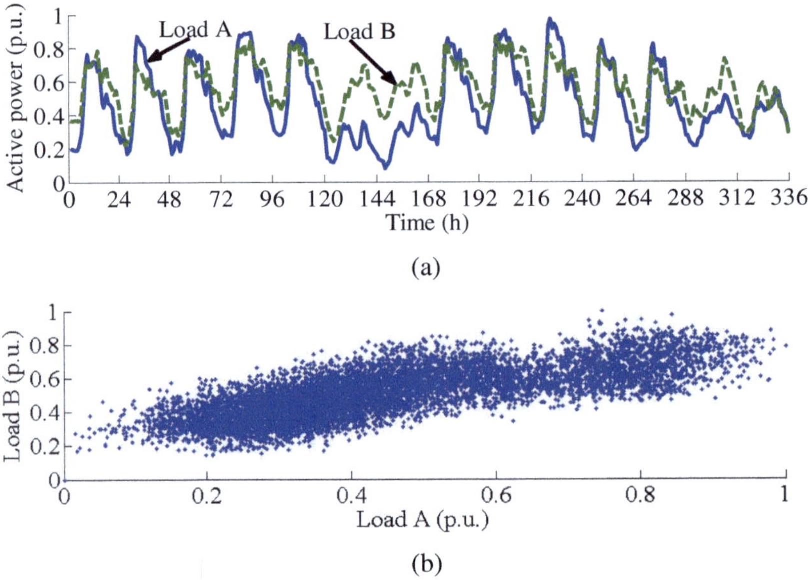

Fig. 11. Load A vs. Load B: (a) time-domain plot, (b) scatter plot

4.2 Results

Simulation that uses the hybrid optimization algorithm is carried out on the above-described distribution system with wind power and load time series. It is assumed that WTs of three different capacities are chosen by the DNO. These capacities are 20 kW, 50 kW and 100 kW. Maximum five WTs of each type are allowed at a given location. This requirement may be set by the available land for building WTs. For another distribution network with a different load level, WTs with different capacities may be considered. Consequently, GA is used to search for the optimal number of WTs of each type at the candidate locations. It is also assumed that the power factor is the same for all WTs connected to the same bus.

The entire method has been implemented in the Matlab® environment, incorporating some features of the MATLAB toolbox for GA (MathWorks, 2004). The basic parameters of the GA are summarized as follows. The total control variables are 27 (= 3x9), corresponding to the number of three types of WTs at the nine candidate locations. The population size of each generation is 20. The initial population is generated at random between zero and five. The GA will stop if any of the following conditions is reached: 1) the maximum generation number exceeds 100, 2) there is no improvement in the objective function for 5 consecutive generations, and 3) the cumulative change in the fitness function value over 5 generations is less than 1e-6. Sensitivity analyses have been carried out to consider different values for the GA parameters such as stop criteria, population size and genetic operators. From these analyses, it was shown that the used values guarantee the convergence of the algorithm to a satisfactory solution in this case.

Table 1 lists the initial values of the number of WTs at different buses for the GA. These initial values are the best individual selected from the initial population (with 20 individuals) at the first generation. The corresponding total capacity of the WTs at each bus is also summarized in Table 1. Table 2 lists the optimal numbers of WTs at different locations found by the GA. The total capacity of all the WTs amounts to 1.7 MW, including ten 20-kW WTs, eight 50-kW WTs and eleven 100-kW WTs. As the peak active load of the network is 2.9 MW, the wind power penetration level (i.e. total WT capacity divided by peak active load) in the network is around 59%. In particular, in area A, the total WT capacity is 0.71 MW and the total peak load is 1.17 MW, corresponding to a 61% wind power penetration level. Whereas in area B, the total WT capacity is 0.99 MW and the total peak load is 1.73 MW, corresponding to a 57% wind power penetration level. At bus 22, three 50-kW WTs are determined. In reality, due to the space limitation, DNOs may prefer to connect one 100-kW and one 50-kW WT instead. However, this should be judged by the DNO according to the actual situation.

Table 1. The initial values of the number and power factor of WTs for the hybrid optimization method

Bus no.	20 kW	50 kW	100 kW	Total Capacity (kW)	Power Factor
7	2	4	4	640	1.0
15	2	1	3	390	1.0
22	5	1	0	150	1.0
29	5	3	2	450	1.0
38	3	4	4	660	1.0
43	0	0	2	200	1.0
50	3	5	5	810	1.0
56	4	2	4	580	1.0
64	2	2	2	340	1.0

Table 2. The optimal values of the number and power factor of WTs found by the hybrid optimization method

Bus no.	20 kW	50 kW	100 kW	Total Capacity (kW)	Power Factor
7	1	1	1	170	0.91
15	2	0	1	140	0.92
22	0	3	0	150	0.97
29	0	1	2	250	0.80
38	0	1	1	150	0.89
43	0	0	2	200	0.86
50	2	1	1	190	0.80
56	1	0	2	220	0.80
64	4	1	1	230	0.86

All wind turbines generate reactive power.

Furthermore, the corresponding optimal power factor values of the WTs are also listed in Table 2. The optimal power factor values are found by the gradient-based constrained nonlinear optimization algorithm. The initial power factor values were all set to unity for the optimization algorithm as shown in Table 1. The final optimal power factor values indicate that all the WTs generate reactive power to the network. This is to partly supply the local reactive load in order to minimize the network losses due to the reactive power flow.

Figure 12 shows both the best and mean fitness values found by the GA at each generation. A fitness value of the GA is equal to the average system power losses. Both the best and mean fitness values converge very fast. The GA stops due to the third stop criterion, which indicates that the cumulative change in the fitness function value over five generations is less than 1e-6. The GA obtains an optimal solution after 35 generations. The best fitness value found is around 13.7 kW, which is the minimum average network power loss. Recall that the average network power loss was 25 kW without the connection of WTs. On the basis of the optimal location, capacity and power factor of WTs, the power losses of the network are reduced by more than 45% with the connection of WTs to the network.

Fig. 13 shows the probability density function and cumulative distribution function of the voltage at bus 29, with the given capacity, location and power factor of these WTs. Bus 29 has the largest WT capacity (250 kW). Recall that the voltage constraints require that the 97.5%-quantile voltage should be lower than the maximum voltage (1.06 p.u.) and that the 2.5%-quantile voltage should be higher than the minimum voltage (0.94 p.u.). These two requirements are both fulfilled as shown in Fig. 13. The actual voltage at bus 29 is between 0.96 p.u. and 1.06 p.u., with the mean value of 1.01 p.u.

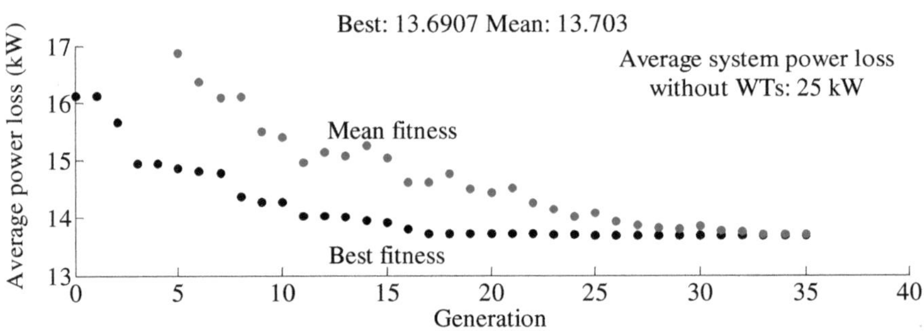

Fig. 12. Best and mean fitness found by the genetic algorithm at each generation

4.3 Discussions

The DNOs' decisions on the optimal locations and capacities of WTs are dependent on the actual or perceived costs or benefits associated with the connection of WTs. In this case, the DNO is interested in reducing the operating costs by simply

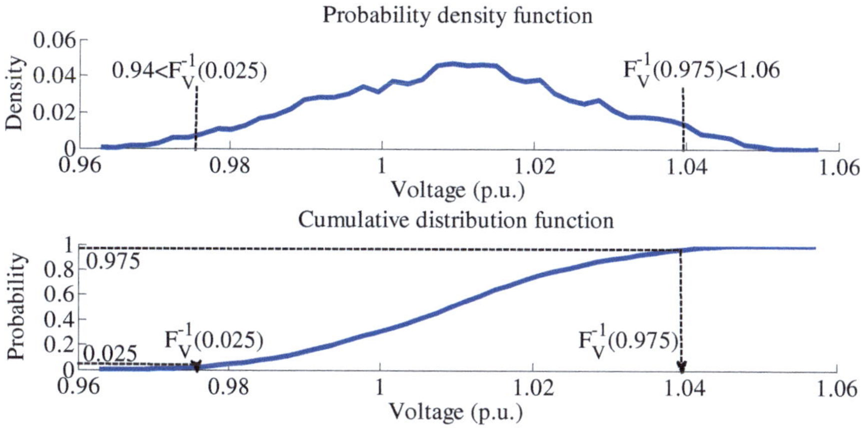

Fig. 13. Probability density function and cumulative distribution function of voltage at bus 29

minimizing network power losses. The proposed method allows the DNO to achieve the power losses minimization by strategically connecting a chosen number and types of WTs, among a large number of potential combinations, to the selected locations in a distribution network.

In order to determine the optimal locations and capacities of WTs, the common practice of DNOs is to assume the worst-case situation of maximum generation at minimum load. This worst-case provides the largest reverse power flows and voltage rises (Harrison et al. 2007a, Harrison et al. 2008), even though the situation may occur at a very low probability or even not occur at all. The proposed hybrid optimization method adopts the sequential MCS instead of the worst-case approach in order to account for the stochastic variation of WPG and load. The proposed method considers not only the probability distribution of WPG and load, but also different possible combinations of WPG from adjacent wind farms and of load from different areas. Consequently, the proposed hybrid method provides a more realistic evaluation of the system and thus a more reliable optimal solution.

In particular, when evaluating system power losses with stochastic WPG, it is important to consider the cross-correlation of WPG between wind farms. An assumption of independence usually leads to an underestimation of the system losses. It is shown in (Chen et al. 2009c) that the system losses may be underestimated by 10% under the assumption of the independence of WPG. On the other hand, the system losses may be overestimated if a perfect positive linear cross-correlation is assumed for wind farms that are distant from each other.

The main drawback of the proposed hybrid optimization algorithm is that the simulation time is very long. This is due to the evaluation of the fitness function which each time calls for a lengthy MCS. However, this drawback can be tolerated as simulation time is not a major concern for long-term system planning. In addition, a more powerful computer may improve the simulation speed to a certain extent.

4 Conclusions

This chapter introduces a hybrid optimization method to find the optimal sitting, sizing and power factor setting of WTs in a distribution system in order to minimize the network power losses. The method combines the GA, gradient-based constrained nonlinear optimization and the sequential MCS method, which takes into account the stochastic behaviour of WPG and load. The optimization algorithm considers a 95%-probability of fulfilling the bus voltage and line/transformer thermal limits. With the proposed optimization algorithm, a significant reduction of system power losses is achieved as a result of the integration of wind power. Therefore, the described hybrid optimization method can be used to assist the network operators to assess the system performance and to plan future integration of WTs in an effective and practical way.

References

Box, G.E.P., Jenkins, G.M., Reinsel, G.C.: Time Series Analysis: Forecasting and Control, 3rd edn. Prentice-Hall, New Jersey (1994)

Celli, G., Ghiani, E., Mocci, S., Pilo, F.: A multiobjective evolutionary algorithm for the sizing and siting of distributed generation. IEEE Trans. Power Systems 20, 750–757 (2005)

Chen, P., Bak-Jensen, B., Chen, Z.: Probabilistic load models for simulating the impact of load management. IEEE PES General Meeting 2009, Calgary, Canada (2009a)

Chen, P., Pedersen, T., Bak-Jensen, B., Chen, Z.: ARIMA-based time series model of stochastic wind power generation. IEEE Trans. Power Systems 25, 667–676 (2010)

Chen, P., Siano, P., Bak-Jensen, B., Chen, Z.: Stochastic optimization of wind turbine power factor using stochastic model of wind power generation. IEEE Trans. Sustainable Energy 1, 19–29 (2010)

Das, D.: A fuzzy multiobjective approach for network reconfiguration of distribution systems. IEEE Trans. Power Del. 21, 202–209 (2006)

El-Khaltam, W., Bhattacharya, K., Hegazy, Y., Salama, M.M.A.: Optimal investment planning for distributed generation in a competitive electricity market. IEEE Trans. Power Systems 19, 1674–1684 (2004)

EN 50160, European standard for voltage Characteristics of electricity supplied by public distribution systems. In: European Committee for Electrotechnical Standardization (CENELEC), Brussels, p. 14 (1999)

Greatbanks, J.A., Popovic, D.H., Begovic, M., Pregelj, A., Green, T.C.: On optimization for security and reliability of power systems with distributed generation. In: Proc. IEEE Power Tech. Conf. (2003)

Harrison, G.P., Piccolo, A., Siano, P., Wallace, A.R.: Distributed Generation Capacity Evaluation Using Combined Genetic Algorithm and OPF. International Journal of Emerging Electric Power Systems 8, 1–13 (2007a)

Harrison, G.P., Piccolo, A., Siano, P., Wallace, A.R.: Exploring the Trade-offs Between Incentives for Distributed Generation Developers and DNOs. IEEE Trans. on Power Systems 22, 821–828 (2007b)

Harrison, G.P., Piccolo, A., Siano, P., Wallace, A.R.: Hybrid GA and OPF evaluation of network capacity for distributed generation connections. Electrical Power Systems Research 78, 392–398 (2008)

Keane, A., O'Malley, M.: Optimal allocation of embedded generation on distribution networks. IEEE Trans. Power Systems 20, 1640–1646 (2005)

Keane, A., O'Malley, M.: Optimal utilization of distribution networks for energy harvesting. IEEE Trans. Power Systems 22, 467–475 (2007)

Kim, K.H., Lee, Y.J., Rhee, S.B., Lee, S.K., You, S.K.: Dispersed generator placement using fuzzy-GA in distribution systems. In: IEEE PES Summer Meeting, Chicago, USA, pp. 1148–1153 (2002)

Leon-Garcia, A.: Probability, Statistics, and Random Process for Electrical Engineering, 3rd edn. Pearson Prentice Hall, New Jersey (2009)

Masters, C.L.: Voltage rise: the big issue when connecting embedded generation to long 11 kV overhead lines. Power Eng. J. 16, 5–12 (2002)

MathWorks, Genetic Algorithms and Direct Search Toolbox: User Guide (2004)

Nara, K., Hayashi, K., Ikeda, K., Ashizawa, T.: Application of tabu search to optimal placement of distributed generators. In: Proceedings of the IEEE PES Winter Meeting 2001, pp. 918–923 (2001)

Papaefthymiou, G., Kurowicka, D.: Using copulas for modeling stochastic dependence in power system uncertainty analysis. IEEE Trans. Power Systems 24, 40–49 (2009)

Piccolo, A., Siano, P.: Evaluating the Impact of Network Investment Deferral on Distributed Generation Expansion. IEEE Trans. Power Systems 24, 1559–1567 (2009)

Rau, N.S., Wan, Y.H.: Optimum location of resources in distributed planning. IEEE Trans. Power Systems 9, 2014–2020 (1994)

Ubeda, J.R., Allan, R.N.: Sequential simulation applied to composite system reliability evaluation. IEE Proc. Gen., Trans., Distrib. 139, 81–86 (1992)

Vovos, P.N., Harrison, G.P., Wallace, A.R., Bialek, J.W.: Optimal Power Flow as a tool for fault level constrained network capacity analysis. IEEE Trans. Power Systems 20, 734–741 (2005)

Wei, W.W.S.: Time Series Analysis: Univariate and Multivariate Methods. Addison-Wesley, Redwood City (1990)

Optimal Conductor Size Selection in Distribution Systems with Wind Power Generation

Hamid Falaghi and Chanan Singh

Abstract. Optimal conductor size selection is an important part of the distribution system planning process. Wind power generators installed in the distribution systems have unknown operating cycle and undergo different scenarios according to the wind speed time variation characteristics. This randomness increases the complexity of many of the existing distribution systems planning practices. This chapter proposes a probabilistic approach for conductor sizing in electric power distribution systems in the presence of wind power generators. The objective function developed to determine the optimum conductor profile strikes a balance between the conductor cost and the cost of energy and power losses. Voltage constraints and thermal loading capacity of the conductors are also incorporated. In the proposed approach, the probabilistic evaluation of a solution related to the behavior of the wind power generators is embedded in Genetic Algorithm engine for the search of the optimal conductor planning solutions. The performance of the proposed approach is assessed and illustrated by case studies on typical test distribution systems.

1 Introduction

The primary goal in electric power systems is to supply customer demand in the most economical, reliable, and safe manner possible. Various plans to meet this goal, need to be studied for all major components of the electrical systems namely,

Hamid Falaghi
Department of Electric Power Engineering,
The University of Birjand,
Birjand, Iran

Chanan Singh
Department of Electrical and Computer Engineering,
Texas A&M University,
College Station, TX, USA

L. Wang et al. (Eds): *Wind Power Systems*, Green Energy and Technology, pp. 25–51.
springerlink.com

generation, transmission, and distribution. Power distribution systems are the last link in the chain of production and transport of electric energy and comprise a major part of a power system.

As can be seen from the investment pattern of several electrical utilities, the capital investment in the distribution systems constitutes a significant portion of the total amount spent in the entire power system. Due to the recent power system restructuring, this portion may become even larger. Thus, the optimum planning of power distribution system is one of the most important research fields for electrical engineers. Distribution system planning has been stated as an optimization problem, in which the objective is minimization of total cost, including the investment and operational costs related to the distribution system, subject to a set of technical constraints associated with the characteristics of the electric services and equipment. Because of the various technically feasible alternatives, distribution system planning is a complex problem having a large number of variables and constraints. Therefore powerful optimization techniques must be employed which can lead to remarkable saving for electric utilities and investors.

Due to complexity and difficulty in planning of distribution systems, the planning process is usually divided into sub problems one of which is optimization of conductor profile of the distribution feeders. An ideal conductor set should have the most economic cost characteristics, sufficient thermal capacity and provide acceptable voltage profile at all load points under different operating conditions.

The problem of conductor size selection for distribution systems has been addressed by many authors in the scientific literature. The formulations differ from one and other due to the representation of problem characteristics and the use of various solution algorithms. In [1], effect of voltage regulation on the conductor sizing is studied and an enumeration based technique is proposed for determining economical ACSR conductor size in distribution systems with uniform load. A dynamic cost model and its application in optimal conductor sizing is proposed in [2]. In [3], a multi-stage decision dynamic programming method is suggested for optimal selection of conductor cross-section in radial feeders. A solution algorithm considering the model proposed in [3] is developed in [4]. In [5], an algorithm is suggested for conductor size selection based on some realistic assumptions and specific requirements of voltage and losses. The parameters that affect the selection of cable sizes and the overall economics of the system are studied in [6] using sensitivity analysis. In [7], a very general idea is given about the line economics and the various factors that affect the conductor selection. In [8], a method for the techno-economical long-term optimization of currently operating distribution systems is developed. An algorithm based on economical current density-based method and heuristic approach for conductor sizing is developed in [9]. In [10], a method for selection of optimal set of conductors is presented in which financial and engineering factors are considered. An iterative algorithm for selecting the optimum size of conductors of feeder segments of radial distribution systems is suggested in [11]. A heuristic based solution approach for the problem is developed in [12]. In [13], a method is proposed for improving the maximum allowable loading of radial distribution feeders for different types of load models without violating the maximum current carrying capacity of branch conductors by optimum conductor size selection. A hybrid fuzzy-evolutionary algorithm is proposed in [14] for multiobjective conductor planning in distribution systems. In [15], a generalized method for optimal conductor sizing is presented in which the search space is enumerated systematically using a set of skipping rules.

These days, electric power utilities are concerned with issues related to renewable power generation. Among the renewable energy resources used to generate electricity, wind power has become the fastest growing technology in the last decade. Increased social awareness of harmful environmental, effects of greenhouse gases and energy policies of the governments have caused wind power to be considered seriously as an alternative for electric power generation. Because of variable and stochastic power injections, presence of wind power generation and their increased degree of penetration pose challenges to power system operators and planners. Therefore, its integration into an electric power system has motivated researchers to develop approaches which can evaluate wind power effects on planning and operation of the power systems.

Wind power generators (WPGs) installed in the distribution systems have unknown operating cycle and undergo different scenarios according to the wind speed time variation characteristics and some uncertainties are introduced in the system operation. This randomness increases the complexity of many of the existing distribution systems planning practices which assume that there is no stochastic generation source in the system and that electric power is simply imported from substations into the distribution feeders and delivered to the load points radially. Power injections from WPGs may change magnitude and even direction of network power flows. This causes an impact on planning practices and operation of distribution utilities with both technical and economic implications [16]. Due to the expected high penetration of WPGs in distribution systems, new strategies and methods for distribution systems design and planning need to be developed to accommodate this challenge.

Optimal conductor size selection is one of the problems that needs to be revised in the presence of WPGs. All research done in the previous proposed approaches [1]–[15] is based on the assumption that operating condition of the distribution system is known and there is no WPG in the system.

In this chapter a probabilistic approach for conductor sizing in power distribution systems in the presence of WPGs is presented. The proposed approach allows the optimal economic selection of conductor sizes in distribution systems when stochastic WPGs are running in parallel within the system. The objective function developed to determine the optimum conductor profile strikes a balance between the conductor cost and the cost of energy and power losses. Voltage constraints and thermal loading capacity of the conductors are also incorporated. In the developed approach, the probabilistic evaluation of a solution related to the behavior of the WPGs is embedded in Genetic Algorithm engine for the search of the optimal conductor planning solutions. The performance of the proposed approach is assessed and illustrated by case studies on a typical distribution system.

2 System Modeling

2.1 Load Model

Impacts of WPGs on distribution system depend on system loading. Therefore, the evaluation of the system requires the knowledge of loading at each load point. In order to reduce the problem size and limit the computational burden, the loading at each load point is assume to follow the stepwise load duration curve shown in Fig. 1. On the basis of this load model, the impact of probabilistic operation state of the system can be evaluated in one year. It should be noted that the presented conductor sizing approach can be extended and used easily for more load levels.

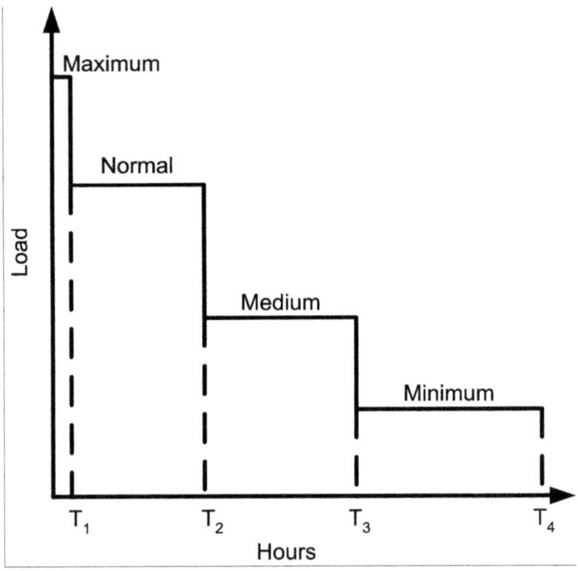

Fig. 1. Load duration curve

2.2 WPG Model

The characteristics of WPGs are very different from those of conventional generators. The generated power of each WPG varies with the wind speed at its site. Consequently, the evaluation of this behavior requires proper modeling of the operation state of the WPG. The power output of a WPG can be determined from its "speed-power" curve, which is a plot of output power against wind speed as shown in Fig. 2. A WPG is designed to start power generation at the cut-in speed (V_{ci}) and to shut down for safety reasons at the cut-out speed (V_{co}). Rated power (PW_r) is generated when the wind speed is between the rated speed (V_r) and the cut-out speed. There is a nonlinear relationship between the generated power and the wind speed when the wind speed lies within the cut-in speed and the rated speed as shown in Fig. 2.

The generated power, PW, which corresponds to a given wind speed, x, can be obtained from Eq. (1).

$$PW = \begin{cases} 0 & x < V_{cin} \\ P_r \cdot (A + Bx + Cx^2) & V_{ci} \le x < V_r \\ P_r & V_r \le x < V_{co} \\ 0 & x \ge V_{co} \end{cases} \tag{1}$$

where, the constants A, B, and C are as follows [17].

$$A = \frac{1}{(V_{ci} - V_r)^2} \left\{ V_{ci}(V_{ci} + V_r) - 4V_{ci}V_r \left[\frac{V_{ci} + V_r}{2V_r} \right]^3 \right\} \tag{2}$$

$$B = \frac{1}{(V_{ci} - V_r)^2} \left\{ 4V_{ci}V_r \left[\frac{V_{ci} + V_r}{2V_r} \right]^3 - (3V_{ci} + V_r) \right\} \tag{3}$$

$$C = \frac{1}{(V_{ci} - V_r)^2} \left\{ 2 - 4 \left[\frac{V_{ci} + V_r}{2V_r} \right]^3 \right\} \tag{4}$$

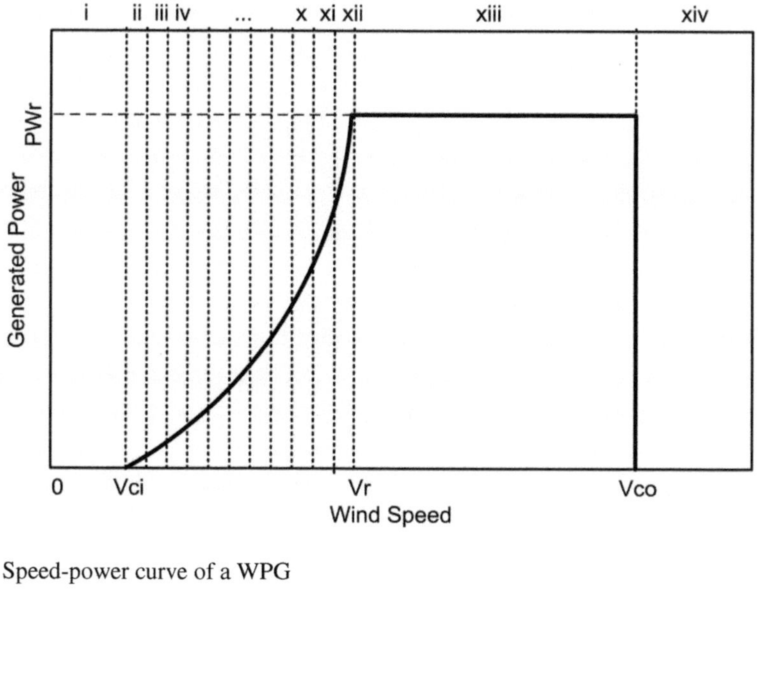

Fig. 2. Speed-power curve of a WPG

The wind speed range of interest can be divided into a finite number of states that do not necessarily have to be equally spaced. Each wind state corresponds to a certain amount of wind power generation derived from speed-power curve of the WPG.

In this work, fourteen wind speed states are defined, i.e., wind states *i–xiv* as shown in Fig. 2. The wind state *i* is the range below cut-in speed, and the wind state *xiv* is the range above the cut-out speed. Wind state *xiii* is the range with the rated power output. A probability is associated with each wind state, which can be derived from historical data or probability density of wind speed at the WPG location. Using these wind states and the speed-power curve, the WPG can be modeled as a multilevel generation source with a probability related to each generation level. In other words, probability density of power generation derived from the WPG can be calculated.

2.3 Load Flow Calculation

A significant amount of calculation burden in conductor size selection is related to load flow in distribution system. By knowing active and reactive load and generated power of WPGs, the load flow equations can be formulated and solved for a conductor profile. With respect to radial configuration and R/X ratio, conventional load flow methods based on Gauss–Siedel and Newton–Raphson techniques are inefficient in solving distribution system problems [18]. So, direct solution based load flow methods, usually apply for solving such systems. The solution technique used here is based on forward and backward propagation to calculate current in each feeder section and voltage at each node [18]. In general, computational steps of the load flow in distribution systems with WPGs are as follows.

1. All the node voltages are initialized at 1 PU as below.

$$V_i = 1 + j0, \quad \forall i \in SN \tag{5}$$

   where, SN is the set of system nodes.
2. Load equivalent currents in all nodes of network are calculated as below.

$$IL_i = \frac{(PL_i - PW_i) - j(QL_i - QW_i)}{V_i^*}, \quad \forall i \in SN \tag{6}$$

   where, PL_i and QL_i are active and reactive load at node i; PW_i and QW_i are active and reactive generated power by WPG at the i-th node (if available). V_i is voltage at node i. In this work, WPGs are modeled as constant power sources in the load flow calculation.
3. Current of each feeder section is calculated using the following equation. This procedure starts from ending feeder sections to the source node (Backward propagation).

$$IS_{(i,j)} = IL_j + \sum_{(j,k)\in F_j} IS_{(j,k)}, \quad \forall (i,j) \in SF \tag{7}$$

where, $IS_{(i,j)}$ is current of feeder section (i,j); F_j is the set of feeder sections connected to the node j of section (i,j) and SF is the set of all feeder sections.

4. The node voltages are calculated using Eq. (8), starting from the source node and proceeding along the feeder sections to ending nodes (forward propagation).

$$V_j = V_i - Z_{(i,j)}\, IS_{(i,j)}, \quad \forall\, j \in SN \tag{8}$$

where, $Z_{(i,j)}$ is impedance of feeder section connecting nodes i and j.

5. The steps outlined for backward and forward propagation are invoked during each iteration of load flow computations. The convergence is based on the difference of voltage at each node with its previous iteration value. If the convergence criterion is satisfied the propagation is stopped. Otherwise iterative process is repeated.

Once the load flow solution is converged, all the feeder sections currents and node voltages are known and the total power losses can be calculated, accordingly.

2.4 Probabilistic Evaluation

The procedure carried out to perform probabilistic evaluation of distribution system containing WPGs summarized as below:

1. Read input data and formulate load flow equations.
2. Select a load level t from the given load levels.
3. Select a generation level g for the WPGs in the system from the given generation levels.
4. Update Eq. (6) according to load at load level t and generation level g.
5. Solve the updated system load flow equations to calculate power loss, voltage profile and section loading at the load level t and generation level g.
6. Repeat steps 3 to 5 for all generation levels in order to calculate the power losses, voltage profile and loading of the system at the load level t.
7. Repeat steps 2 to 6 for different load levels in the year and obtain the voltage profile and loading of the system at different operating condition in the year and also, total annual system losses.

According to the voltage profile of load points and loading of the feeder sections calculated at different load and generation levels, voltage and capacity constraints violations in the system can be evaluated. Also, system power and energy losses can be calculated using the above process as follows.

$$PLOSS_{t-g} = \sum_{(i,j)\in SF} 0.003 \cdot R_{(i,j)} \cdot IS^2_{(i,j)-t-g} \tag{9}$$

$$\mathrm{PLOSS_{sys}} = \sum_{g\in SG} \mathrm{PLOSS_{t-g}} \cdot \mathrm{prob_g} \qquad \text{for t = peak loading} \tag{10}$$

$$ELOSS_{sys} = \sum_{t\in ST} \left(T_t \cdot \sum_{g\in SG} PLOSS_{t-g} \cdot \mathrm{prob}_g \right) \tag{11}$$

In the above equations, $PLOSS_{t-g}$ is system power loss at load level t and generation level g; $R_{(i,j)}$ is resistance of feeder section connecting nodes i and j (in ohm); $IS_{(i,j)-t-g}$ is current of feeder section (i,j) at load level t and generation level g (in Amp.); $\mathrm{PLOSS}_{\mathrm{sys}}$ is the annual system power loss (in kW); prob_g is the probability related to the generation level g; T_t is time duration of load level t (in hour), and $ELOSS_{\mathrm{sys}}$ is annual energy losses of the system load (in kWh). In the above equations, *SF*, *SG* and *ST* are set of all feeder sections, generation levels and load levels, respectively.

3 Optimization Problem

The objective of distribution system conductor planning is to select conductor size of feeder sections while the costs corresponding to the conductors as well as costs associated with the power and energy losses of the system are minimized. This may be formulated as an optimization problem in which decision variables are conductor size of each feeder section. In the following subsections, the objective function and related constraints are explained in detail.

3.1 Objective Function

Mathematical formulation of the objective function is as below.

$$\text{Min} \quad \text{TC} = \text{CC} + \text{EC} + \text{DC} \tag{12}$$

In Eq. (12), *TC* is the overall conductor planning cost (in \$). The cost items are described below.

A. Conductor CC

This cost item represents the investment cost of conductors and their maintenance cost during a useful lifetime, defined as

$$CC = \sum_{(i,j)\in SF} LS_{(i,j)} \cdot IC_{(i,j)} + f_{\mathrm{PW}} \sum_{(i,j)\in SF} LS_{(i,j)} \cdot MC_{(i,j)} \tag{13}$$

and

$$f_{\mathrm{PW}} = \sum_{y=1}^{ny} \left(\frac{1}{1+d} \right)^{y} \tag{14}$$

where, $LS_{(i,j)}$ is the length of feeder section (i,j) (in km); $LC_{(i,j)}$ is the investment costs related to the conductor type used in feeder section (i,j) (in \$/km);

$MC_{(i,j)}$ is annual maintenance cost associated to (*i,j*)-th feeder section conductor type (in \$/km/year); f_{PW} is total present worth factor; d is the annual discount rate, and ny is operational planning period (in year).

B. Energy Cost EC

The cost of energy losses in Eq. (12) is expressed in present value over the planning period referred to the base year as

$$EC = f_{\text{PW}} \cdot C_E \cdot ELOSS_{\text{sys}} \tag{15}$$

where, C_E is the cost of energy losses (in \$/kWh).

C. Demand Cost –DC

The demand cost in Eq. (12) represents the cost of useful system capacity lost and can be expressed in present value over the planning period referred to the base year as

$$DC = f_{PW}\, C_P\, PLOSS_{sys} \tag{16}$$

where, C_P is the levelized annual demand cost of loss (in \$/kW).

3.2 System Constraints

The objective function (12) is minimized subject to a set of constraints which have to be imposed in the conductor size selection of a distribution system. The objective of these constraints is to operate the system within its allowable limits. These constraints are thermal capacity of conductors and voltage constraints.

A. Capacities of the Conductors

Current following through feeder section (*i,j*) at different operating conditions should be less than the maximum current carrying capacity of its conductor, $IS_{(i,j)}^{\max}$, i.e.

$$IS_{(i,j)-t-g} \le IS_{(i,j)}^{\max}, \quad \forall (i,j) \in SF, \quad \forall t \in ST, \text{and} \quad \forall g \in SG. \tag{17}$$

B. Over and under Voltage Magnitude Limits

The voltage at every node in the feeder must be within the acceptable voltage bound, i.e.

$$V^{\min} \le V_{i-t-g} \le V^{\max}, \quad \forall i \in SN, \quad \forall t \in ST, and \quad \forall g \in SG. \tag{18}$$

In Eq. (18), V_{i-t-g} is calculated i-th node voltage magnitude at load level t and generation level g; $V^{\min}$ and $V^{\max}$ are minimum and maximum allowed operation voltage (in volt).

4 Genetic Based Optimization

The Genetic Algorithm (GA) is a numerical search tool based on the mechanics of natural selection and natural genetics. Its theory is based on "a better chance of surviving for a population with a better fitness". The GA can be used for the estimation of global minimum or maximum values of a function even in the presence of some local extermum. In other words, it will not be trapped in the local extermum and searches for the global one. The GA begins with a randomly generated population of individuals (chromosomes); representing decision-making variables. The population which represents the candidate solutions of the problem is evolved generation after generation to search an optimal solution and historical information is then exploited to speculate on new search points with expected performance during the iteration [19]. The GA searches the optimal solution via three major genetic operators, namely, selection, crossover, and mutation. First, the selection operator selects the fittest individuals of the previous generation to be the parents of the new generation. This is used to preserve the better historical information to survive at the new generation. After the selection, the crossover operator generates new individuals by crossing pair parents of the old. The new individuals succeed and exchange the best information of parents to be the new individuals. To avoid the loss of some important genes and increase the variation of the individuals, mutation operator is imported to add new information occasionally. Different applications of GA in biology, computer science, image processing, and social sciences have been reported [19].

The flowchart of the GA implementation for the problem of optimal conductor size selection is shown in Fig. 3 and discussed in the following subsections.

4.1 Chromosome Structure

The first issue that should be defined is the type of codification to be used, so that a chromosome represents solution of the candidate problem. In the proposed method each chromosome is a string which has *nf* genes where *nf* represents number of feeder sections. Each gene is related to one feeder section and contains an integer number showing conductor size number of the corresponding feeder section.

For example, in a conductor sizing problem if $nf=10$ and there are 4 conductor type in the list of available conductors to be used, then the chromosome is as following: "4 2 1 3 3 2 1 4 1 4".

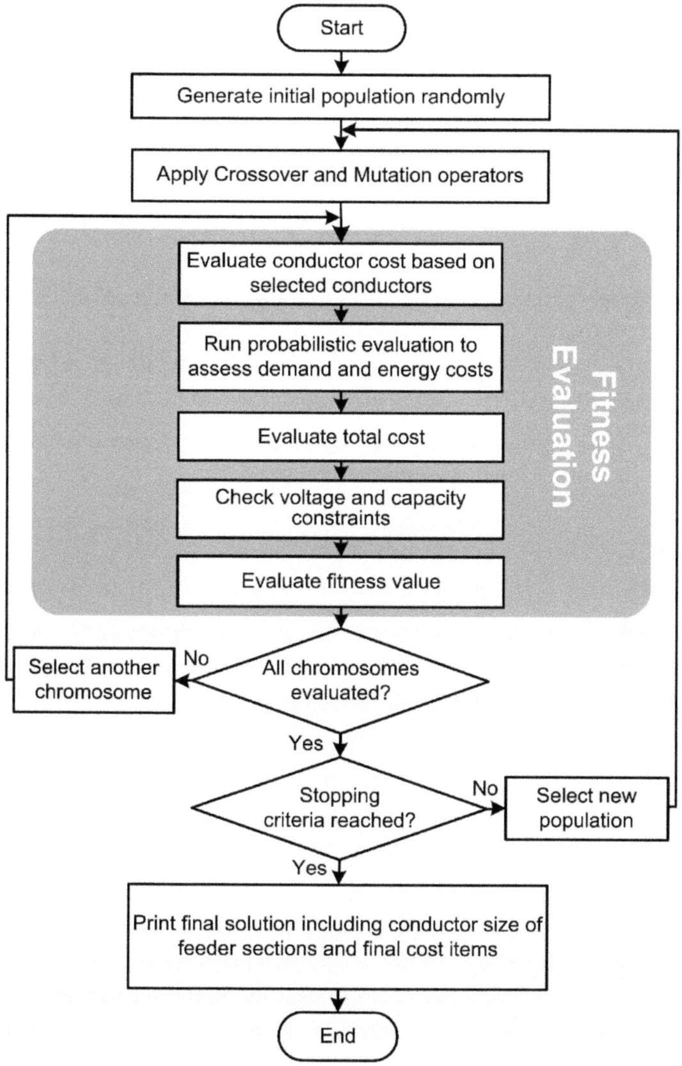

Fig. 3. Flow chart of GA based optimization

4.2 *Crossover Operator*

The crossover operator is used for random recombination of chromosomes to create new individuals called children. The crossover operation consists of two stages. The first, two chromosomes are selected randomly from the population as parents. Then, a uniform random number is generated into the interval [0, 1]. If this number is lower than the crossover rate, *pc*, then the crossover operator will perform on the selected parents. The crossover rate represents the percentage of the population on which the crossover is performed. Children are generated using the following procedure:

Random integer *c* is generated in the interval [1, *nf*]. The *c*-th first genes of the children are the same components as the respective parents (i.e. the first child from the first parent and the second child from the second parent). The remaining genes are selected according to the following rules:

1. The $(c+i)$-th gene of the first child is replaced by the $(nf - i + 1)$-th gene of the second parent (for $i = 1, 2, \ldots, nf - c$).
2. The $(c+i)$-th gene of the second child is replaced by the $(nf - i + 1)$-th gene of the first parent (for $i = 1, 2, \ldots, nf - c$).

For example, by applying the proposed operator for the following parents, and assuming *c=6*, the following children are obtained:

Parents	Children
123412|3412	123412|4131
432114|1314	432114|2143

It is to be noted that the proposed operator generates chromosomes with more variety in comparison with the standard crossover operator, because this operator can generate different children from similar parents.

4.3 Mutation Operator

The mutation diversifies the search and prevents the premature convergence that leads to nearly the same individual within a population after several generations [19]. Mutation operation in the simplest form is a random alteration of one or more genes of a chromosome. In this step, at first, one chromosome is selected randomly from the population. Then, a uniform random number is generated into the interval [0, 1]. If this number is lower than the mutation rate, *pm*, then the mutation operator will perform on the selected chromosome. The mutation rate represents the percentage of the population on which the mutation is performed. In order to perform mutation, a uniform integer random number *u* is generated in the interval [1, *nf*]. For generating the new chromosome the *u*-th component is swapped for another random value within a specific interval i.e. randomly selected another conductor size number.

The *pm* must be sufficiently small to ensure not only that the crossover is the primary means of creating new chromosomes, but also that the GA is not reducing to a random search. However, too small a mutation rate can not avoid premature convergence. In this work, to enhance the local search around the optimum solution, the mutation rate is a dynamic value that is minimal at the early stages of evolution and grows with each generation.

4.4 Fitness Evaluation

The individuals evolve according to their fitness to the environment. The fitness function of the problem is defined as the inverse of the total cost (12) plus a penalty factor for the infeasible solutions (i.e., the ones violating the constraints). To speed up the convergence properties of the algorithm and at the same time, to use the information that may still be useful in rejected chromosomes, this penalty factor is linearly increased (through iterations) from zero toward a very high value.

4.5 Selection and Elitism

If the number of generated races is sufficient then we proceed to the next step, otherwise we select new population among the present population and the chromosomes that are generated in this iteration. The criterion to select the chromosomes is based on their fitness values [19]. Individuals with the larger fitness value, i.e. better solutions to the problem, receive corresponding larger numbers of copies in the mating pool.

There are several proposals on how to implement selection but most of them work in a similar way in a planning problem. Therefore, the fast and efficient method called tournament selection is used. This proposal carries out *np* games for a population size *np*. In each game, a set of *k* chromosomes is randomly chosen and the winner chromosome is the one with the best fitness value. The value *k* is generally small, usually $k \in \{2, 3, 4, 5\}$.

Elitism is an effective means of saving early solution by ensuring the survival of the fitness string in each generation. The elitism puts the best individual of the current population into the new population to further improve the convergence performance of the GA.

5 Numerical Results and Discussion

The proposed conductor sizing approach for distribution systems with WPGs was applied on a typical 11-kV distribution system [20]. The single line diagram of this 30-node distribution system is shown in Fig. 4. The relevant peak load and section length data are listed in Table 1. Technical and economical characteristics of conductor types considered in the case studies are given in Table 2. The loading at each node is assumed to follow the load duration curve in Fig. 1 and the corresponding loading levels and duration are given in Table. 3. In this table different loading levels of the system are modeled as percentage of the peak load. Although it is assumed that all load points follow the same loading levels, this assumption is not mandatory for the proposed approach. Different loading levels per load point could also be used without the need of any modifications in the presented conductor sizing approach. Other technical and cost parameters are listed in Table 4.

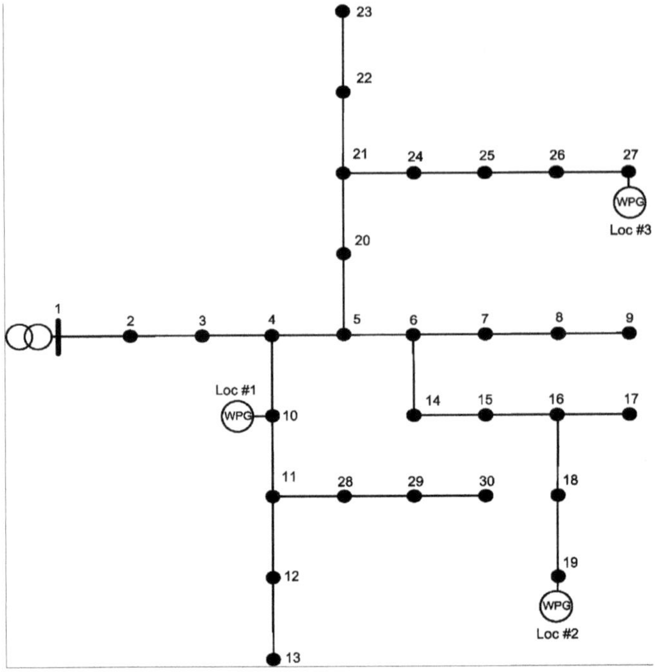

Fig. 4. Single line diagram of the distribution system under study

To demonstrate the efficiency of the proposed approach and at the same time to analyze the impact of connection point of WPGs, three locations are assigned to connect a WPG which are illustrated in Fig. 4. Three tests are conducted to investigate the impact of capacity and location of WPGs on optimal conductor sizing in the distribution system. In the first test, a WPG is assumed to be installed at Loc#1 and optimal conductor sizes for the system are determined using proposed approach for different capacities of the WPG changing from 0 to 1000 kW in steps of 100 kW. In the second and the third tests, the location of the WPG is changed respectively to Loc#2 and Loc#3 and the optimal conductor sizes are determined for these different capacities of the WPG as the first test.

The wind speed probability distribution used in all of the above tests is illustrated in Fig. 5. The cut-in, rated, and cut-out wind speeds of the WPG are 4, 15, and 25 m/sec, respectively. Using the presented modeling approach, the probabilistic multilevel generation model of the WPG is determined and illustrated in Fig. 6.

For comparison purpose, other input data in all the above tests is fixed and the same.

The proposed approach is used to determine optimal conductor sizes in the system for the above tests.

Table 1. Section and load data of the distribution system under study

Sending end node	Receiving end node	Length [km]	Load at receiving end node	
			PL[kW]	QL[kVAr]
1	2	1.5	35	30
2	3	1	25	15
3	4	0.5	10	6
4	5	2.5	25	15
5	6	0.5	38	28
6	7	0.5	10	5
7	8	1	15	12
8	9	1.5	30	30
4	10	2	40	30
10	11	0.5	54	30
11	12	0.25	30	15
12	13	0.25	10	5
6	14	0.5	25	20
14	15	1	40	30
15	16	1	40	30
16	17	0.5	100	90
16	18	0.5	60	30
18	19	0.5	30	30
5	20	0.5	10	5
20	21	1	40	40
21	22	1.5	40	30
22	23	0.5	27	20
21	24	1	80	70
24	25	1.1	45	39
25	26	1	25	20
26	27	1	20	10
11	28	0.5	40	30
28	29	0.5	30	20
29	30	1.5	22	12

Table 2. Technical and economical characteristics of conductors available for the case studies

Type of Conductor	R [Ω/km]	X [Ω/km]	Thermal capacity [Amp.]	Investment cost [$/km]
Squirrel	1.3760	0.3896	115	2600
Weasel	0.9108	0.3797	150	4000
Rabbit	0.5441	0.3973	208	6400
Raccoon	0.3675	0.3579	270	9600

Table 3. System load levels

Load band	Load [% of peak]	Duration [hour]
maximum	100	73
normal	80	2847
medium	60	2920
minimum	40	2920

Table 4. Technical and cost parameters

Parameter	Value
Planning period [year]	30
Discount rate [%]	9
Annual maintenance cost of conductors during life time	2% of investment cost
Load power factor	0.8
WPG operation power factor	0.9 lagging (producing reactive power)
Maximum allowed operation voltage [PU]	0.95
Minimum allowed operation voltage [PU]	1.05
Cost of energy losses [$/kWh]	0.06
Cost of power losses [$/kW]	300

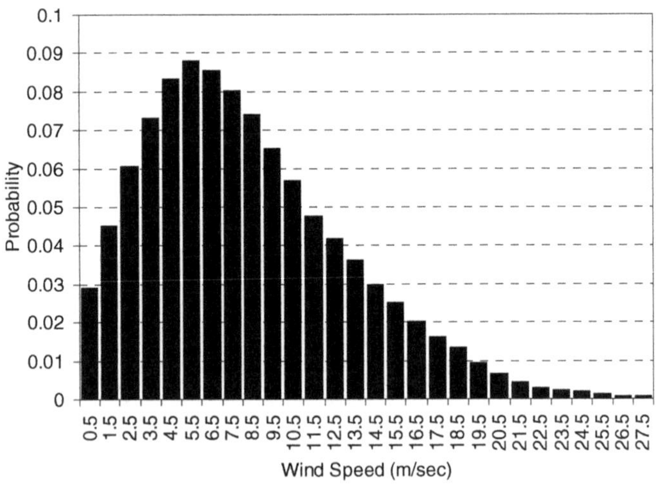

Fig. 5. Probability distribution of wind speed

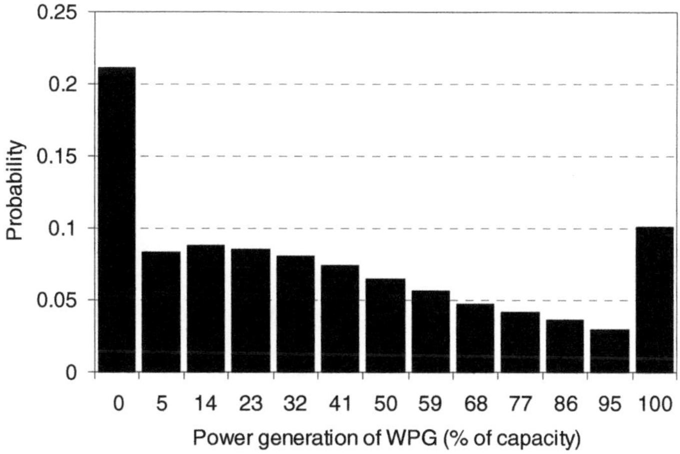

Fig. 6. Probabilistic multilevel generation model of the WPG

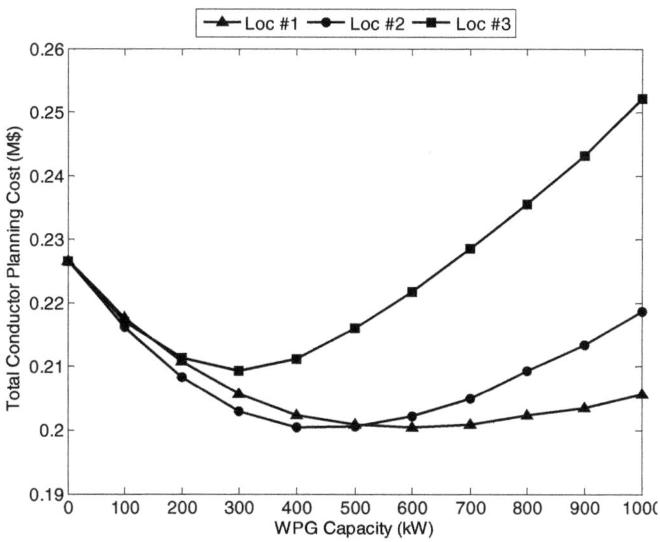

Fig. 7. Variation of total conductor planning cost vs. WPG capacity

Fig. 7 shows a comparison among the obtained total conductor planning costs for different WPG capacities connected in Loc#1, Loc#2, and Loc#3. Result of optimal conductor sizes of the system without WPG is shown in Fig. 8. Optimal conductor planning results for selected WPG capacities in the above tests are

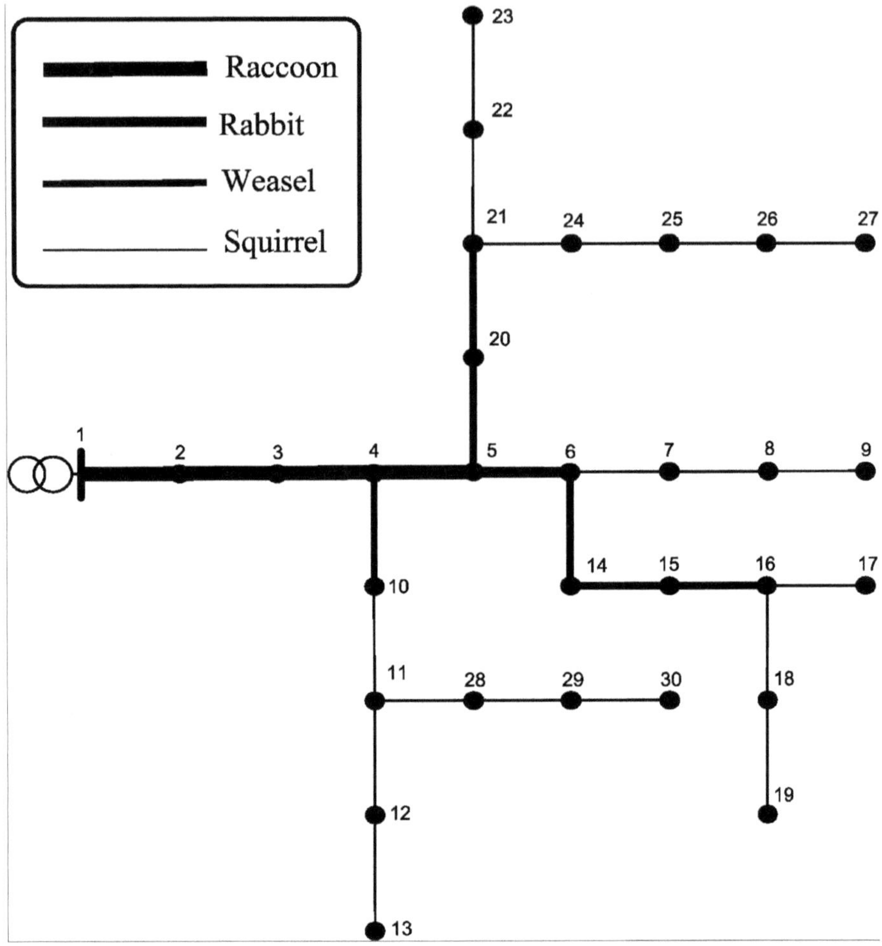

Fig. 8. Conductor planning result for the system without WPG

compared in Figs. 9–11. The related cost items in these selected cases are listed in Table 5.

The results obtained show that different locations and capacities of WPGs give different conductor planning costs and results. It is to be mentioned that for a certain WPG location, the total cost of conductor planning begins to decrease when connecting small amount of WPG capacity until it achieves its minimum level. Once this minimum level is reached, if WPG penetration level still increases, then total cost begins to rise too. It is worth pointing that at high WPG penetration levels, total cost of conductor planning can becomes larger than that without WPG connected. It means that there is a beneficial penetration level for wind based

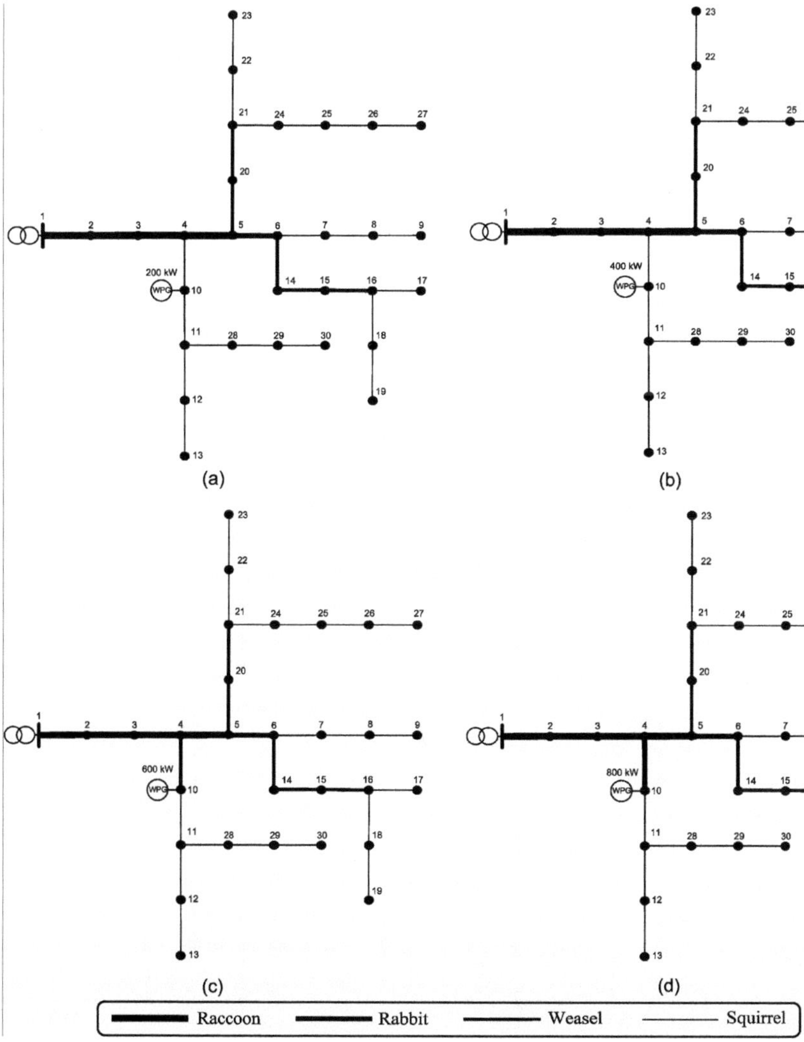

Fig. 9. Conductor planning results for WPG at Loc#1 with capacity equal to: **a)** 200 kW, **b)** 400 kW, **c)** 600 kW, and **d)** 800 kW

power generation from system conductor planning view point. Fig. 7 also shows that the effect of the WPG penetration level on the total cost of conductor planning is site specific and the beneficial penetration level for different WPG locations are not identical. In our case study, the most beneficial penetration level when the WPG is connected to the Loc#1 is about 600 kW while it is 400 kW for Loc#2 and 300 kW for Loc#3.

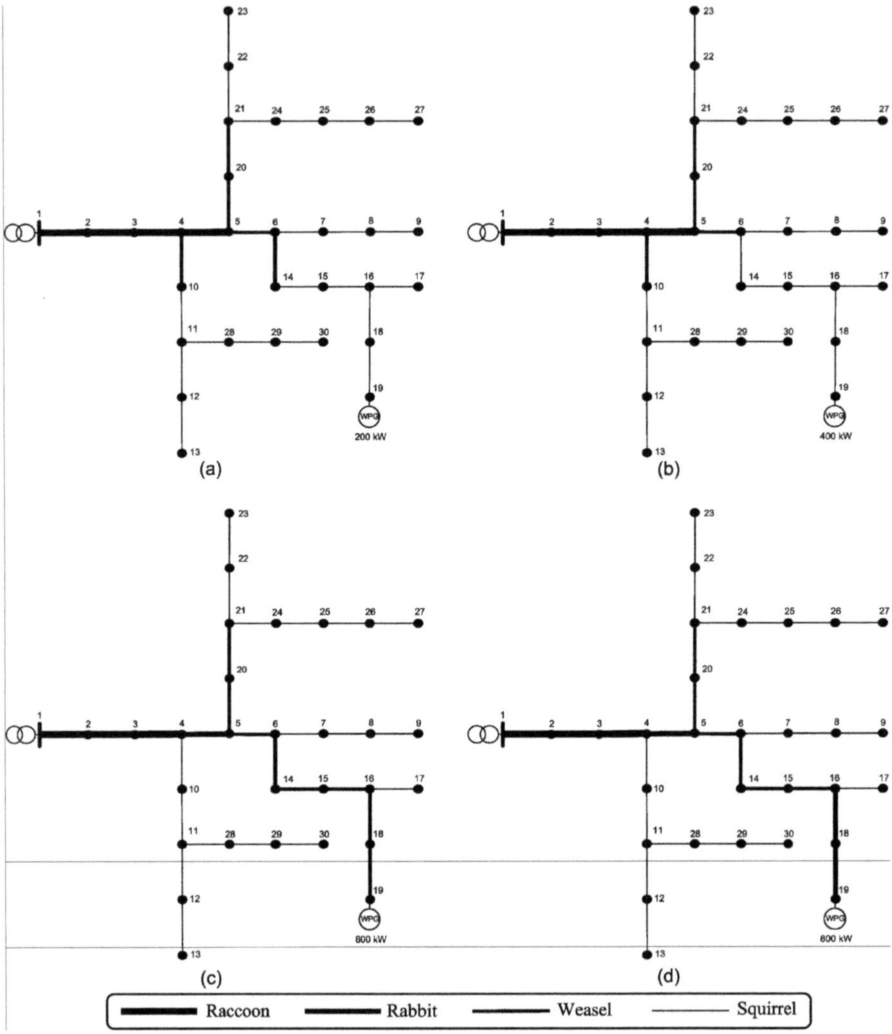

Fig. 10. Conductor planning results for WPG at Loc#2 with capacity equal to: **a)** 200 kW, **b)** 400 kW, **c)** 600 kW, and **d)** 800 kW

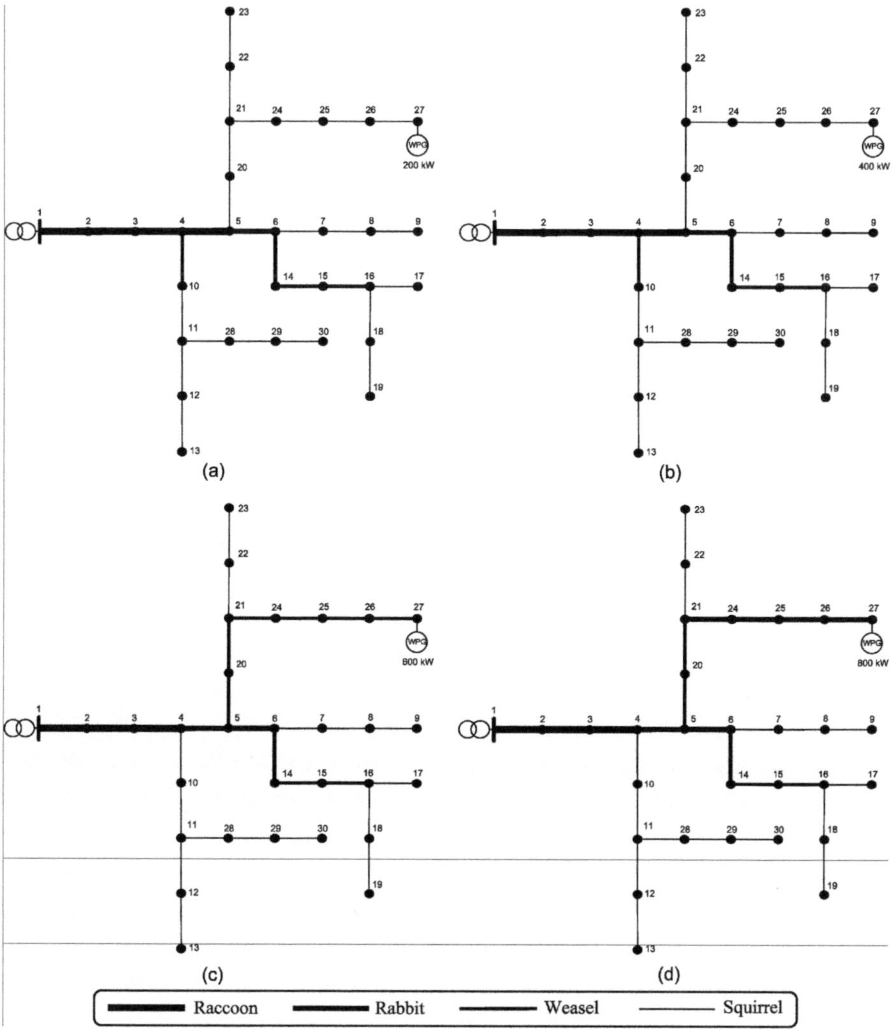

Fig. 11. Conductor planning results for WPG at Loc#3 with capacity equal to: **a**) 200 kW, **b**) 400 kW, **c**) 600 kW, and **d**) 800 kW

Table 6. Comparison of conductor planning cost items for selected WPG capacities

WPG Location	Cost item	WPG Capacity (kW) 200	400	600	800
Loc#1	CC ($)	111,217	111,217	114,592	120,379
	EC ($)	38,816	35,893	34,822	34,146
	DC ($)	60,695	55,222	51,028	47,867
	TC ($)	210,729	202,332	200,443	202,391
Loc#2	CC ($)	109,770	108,927	110,494	113,387
	EC ($)	38,288	36,859	39,022	44,321
	DC ($)	60,204	54,687	52,623	51,601
	TC ($)	208,263	200,472	202,139	209,309
Loc#3	CC ($)	112,061	112,061	117,172	129,034
	EC ($)	39,235	42,196	47,314	50,619
	DC ($)	60,054	57,033	57,287	55,892
	TC ($)	211,350	211,290	221,774	235,545

The voltage profile corresponding to the final conductor sizes for selected WPG capacities at the three locations are shown in Figs. 12–14. For each location, voltage profiles at minimum and maximum system loading conditions are given. From these figures it is observed that by increasing WPG capacity, voltage drop at different nodes for maximum loading condition is reduce and voltage rise at system nodes for minimum loading condition increases.

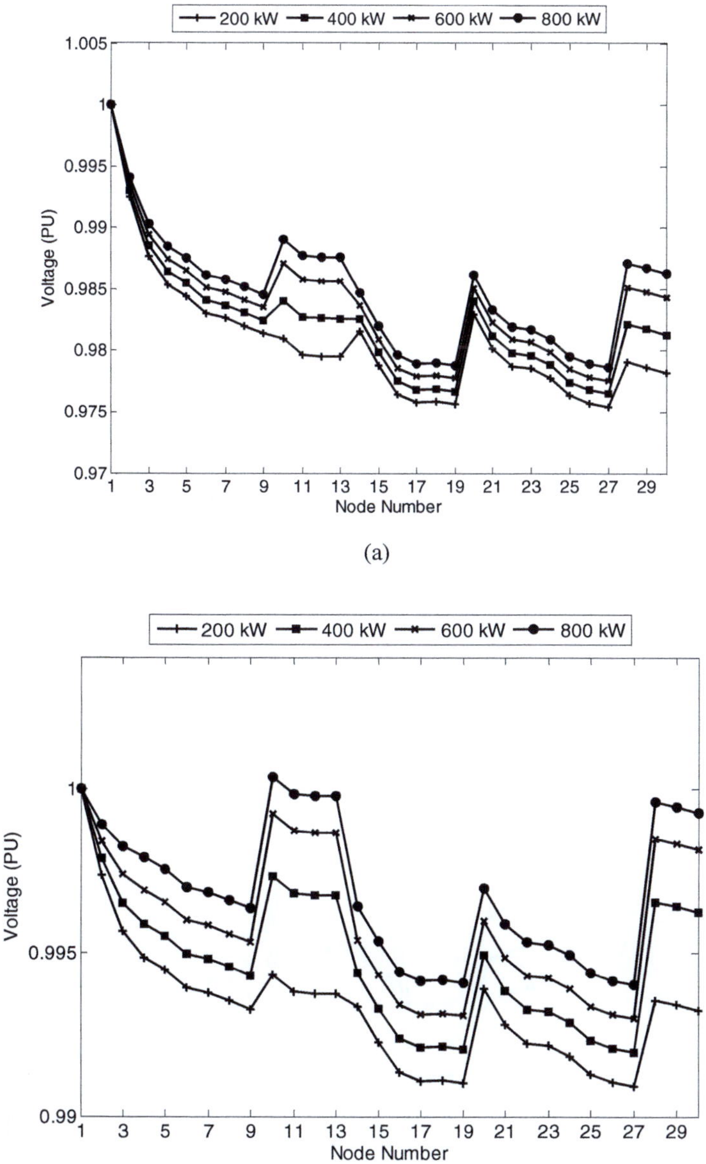

(a)

(b)

Fig. 12. The voltage profile corresponding to the final conductor sizing result for selected WPG capacities at Loc#1 at the maximum (a), and the minimum (b) loading conditions

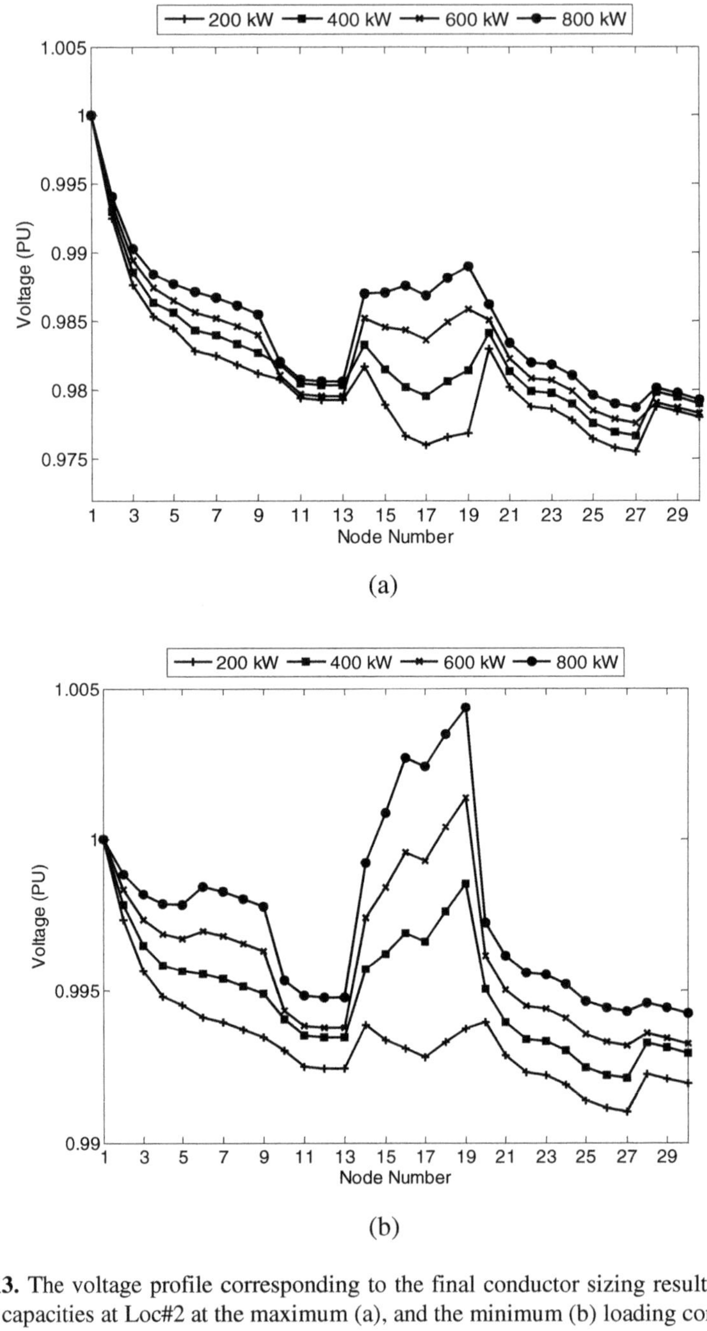

Fig. 13. The voltage profile corresponding to the final conductor sizing result for selected WPG capacities at Loc#2 at the maximum (a), and the minimum (b) loading conditions

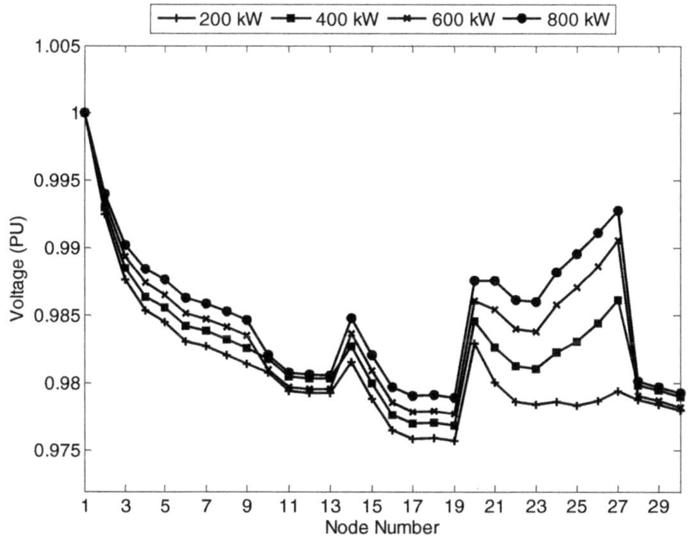

(a)

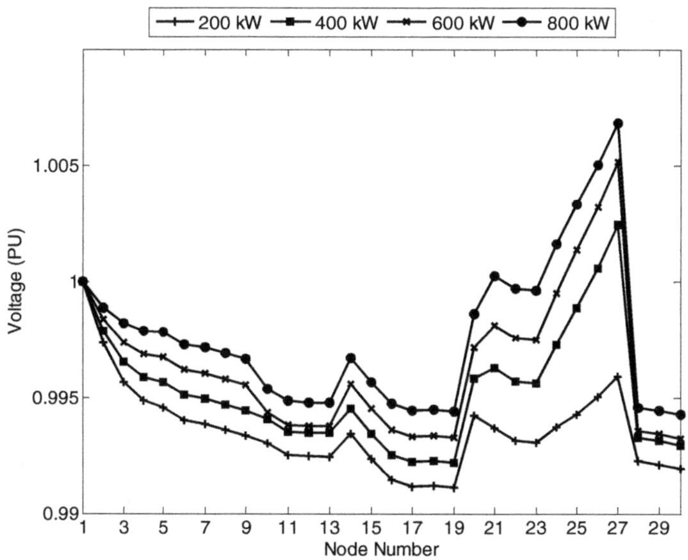

(b)

Fig. 14. The voltage profile corresponding to the final conductor sizing result for selected WPG capacities at Loc#3 at the maximum **(a)**, and the minimum **(b)** loading conditions

6 Conclusion

This chapter has presented a novel GA based approach for the optimal conductor size selection problem in distribution systems when WPGs are running within the system. The approach considers intermittent and probabilistic power injected by each WPG which is the main source of uncertainty in the operation of the system. The proposed approach employs probabilistic evaluation in conjunction with GA as a solution tool to model the varying behavior of the system and search the optimal conductor sizing solution. The approach takes into account the load variation and properly handles thermal capacity of conductors and voltage constraints. The capability and the performance of the proposed model have been demonstrated using different case studies. The results obtained show how the varying operation of WPGs has a major impact on result of the conductor sizing process. Results obtained through the presented approach help to understand the site specific influence of different WPGs penetration on the cost of conductor planning in distribution systems. Use of the proposed conductor sizing approach can enable distribution utilities to design their system in the presence of intermittent WPGs.

References

1. Funkhouser, A.W., Huber, R.P.: A method for determining economical ACSR conductor sizes for distribution systems. AIEE Transactions on Power Apparatus and Systems PAS 74, 479–484 (1995)
2. Kiran, W.C., Adler, R.B.: A distribution system cost model and its application to optimal conductor sizing. IEEE Transactions on Power Apparatus and Systems PAS 101(2), 271–275 (1982)
3. Ponnavaikko, M., Rao, K.S.P.: An approach to optimal distribution system planning through conductor gradation. IEEE Transactions on Power Apparatus and Systems PAS 101, 1735–1741 (1982)
4. Rao, P.S.N.: An extremely simple method of determining optimal conductor selection for radial distribution feeders. IEEE Transactions on Power Apparatus and Systems 104(6), 1439–1442 (1985)
5. Tram, H.N., Wall, D.L.: Optimal conductor selection in planning radial distribution systems. IEEE Transactions On Power Systems 3, 200–206 (1988)
6. Anders, G.J., Vainberg, M., Horrocks, D.J., Foty, S.M., Moths, J.: Parameters affecting economic selection of cable sizes. IEEE Transactions On Power Delivery 8, 1661–1667 (1993)
7. Willis, H.L.: Power Distribution Planning Reference Book. Marcel Dekker Press, New York (1997)
8. Salis, G.J., Safigianni, A.S.: Long-term optimization of radial primary distribution networks by conductor replacements. International Journal of Electrical Power and Energy Systems 21, 349–355 (1999)
9. Wang, Z., Liu, H., Yu, D.C., Wang, X., Song, H.: A practical approach to the conductor size selection in planning radial distribution systems. IEEE Transactions On Power Delivery 15(1), 350–354 (2000)
10. Mandal, S., Pahwa, A.: Optimal Selection of Conductors for Distribution Feeders. IEEE Transactions On Power Systems 17(1), 192–197 (2002)

11. Sivanagaraju, S., Sreenivasulu, N., Vijayakumar, M., Ramana, T.: Optimal conductor selection for radial distribution systems. Electric Power Systems Research 63, 95–103 (2002)
12. Falaghi, H., Ramazani, M., Haghifam, M.R., Roshan Milani, K.: Optimal selection of conductors in radial distribution systems with time varying load. In: CIRED 2005, paper no. 423 (2005)
13. Satyanarayana, S., Ramana, T., Rao, G.K., Sivanagaraju, S.: Improving the maximum loading by optimal conductor selection of radial distribution systems. Electric Power Components and Systems 34, 747–757 (2006)
14. Anjan, R., Venkatesh, B., Das, D.: Optimal conductor selection of radial distribution networks using fuzzy adaptation of evolutionary programming. International Journal of Power & Energy systems 26(3), 226–233 (2006)
15. Kaur, D., Sharma, J.: Optimal conductor sizing in radial distribution systems planning. International Journal of Electrical Power and Energy Systems 30(4), 261–271 (2008)
16. El-Khattam, W., Salama, M.M.A.: Distribution system planning using distributed generation. In: IEEE Canadian Conference on Electrical and Computer Engineering, vol. 1, pp. 579–582 (2003)
17. Masters, C.L., Mutale, J., Strbac, G., Curcic, S., Jenkins, N.: Statistical evaluation of voltages in distribution systems with embedded wind generation. IEE Proceedings on Generation, Transmission & Distribution 147(4), 207–212 (2000)
18. Cheng, C.S., Shirmohammadi, D.: A three phase power flow method for real time distribution system analysis. IEEE Transactions On Power Systems 10(2), 671–679 (1995)
19. Goldberg, D.E.: Genetic Algorithms in Search, Optimization and Machine Learning. Addison-Wesley Longman Publishing Co., Boston (1989)
20. Das, D.: Maximum loading and cost of energy loss of radial distribution feeders. International Journal of Electrical Power and Energy Systems 26(4), 307–314 (2004)

Global Optimization of Wind Farms Using Evolutive Algorithms

Angel G. Conzalez-Rodriguez, Javier Serrano-Conzalez, Jesus M. Riquelme-Santos, Manuel Burgos-Payán, Jose Castro-Mora, and S.A. Persan

Abstract. The design of a facility for wind power generation is a complex and multidisciplinary problem. The complexity of the problem derives mainly from the many interrelated variables and constraints or restrictions involved. Thus, the solution is usually obtained by heuristics after several cycles of trial and error, and it is heavily based on previous experience of the team planner. Evolutionary algorithms are efficient optimization techniques to tackle the problem of global optimization for wind farms by considering the turbines layout and electrical and civil infrastructure as a whole. The algorithm should evaluate each potential solution based on their economic returns over the entire period production of the wind farm, providing economic and financial information useful for prospective developers. Therefore, the algorithm needs to be driven by a thorough cost wind farm model that considers both the initial costs of acquisition and installation of equipment (initial investment) and the yearly cash flow. This cash flow is calculated as the difference between the incomes due to the energy selling and the ordinary maintenance and operation costs, along the whole lifespan of the wind farm. A final cost for the installation decommissioning and a residual value, after the facility production period, should also be considered.

The content of this chapter is organized into five main sections. After an initial introductory section, the problem of wind farm design and planning is formulated. Then there is a brief section on the basics of evolutionary algorithms. Having discussed the problem and the optimization technique, the next section is devoted to integral wind farm optimization through evolutionary algorithms. This section

---

Angel G. Conzalez-Rodriguez
Department of Electronic and Control Engineering of the University of Jaen, Spain
e-mail: agaspar@ujaen.es

Javier Serrano-Conzalez · Manuel Burgos-Payán · Jose Castro-Mora · S.A. Persan
Department of Electrical Engineering University of Seville, Spain
e-mail: javierserrano@us.es, jsantos@us.es, mburgos@us.es, jcastro@persan.es

L. Wang et al. (Eds): *Wind Power Systems*, Green Energy and Technology, pp. 53–104.
springerlink.com

includes a comparison with published works and a collection of new cases to test this new tool performance. The chapter ends with conclusions and references.

1 Introduction

Amongst all renewable energies, wind power is the one that has achieved a quickest and highest degree of development and technological maturity. At the beginning of 2009 the worldwide operational wind power capacity reached 120.6 GW. The regional leader is Europe with 65.9 GW wind power capacity and the country leader is USA with 25.4 GW [1]. The rate of yearly growing of installed wind power over the last years maintains a high stable value that exceeds 25%. This increase comes parallel with the size of recently installed or projected winds farms, rendering the development of a systematical tool for optimal design and installation of wind farms, a line of work of special relevance that is at the forefront, both from a technical and an economical point of view.

The initial task of identifying a suitable location for a wind farm requires to consider three key factors: availability and quality of the wind; availability and access to the electric power distribution network; availability and access to the terrain. Apart from these factors, it is advisable for the terrain to have a high bearing capacity, it should not be located in an area with risk of hurricanes or other natural disasters, and it should have a low level of thunder and lightning activity.

Other factors of equal relevance are the eventual barriers of a social or an administrative kind, or any sort of environmental protection, since any of these points could obstacle, delay or even prevent the obtention of the necessary administrative permissions.

Traditionally, once the area with potential enough to install a wind farm has been located, the geographical individual location for each turbine is projected [2-6]. With this preliminary setting out in mind, the wind farm designer habitually uses some commercial package for micro-location [7-9] to calculate the amount of energy potentially generable using that geographical distribution of the wind turbines, taking into account the topographical characteristics of the terrain and the wind available in the area. After a few trial-and-error iterations [2,3], the designer reaches a layout from which a certain potentially collectable energy is expected [4].

This chapter introduces the development of an aid tool for global optimization of wind farms based in genetic algorithms [10-12]. The proposed method combines a model of wind farm costs based in the life cycle of the facility and a method for searching optimal turbine location and wind farm configuration based in genetic algorithms.

The algorithm also manges the possibility of defining spatial constraints affecting the location of turbines or the presence of electric lines, like existing physical obstacles or protected areas.

The complexity of this problem arises not only from a technical point of view, due to strong links between its variables, but also from a purely mathematical point of view. The problem consists of both discrete and continuous variables, being therefore an integer-mixed type problem. The problem exhibits manifold optimal solutions (convexity), it can not be completely described in an analytical form, some variables have a range of non allowed values (solutions space not simply connected) and others are integers. This fact makes the problem non-derivable, preventing the use of classical analytical optimization techniques.

As we will see in section two, evolutive algorithms constitute an optimization tool that doesn't require the use of derivative calculations and turns out to be a suitable tool for this kind of problems.

In a first stage, and to outline the features for each optimization problem, the problem of global optimization of a wind farm is split up into two subproblems [13-15]: individual optimization of the location (or place) of the wind turbines, introduced in section three, and optimal design of the electrical infrastructure of the wind farm, that will be introduced in section four.

Section five introduces the problem of global optimization of the farm, analyzing in an integrated and jointly way the problem of individual location of the turbines and the design of the electrical infrastructure of the wind farm and its corresponding interaction.

The chapter finishes with a summary of the main conclusions of this work.

2 A Brief Overview of Genetic Algorithms

Genetic algorithms are robust optimum search techniques that find the minimum or the maximum of a function based on principles inspired from the natural genetic and evolution mechanisms observed in the nature [16-18]. These algorithms use multiple paths of search instead of single point, using encoded solutions to the problem (variable values or genotypes), instead of their real values. Their main principle is the maintenance of a set of encoded solutions (population) that evolves along the time, guiding the population towards the optimum solution. The evolution (optimization) process is based on the continuous Darwinian improvement cycle of evaluation, selection and reproduction of the best individuals, according to an objective function suited to the specific problem, as shown in Fig. 1.

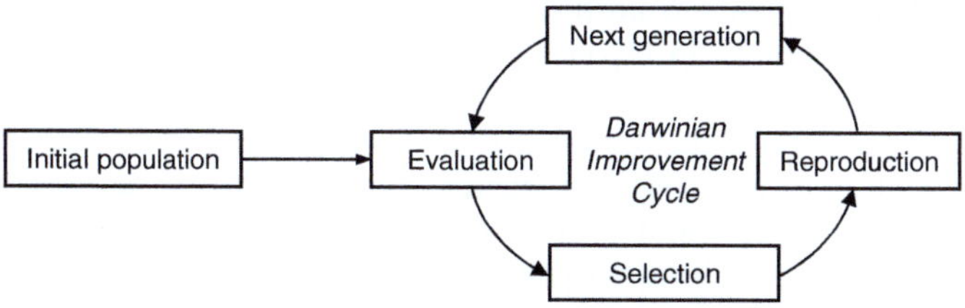

Fig. 1. Darwinian improvement cycle

The initial population can be settled both randomly or heuristically. Genetic reproduction is performed by means of a few basic genetic operators, mainly crossover and mutation that recombine highly fit individuals (best solutions). The evaluation of the population (solutions) is performed by means of a specific objective function (fitness) that depends on each particular optimization problem. The objective function plays a paramount role as this function guide the evaluation-selection-optimization process. The role of the objective function is similar to the gradient or the derivative in the conventional optimization methods. Individual selection is performed according to a selection strategy that chooses parents with probability proportional to their relative worth (fitness) and explores the solution space looking for better and better solutions.

As can be seen, genetic algorithms are, basically, mechanisms of search based on the species evolution, on the natural selection and survival of the best adapted individuals, objectively evaluated with a suited fitness function.

3 Problem Description

The relative low values of the wind turbine rated capacities available nowadays, compared to conventional power station units, means that a high number of turbines must be installed in a single site, a wind station or wind farm, in order to reach an installed capacity similar to a conventional power station. This wind turbine cluster disposition, more or less packed, offers some economic advantages related to the investment and to the plant operation and maintenance costs. But the wind turbine compactness degree is limited by spacing constrains due to wind shadow or wake decay effects. That is, when two wind turbines are placed too close one behind the other in the prevailing wind direction, the total amount of generated power (or energy) is less than the initially expected individual power sum (at the free air stream) because the wind power in the air stream available for the downwind turbine is reduced due to the wind power extracted by the upwind rotor turbine. As a consequence, the layout or specific individual wind turbine position determines the overall potential wind energy extraction efficiency of a wind farm.

Most sites with the best wind conditions are already in exploitation to guarantee fast revenue of the investment. So, nowadays the wind energy potential of the non exploited sites are not so exceptional but due to new research applications and economy of scale (mass production), the unitary cost of the installed wind power kilowatt is now lower and wind turbines are bigger and more efficient today. In this scenario, the wind farm layout design must be carefully analyzed, choosing the best solution that provides the higher profit for a certain investment. There are sophisticated programs that, starting from the available wind speed distribution, topography and roughness characteristic data of a specific site, are able to estimate the performance of a potential wind turbine layout. Some of these packages also provide visualization facilities to generate noise maps or photomontages of the wind farm to predict the visual impact. However, in spite of the huge growing experienced by this technology still nowadays there is scarce significant literature about the optimization of wind turbines positioning problem, probably due to its complexity. The problem exhibits manifold optimal solutions (convexity), cannot be completely described in an analytical form, some variables have a rank of non allowed values (solution space no simply connected) and some others are discrete variables. This makes the problem non-derivable, preventing the use of any of the classic analytical optimization techniques.

3.1 Previous Approaches

Evolutive algorithms have been used in the past in similar applications of optimal solution search, both for the design of turbines layout in a wind farm and for the design of its electrical infrastructure. The number of contributions in both fields is uneven.

The first *concern*, that of searching for the layout of optimal turbines using genetic algorithms, has been raised only by two relevant works:

- The first one by G. Mosetti, C. Poloni, and B. Diviacco [19], dated in 1994, uses a very simplified model of simulation of a wind farm based in scale-economies and overlapping of wakes.
- One decade later, in 2005, S.A. Grady, M.Y. Hussaini and M.M. Abdullah [20], using the same model and the later structure, achieved some quantitative improvements based on the same assumptions.

The authors agree in just wanting to proof the applicability of the proposed method, allowing them to justify some simplifications introduced to the models.

Regarding the optimization of the design of the electrical infrastructure, some techniques and results can be used that derive from generic research on planning of electrical networks, that could also be extrapolated both to the design of the electrical infrastructure of wind farms and to the selection of turbine location.

To solve this type of non-linear problems several optimisation algorithms have been developed that, according to its resolution technique, can be classified into two different categories:

- Mathematical optimization techniques. They use a calculation procedure to solve the exact mathematical formulation of the problem. The response deteriorates in systems with strong non-linearities or where discrete variables are used.
- Heuristical optimization techniques. They represent the current alternative to mathematical methods. Heuristical methods generate, evaluate and select possible options for expansion and modification of the installation. To achieve this, they perform local searches following either empirical rules or rules based on sensitivities, to generate and classify possible solutions during the search. The process ends when the algorithm cannot find a better solution.

Some recent approaches focus into modern heuristical techniques. Amongst other advantages, heuristical methods can generate a number of alternative solutions, there are no mathematical constraints in the formulation of the problem, and they are relatively easy to program and numerically robust.

The only inconvenient of these models is that they cannot guarantee converging to an absolute optimal, since they cannot evaluate all possible combinations, a massive set for this type of combinatorial problems.

The heuristical techniques of optimization most frequently used when solving planning problems are genetic algorithms [21-25], simulated annealing [26], ant colonies systems [27-30] and tabu search [31-35].

Apart from more specific bibliographic notes, a global review on wind installations can be seen in [4,6,36-45], and on genetic algorithms in [16-18]. Wake models can be found in [46-52], whereas in [53-60] you will find aspects related with the integration of wind farms in the distribution and transport electric power network. Finally, other elements can be found in [61-72], such as cost models, estimate of parameters or applications of genetic algorithms related with wind power technology, amongst others.

3.2 Application of Genetic Algorithms to Wind Farms

The problem of positioning wind turbines in a wind farm is a typical discrete one that, in the same way as the travelling salesman problem, is unviable to be solved exactly beyond a given size of the problem. On the other hand, methods based in the gradient of the objective function are not easily applicable, mainly in non-connected spaces.

Nevertheless, a search genetic algorithm, like the ones originally developed by Holland [17], could be directly and easily applied to solve the problem of designing a wind farm, including both the optimisation of location and the electrical infrastructure.

The resolution of the global problem has been initially approached searching separately for the optimum for these two subproblems: location, and electric infrastructure. At a later stage, the problem has been solved jointly, without splitting it in two subproblems. The analysis of pros and cons of both methods has allowed to develop a hybrid method that, in most cases, results in the same solution provided by the integrated search, but in a significant lower amount of time.

3.3 Using NPV as a Measure of Profitability

Table 1 shows the cost distribution of a wind farm adapted from [73]. As can be seen, almost three fourth parts of the initial cost corresponds to the wind turbines. The remaining chapters correspond to the electric infrastructure, the civil work, components erection and installation and other costs. Besides that, the yearly operation and maintenance costs (about 3% of the initial installation cost) and the wind farm unavailability (typically 2%) must be considered. Finally, at the end of the wind farm production life, the decommissioning cost and the equipment residual value must be considered. The present removing cost and the residual value are similar and equivalent to the 1-3% of the initial installation cost.

Table 1. Typical Initial Cost Structure of a Wind Farm

Item	%
Wind turbines	65-75
Substation and electrical infrastructure	10-15
Civil work	5-10
Component installation	0-5
Other	5
Overall cost (€/kW)	800-1100

The criterion for optimising the layout of wind farm turbines will be that of getting maximum investment profitability, once the technical constraints or any other kind of applicable constraints have been met. As a measure of economic profitability of the investment required installing and commissioning a wind farm, the Net Present Value (NPV) through the life cycle of the wind farm has been chosen, being this parameter usual in decision making that involves investment alternatives.

Therefore, a wind farm with a certain turbine configuration, x, requires an initial capital investment to build and put the installation into production, $I_{WF}(x)$. This initial investment is necessary mainly to face wind turbines acquisition, wind turbine civil work and electrical infrastructure costs. On the other hand, the wind farm delivers a stream of both financial benefit (profits from the energy selling), $P_{WF}(x)$, and ordinary operation and maintenance costs, $C_{O\&M}(x)$, year after year, during the installation life time (production period), LT. A final present cost for the installation decommissioning, $C_D(x)$, and a present residual value, $V_R(x)$, after the production live, must also be considered. So, the net present value of the wind farm initial capital investment, $I_{WF}(x)$, for an installation live spam of LT years with an equivalent discount rate, r, can be written as:

$$NPV(I_{WF}(x)) = -I_{WF}(x) - C_D(x) + V_R(x) + \\ + \sum_{k=1}^{LT} \frac{E(x)\, p_{kWh}(1+\Delta p_{kWh})^{k+1} - C_{OM}(x)(1+\Delta C_{OM})^{k+1}}{(1+r)^k} \tag{1}$$

Fig. 2 shows a simplified block diagram structure of the wind farm global/layout optimization algorithm. As can be seen, there are four main modules –the initial wind farm cost model, the wind farm production model, the wind farm operation and maintenance cost model and the wind farm removing cost model. Each of them integrates two to four submodels.

From the analysis of the previous expression it can be concluded that, in order to minimize the investment and operation costs, it is desirable to reduce the distance between wind turbines (*compact wind farm*) and to select the turbines having a lower acquisition cost. The downside is that choosing the compact wind farm solution, integrated by the most economical wind turbines (generally less efficient), reduces the investment costs at the expense of a double reduction (wakes and performance) of the return obtained by sale of the generated electrical power. On the other hand, a layout with a higher spacing between the wind turbines and more efficient turbines would have the opposite effect. Therefore, the role of the wind farm design team is to get a configuration/layout x that maximizes the preceding economic figure.

Considering the structural complexity of the problem, both from a technical-economic and the purely mathematical point of view, and as a first approach, the problem of the global wind farm optimization has been separated in two subproblems [13-15]:

- Optimization of the location (or siting) of individual wind turbines (turbine type and height, location and configuration of the auxiliary road web)

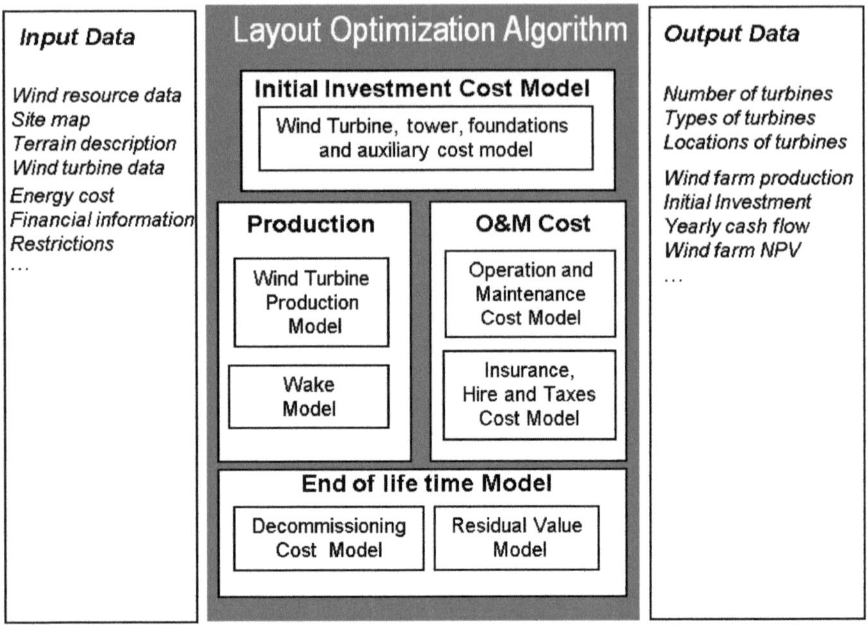

Fig. 2. Block diagram of the wind farm cost model considered by the optimization evolutive algorithm

- Optimization of the configuration of the wind farm electrical network (type, location and placement of the components of the wind farm inner distribution network, substation and HV evacuation line)

This problem division leads to a simplification derived from the decoupling of the problem and it can be explained in terms of the economic relevance of each of them (Table 1).

From a purely economic point of view, the first of the problems, the location of the individual wind turbines in the park, is the most significant since it is responsible for between two thirds and three quarters of the total investment required and is the most directly affect the annual production of electricity, i.e. the posterior return of the investment (ROI) [13].

The second problem, the design of the wind farm electrical infrastructure, is very similar to the design of a new radial network [25-28] and has less significance, in terms of initial investment. But the configuration of the electrical infrastructure also affects the yearly net production (energy) of the wind farm because the electrical losses.

Both problems can be analyzed using economic functions that allow analysis and comparison of different cost components related to the design and operation of a wind farm. Obviously, this division of the problem leads to suboptimal solutions, but rather close to the global optimum, as will be shown later.

This kind of optimum sequential solution, which is simply the result of combining the optimum site solution and the optimum network configuration, is potentially superior to the solution derived from the usual heuristic wind farm methods. This is due to the fact that sequential optimum solution guaranties both the optimum location of the turbines and the complete electrical infrastructure, taking into account the inner distribution network, the substation and the HV evacuation line.

Later, as a further improvement, the problem of the wind farm global optimization, is faced. Now, both problems, the wind turbine sitting and the network configuration, are considered as an integrated problem. This approach leads to an interaction between the location of individual turbines and the wind farm electrical infrastructure design, not allowed in the sequential approach.

4 Problem of Location of Wind Turbines

This chapter introduces the problem of finding an optimal layout of the wind turbines on the terrain, as well as selecting the type and height of a wind turbine that is most suitable to the conditions of the problem. A genetic algorithm is specifically developed for its determination.

4.1 Model of Initial Costs of a Wind Farm

In broad strokes, the necessary investment to commission a wind farm can be broken down into three entries: investment in wind turbines, investment in civil works and investment in electrical installation, both in internal distribution and in substations and electric lines of power disposal. A typical distribution of the wind farm installation costs approximately corresponds to 70% for wind generators, 15% for electrical installation and the remaining 15% for civil works and additional entries.

In order to not take the system into account in the first stage, during the search for an optimal location the costs linked to electrical installation will be approximated, considering them numerically identical to the costs linked to civil works.

Investment in Wind Turbines

Total investment in wind turbines in a wind farm equals the sum of acquisition costs of the turbines plus the costs of the towers.

The most important feature of a wind turbine, with respect to characterizing the efficiency of conversion of the wind kinetic energy in electricity, is its power curve. This curve describes the evolution of the net electrical power generated by the wind turbine as a function of the wind speed, within the range going from 0 to 25 m/s at 1 m/s intervals.

Without getting into further detail for the analysis and classification of the types of wind turbines, let's just say that some wind turbines present a wider working speed range. A greater variability in the rotor speed means a better adaptation to wind conditions and therefore a higher use of the wind, however requiring more expensive turbines.

Apart from the investment needed to buy the turbine, an extra investment is required for the acquisition of the mounting tower. Since the speed of the wind greatly increases with the height over the level of the terrain (*shear effect*), the higher the height of the hub, the higher the amount of usable energy, at the expense of higher costs for the tower and foundations.

As a summary, an analysis is needed on the energy collectable for each model and turbine height, for the current wind conditions and for the location being studied.

Investment in Civil Works

As previously stated, civil works costs account for roughly 15% of the total investment and it is mostly made up of: costs for the foundation of the machines, costs to transport the turbines and tower sections, and costs of execution of the service roads that connect the main road to the final location of the wind turbines.

It is also necessary to take into account the orography of the terrain and the existence of areas where installing wind turbines could be extremely difficult, or not even possible, due to existing rivers, dams, environmentally protected areas, proximity to populated areas, etcetera.

Foundations costs for each machine mainly depend on the height of the tower used, the diameter of the machine (related to its power) and the bearing capacity of the terrain. The cost model uses average values for each machine, corresponding to terrains with a high bearing capacity, and then adds an increased cost factor for terrains with a low bearing capacity.

Transport costs for the nacelle, the blades, and the sections of the tower, as well as the part corresponding to the wind turbine mounting, will be considered fixed for each machine, no matter which position in the wind farm they are located in. A variable cost for each machine is also considered depending on its distance to the main road.

Service roads costs are proportional to their length. Therefore, it is necessary to calculate the network of service roads connecting the wind turbines to the main road. To design the network of service roads the following procedure can be used (Fig. 3):

1. Calculate the distance from each wind turbine to the main road/s. The wind turbine with the minimum distance is then connected through the service road. This wind turbine will now remain *connected.*
2. Calculate the distances from each one of the non-connected wind turbines to the wind turbines that are connected, and also to the main roads.
3. The lower distance identifies the next turbine to be connected, next go back to step 2.

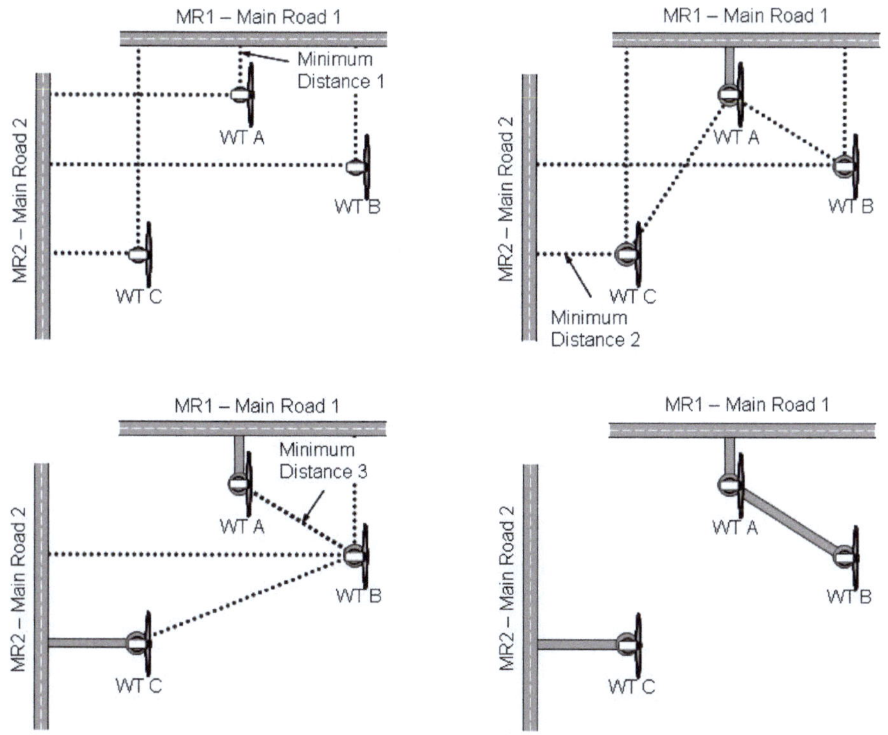

Fig. 3. Auxiliary road design for a set of three turbines and two main roads

4.2 Model of Evaluation of Electric Energy Production

It is necessary to calculate or estimate the energy produced by the set of turbines that compose a given layout, and also the operation and maintenance costs.

4.3 Annual Energy Generated by a Wind Farm

To estimate the energy generated by a set of wind turbines composing a wind farm for each year of the working life of the facility, it is necessary to have the available information on wind parameters such as speed, frequency and direction. Direction is not important for the location of isolated turbines, but it is one of the most relevant factors in selecting and optimizing wind farms, due to energy losses produced by wakes. Let's start estimating the energy that could be obtained for each turbine if it were isolated, next we will estimate the energy that could be generated by a wind farm with a specific turbine layout.

Evaluation of Energy Generated Per Turbine

The energy captured by an isolated turbine can be estimated by multiplying the number of hours the wind is expected to reach a given speed by the power generated by the turbine at that speed. To get the number of hours the wind blows at a given speed two expressions are generally used: the Weibull distribution and the wind shear expression.

Weibull Parameters

The analysis of measurements made during the observation period allow us to calculate the shape parameter K and the scale parameter C for the Weibull distribution, both being used to estimate the probability or frequency at which a wind speed, v, appears:

$$p(v) = \frac{K}{C} \cdot \left(\frac{v}{C}\right)^{K-1} \cdot e^{-\left(\frac{v}{C}\right)^{K}} \tag{2}$$

The scale parameter, C, like the average wind speed, shows how windy, on average, is the location. The shape parameter, K, indicates how pointed the distribution is.

Wind Shear

Once the behaviour of the wind has been estimated at a given reference height (that of the measurement equipment, typically 50 meters), to calculate the energy produced by a wind turbine it is necessary to know the wind behaviour at the height of the hub. Wind speed increases with height, this is known as wind shear. Hence, when the wind speed $v(z_{ref})$, measured at a reference height, z_{ref}, is known, the corresponding wind speed at a different height, z, can be calculated by means of an exponential function like that of (3), where z_0 is the length of roughness for the terrain.

It can be demonstrated that the shape constant K of the Weibull distribution is not affected by height, whereas the scale parameter, C, is affected in the same way as the speed field (3).

$$v(z) = v(z_{ref}) \cdot \frac{Ln\frac{z}{z_0}}{Ln\frac{z_{ref}}{z_0}} \Rightarrow \begin{cases} K \neq f(z) \\ C(z) = C(z_{ref}) \cdot \frac{Ln\frac{z}{z_0}}{Ln\frac{z_{ref}}{z_0}} \end{cases} \tag{3}$$

Thus, using the data for speed, roughness and direction, it is possible to obtain a Weibull function calculated for the height of the hub.

Evaluation of the Energy Generated by a Wind Farm

In wind farms with multiple turbines each one of the turbines has the potential to generate wakes that reduce the energy available downstream in the direction of the incident wind. Apart from the previous measurements, it is necessary to know with which frequency the winds blows for each direction. A prevailing direction will set a trend to install the turbines in lines perpendicular to that direction.

Wind Rose

The wind rose is a chart that shows the frequency the wind blows for each direction, and the average speed achieved when doing so (or quadratic average speed). Wind roses come with 8, 12, 16, 32, … sectors.

Likewise, it is more accurate to obtain the Weibull parameters and the length of the roughness separately for each one of the sectors. For example, in locations close to the sea, the behaviour of the wind (speed) is different when the wind blows from that direction compared to blowing from inland.

Wakes Model

To calculate the annual total energy of a wind farm it is necessary to evaluate the influence that a wind turbine has on the turbines located downstream (leeward) in the wind direction, also known as wake. The collection of the wind energy performed by the turbines reduces the speed of the wind through it, causing a reduction in the kinetic energy available for the turbines located downstream in the direction of the incident wind.

Figure 4 schematizes the evolution of the wind speeds of the stream going through the rotor. The speed of the incident wind, U_0, considered in the first instance equal to the speed of the wind in free flow, reduces its axial component after passing through the area formed by the moving rotor, U_a.

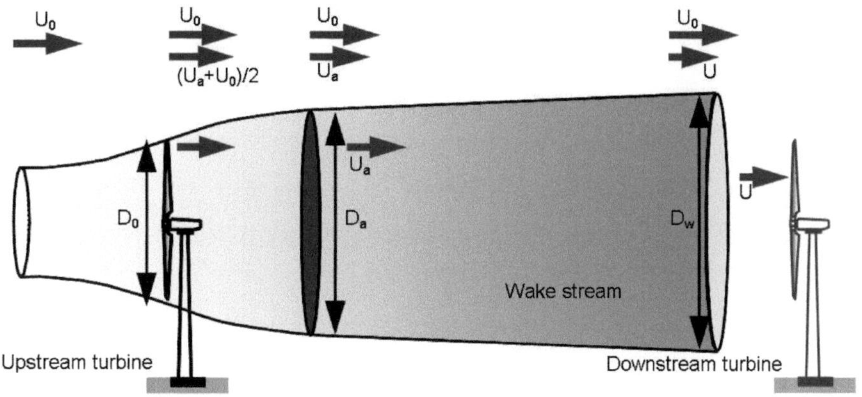

Fig. 4. Evolution of the wind speed field in a wake

For a point located in the wake stream of a turbine following the initial expansion area, the reduction of the wind speed can be calculated assuming the kinetic momentum of the air mass remains unchanged. Frandsen suggests in [50] that the ratio of speed in points of the wake stream compared to the speed of free flow can be expressed by (4).

$$\frac{U}{U_0} = \frac{1}{2} + \frac{\sqrt{1 - 2C_T \frac{D_1^2}{D^2}}}{2} \tag{4}$$

where D_1 and D are the diameters of the areas swept by the rotor in a point of the wake stream at an x distance, respectively, being D:

$$D(x) = \left(1 + 0.2\frac{x}{D_1}\right)\sqrt{\beta}D_1 \tag{5}$$

However, considering this new speed in the calculation of the energy of a turbine located downstream would be equivalent to consider the whole area swept by that wind turbine to be affected by the wake stream and, in principle, this would only be correct for turbines located in the same incident direction of the wind, but not for the ones that are slightly apart from that direction.

The percentage of the rotor area affected by the wake stream is calculated geometrically using the diagram shown in Figure 5.

Hence, the overlap area is:

$$A_{overlapped} = R^2\left(\gamma_R - \frac{\sin(2\gamma_R)}{2}\right) + \tag{6}$$

$$+r^2\left(\gamma_r - \frac{\sin(2\gamma_r)}{2}\right) \quad \forall X : R - r \le X \le R + r$$

$$A_{overlapped} = 0 \quad \forall X \ge R + r$$

$$A_{overlapped} = \pi r^2 \quad \forall X \le R - r$$

where R and r are the radiuses of the wake and rotor girths being analyzed, respectively, and the angles γ_R y γ_r are:

$$\gamma_R = \cos^{-1}\left(\frac{R^2 + X^2 - r^2}{2XR}\right) \tag{7}$$

$$0 \le \gamma_R \le \frac{\pi}{2} \quad \forall X : R - r \le X \le R + r$$

$$\gamma_r = \cos^{-1}\left(\frac{R^2 - X^2 - r^2}{2Xr}\right) \tag{8}$$

$$0 \le \gamma_r \le \frac{\pi}{2} \qquad \forall X : \sqrt{R^2 + r^2} \le X \le R + r$$

$$\frac{\pi}{2} \le \gamma_r \le \pi \qquad \forall X : R - r \le X \le \sqrt{R^2 + r^2}$$

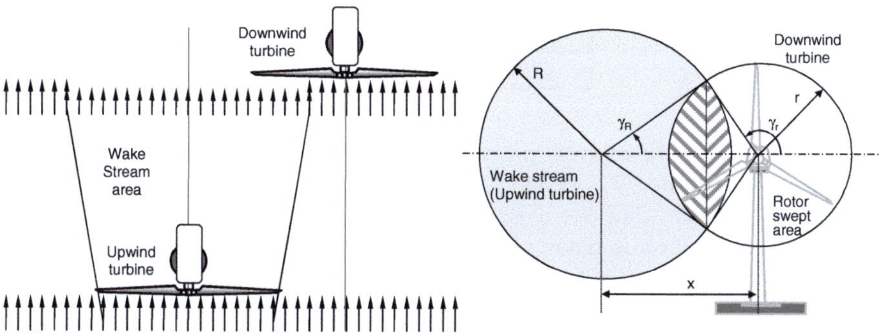

Fig. 5. Partial wake. Diagram for calculating the section of the turbine rotor area partially affected by a wake.

Besides, in wind farms there are generally perturbations produced by more than just one wind turbine, with existing interferences between different wakes causing a final reduction in the speed you need to calculate. Katic and Jensen in [47] propose that the deficit of equivalent speed can be calculated by using the average quadratic sum of the deficits produced by each turbine separately (9).

$$(U - U_0)^2 = \sum_i \frac{A_{overlapped\_i}}{\pi \cdot r^2} (U_i - U_0)^2 \tag{9}$$

where U is the speed to be considered in the turbine downstream, and U_i is the speed of the air due to the wake of turbine i.

This model must be completed with the effect on turbines located in the direction transversal to the wind incidence, due to the creation of a backwater zone in the speed field just in front of the area swept by the rotor. The result is that the wind speed is linearly reduced to transversal positions located four times away the diameter (4*D*).

4.3 Proposed Method

Next we will explain the coding used to solve the problem as well as the specific operators developed for this goal.

Codification

Each potential solution (individual) corresponds to a solution of the problem of selection of a number, type, height and position of the turbines of a wind farm. Therefore, each individual will be codified in such a way that the involved parameters are defined univocally, as shown in Table 2.

Table 2. Array representation of an individual

X_1	X_2	...	X_n
Y_1	Y_2	...	Y_n
T_1	T_2	...	T_n
H_1	H_2	...	H_n

Where you can see the relevant properties: position for each generator, X_k and Y_k , type of wind generator, T_k , height of tower for each generator, H_k, and number of turbines, n.

Since each solution is made out of a variable number of wind turbines, when crossing individuals you must allow for different sizes. Furthermore, for the algorithm to work properly it is necessary to guarantee diversity in sizes.

In order to codify the position of each generator the terrain is discretised in cells, whose dimension or number will depend on the desired accuracy.

The height of the towers has been discretised in 5 m modules, and the type of machine will match its index in the machine database being used.

Population Evaluation

The *fitness* of each individual is obtained from the profitability of the layout that the individual proposes as a solution to the problem of location and selection of turbines. Fitness can be measured with different economical indicators, like for case the *Net Present Value* (NPV).

Individuals with negative values of NPV correspond to unviable projects and will have little effect in future generations. The higher the NPV, the higher the probability for an individual to be chosen to generate the population of the next generation.

When the best individual of the population repeats during a particular number of generations, the algorithm takes it as a sign of having reached the optimum and the evolution stops (criterion of convergence).

The proposed evolutive algorithm accepts two simultaneous global schemes of execution:

- Obtaining the optimal solution for a wind farm with a maximum number of predetermined wind generators.
- An optimal solution for a wind farm that maximizes the profit for a maximum limit of available investment.

Next we will introduce the operators specifically developed for this application.

Selection operator

The selection of the individuals on which the crossover and mutation will be applied is performed in a pseudorandom way, being the probability of an individual to be selected proportional to its fitness. Thus, the population evolves with the features of the fittest individuals. The method employed is also known as roulette method, where the individuals with the higher *fitness* get a greater number of sectors, therefore the probability of the roulette to stop in one of these positions is proportionally higher (Fig. 6).

To prevent an individual from being selected a high number of times a coefficient is used that penalises its fitness every time it is selected. This prevents the population from stagnating in the initial stages of evolution.

Individual	Fitness
A	25 (44%)
B	15 (27%)
C	10 (18%)
D	5 (9%)
E	1 (2%)

=▶

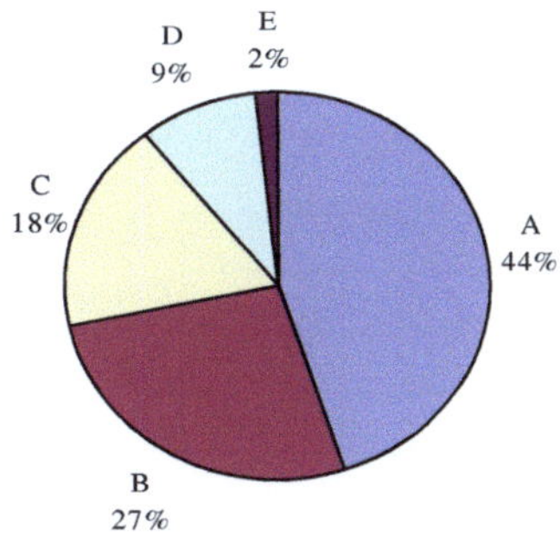

Fig. 6. Roulette method

Crossover Operators

The crossover operation is the main tool that allows the population to evolve towards an optimal solution. It recombines the genes of two individuals to get two new ones.

Individuals are selected randomly according to their fitness, depending on their *selection* operator. A parameter has been added, called *crossing probability*, so that once the parents have been selected the operation will be carried out only if a randomly selected number exceeds that parameter. Five different types of crossings have been implemented, that are randomly applied, in order to obtain a higher diversity for each generation and therefore speed up the convergence process.

Any inconsistencies got in the childs, like duplication or out of range values, are fixed by the operator *correct* that will be described later.

Operator Mutation

Mutation allows evolution when the population has stagnated around a local maximum, allowing to carry on the exploration of the solutions space in search for a global maximum.

Just like in the crossover operation, the *selection* operator chooses the individual amongst those having a higher fitness. An operator called *probability of mutation* has also been introduced to control the number of mutations. The use of mutation should be moderate, since an intensive use of this operator would result in a random search.

Once it is selected, a pattern governed by a random mask is generated, that will determine which genes will be mutated, as shown in Figure 7. Mutation changes the value for each quality or gene of the affected individual. The degree of alteration is, within the range permitted for that quality, also random.

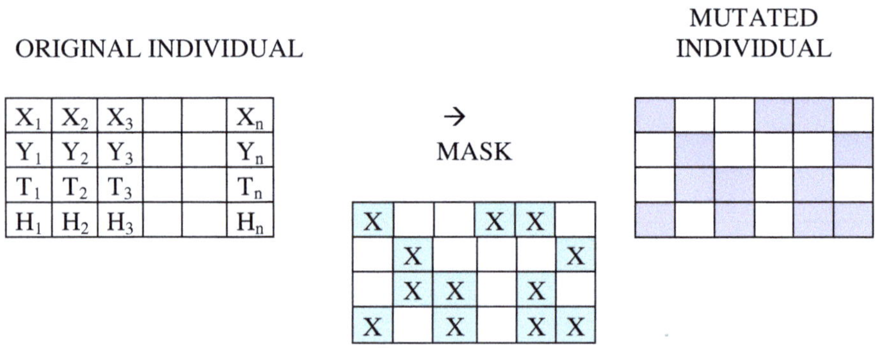

Fig. 7. Mutation

Operator Correct and Crop Individual

Individuals obtained applying the crossover and mutation operators can be non valid individuals in that they might include out of range values of position, type or height; or in that they might contain more than one machine in the same position. The operator *correct* reviews and, if needed, modifies the genes of the individuals to make them valid.

Next, the operator *correct* verifies that the number of machines for each individual complies with the *maximum number of machines* constraint, in case there is one. If this is the case, it executes a procedure that keeps the group of machines that provide a higher profit. It also verifies whether the investment limit has been exceeded or not. If this is the case, it evaluates options such as a reduction in the height of the tower, a shift to more economical turbines or an elimination of turbines, keeping the option of a lower NPV reduction.

Local Search Operator

As in the case of genetic algorithms applied to other problems (like the travelling salesman problem), a local optimization operator appears to be convenient to reduce the computational effort. In the proposed algorithm, after a number of generations without changing the best individual, a local search operator has been applied to the best individual and also to a randomly selected one. The local search operator moves every turbine to each one of the free positions of the farm, evaluating whether the NPV has improved or not. The best NPV improvement generates a new individual as seen in Figure 8.

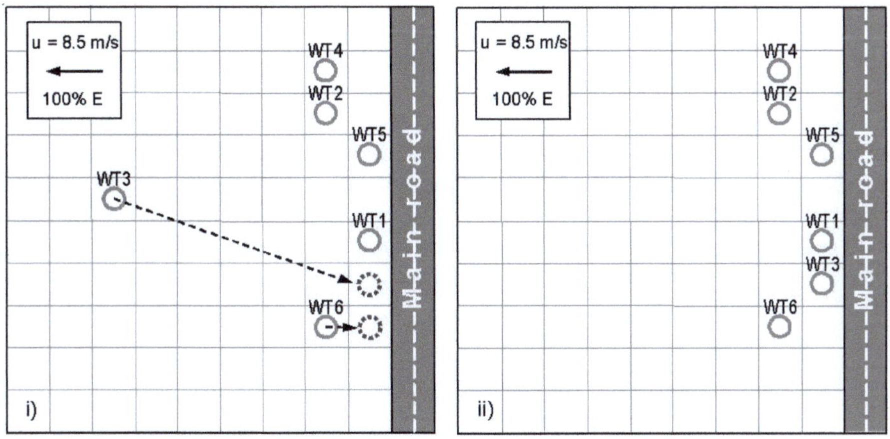

Fig. 8. Local search operator

Constraints and Generation Evolution

In the location of the wind turbines a consideration must be made on the existence of areas where the location of machines is forbidden, due to physical obstacles, buildings or environmentally protected areas. The layouts (individuals) with machines located in any of these forbidden positions are penalised with a highly negative NPV, and will be candidates to be deleted from the population without participating in crosses or mutations.

On the other hand, once the population considered has generated the population resulting from the crossover and the population resulting from the mutation, and after being subject to the *correct* and *crop individual* operators, joins these two new populations and acts as the basis for the next generation. Once the duplicated individuals have been deleted, the final population will be made out of the best t_p individuals, being t_p the *size of the population*, which is adjusted depending on the complexity of the case to be solved. If, on the contrary, once the duplicated individuals have been deleted the resulting population still doesn't reach the indicated size, new individuals will then be generated by mutation.

The bigger the size of the population, the bigger the variety of individuals, becoming more likely to find the global maximum in few iterations. On the contrary, the computing time for generation will increase.

Example

This section will put to the test the fitness of the algorithm introduced by resolving a general instance.

For this instance it has been considered a 4000 m long on each side square flat area, that has then been discretised using a 10 x 10 cells array. The terrain possesses a good bearing capacity in every point. The central area of the terrain is crossed West to East by the main road (see Table 3).

Regarding the existing wind the assumptions are a scale factor $C = 12$ m/s and a shape factor $K = 2$. The roughness coefficient (given by its height) is 0.0055 m. These values are identical for every sector.

For the wind direction an 8-sector wind rose has been used with a 50% probability for the wind to blow from the North-South direction, 25% of blowing from the Northwest-Southwest direction and the remaining 25% doing it from an East-West direction.

There are two areas where installing the wind turbines is forbidden. Said areas are located on both sides of the main road, being shown in Figure 9 in dark colour.

12	12	12	12	12	12	12	12	12	12
12	12	12	12	12	12	12	12	12	12
12	12	12	12	12	12	12	12	12	12
12	12	12	12	12	12	12	12	12	12
12	12	12	12	12	12	12	12	12	12
12	12	12	12	12	12	12	12	12	12
12	12	12	12	12	12	12	12	12	12
12	12	12	12	12	12	12	12	12	12
12	12	12	12	12	12	12	12	12	12
12	12	12	12	12	12	12	12	12	12

Fig. 9. Values of the scale factor, *C*, of the Weibull function, main road (light colour) and forbidden areas (dark colour)

Let's consider three kind of turbines with the same power, 2000 kW, similar prices and features, but being designed for different wind conditions. Machine 1 is class -A (according to IEC-61400-1), optimized for locations with average speed $v_{av} = 7.5$ m/s. Machine 2 is class II, with $v_{av} = 8.5$ m/s and machine 3 is class I-A, with $v_{av} = 10$ m/s. Assuming a shape factor $K = 2$, the previous average speeds occurred for scale factors of 8.4 m/s, 9.5 m/s and 11.2 m/s, respectively. The maximum number of turbines has been limited to 10.

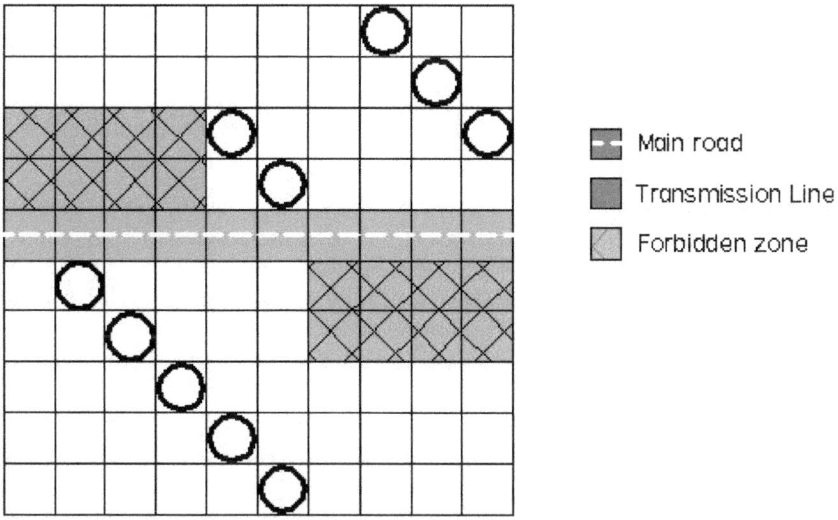

Fig. 10. Optimal solution of the described problem

The solution reached by the algorithm is shown in Fig. 10. As expected, the resulting layout proposes a Northeast orientation for the wind farm, hence minimizing the influence of wakes. On the other hand, the rows of wind turbines branch off the main road, therefore attaining a minimal length for the service roads.

The chosen machine corresponds to machine 3, with the maximum available mounting height, 100 m. This is a consistent solution as this turbine is the most suitable for high winds.

5 Electrical Infrastructure for the Wind Farm

This section introduces the design of the path of the electric power distribution network. For an electrical network, a wrong design can cause unreasonable losses due to Joule effect, reducing the available energy that can be sold. Therefore, the design of the electrical infrastructure needs to be done also "balancing" the initial cost to the investment, along with an annual extra cost due to Joule losses in case of not choosing the optimal location. Investment NPV will again be used as a selection criterion of the optimal solution.

5.1 Problem Approach

The electrical installation of a wind farm consists of two components:

- Internal installation of medium voltage distribution (1 to 35 kV according to ANSI/IEEE 1585-2002 and IEEE Std 1623-2004). This is the energy collecting installation that connects wind turbines one to each other and with the boosting substation-s.

- Installation for disposal of high voltage energy. This is the line of energy disposal connecting the substation-s with the line of the electrical distribution system previously existent, that allows the use of the electric energy generated by the farm. Although not being the most usual, we will consider the most general case of several high voltage lines.

The problem of designing the electrical infrastructure of the wind farm comes down to determine the number of substations to be used and their location, as well as the specific layout of the high and medium voltage collecting networks.

For the resolution of this problem the initial data are the position of the wind turbines and the path of the existing transport lines. It also has to take into account certain constraints like forbidden areas or those of a technical kind (limits of energy disposal on the high voltage lines, maximum admissible current or tension drops).

As a result, the design of the electrical installation brings in a new optimisation problem solved by an evolutive algorithm with specific functionalities. Whenever the algorithm reaches the convergence criterion, defined by the optimal solution repeating for a given number of generations, it stops and shows the reached solution.

5.2 Operator and Algorithm Specific Tools

Codification

To speed up the evaluation of individuals, it has been preferred not to codify the location of substations, therefore reducing the individuals codification to represent the element (turbine or substation) to which each turbine is connected. Calculation of the exact substation coordinates is only necessary for the exact evaluation of costs and will be described in detail later.

In this sense, each individual represents a possible solution to the proposed problem and it is represented by a vector, A_i, with a $2\times np$ size, np being the number of wind turbines.

The first np elements of this vector correspond to a permutation of the np wind turbines or generation points.

The next np elements refer either to other wind turbines or to the substations to which the previous generation points are connected. To differentiate when a connexion is done to the point of generation k or to substation k, positive numbers will be used for connexion to the substation and negative numbers for connexions to other points. Besides, in the connexion between 2 points the negative numbers will indicate the position occupied by the point to which the associated wind turbine is connected. Thus, a negative value, for instance, $a_{i\,j+np} = -4$, would indicate that the generation point in position j of the permutation is connected to the point occupying position 4. On the contrary, $a_{i\,j+np} = 4$ would indicate that the generation point in position j of the permutation is connected to substation 4.

By way of illustration, a possible solution for a problem involving three points of generation (wind turbines) and one substation would be:

$$\text{Solution } i = [2\ \ 1\ \ 3\ \ 1\ \ 1\ \ -2]$$

whose interpretation corresponds to the following solution (Fig. 11):

Point of generation *WT2* ($a_{i\,1} = 2$) is connected to substation 1 ($a_{i\,1+3} = 1$)

Point of generation *WT1* ($a_{i\,2} = 1$) is connected to substation 1 ($a_{i\,2+3} = 1$)

Point of generation *WT3* ($a_{i\,3} = 3$) is connected to substation *WT1* (located in position 2 in the vector of generation points, $a_{i\,3+3} = -2$).

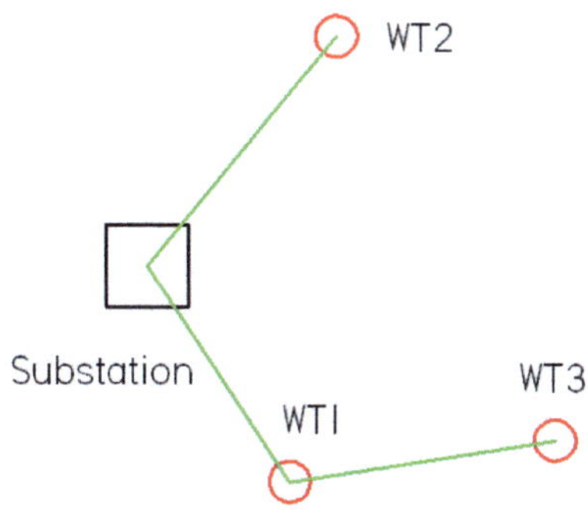

Fig. 11. An Case of electrical individual codification

Being *ns* the number of substations, with the adopted codification the possible values of $a_{i\,j+np}$ are limited to the set:

$$\{-j+1, -j+2, ..., -1, 1, 2, ..., ns\}.$$

This codification allows to maintain the tree structure in the network and to limit connexions between points of generation in such a way that each point (wind turbine) can only be connected to previous points of its permutation.

Because of simplicity, in the proposed examples it has been established that only one station could be connected to the high voltage line, allocating the same index to both the substation and the line.

Costs of the Wind Farm Electrical Network

The calculation of the costs of the internal energy distribution and disposal network of the wind farm, and the optimal electrical network can be divided, in principle, in two vast chapters:

- Unitary costs (by unit of length) of the lines, both for the transport line and for the internal distribution.

- Transforming substation costs, including transformers and elements needed for manipulation, protection and measurement. It is defined by voltages of transformation, nominal design power and the number of medium voltage lines branching off a given substation.

In both cases the fixed costs will be determined by investment costs, including equipment, materials and associated civil work. The variable costs are updated cost of energy lost by Joule effect that cannot be sold. Variable costs depend on the square of the circulating power P on the conductor or substation. This power P is pro rated as root mean square on a year and depends on the wind farm configuration, on the wind rose and on the Weibull parameters.

Assuming there is an extensive catalogue for the conductors of the medium voltage network (with different sections), and once the technical criteria have been met, you can check that the envelope of the different cost functions for each conductor by unit of length is fairly well approximated by using the fitting straight line of the intersection points between cost curves corresponding to consecutive sections, known as *economical breakeven points* (see Fig. 12).

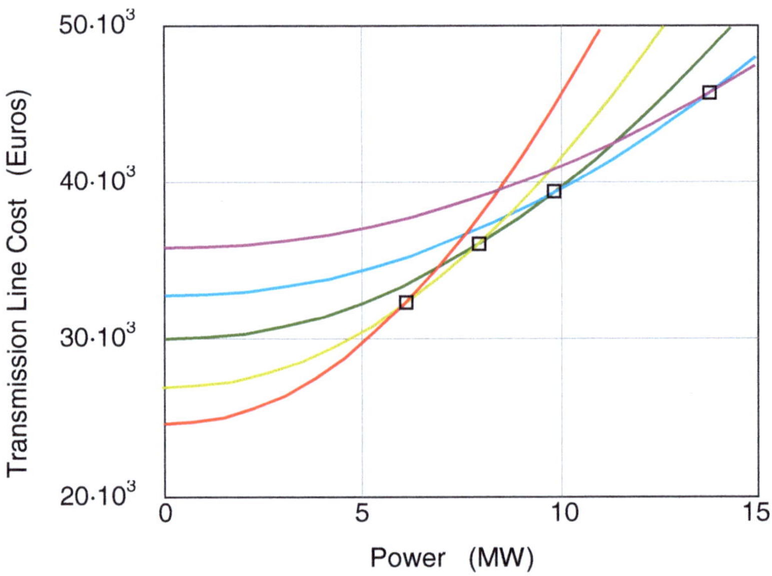

Fig. 12. Determination of economical breakeven points

The straight line will follow the expression:

$$C_m(P) = C_{mv}P + C_{mf} \tag{10}$$

The slope, C_{mv}, multiplied by the power to be carried, P, represents the variable updated cost of the line per length unit. The ordinate at the origin, $b = C_{mf}$, represents the fixed cost per length unit, updated too.

Logically, the space of the available conductors sections is not continuous, but this proxy allows a significant simplification of the calculation with an acceptable accuracy, as it relates directly and lineally the power to be carried with the minimum costs (those of the *optimal* conductor). Once this simplification has been adopted to obtain in the first step the approximate value of the total cost, the actual value of fixed and variable costs can be calculated in a more accurate way once the power to be carried is known.

This same reasoning can be applied to high voltage lines and to substations, however in this case the cost would not be by length unit.

Position of Substations

To evaluate the cost associated to each solution it is required to set the optimal substations placement from the information generated by the genetic algorithm. The coordinates of the substation k, (x_{0k}, y_{0k}), are chosen in order to minimize the following cost,

$$Z_k = \left(C_{HVv}\ P_k + C_{HVf}\right)L_k + \sum_{j=1}^{nr}\left(C_{LVv}\ P_j + C_{LVf}\right)L_j \tag{11}$$

C_{HVv}, C_{HVf}: Fixed and variable costs related to HV
C_{LVv}, C_{LVf}: Fixed and variable costs related to LV
P_k: Power transformed by the substation
P_j: Power injected at node j

$L_k = \dfrac{\left|a_k\ x_{0k} + b_k\ y_{0k} - c_k\right|}{\sqrt{{a_k}^2 + {b_k}^2}}$: Distance from the substation coordinates (x_{0k}, y_{0k}), to the HV line $(a_k x + b_k y = c_k)$.

$L_j = \sqrt{\left(x_{0k} - x_j\right)^2 + \left(y_{0k} - y_j\right)^2}$: Distance from substation to wind generator/node j.

As can be seen, the whole LV route is not needed. Only the generators connected to the substation and its power are needed. It must be observed that the function cost includes the fixed and variable cost related to the HV line (from the HV line to the substation) and the cost related to the LV line (form substations to wind generators). The costs of every transformer substation bus are included in the main algorithm. The fixed costs related to the beginning and end of the HV and LV lines are not considered because they are constant and do not determine the substation placement.

The analitical obtention of optimal x_0, y_0 requirees a certain amount of computation time, especially when it must bee repeated a high number of times. So, this coordinates are only computed in the last generation of the genetic algorithm. This is the reason why, apart from the exact analitical solution, three approximate solutions taking into account the center of gravity (x_{gc}, y_{gc}).

$$x_{gc}=\sum_i w_i x_i \quad y_{gc}=\sum_i w_i y_i \quad w_i=\frac{(P_i\,C_{LVv}+C_{LVf})}{\sum_i (P_i\,C_{LVv}+C_{LVf})} \tag{12}$$

P_i: Power injected at node i
C_{LVf}: LV fixed costs related to the conductor path and poles
C_{LVv}: LV variable costs related to line power losses

The three following possible substation placements are considered:

a) The intersection point between the HV line and the perpendicular from the gravity centre, (x_L,y_L), as shown in Fig. 13 left.
b) The same gravity centre coordinate, (x_{gc},y_{gc}), as in Fig. 13 midle.
c) A point between both, (x_I,y_I), as in Fig. 13 right. The position is calculated as a result of the following weighting between the HV and the LV costs.

$$x_I=x_L\,\lambda+x_{gc}\,(1-\lambda) \quad y_I=y_L\,\lambda+y_{gc}\,(1-\lambda)$$

$$\lambda=\frac{C_{HVv}\,P_i+C_{HVf}}{(C_{HVv}+C_{LVv})\,P_i+C_{HVf}+C_{LVf}\,Nb}$$

P_i: Power injected at node i
Nb: Number of branches
C_{LVf}: LV fixed costs related to the conductor path and poles
C_{LVv}: LV variable costs related to line power losses
C_{HVf}: HV fixed costs related to the conductor path and poles
C_{HVv}: HV variable costs related to line power losses

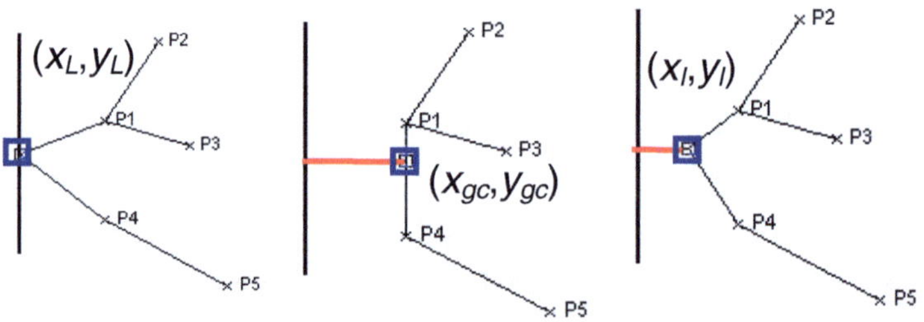

Fig. 13. Substation placement

The placement with the minimum value of Z_k is chosen. This computation is repeated for each active substation. When a main line is overloaded, exceeding its maximum transmission power flow, a penalty is added to the total cost.

Crossover Operator

Considering the special features of the codification used, an specific crossover operator has been designed that, depending on the crossing point, manages the crossing in a particular way.

The crossover process is defined as follows: a random number n_{rand} is generated, its value being between 1 and 2×np-1 (twice the number of generation points minus one). This number gives the position, in the vector representing the solution, beyond which the elements will be exchanged according to the following considerations:

1. When $n_{rand} < n_p$ the points exchanged will be the ones located in these positions for both parents.
2. When $n_{rand} \geq n_p$ two "child" solutions will be generated:

    a. The first one will be a replicate of the first elements till the crossing position of the first “father”. The rest of elements of this solution will be a copy of the last elements (from the crossing position, to the end) of the second “father”.
    b. The elements of the second "child" solution will be those not used in the first one.

One example of this crossover procedure is showed in Fig. 14, where the randomly selected position is number 5. It is possible to see than the children have characteristics from the parents and one's own thanks to the crossing.

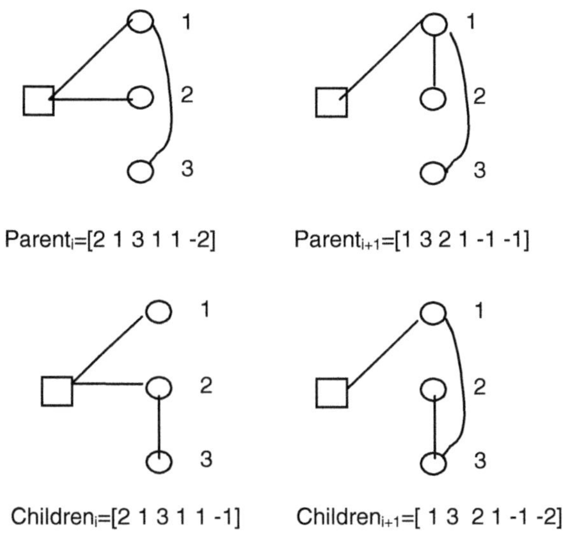

Fig. 14. Individual codification and crossover

Mutation Operator

Following the crossover operations, the resulting population is subjected to the mutation operation. The number of mutations is controlled by the *"Rate of mutation "* parameter. Two random numbers are chosen to perform these mutations. The first random number determines the individual to be mutated, whereas the second number n_{gen} establishes which gene of the individual will be mutated.

1. When $n_{gen} < n_p$ the mutation respects the permutation and a *special mutation* is performed. A third number will then be randomly chosen within the range (1, *np*) and the points located in these positions will be exchanged.

$$[2 \ 1 \ 3 \ 1 \ \ 1-2] \Rightarrow \xrightarrow{n_{gen}=3 \quad exchange=1} [3 \ 1 \ 2 \ 1 \ \ 1-2]$$

2. When $n_{gen} \geq n_p$ If the gene belongs to the group that set the connections of the wind generators, a guided mutation is performed. To do that, each possible mutation fitness is computed in an approximate way. The approximate cost is calculated:

   a) Connecting the wind turbine to each one of the transforming substations.
   b) Connecting the wind turbine to each one of the preceding wind turbines in the permutation.

A lower cost or a better fitness determines the new value of the mutated gene. This way, certain favourable branches are given a higher probability.

In order not to lose the best solution of each generation, one of the copies of the best individual is not subjected to mutation.

Repetition Operator

Due to the nature of the problem, branches can only be included from the farthest points to the nearest points. In consequence, the solution is very influenced by the codification of wind generators' position (the n_p digits in the individual codification. If the Figure 15 left were the best solution for an iteration and the Figure 15 right was the optimum solution, it will difficult for the algoritm to evolve from left to the right. To avoid that situation, the best individual is copied changing the codification form. In the first solution, the points are ordered depending on the distance form the point to the nearest substation, starting from the first substation and finishing by the last ([1 2 3 5 4 1 -1 1 2 2]). In the second solution, the points are ordered in the same way, but this time inverting the previous order ([5 4 1 2 3 2 2 1 -3 1]. This way, each wind turbine has the same probability to change of substation.

The population resulting of applying the operator to the initial population of the iteration plus the population generated applying the crossover operator will join them to make the starting population for the next iteration (generation).

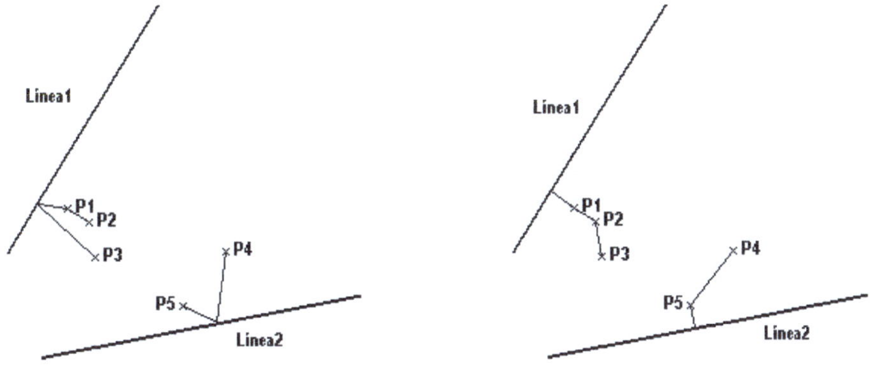

[1 3 4 2 5 1 1 2 −1 2]   [1 2 3 5 4 1 −1 −2 2 −4]

Fig. 15. The best solution for an iteration and the optimum solution

Detour of Forbidden Areas

The algorithm contemplates the possibility of existing restricted-use areas in the terrain, incompatible with the electrical installation. They are called forbidden areas.

There are two distinct cases in the treatment of forbidden areas:

- Medium voltage lines. For a medium voltage line, both the source and the destination point are known and invariable, therefore the tool will only detour from the forbidden area. This also happens when the point of connexion to the HV line is imposed by the distribution company.
- High voltage line. In this case, the path attempts to connect a specific known point (that of the substation) to an undetermined point belonging to the high voltage line (of the transport network) to which you want to connect to. The procedure is akin to the previous one with the initial condition being now to find a perpendicular to the high voltage line starting in the substation or from the apex being detoured (see Fig. 16).

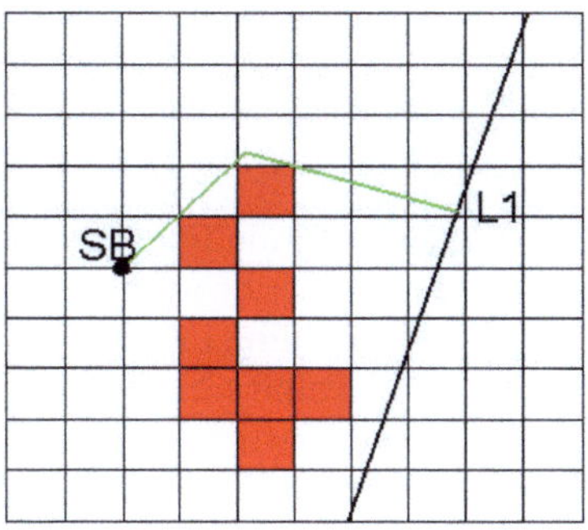

Fig. 16. Detour of forbidden areas

The graph composed of all the vertex (and the start and goal points) and the links joining them that do not cross any obstacle is called the *Visibility graph.* A compilation of different methods for the search of the shortest paths in visibility graphs can be found in [74].

In order to compatibilise the optimal point search and the electrical linking procedures, it has been developed an algorithm focused in a discretisation of the definable terrain that restrain the solutions space to the minimum space unit considered, that is, the cell.

Example

The ability of the algorithm to solve a complete case will now be put to the test.

Figure 17 shows the distribution plan view of the wind farm. As you can see, the available terrain for the wind farm is a square area 20 km long on each side, discretised using a grid of 20x20 square cells.Two HV evacuation lines (L1,L2) are consider. The line capacity L1 is limit to 5 MW. All the wind turbines to be considered (*A1…A10*) are identical, with a nominal power of 1 MW. Two forbidden areas have been added to check the response of the detour procedure on the high and medium voltage electrical network.

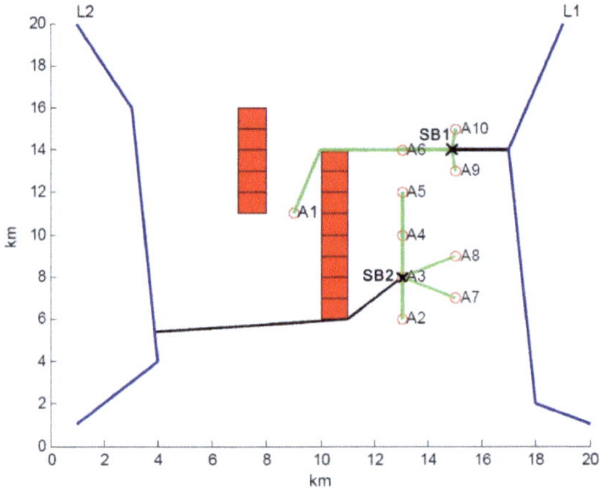

Fig. 17. Optimal solution for detour example

Figure 17 also shows the optimal solution found by the algorithm. It can be noticed in the proposed solution that the set of wind turbines composed by *A1*, *A6*, *A9* and *A10* is linked to the high voltage line *L1*. On the other hand, the remaining wind turbines are linked to the high voltage electrical line *L2* through substation *SB2,* that envisages the position close to wind turbine *A5* as the one that optimises costs. The high voltage link does a detour of the forbidden area through the lower area.

6 Global Planning of a Wind Farm

The proposed method involves the use of an evolutive algorithm that combines the algorithms developed for each one of the parts of the design problem. This global optimisation algorithm basically performs the following operation in each iteration:

1. The algorithm of location of wind turbines is executed.
2. The problem of electrical connexion for each one of the individuals of the generation is solved.
3. The NPV for each individual is calculated.
4. Finally, the individuals are arranged, the best ones being selected to participate in the next iteration.

Example

In order to examine the operation and potential of the proposed algorithm a case has been solved, as an example, that will let us assess its performance.

Let's consider a square plot of terrain with the following features (see Figure 18) there is a wind with a sole North-South direction, with an inner area

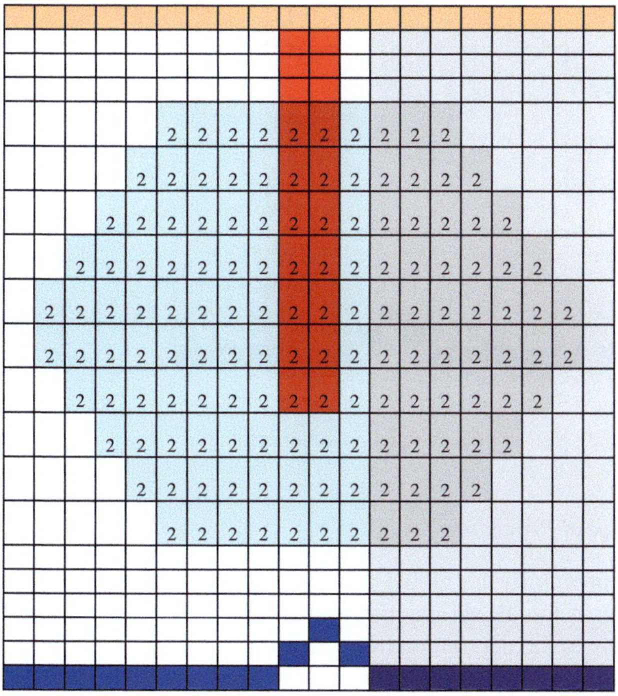

Fig. 18. Main road, high voltage line and values of scale factor, C, for the Weibull function

where the average speed is much higher; the terrain plot includes a main road that runs West to East along the North boundary; there is a high voltage line in the same direction but along the South boundary of the terrain, with a point closer to the potential location of the turbines; there is a forbidden area and another area with a low bearing capacity; an investment limit of 4.200.000 € is established; and only a type of machine, with a height between 60 and 100 m, will be considered.

Sequential Procedure

A first approach to the problem of global optimisation, that could also be called *sequential*, entails the separate resolution of two subproblems: that of locating the wind turbines on the wind farm, and that of calculating the related optimal electrical installation.

The solution found using the sequential procedure is shown in Figure 19, and it is made up of 5 wind turbines.

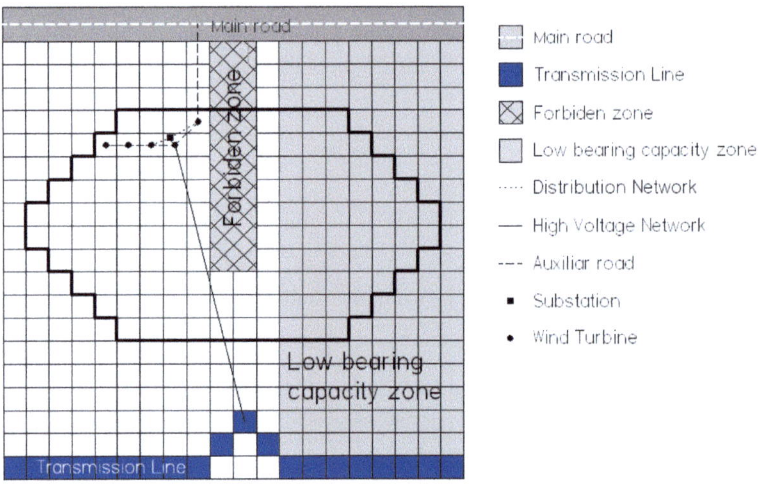

Fig. 19. Optimal solution of sequential procedure example

Complete Global Procedure

A second approach to the global problem, that could be called *complete*, entails the execution of the global algorithm to compute the electrical installation of all the individuals of the population in all iterations.

Figure 20 shows the solution reached by this procedure, that involves a sixth turbine without exceeding the investment limit.

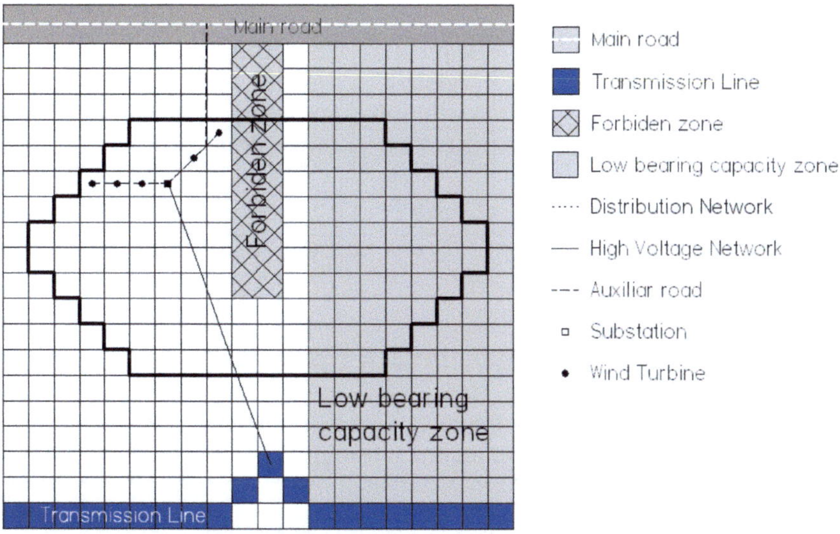

Fig. 20. Optimal solution of global procedure example

The preceding sequential procedure had wrongly limited the number of wind turbines to 5 since an excessive estimate of the costs of electrical infrastructure would have exceeded the investment limit in case 6 wind turbines were used. Actually this would have been a worse solution.

Accelerated or Hybrid Global Procedure

A third and last approach, called *accelerated* or *hybrid*, starts the execution of the algorithm by solving just the setting out problem (estimating the costs of the electrical installation at the same value as the costs of civil works) till it leaves the initial search random stage. The first generations gone, the resolution of the problem of the electrical installation is added. In this sense, it's been made obvious the convenience of differentiating between an approximate solution, which is quicker, and a more exact one. The difference between them relies on how they calculate the position of the substation, that for the approximated case is computed by ponderating the costs of medium and high voltage lines according to one otf the three possible cases considered electrical gravity center, whereas in the exact procedure said position is analiticaly computed using the cost equation (11).

The precise timing of inclusion of the resolution of the electrical installation, first in a approximated way and later in an exact way, are 40% and 90% measured in terms of percentage of the convergence criterion. In most cases this procedure reaches the same solution as the complete global procedure, but with a significant reduction in computing time (Table 3).

Table 3. Economic summary of proposed methods

Concepts	Sequential global optimum	Complete global optimum	Accelerated global Optimum
NPV (€)	20.66 M€	24.96 M€	24.96 M€
Total investment(€)	3.50 M€	4.14 M€	4.14 M€
Computational cost *	985 s	$9.74 \cdot 10^4$	$5.51 \cdot 10^4$

(*) Computing time using a PC with Intel Core Duo T2400 processor (16000 MIPS).

7 Test Peformance and Benchmark Cases

In order to show the potential of the proposed algorithm a series of examples have been solved where, starting from a common initial approach, different restraints and difficulties will be incorporated, that will allow us to evaluate the optimal solution reached by the algorithm and evaluate its fitness (optimisation). The Table 4 shows a summary of the features and restraints included for each one of the sample cases tested.

Table 4. Main charateristic of each case

	Case			
	1	2	3	4
Presence of forbidden areas	X	X	X	X
Zones with different bearing capacity		X	X	X
Variable direction of the wind			X	X
Computation of the electrical installation				X

The Figure 21 shows the terrain chosen to materialize the examples, a rectangular plain 3500 m high by 5250 m wide where, as it can be seen, there are two forbidden areas, one in the central area where it is neither possible to locate wind generators, nor can it be crossed by service access roads, and a second forbidden area in the most northern part of the terrain close to the main road, where no wind turbines can be located, but that can be crossed by a service road. On the other hand, there is also an area where the bearing capacity of the terrain is low and, therefore, it is necessary to face an increase in foundations cost. As shown in the Figure 21, the terrain has been discretised as an array of 10x15 square cells, with dimensions of 350m x 350m.

The main road runs on the North side of the surface, according to this fact execution costs of 100 € /meter will be allocated for service roads.

The wind is characterisedby a scale factor C = 6.5 and K=2 identical for all cells. For examples 1 and 2 we will consider that the prevailing wind comes from

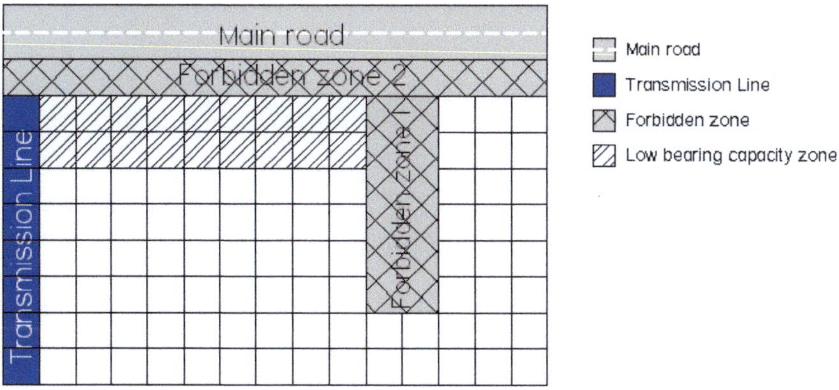

Fig. 21. Case scenario

the North direction, and for examples 3 and 4 that the direction of the wind is variable. The coefficient of roughness is 0.0055 for the whole terrain.

The algorithm has the ability to select the fittest wind turbine in each situation, whose main features appear in the Table 5.

Table 5. Main charasteristic of the considered wind turbine

	WT A	WT B	WT C	WT D	WT E
Rated capacity (*MW*)	1.3	1.67	1.67	2	2
Min. height (*m*)	60	60	60	60	60
Max. height (m)	80	80	80	100	100
Price/Cost (*M€*)	0.748	1.253	1.253	1.65	1.5
Tower cost (*k€/m*)	1.5	1.5	1.5	1.5	1.5
Foundation cost (k€)	80	80	80	80	80

Power curves for each wind generator are shown in Fig. 22.

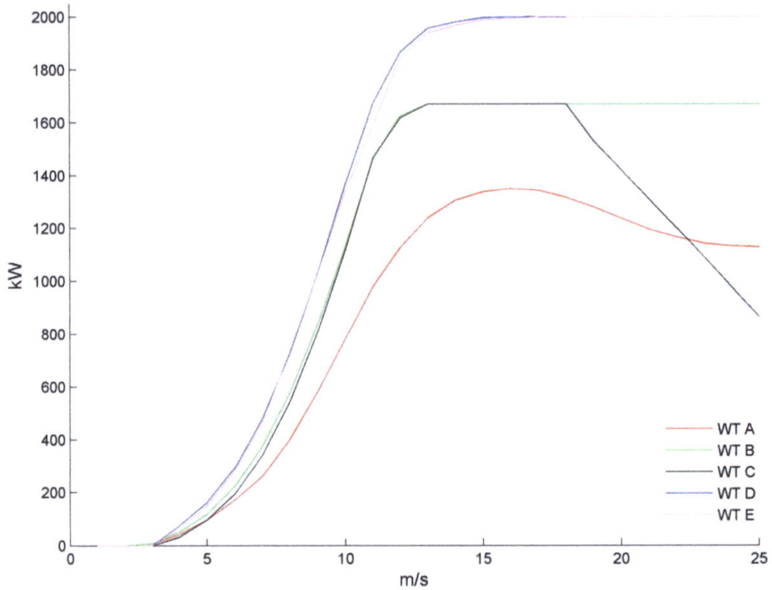

Fig. 22. Wind turbines power curves

The Table 6 below shows the main parameters and economical data used to run these examples.

Table 6. Main input agorithm parameters

Algorithm parameters		*Economical information/data*	
Size of population	200	Life time (*years*)	20
Initial number of turbines	100	Interest rate (%)	6
Initial solution considered	No	Price of energy (*€/kWh*)	0.07
Maximum number of turbines	7	Increase of energy price (%)	3
Crossing probability (%)	30	Availability factor (%)	95
Mutation probability (%)	80	Present cost of decommission (%)	3
Number of repetitions to finish	40	Present residual value (%)	3

7.1 Results

Case 1

The Figure 23 shows results obtained for case 1, previously described.

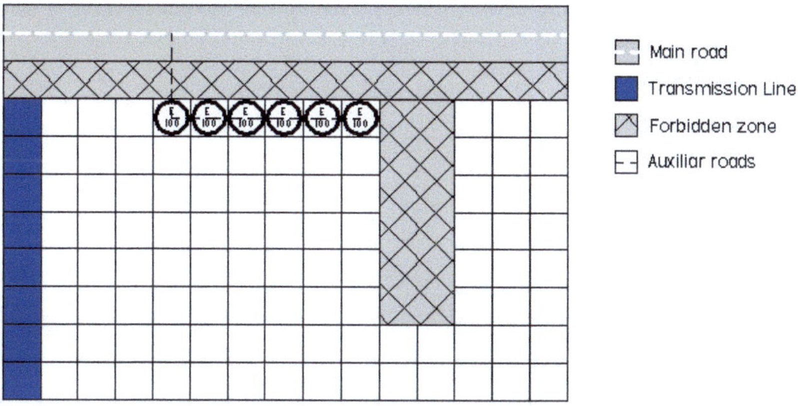

Fig. 23. Case 1. Optimal configuration.

The algorithm reaches an optimum in 54 generations, obtaining a solution where all wind turbines are of type E, with a tower height of 100m. All turbines have been located in the cell being closest to the main road, in order to minimise the global distance between the service roads to the main road, and not let the energy production be affected by effect of wakes due to the existence of close wind turbines in the source direction of the wind.

Table 7. Results for Case 1

NPV(k€)	21991
Investment(k€)	13790
Turbines Cost (k€)	11550
Civil Work Cost (k€)	1120
Electrical infrastructure cost (k€)	1120
Average Power (MW)	3.538
Yearly produced energy (GWh)	30.99

The Table 7 shows the main economical data for this example. It can be noticed that most of the initial investment, 84%, goes to acquisition of turbines, whereas the remaining fraction are costs in civil works and electrical installation (that, in this example, will be considered equivalent to the cost in civil works, as its exact calculation is not computed).

Case 2

In this case we will consider the existence of an area of terrain presenting a low bearing capacity and therefore, a higher investment in foundation works will be necessary. Said low bearing capacity will be located in the cells closer to the road in the western part of the terrain. The Figure 24 and the Table 8 show the optimal solution reached by the algorithm.

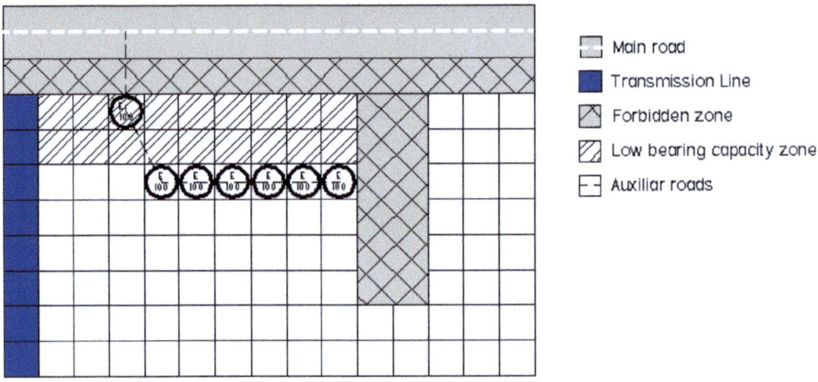

Fig. 24. Case 2. Optimal configuration.

Table 8. Results for Case 2

NPV(k€)	21864
Investment(k€)	13917
Turbines Cost (k€)	11550
Civil Work Cost (k€)	1183
Electrical infrastructure cost (k€)	1183
Average Power (MW)	3.538
Yearly produced energy (GWh)	30.99

All the wind turbines have been located forming a row in the cell of high bearing capacity that is closest to the main road, apart from a wind turbine located in a cell adjacent to the road with a low bearing capacity of the terrain. This configuration minimises the total costs derived from access roads and foundation costs, since locating the wind turbine in an area of low bearing capacity reduces the length of the roads (at the expense of increasing the foundation costs of a wind turbine) compared to a solution where all the wind turbines were lined in the cells with high bearing capacity near the main road; on the other hand, it would be possible to additionally reduce the distance of the service roads by moving the most Northern wind turbine one cell to the East, but as a consequence such configuration would cause a reduction in the energy produced by the wind turbine located downstream in the source direction of the wind caused by the wakes, resulting in a loss of profitability for the wind farm.

Case 3

This case has been solved replicating the conditions of case 2, apart from the fact that the source direction of the wind is variable with occurrence probabilities of 25 % for North direction, 50% for Northwest direction, and 25 % for West direction, being nil for the rest of directions.

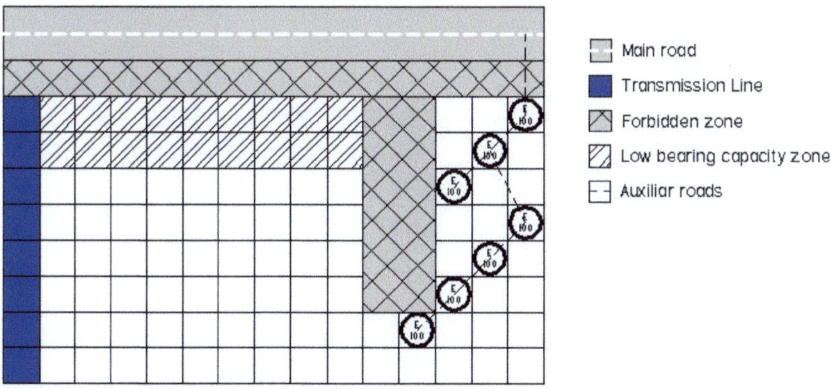

Fig. 25. Case 3. Optimal configuration.

The algorithm reaches an optimal solution in 54 generations, being the selected wind turbines of the same type and height than in Cases 1 and 2.

In this case (Fig. 25) the adopted layout contains two diagonal staggered rows, with a spacing distance that doesn't affect the wakes. These rows follow a Southwest-Northwest direction and are located in the most oriental part of the terrain, so that the total length of access roads is minimal while preventing the location of wind turbines in cells where the bearing capacity of the terrain is low, therefore minimising the global cost in civil works.

The figures 26 – 27 show two possible solutions to the example, and later some detail is provided on the derived costs for construction of service roads, as well as the cost of foundation for both situations.

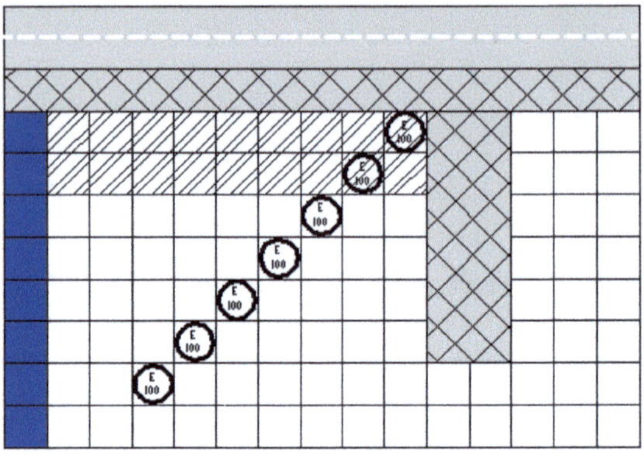

Fig. 26. Case 3. Proposed solution 1.

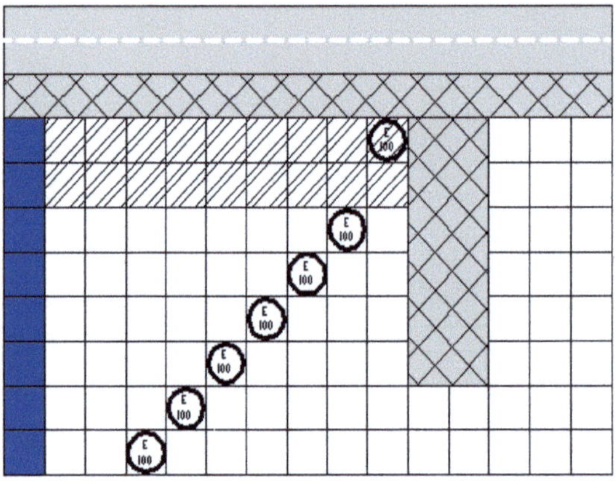

Fig. 27. Case 3. Proposed solution 2.

The Table 9 shows the economical data and the objective function for the three solutions.

Table 9. Comparasion betwen optimal solution and proposed solutions

	Optimal Solution	Proposed Solution 1	Proposed Solution 2
NPV (*k€*)	21760	21737	21720
Investment (*k€*)	14021	14044	14061
Turbines Cost (*k€*)	11550	11550	11550
Civil Work Cost (*k€*)	1236	1247	1256
Foundations Cost (k€)	*840*	*880*	*860*
Auxiliary roads cost (k€)	*396*	*367*	*396*
Electrical infrastructure cost (*k€*)	1236	1247	1256
Average Power (*MW*)	3.538	3.538	3.538
Yearly produced energy (*GWh*)	30.99	30.99	30.99

It can be noticed that the three solutions only differ in the cost of civil works, that can be broken down into foundation costs and costs for construction of service roads. It is to be noticed that it is more profitable to increase the length of the service roads (optimal solution), with respect to proposed solution 1, in order to prevent foundation in low bearing capacity areas. On the other hand, it is also noticeable how in proposed solution 2 the cost of service roads is the same as that of the optimal solution, however the foundation cost is higher that in the later one, as it is necessary to locate a wind turbine in a low bearing capacity area.

Case 4

Finally, the global problem will be solved calculating the exact electrical infrastructure using the case described in Case 3. A high voltage line runs North to South, on the western side of the terrain.

The data and parameters required to solve the positioning problem are those of previous examples. The table 10 shows the most relevant parameters to solve the calculation of the electrical installation.

Table 10. Main input of electrical infraestructure aglorithm

Number of repetitions to start the electrical exact calculation	10
Size of population	100
Number of repetitions to finish	40
Crossing probability (%)	75
Mutation probability (%)	5
Number of the optimal solution repetitions	6

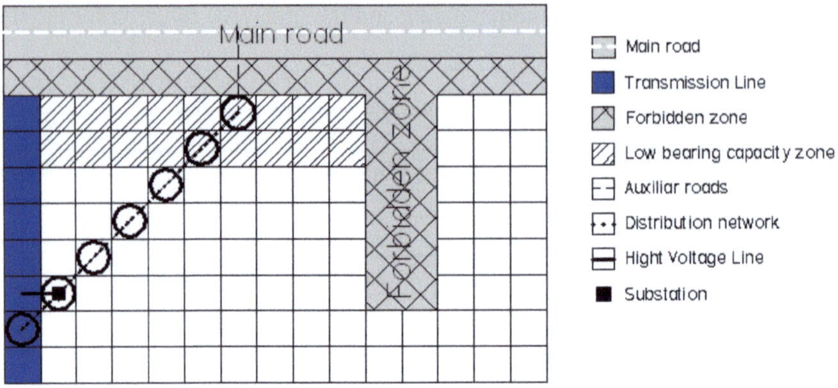

Fig. 28. Case 4. Optimal Solution.

The Figure 28 shows the values obtained by the algorithm. This time the wind turbines have been located in the western side of the terrain aiming at minimising the global length of conductors. This will be an optimal solution, in spite of locating two wind turbines in a low bearing capacity area of the terrain, since it minimises the global length of roads and conductors.

The algorithm has reached an optimum in 103 generations. The path of the service roads and the medium voltage installation are coincidental along most of the path, with the exception of the link of the roads to the main road (link between the most northern wind turbine and the main road) and the interconnexion between the substation and the high voltage line, as this is the position of the substation that makes the total cost of the electrical installation to be minimal.

The Table 11 shows the most relevant values for this example, the first column representing the results obtained employing the coupled global method, the second

Table 11. Compasion between results of global and sequential procedure

	Global	Sequential
NPV (k€)	22605	22499
Investment (k€)	13176	132283
Turbines Cost (k€)	11550	11550
Civil Work Cost (k€)	1247	1236
Electrical Cost (k€)	297	394
Distribution Network Cost	88	99
High Voltage Cost	86	95
Substations Cost	200	200
Network losses	83	103
Average Power (MW)	3.538	3.538
Annual produced energy (GWh)	30.99	30.99

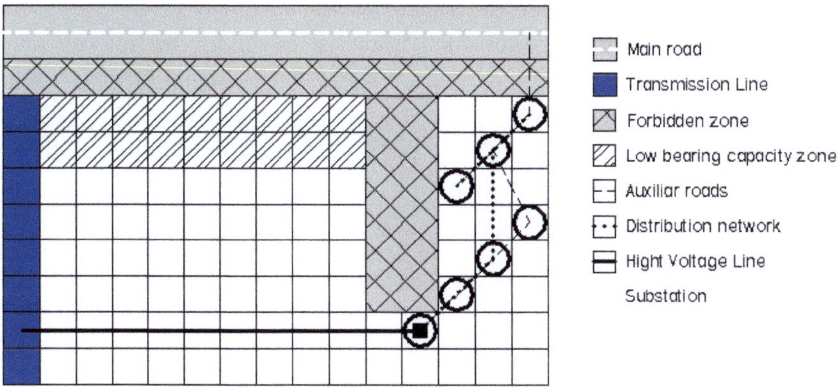

Fig. 29. Case 4. Sequential procedure solution.

corresponding to results obtained using the sequential method, where the layout of the wind turbines is the one obtained in Case3, whose configuration is shown in the Figure 29.

Repetitivity

To expose the robustness of the proposed method a repetitivity test has been run, executing 50 times the sample 4, and obtaining the results shown in the Table 12.

Table 12. Results of repetitivity test

	Frequency	NPV (k€)	Relative Error (%)
Optimal Solution	24	22605	-
2nd Solution	4	22594	0.04
3erd Solution	21	22589	0.07
4th Solution	1	22557	0.21

These results show that evolutive algorithms, although not always converging to an optimum, reach the optimum most of the time, the rest of solutions also being close to the optimum (relative errors lower than 0.21%). The proposed method is therefore suitable, given the complexity of the problem introduced in this chapter.

Sensitivity

In this section a sensitivity test to the parameters will be done for the conditions described in Case 4. The goal of this test is to observe to what degree the optimal solution is affected by eventual variations that may exist in the raw data entered in the algorithm, like the interest rate and the price for kWh.

Sensitivity to Interest Rates

Starting with Case 4 the algorithm has been run modifying the interest rate for values ranging from 1 % to 15 %. Fig. 30 shows the evolution for the most relevant economical results.

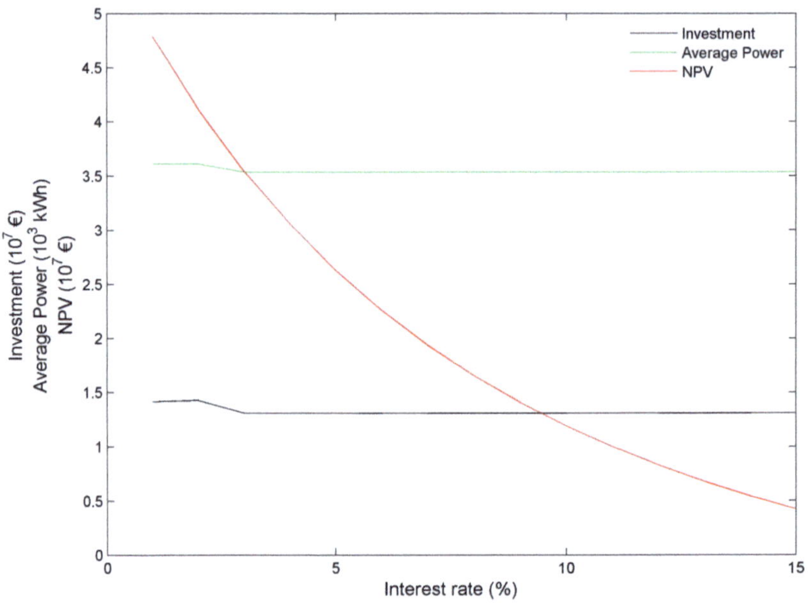

Fig. 30. Solution sensitivity to variation of the interest rate

You will notice that as interest progresses there is a drop in the project profitability. For interest rates lower than 3%, it is more profitable to increase the investment, mainly in wind turbines, selecting for these interest rates type-D turbines that are slightly more expensive, however allowing an increase in annual production.

Fig. 31 shows the evolution for the most relevant factors regarding the electrical installation. It can be noticed how an increase in the interest rate reduces the costs derived from electrical losses while, at the same time, it becomes more profitable to make a lower investment in electrical installation, either reducing the investment in high tension, or reducing the investment in medium tension.

Sensitivity to Energy Price (€/kWh)

In this test the energy price (€/kWh) will vary from 1 Euro cent to 15 Euro cents. Fig. 32 shows the evolution for the most relevant results.

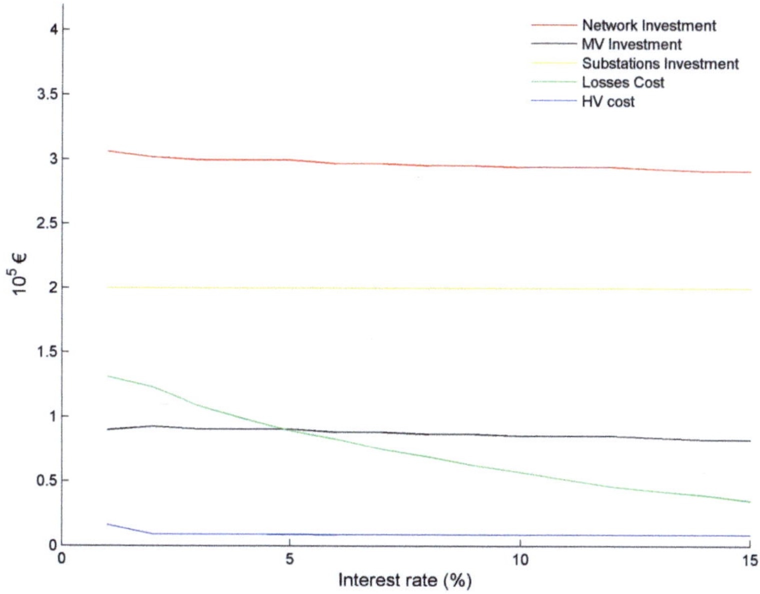

Fig. 31. Solution sensitivity to variation of the interest rate

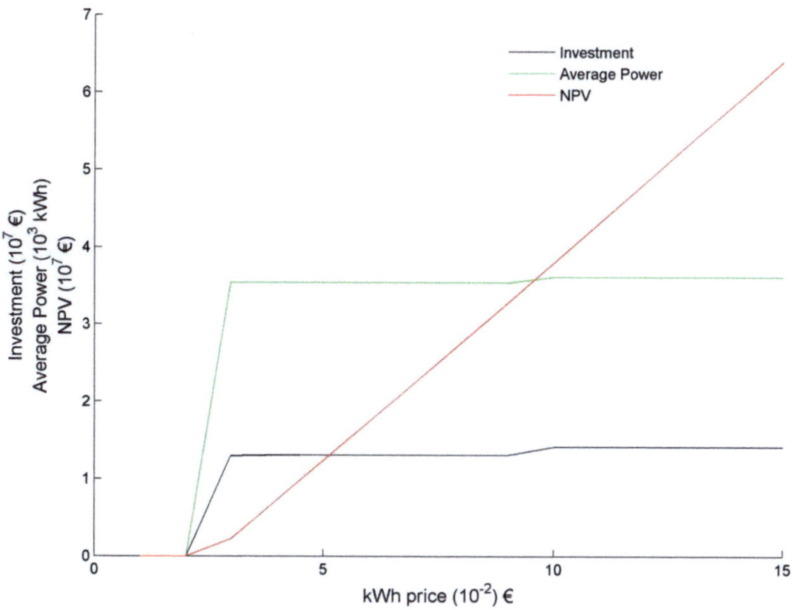

Fig. 32. Solution sensitivity to variation of the energy price

It can be seen that for energy prices lower than 3 cent € the wind farm project is not profitable, on the other hand it is to be noticed that there is a progressively higher profitability as energy price increases, whereas for prices higher than 9 cent € it becomes more convenient to increase the investment in wind turbines, selecting type D for all of them, therefore achieving an increase in the global energy generated by the wind farm.

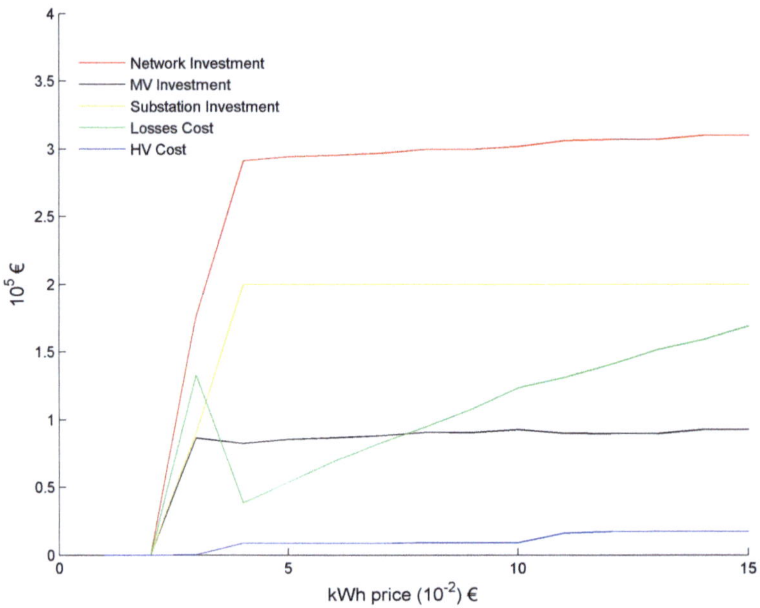

Fig. 33. Solution sensitivity to variation of the energy price

The Figure 33 shows that for increases in the energy price it is more convenient to increase the investment in electric installation, a trend caused by the reduction of electrical losses due to an increased section of conductors. The curve of cost of electric losses shows the effect of increasing the section of conductors by means of decreasing the slope of the curve. On the other hand, it can also be noticed that, for an energy price of 3 cent €, where the profitability of the wind farm is relatively low, it is preferable to make the lowest possible investment in electrical infrastructure, locating the substation in the line (minimising the investment in high voltage installations), at the expense of increasing the length of medium voltage conductors, cheaper than the high voltage ones, but with a subsequent increase in electrical losses.

8 Conclusions

The design of a wind farm aimed at the generation of electric energy is an extraordinarily complex and multidisciplinary task where the most diverse areas of knowledge converge. The complexity of the problem comes, amongst other causes, from the high number of mutually dependent variables and underlying constraints. This makes the solution achievable by means of heuristical methods, following several test and error cycles, and being strongly based in the previous experience of the design team.

In practice, and in order to simplify the solution, the problem of global design of a wind farm is usually split into two separate subproblems that, as a consequence, are easier to approach. On the one hand, you study the (sub) problem of the individual location of turbines and, on the other, the (sub) problem of design and configuration of the wind farm electrical network. This 'decoupling' is justified by the fact that 75% of the initial investment of the wind farm corresponds to costs calculated in the first subproblem, playing in addition a key role in the annual production of electric energy, that is, in the return of the investment.

To optimise the problem of the location and selection of turbines it has been shown a genetic algorithm that makes use of a cost model as an instrument of assessment that allows the algorithm to perform a gradual remodelling of the global configuration of the wind farm till it reaches a maximum return of investment (NPV) through all the productive cycle of the wind farm.

Later, using the previously established setting out as a starting point, we have approached the problem of the design and configuration of the electrical infrastructure of the wind farm.

In both cases the proposed optimisation algorithm also deals with constraints such as the existence of access roads close to the wind farm and of one or two energy disposal lines with a limited evacuation capacity, as well as forbidden areas (either for the installation or for the electrical network), a terrain with a low bearing capacity, aside from the technical constraints of the electrical network (cable ampacity, voltage drop or ability to withstand shortcuts).

As it could be expected, a decoupled solution for the problem of global configuration of the wind farm does not always lead to the best global solution of the problem, especially in those cases where there is an economic cap for the investment. For this reason, the approach has been to focus in achieving a resolution for a global optimisation of the wind farm, considering jointly the setting out and the electrical infrastructure in an integrated way, as parts of a single global problem. The proposed algorithm combines the genetic algorithms both for the selection of type and location of the turbines, and for the design of the electrical infrastructure.

Finally throuht series cases, the algorithm capacity to consider the practical and geographical constrainst (forbidden areas, capacity limit, voltage drop...) has been shown. Besides, an analysis of the sensibility of the optimum solution to the main parameters has been performed and the robustness and quality of the optimum solutions found by the algorithm has been favorably demonstrated with a series of repetition tests on the same case.

In summary, the wind farm design could be optimize using an algorithm that integrates a realistic economic model working with an evolutive algorithm that, generation after generation, guides the wind farm integral configuration towards the global optimum.

References

[1] Windicator-Wind Power Monthly, available at http://www.windpower-monthly.com/WPM:WINDICATOR
[2] Jenkins, N.: Electrical design of wind farms. In: Proc. IEEE/NTUA Athens Power Tech Conference, Athens, Greece, September 5-8, pp. 990–994 (1993)
[3] Jenkins, N.: Engineering wind farms. Power Engineering Journal 7(2), 53–60 (1993)
[4] Walker, J.F., Jenkins, N.: Wind energy technology. John Wiley and Sons, Ltd./Inc., New York (1997)
[5] Bansal, R.C., Shatti, T.S., Kothari, D.P.: On some of the design aspects of wind energy conversion systems. Energy Conversion and Management 43(16), 2175–2187 (2002)
[6] Johnson, G.L.: Wind energy systems (Electronic Edition, Revised January 29, 2004), `http://www.rpc.com.au/products/windturbines/wind-book/WindTOC.html`
[7] WAsP, `http://www.wasp.dk`
[8] WindFarm, `http://www.resoft.co.uk/Spanish/index.HTM`
[9] WindPro, `http://www.emd.dk/`
[10] Goldberg, D.E.: Genetic algorithms in search, optimization and learning. Addison-Wesley Pub. Co. Inc., Reading (1989)
[11] Spillman, R.: Genetic algorithms, nature's way to search for the best. Dr. Dobb's J., 26–30 (1993)
[12] Grefenstette, J.J.: Optimization of control parameters for genetic algorithms. IEEE Trans. On Systems Man, And Cybernetics SMC-16, 122–128 (1986)
[13] Riquelme Santos, J., Burgos Payan, M., Calero, J.M., Castro Mora, J.: An evolutive algorithm for wind farm optimal design. Neurocomputing 70(16-18), 2651–2658 (2007)
[14] Martínez Ramos, J.L., Castro, J., Riquelme Santos, J., Burgos Payán, M.: A hybrid evolutive algorithm for wind farm optimum network design. In: Artificial Intelligence in Energy Systems and Power Aiesp 2006, Madeira, Portugal, pp. 1–5 (2006)
[15] Castro Mora, J., González Rodríguez, Á.G., Riquelme Santos, J.M., Burgos Payán, M.: A mixed hybrid algorithm for integral wind farm optimum design. In: International Conference on Renewable Energy and Power Quality – ICREPQ 2007, Sevilla, Marzo de (2007)
[16] Hajela, P.: Genetic search - an approach to the nonconvex optimization problem. AIAA J. 28(7) (1990)
[17] Holland, J.H.: Adaptation in natural and artificial systems. Univ. of Michigan Press, Ann Arbour (1975)
[18] Davis, L., Steenstrup, M.: Genetic search and simulated annealing. Lawrence Davis and Morgan Kauffman, Los Altos (1987)

[19] Mosetti, G., Poloni, C., Diviacco, B.: Optimization of wind turbine positioning in large wind farms by means of a genetic algorithm. Journal of Wind Engineering and Industrial Aerodynamics 51(1), 105–116 (1994)
[20] Grady, S.A., Hussaini, M.Y., Abdullah, M.M.: Placement of wind turbines using genetic algorithms. Renewable Energy 30(2), 259–270 (2005)
[21] Carrano, E.G., Soares, L.A.E., Takahashi, R.H.C., Saldanha, R.R., Neto, O.M.: Electric distribution network multiobjective design using a problem-specific genetic algorithm. IEEE Trans. Power Syst. 21(2), 995–1005 (2006)
[22] Ramírez, I.J., Bernal, J.L.: Genetic algorithms applied to the design of large distribution systems. IEEE Trans. Power Syst. 13(2), 696–703 (1998)
[23] Miranda, V., Proenca, L.M.: Probabilistic choice vs. risk analysis - conflicts and synthesis in power system planning. IEEE Trans. Power Syst. 13(3), 1038–1043 (1998)
[24] Yeh, E.C., Venkata, S.S., Sumic, Z.: Improved distribution system planning using computational evolution. IEEE Trans. Power Syst. 11(2), 668–674 (1996)
[25] Miranda, V., Ramito, J.V., Proenca, L.M.: Genetic algorithms in optimal multistage distribution network planning. IEEE Trans. Power Syst. 9(4), 1927–1933 (1994)
[26] Parada, V., Ferland, J.A., Arias, M., Daniels, K.: Optimization of electrical distribution feeders using simulated annealing. IEEE Trans. Power Syst. 19(3), 1135–1141 (2004)
[27] Dorigo, M., Maniezzo, V., Colorni, A.: The ant system: Optimization by a colony of cooperating agents. IEEE Trans. Syst. 26, 29–41 (1996)
[28] Gómez, J.F., Khodr, H.M., De Oliveira, P.M., Ocque, L., Yusta, J.M., Villasana, R., Urdaneta, A.J.: Ant colony system algorithm for the planning of primary distribution circuits. IEEE Trans. Power Syst. 19(2), 996–1004 (2004)
[29] Teng, J.-H., Liu, Y.-H.: A novel ACS-based optimum switch relocation method. IEEE Trans. Power Syst. 18(1), 113–120 (2003)
[30] Teng, J.-H., Liu, Y.-H.: Application of the ant colony system for optimum switch adjustment. IEEE Trans Power Syst 2(6-10), 751–756 (2002)
[31] Ramírez, I.J., Domínguez, J.A.: New multiobjective tabu search algorithm for fuzzy optimal planning of power distribution systems. IEEE Trans. Power Syst. 21(1), 224–233 (2006)
[32] Ramírez, I.J., Domínguez, J.A.: Possibilistic model based on fuzzy sets for the multiobjective optimal planning of electric power distribution networks. IEEE Trans. Power Syst. 19(4), 1801–1810 (2004)
[33] Mori, H., Yamada, Y.: Two-layered neighborhood tabu search for mutiobjective distribution network expansion planning. IEEE Circuits and Systems, 21–24 (Mayo 2006)
[34] Mori, H., Yamada, Y.: An improved tabu search approach to distribution network expansion planning under new environment. IEEE Power Syst. Tech., 981–986 (November 2004)
[35] Ramírez, I.J., Domínguez, J.A., Yusta-Loyo, J.M.: A new model for optimal electricity distribution planning based on fuzzy set techniques. IEEE, Power Engineering Society Summer Meeting 2, 1048–1054 (1999)
[36] Jarass, L., Hoffmann, L., Jarass, A., Obermair, G.: Wind energy. Springer, Berlin (1981)
[37] Masters, G.M.: Renewable and efficient electric power systems. John Wiley and Sons, Ltd./Inc., New York (2004)

[38] Patel, M.R.: Wind and solar power systems. CRC Press, Boca Raton (1999)
[39] Pohlheim, H.: GEATbx: Genetic and evolutionary algorithm toolbox for use with MATLAB (1999)
[40] Heier, S.: Grid integration of wind energy conversion systems. John Wiley and Sons, New York (1998)
[41] Gipe, P.: Wind energy comes to age. John Wiley and Sons, Ltd./Inc., New York (1995)
[42] European Wind Energy Association, Wind energy in Europe, EWEA, Greenpeace, Wind Force 12. A blueprint to achieve 12% of the worlds electricity from wind power by the year 2020, European Wind Energy Association, p. 52 (2003), `http://www.ewea.org/src/europe.htm2003`
[43] Danish Wind Turbine Manufacturers Association, Guided tour on wind energy (accessed July 2009), `http://www.windpower.org/en/core.htm`
[44] Ackermann, T.: Wind power in power systems. Wiley, Chichester (2005)
[45] van Kuik, G.A.M., Bierboos, W.A.A.M.: The status of wind energy. In: Nàprstek, J., Fischer, C. (eds.) EACWE4 – The Forth European & African Conference on Wind Engineering, ITAM AS CR, Prague, Paper #260, July 11-15 (2005)
[46] Jensen, N.O.: A note on wind generator interaction. Risø National Laboratory, Roskilde (1993)
[47] Katic, A., Hojstrup, J., Jensen, N.O.: A simple model for cluster efficiency. In: Proceedings of the European Wind Energy Association Conference and Exhibition EWEC 1986, Rome, Italy, October 7-9, pp. 407–410 (1986)
[48] Frandsen, S.: On the wind speed reduction in the center of large clusters of wind turbines. In: Proceedings of the European Wind Energy Association Conference and Exhibition EWEC 1991, Amsterdam, The Netherlands, pp. 375–380 (1991)
[49] Frandsen, S., Barthelmie, R., Pryor, S., Rathmann, O., Larsen, S., Højstrup, J., Thøgersen, M.: Analytical modelling of wind speed deficit in large offshore wind farms. In: Wind Energy 2006, vol. 9, pp. 39–53 (2006)
[50] Frandsen, S.: Turbulence and turbulence-generated structural loading in wind turbine clusters, Risø-R-1188(EN), Risø National Laboratory, Roskilde, Denmark (January 2007)
[51] Beyer, H.G., Pahlke, T., Schmidt, W., Waldl, H.P., de Uit, U.: Wake effects in a linear wind farm. Journal of Wind Engineering and Ind. Aerodyn. 51, 303–318 (1994)
[52] Vermeer, L.J., Sørensen, J.N., Crespo, A.: Wind turbine wake aerodynamics. Progress in Aerospace Sciences 39, 467–510 (2003)
[53] Tande, J.O.G.: Grid integration of wind farms. Wind Energy 6, 281–295 (2003)
[54] González Rodríguez, Á.G.: Improvement of a fixed-speed wind turbine soft-starter based on a sliding-mode control Tesis doctoral, Mención Doctor Europeo, Escuela Superior de Ingenieros, Universidad de Sevilla (mayo 2006) `http://www4.ujaen.es/~agaspar/Thesis`
[55] González Rodríguez, Á.G., Burgos Payán, M., Izquierdo Mitchell, C.: PSCAD based simulation of the connection of a wind generator to the network. In: IEEE International of Power Tech Proceedings, 2001 IEEE Porto., Oporto, Portugal, vol. (8), pp. 307–312 (2001) ISBN: 0-7803-7139-9
[56] González Rodríguez, Á.G., Burgos Payán, M., Izquierdo Mitchell, C.: Improving soft-starting of induction generators. In: 15th International Conference on Electrical Machines ICEM 2002, Brujas, Bélgica, pp. 1–6 (2002)

[57] González Rodríguez, Á.G., Burgos Payan, M.: Impacto del arranque de un generador eólico. Energía 29(173), 104–108 (2003)

[58] González Rodríguez, Á.G., Burgos Payán, M., Trigo García, Á.L., Roldán Fernández, J.M., Izquierdo Mitchell, C.: Improvement of induction generator soft-start. In: 8^0 Congresso Luso Espanhol de Engenharia Electrotécnica, Vilamoura (Portugal), pp. 4.115–4.120 (2003) ISBN: 972-8822-00-6

[59] Cano Marín, R., Gómez, A., Burgos, M.: Wind energy integration in distribution networks: a voltage-stability constrained case study. In: Bulk Power System Dynamics and Control - IREP 2004, Cortina d'Ampezzo (Italia), Agosto de (2004)

[60] Roldán Fernández, J.M., Caballero Gutiérrez, M.A., del Pino, J.C., Gonzélez Rodríguez, A.G., Burgos Payán, M.: Monitoring and control of a laboratory scale wind farm. In: International Conference Conference on Renewable Energy and Power Quality - ICREPQ 2007, Sevilla, Marzo de (2007) ISBN: 978-84-611-4707-6

[61] Diveux, T., Sebastian, P., Bernard, D., Puiggali, J.R., Grandidier, J.Y.: Horizontal axis wind turbine systems: optimization using genetic algorithms. In: Wind Energy, vol. 4, pp. 151–171. John Wiley & Sons Ltd., Chichester (2001)

[62] Middelgrunden Wind Turbine Co-Operative Website, `http://www.middelgrunden.dk/MG_UK/economy/budget.htm`

[63] Wood, D.H.: Determination of the optimum tower height for a small wind turbine. Int. J. REE 3(2), 356–359 (2001)

[64] Wolsink, M.: Wind power implementation: The nature of public attitudes: Equity and fairness instead of 'backyard motives'. Renewable and Sustainable Energy Reviews 11, 1188–1207 (2007)

[65] González Rodríguez, Á.G., González Rodr'iguez, A., Burgos Payán, M.: Estimating wind turbines mechanical constants. In: International Conference on Renewable Energy and Power Quality - ICREPQ 2007, Sevilla, Marzo de (2007)

[66] Roldán Fernández, J.M., Caballero Gutiérrez, M.A., González Rodríguez, A.G., Burgos Payán, M.: Wind farm optimun efficiency by voltage control. In: International Conference on Renewable Energy and Power Quality - ICREPQ 2007, Sevilla, Marzo de (2007)

[67] Beyer, H.G., Lange, B., Waldl, H.P.: Modelling tools for wind farm upgrading. In: Proceedings of the European Union Wind Energy Conference EUWEC, Göteburg, Sweden, pp. 1069–1073 (1996)

[68] Shahirinia, A.H., Tafreshi, S.M.M., Gastaj, A.H., Moghaddomjoo, A.R.: Optimal sizing of hybrid power system using genetic algorithm. In: International Conference on Future Power Systems, November 2005, pp. 16–18 (2005)

[69] Elkinton, C., Manwell, J., McGowan, J.: Offshore wind farm layout optimization (OWFLO) project: preliminary results. In: Proc. 44th AIAA Aerospace Sciences Meeting and Exhibit, Reno, Nevada, January 9-12 (2006)

[70] Hopewell, P.D., Castro-Sayas, F., Bailey, D.I.: Optimising the design of offshore wind farm collection networks. In: Proceedings of the 41st International Universities Power Engineering Conference, UPEC 2006, September 2006, vol. 1, pp. 84–88 (2006)

[71] Zhao, M., Chen, Z., Hjerrild, J.: Analysis of the behaviour of genetic algorithm applied in optimization of electrical system design for offshore wind farms. In: 32nd IEEE Annual Conference on Industrial Electronics, IECON 2006, pp. 2335–2340 (November 2006)

[72] Carrasco, J.M., Garcia Franquelo, L., Bialasiewicz, J.T., Galván, E., Portillo Guisado, R.C., Martín Prats, M.Á., León, J.I., Moreno Alfonso, N.: Power-Electronic Systems for the Grid Integration of Renewable Energy Sources: A Survey. IEEE Trans. on Industrial Electronics 53(4), 1002–1016 (2006)

[73] Junginger, M., Faaij, A., Turkenburg, W.: Cost reduction prospects for offshore wind farms. Wind Eng. 28(1), 97–118 (2004)

[74] LaValle, S.M.: Planning Algorithms. Cambridge University Press, Cambridge (2006), http://planning.cs.uiuc.edu/

Capacity Benefit Margin Evaluation in Multi-area Power Systems Including Wind Power Generation Using Particle Swarm Optimization

Maryam Ramezani, Hamid Falaghi, and Chanan Singh

Abstract. Available transfer capability (ATC) is an index showing the measure of transfer capability remaining in the physical transmission network over and above already existing transactions. To determine ATC between two areas in a multi-area power system, different parameters such as total transfer capability (TTC), transmission reliability margin (TRM), and capacity benefit margin (CBM) should be calculated. CBM ensures security of system operation when the system faces generation deficiency in some areas. The presence of wind turbine generators (WTGs) in multi-area power systems creates new challenges in CBM calculation process. In this chapter, three different methods are proposed for CBM evaluation considering WTG which reflect different objectives. In the proposed methods, CBM determination is formulated as an optimization problem and Particle Swarm Optimization method is used to solve the problem. The numerical results for modified IEEE reliability test system are presented to demonstrate the effectiveness of the proposed approaches.

1 Introduction

Increased social awareness of harmful environmental effects of greenhouse gases and energy policies of the governments have caused wind farms (WFs) to be considered seriously as an alternative for electric power generation. Currently, in some countries having potential of wind energy utilization, WFs produce as much as 20% of the total electric power [1] and some countries intend to produce

Maryam Ramezani · Hamid Falaghi
Department of Electric Power Engineering, The University of Birjand, Birjand, Iran

Chanan Singh
Department of Electrical and Computer Engineering, Texas A&M University, College Station, TX, USA

L. Wang et al. (Eds): *Wind Power Systems*, Green Energy and Technology, pp. 105–123.
springerlink.com

between 5% and 25% of electric power from renewable energy sources by 2010–2015 [2]. Significant part of this is expected to be supplied by WFs. Presence of WFs and their increased degree of penetration have motivated researchers to develop approaches which can evaluate WF effects on planning and operation of the power system. Several studies have been performed about effects of wind power and intermittent or fluctuating sources in general on reliability evaluation [2], [3]. The reinforcement of the transmission network for connecting WFs has also been studied [4]. The effects of WF on thermal generation unit commitment and dispatch have been investigated in [5] and [6], respectively. With respect to the use of WFs in the multi-area power systems, it is necessary to provide the possibility of evaluation, control, and management of power system in the new environment.

Available transfer capability (ATC) is a measure of the transfer capability remaining in the physical transmission network for future commercial activities over and above already committed uses [7]. It is a key index to determine allowable power transaction between areas. Mathematically ATC is expressed as follows:

$$\text{ATC}=\text{TTC}-\text{TRM}-\text{CBM}-\text{Existing transmission commitments.}$$

Several technical terms have been defined in [7] for ATC calculation. Total transfer capability (TTC) is defined as the amount of electric power that can be transferred over the interconnected transmission network in a reliable manner. Transmission reliability margin (TRM) is defined as the transfer capability necessary to ensure that the interconnected transmission network is secure under a reasonable range of uncertainties in system conditions. Capacity benefit margin (CBM) is defined as the amount of transfer capability reserved by load serving entities to ensure access to generation from the interconnected system to meet generation reliability requirements [8]. Thus, CBM is an important index reflecting reliability of the system and needs to be evaluated carefully.

CBM can be determined either by deterministic or probabilistic approaches. Deterministic approaches are approximate and typically centered on maintaining a specified reserve or capacity margin, or may be based upon surviving the loss of largest generating unit. Probabilistic approaches, which have been utilized in some recent research efforts, often use loss of load expectation (LOLE) to include reliability considerations. The basis of all of the proposed approaches is to keep the LOLE less than a target value (for example 2.4 hrs/yr) for the areas in the interconnected power system [8]-[10].

The presence of WFs in multi-area power systems creates new challenges in CBM calculation process due to fluctuating nature of WTG causing by wind speed variations. Thus, the CBM problem should be revised in the presence of WFs in the power systems.

This chapter discusses the issue of CBM determination and allocation in the interconnected power systems with wind power generation. In this chapter, the area that needs to import power from external areas in emergency states is called "supported area" and the area which has enough generation capacity is named "supporting area".

CBM determination policies can differ from one system to another. These differences basically depend on the system operation philosophy and objectives.

Keeping in mind different targets and strategies for system management, three methods are proposed to specify and allocate CBM, thus providing a set of choices for different applications. The proposed methods try to reach special points of the feasible solution space based on system operator's decision. Also, by incorporating CBM allocation in the optimization process, feasibility of the candidate solutions is checked and the best possible solution is obtained finally. In the first method, the main objective is to minimize the summation of LOLE of the areas that can be considered in the pool operation. The second method attempts to minimize the reliability degradation of supporting areas and in the third one, CBM is minimized directly. The second and third methods can prove significant in bilateral markets, because reduction of CBM leads to decreased reservation in the network and more potential for commercial activities. The constraint which has to be satisfied in all of the methods is achieving LOLE less than the target value. In the proposed methods, CBM determination and allocation for all supported areas are done simultaneously in one optimization process. Due to use of LOLE and power flow calculations in the presence of WFs, the proposed methods for CBM determination become combinatorial optimization problems with nonlinear and non-differentiable objective functions. Heuristic or evolutionary optimization techniques are the main class of techniques that can be applied conveniently in these problems. Particle swarm optimization (PSO) is an evolutionary based technique inspired by social systems. It has simple concepts and can be easily applied to the nonlinear and non-continuous optimization problems with continuous variables. PSO is considered robust to control parameters and computationally efficient. PSO technique is expected to generate high-quality solutions within shorter calculation time and has more stable convergence characteristic than other stochastic methods [11], [12].This has motivated investigation and application of PSO technique in the proposed approach.

The advantages of the proposed methods are illustrated by a modified IEEE 24-bus reliability test system.

2 Theoretical Background

As mentioned earlier, to calculate and allocate CBM in the interconnected power system with wind power generation, it is necessary to recognize supported areas according to the target value of LOLE (LOLT). Then, the imported and exported power should be adjusted to obtain available generation of areas to achieve the LOLT. Finally, power transactions are assigned to the tie lines. The following presents the basic context of the problem.

2.1 LOLE Evaluation and Wind Farm Modeling

LOLE indicates the expected risk of loss of load for a specific duration and is obtained by combining system capacity outage probability table and the system load characteristics. Capacity outage probability table is simply an array of possible capacity levels and the associated probabilities. There are several possible load models which can be used resulting in different units for LOLE. In this work, the hourly load duration curve is used in conjunction with the capacity outage probability table to obtain the LOLE. Thus, the units of LOLE are (hrs/yr) and can be calculated by Eq. (1).

$$LOLE = \sum_{i=1}^{Nh} P_i(C_i < L_i) \qquad \text{(h/yr)} \tag{1}$$

where N_h is total number of hours in the period under study; C_i is available generation capacity at hour i; L_i is forecasted load at hour i; $P_i(C_i < L_i)$ is probability of loss of load at hour i.

The generated power of WF varies with the wind speed at its location and can be determined from its "speed-power" curve, which is a plot of output power against wind speed. The output power, PWF, which corresponds to a given wind speed, x, can be obtained from Eq. (2).

$$PWF = \begin{cases} 0 & x < V_{cin} \\ P_r \cdot (A + Bx + Cx^2) & V_{ci} \le x < V_r \\ P_r & V_r \le x < V_{co} \\ 0 & x \ge V_{co} \end{cases} \tag{2}$$

The constants A, B, and C are as follows [13]:

$$A = \frac{1}{(V_{ci} - V_r)^2} \left\{ V_{ci}(V_{ci} + V_r) - 4V_{ci}V_r \left[\frac{V_{ci} + V_r}{2V_r} \right]^3 \right\} \tag{4}$$

$$B = \frac{1}{(V_{ci} - V_r)^2} \left\{ 4V_{ci}V_r \left[\frac{V_{ci} + V_r}{2V_r} \right]^3 - (3V_{ci} + V_r) \right\} \tag{5}$$

$$C = \frac{1}{(V_{ci} - V_r)^2} \left\{ 2 - 4 \left[\frac{V_{ci} + V_r}{2V_r} \right]^3 \right\} \tag{6}$$

where, P_r is the rated power, V_{ci} , V_r , and V_{co} are the cut-in, rated, and cut-out wind speed, respectively. The real electric power from WTGs.

A probability density of wind speed at the WF location can be derived from historical data of wind speed at that site. Using stepwise version of this probability density function and the speed-power curve, the WF can be modeled as a multi-level generation source with a probability related to each generation level.

Therefore, using the above mentioned multilevel generation model, LOLE of the system with wind power generation can be evaluated as follows:

$$LOLE = \sum_{g=1}^{Ng} prob_g \cdot LOLE_g \qquad \text{(h/yr)} \tag{6}$$

where, Ng is number of generation levels of the WF model; $prob_g$ is the probability related to the generation level g and $LOLE_g$ is LOLE at generating level g.

To specify and allocate CBM, after determining available generation of areas, it is necessary to calculate LOLE for the new condition. For this purpose, the following procedure can be used:

1. The imported power to an area is modeled as reduction of that area's load by an equivalent value.
2. The exported power from an area is modeled as increment of that area's load by an equivalent value.

After changing the load of areas, LOLE can be calculated by the mentioned method.

2.2. CBM Description

CBM is the amount of transfer capability reserved for load serving entities on the host transmission system, to enable access to generation from the interconnected power system to meet generation reliability requirements [8]. Reservation of CBM for a load serving entity allows that entity to reduce its installed generation capacity below what may otherwise have been necessary without interconnection to meet its generation reliability requirements. The transmission capability reserved as CBM is intended to be used by load serving entities only in times of emergency generation deficiencies. The planned purchase of energy to serve the network load and meet required generation reserve levels is not to be included in the CBM.

2.3 CBM Allocation

The CBM determination involves calculating the amount of import and export power and allocating these values to the interconnected transmission network. In the proposed approaches, determination of import power from external areas and CBM allocation to tie lines is done simultaneously as a single optimization process. For this purpose, reduction or increment of available generation for each area for the alternative solution should be calculated by comparison between its generation in base case and alternative solution. The base case is an assumed power system operating condition for transfers in the ATC calculation problem. In this condition, the system is secure and all constraints such as line flows and bus voltage magnitudes lie within their operating limits. Then, CBM is allocated to the tie lines by modeling the generation changes as base transfers. Actually, in this method, the amounts of area power generation are modified. If generation change of area is positive (area imports power from external areas for emergency condition), the generation output of units is updated by subtracting. Indeed, the generation change of the area is distributed between its generating units in equal proportion or specified values. If generation change of area is negative (area exports power to shortage areas in emergency condition), this process is repeated by adding a value to generating units. Actually, modeling of power transaction from supporting areas to supported areas which is the goal of this approach is done by increasing generation output of the supporting area by the determined value.

After modifying the generation output of each area, physical constraints are investigated by the AC power flow. If no violation is detected, CBM of the tie lines can be determined by comparison between real power transaction of the lines before and after modification.

$$CBM_k = \sum_{g=1}^{Ng} prob_g (S_{Tk-g}^{after} - S_{Tk-g}^{before}) \quad (7)$$

where, CBM_k is CBM of tie line k; S_{Tk-g}^{before} and S_{Tk-g}^{after} are power transaction from tie line k at wind power generation level g before and after modification, respectively.

The method provides more flexibility and any limitation of power transfer can be modeled in this way. For example, a reservation may not be desired in some part of the network (for some reason) in spite of sufficient transmission capacity. In this situation, the restriction can be modeled by limiting the physical limitations of the lines and obtaining suitable results.

3 Optimization Problem Formulation

In CBM determination, an important objective is to improve the LOLE of areas to be equal or less than LOLT. Since, system provider has responsibility of CBM determination, magnitude of CBM can be affected by the discretion of the transmission provider. Therefore, it is possible that other objectives also need be proposed in addition.

There are several ways to improve the situation and compensate generation deficiency by importing power from external areas. Also, it is possible that there are several areas that can export power to supported areas. On the other hand, LOLE has a non-linear relationship with power generation of the area and reduction of generation in two areas with equal additional power leads to different LOLE for them. Now, the key question is how much power and from which areas it is purchased. Thus, in this chapter three methods are presented to incorporate these targets as described in Sections 3-1 to 3-3. The various steps of the general optimization procedure are summarized as follows:

1. Select a base case.
2. Evaluate LOLE for the base case.
3. Identify the supporting and supported areas.
4. Implement one of the optimization methods described below and specify the available generation for each area and the CBM allocated to the lines as a final result.

The three proposed optimization methods are described below.

3.1 Method I: Minimization of Summation of Area LOLEs

In the pool operation, in addition to satisfying the loads with minimum cost, minimization of damage caused by generation deficiency and reduced LOLE is also important. In this formulation, besides bringing LOLE to allowable level, the procedure looks for transactions that cause reduction of the summation of area LOLE values. The problem formulation is illustrated in Eq. (8).

$$\begin{aligned} &\text{Min } z = \sum_{j\in Ta} LOLE_j \\ &\text{S.t}: LOLE_j \le LOLT \quad \forall j \in Ta \\ &\quad S_{Tk} \le S_{Tk}^{\max} \qquad \forall k \in Tl. \end{aligned} \tag{8}$$

where, z is objective function value; $LOLE_j$ is loss of load expectation for area j; Ta is set of areas in the interconnected network; $S_{Tk}^{\max}$ is maximum power transaction from tie line k and Tl is set of tie lines between supporting and supported areas.

It should be noted that, this method may result in more power transfer between areas because besides observing LOLT constraint, reduction of LOLE of areas is considered in the objective function.

3.2 Method II: Minimization of LOLE Changes of Supporting Areas

Generation deficiencies of the supported areas have to be compensated by importing power from supporting areas. Changing generation of supporting areas will certainly affect their LOLE values. On the other hand, because of the technical characteristic of generating units and the amount of loads, areas present differing behavior with respect to export power. Therefore, minimization of these effects can be taken as the main objective. The formulation of the problem is demonstrated in Eq. (9).

$$\begin{aligned} &\text{Min } z = \sum_{i\in Sa} (LOLE_i^{\text{new}} - LOLE_i^{\text{old}}) \\ &\text{S.t}: LOLE_j \le LOLT \quad \forall j \in Ta \\ &\quad S_{Tk} \le S_{Tk}^{\max} \qquad \forall k \in Tl. \end{aligned} \tag{9}$$

where, $\text{LOLE}_i^{\text{old}}$ and $\text{LOLE}_i^{\text{new}}$ are loss of load expectation before and after optimization for area i. Sa is set of supporting areas.

This method provides a better choice for external areas to export power with minimum damage. Some generation owners may not want to change their initial suitable position or generators in some areas may have unique owners. In this situation, generator owners may want to export power from the area that causes minimum change in its LOLE. Therefore, this objective can be considered in the CBM determination.

3.3 Method III: Minimization of CBM

As mentioned earlier, there is a wide variety of possible supporting policies to eliminate generation deficiency in the interconnected power system. Different supporting policies create various clusters of supporting areas and different CBM in the tie lines.

Maximization of ATC means more flexibility in contract adjustment (in bilateral the market) and more potential for electricity transactions. For this reason minimization of CBM can be considered directly in the objective function as Eq. (10).

$$\begin{aligned} &\text{Min } z = \sum_{i \in Tl} W_i \, CBM_i \\ &\text{S.t} : LOLE_j \leq LOLT \quad \forall j \in Ta \\ &\quad S_{Tk} \leq S_{Tk}^{\max} \qquad \forall k \in Tl. \end{aligned} \tag{10}$$

where CBM_i and W_i are CBM of tie line i and its related weighting factor.

In the interconnected power system, some tie lines may have higher priority with respect to the rest. Use of weighting factors can include a priority order supporting policy which is most commonly used. The tie line with the first priority has maximum value of W_i. A larger weighting factor results in lower CBM (greater ATC) in the corresponding interfaces. Thus, these values of W_i are specified in terms of the relative priority of tie lines.

In the bulk interconnected power system that has a higher number of supporting and supported areas and tie lines, method II and specially method III can reduce reservation in tie lines and therefore increase ATC.

4 PSO Based Optimization

Use of reliability and power flow calculations in the problem casts the proposed CBM determination models as combinatorial optimization problems with nonlinear and non-differentiable objective functions. In this case, mathematical optimization approaches such as linear or nonlinear programming can not be easily implemented [14], [15]. Heuristic or evolutionary optimization techniques have become the main class of techniques for solving such problems. In this study, PSO is used as the optimization tool. PSO is an effective approach to handle different continuous optimization problems with nonlinear objective function. In the following sections, first the PSO algorithm is explained briefly and then stages of applying PSO to the CBM problem are illustrated.

4.1 Overview of the PSO Algorithm

PSO is one of the evolutionary optimization techniques inspired by the behavior of a flock of birds or sociological behavior of a group of people [16]. PSO is a population based optimization technique where the population is called a swarm. A swarm contains a number of individuals that can be candidate solutions for the optimization problem at hand. Similar to other population based evolutionary algorithms, initial swarm is produced randomly. Each individual is formed by different components that determine the position of the individual in the n-dimensional search space [11]. Besides position, each candidate solution is associated with a velocity. In the n-dimensional solution space, the position and velocity of each individual are illustrated as the vectors, $X_i^0 = [P_{i,1}^0, P_{i,2}^0, \ldots, P_{i,n}^0]$, and $V_i^0 = [v_{i,1}^0, v_{i,2}^0, \ldots, v_{i,n}^0]$, respectively.

Similar to the other evolutionary algorithms, PSO is repeated up to a pre-specified number of iterations and it has an evaluation function that assigns the individual's position based on its evaluation value. In each iteration, the best position of individual *i* with respect to previous iterations is determined and represented by $X_{pbest,i}^k$. Then, the best evaluation value obtained in all the previous runs is called global best, X_{Gbest}^k. The velocity and position of each individual are updated using the following expressions [11].

$$V_i^{k+1} = \omega V_i^k + \alpha_1 R_1 \times (X_{pbest,i}^k - X_i^k) + \alpha_2 R_2 \times (X_{Gbest}^k - X_i^k) \tag{11}$$

$$X_i^{k+1} = X_i^k + V_i^{k+1} \tag{12}$$

where V_i^k is velocity of individual *i* at iteration *k*; ω is inertia weight; α_1 and α_2 are acceleration coefficients; R_1 and R_2 are random numbers in the range [0, 1]; X_i^k is position of individual *i* at iteration *k*.

The acceleration coefficients α_1 and α_2 control how far an individual will move in a single iteration. The inertia weight which is used to control the convergence behavior of the PSO is adapted during the training stage *k* according to Eq. (13).

$$\omega = \omega_{max} - \frac{\omega_{max} - \omega_{min}}{k_{max}} \cdot k \tag{13}$$

where, ω_{min} and ω_{max} are initial and final inertia weight; k_{max} is maximum number of iterations.

4.2 Development of the Proposed Approach

In the proposed approach, the procedure for CBM determination using a PSO algorithm is developed to obtain the best power transaction schedule in emergency conditions. This section explains the application of PSO approach to the problem using the following steps and the flowchart shown in Fig. 1.

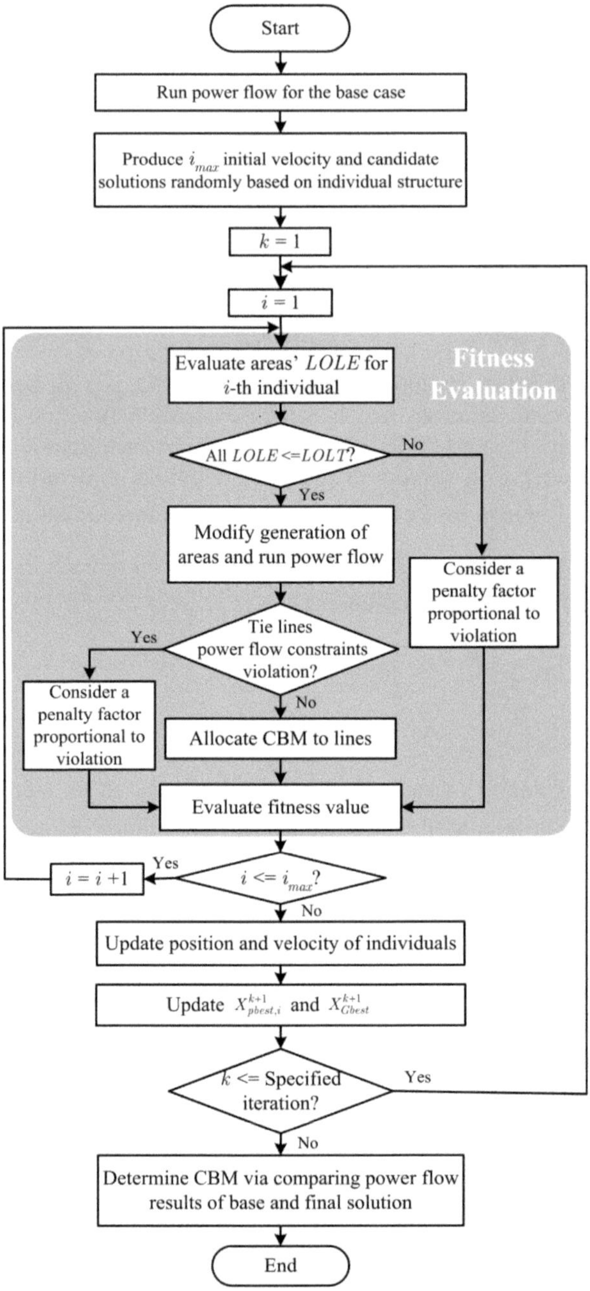

Fig. 1. General optimization procedure for CBM determination

1. Determination of individual's structure: In the CBM specification problem, the determination of import or export power is the basis of the study. So, available area generations are selected as decision variables of the problem. Each component of an individual shows the available generation of its associated area. The number of individual's components is equal to the number of areas that cooperate in the power transaction. The value assigned to each component is equal to total generation of areas plus imported power or minus exported power. Each individual i is described by a vector as $X_i^k = [G_{i,1}^k, G_{i,2}^k, \ldots, G_{i,j}^k, \ldots, G_{i,Ta}^k]$ where $G_{i,j}^k$ is available generation capacity for area j for individual i. The difference between a component's value and installed generation capacity of related area shows import/export power of that area. If the component's value is larger than the installed generation capacity of corresponding area in the base case, power is imported to the area and vice versa.
2. Velocity and position initialization: The velocity and position of the individuals are produced randomly but the total amount of generation in the system is a fixed number. Therefore, the initialization should be done observing this essential point. Initialization of each individual can be described as follows:

   2-i) One of the components j (in other words, one of the areas) of the individual i is chosen in a random fashion.

   2-ii) The amount of generation of the corresponding area, $G_{i,j}^k$, is selected in the range $[Load_j, G_T - \sum_{h=1}^{Sn} Load_h]$ where $Load_j$ is load of area j; G_T is total generation capacity of areas cooperating in improvement of generation reliability; Sn is set of non-initialized components.

   2-iii) Another component is selected randomly from the non-initialized components.

   2-iv) Initial value for selected component is chosen from range $[Load_j, G_T - \sum_{h \in Si} Gen_h - \sum_{h \in Sn} Load_h]$ where Gen_h is allocated generation capacity to area h; Si and Sn are set of initialized and non-initialized components, respectively.

   2-v) If all components but one are initialized, go to (2-vi) otherwise go to (2-iii).

   2-vi) The last component is equal to $[G_T - \sum_{h \in Si} Gen_h]$.

   The velocity of each component is selected in the range $[-V_{max}, +V_{max}]$ randomly. (In this study V_{max} of 4 has been selected for the PSO process).
3. Fitness evaluation: The fitness evaluation for individuals in the swarm is accomplished as follows:

3-i) Power flow implementation and LOLE evaluation for the base case.
3-ii) LOLE evaluation for each area according to available generation that has been represented by corresponding individual's component.
3-iii) If the LOLE of the area is in the allowable range, the generation output of the areas of transfer will be updated as described in Section 2-3 and the power flow will be run for the new condition. Otherwise a penalty factor (K_1) is considered.
3-iv) If tie line power flow constraints are satisfied, the amount of CBM will be calculated by comparison between the initial (base case) and final (candidate solution) power transaction otherwise a penalty factor (K_2) is considered.
3-v) The value of the fitness for each individual is calculated by Eq. (14).

$$fit(X_i^k) = \begin{cases} M - (K_1 + K_2), & \text{if } K_1 + K_2 > 0 \\ M - z(X_i^k), & \text{otherwise} \end{cases} \tag{14}$$

where, M is a large number. K_1, K_2 and M should be large enough and M has to be larger than $K_1 + K_2$. K_1 and K_2 are set to 500 and M is 1500. In the first and second optimization problem as objective values are not large, these values can be used regardless the kind of the network, but in the third method, K_1, K_2 and M should be selected considering the network and probable CBM values.

4. Velocity and position update: Since equality constraint corresponding to constant total generation of the interconnected power system should be satisfied, position updating process is carried out by considering this issue. In this way, after updating velocity and position by Eqs. (11) and (12), respectively, the position is modified by distributing the additional or subtracted power among the component's individual.
5. Updating of $X_{pbest,i}^{k+1}$, X_{Gbest}^{k+1}: $X_{pbest,i}$ of individual i at iteration $k+1$ is updated as follows:

$$X_{pbest,i}^{k+1} = \begin{cases} X_i^{k+1} & if \quad fit(X_i^{k+1}) < fit(X_i^k), \\ P_{best,i}^k & if \quad fit(X_i^{k+1}) \geq fit(X_i^k). \end{cases} \tag{15}$$

Additionally, X_{Gbest}^{k+1} is set as the best evaluated position among $X_{pbest,i}^{k+1}$.

6. Stopping criterion: The optimization process is stopped after a predefined maximum number of iterations.

5 Numerical Results and Discussion

To investigate the proposed methods for determination and allocation of CBM in the presence of wind power generation and to evaluate their efficiency a modified IEEE 24 bus reliability test system (MRTS) is used [8]. The data about initial and additional generating units is shown in Table 1. As shown in Fig. 2, this system

Table 1. Initial and additional generation units in the base case

Area	Initial Generation Units Capacity [MW (number)]	Additional Generation Units Capacity [MW (number)]
1	12(5), 155(2), 400(2)	155(3), 400(1)
2	50(6), 155(2), 197(3), 350(1)	197(1)
3	20(4), 76(4), 100(3)	100(1)

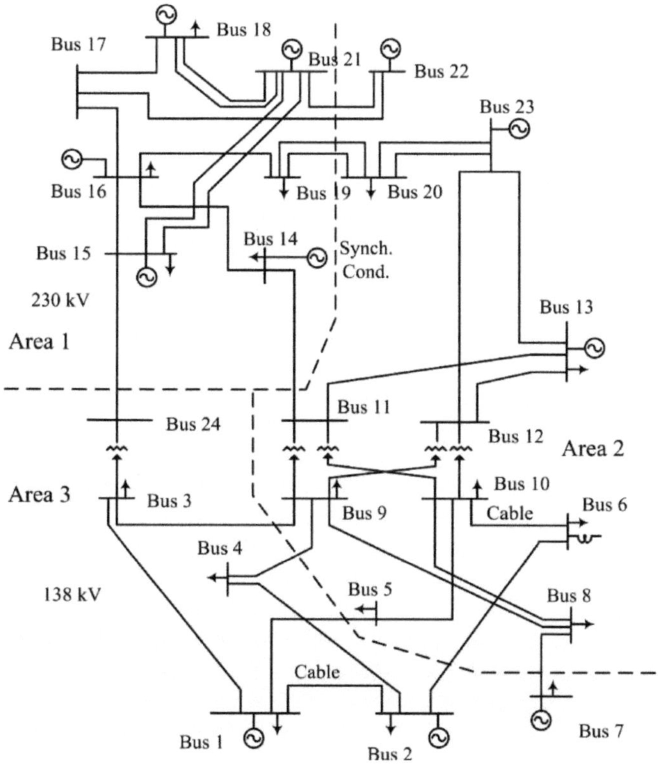

Fig. 2. Control Areas of the IEEE 24 Bus System

has been divided into three areas. In order to study WF effects on CBM, one WF of 60 MW is connected to different areas and replaced by conventional units. Table 2 lists the characteristics of the fixed speed wind turbines used in the WF.

The target value for LOLE for all areas is set at 2.4 h/yr. Detailed information of the system can be found in [17]. Hourly load model [17] and historical wind speed data from Manjil in North of Iran is used [18]. Fig. 3 shows probability density of wind speed at this location.

Table 2. The characteristics of wind turbines used in the WF

Operation power factor	V_{ci} [m/s]	V_r [m/s]	V_{co} [m/s]	P_r [MW]
0.9 lagging (producing reactive power)	4	10	22	2

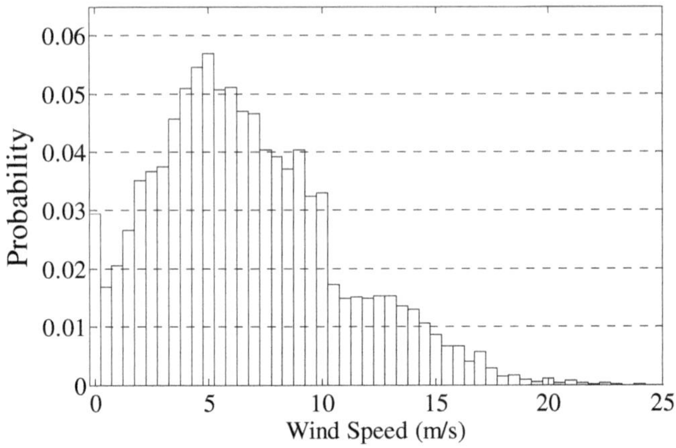

Fig. 3. Probability density of wind speed used in the case study

This study is directed by calculating CBM for MRTS with and without WF. MRTS without WF is considered as base case. Table 3 shows the base case generation information; it lists initial value of generation of each area and associated LOLE. Area LOLE values at initial condition show that the area 1 faces generation deficiency and has to import needed power from external areas. On the other hand, area 2 and area 3 have enough generation sources and have the exporting potential.

As mentioned before, the requested power of area 1 can be provided in several ways that result in different LOLE for areas and various CBM for tie lines. In the following, first, results obtained by the three proposed methods for the base case are studied then MRTS including WF are considered.

Table 3. Initial condition of each area at the base case

Area	Load [MW]	Generation [MW]	LOLE [h/year]
1	1125	2035	4.77
2	1141	1748	0.63
3	584	784	0.66

5.1 *CBM Evaluation of the Base Case (without WF)*

In the first method, the study is performed with the objective to achieve minimum total LOLE in the interconnected power system. The results of the optimization process for the base case, which are presented in Table 4, demonstrate that the LOLE of supported area has been decreased to 1.94 h/year that is lower than LOLT. And, simultaneously LOLE of area 2 has been increased to 2.04 h/year that is higher than the LOLE of the supporting area in the final solution.

The use of method II can control the exporting power to supported area in such a way that the LOLEs of the supporting areas have minimum changes with respect to the first condition. The results of the method II is listed in Table 4. It could be expected that method II tries to reach the LOLE of supported area to 2.40 h/yr.

Table 4. Final result after optimization in the base case (without WF)

Method	Area	Import/Export power [MW]	LOLE [h/year]
I	1	98.46 (import)	1.94
	2	94.96 (export)	2.04
	3	3.50 (export)	0.73
II	1	79.40 (import)	2.40
	2	79.40 (export)	1.74
	3	0	0.66
III	1	79.38 (import)	2.40
	2	36.37 (export)	1.01
	3	43.01 (export)	2.40

Method III reduces CBM in the system according to specified weighting factors. There are three tie lines between areas in the test system so three factors is considered. W_1, W_2 and W_3 refer to tie lines between area 1 and 2 (L_{1-2}), areas 1 and 3 (L_{1-3}) and areas 2 and 3 (L_{2-3}), respectively. The optimization results of method III by W_1 equals to 1 and W_2 and W_3 equal to 0 are demonstrated in Table 4. These weighting factors make attempt to minimize CBM between area 1 and 2. So they lead system into the situation that area 3 exports more power to area 1 until LOLE constraint is satisfied. Table 5 shows CBM for L_{1-2} and L_{1-3} obtained by different methods in the base case.

Table 5. CBM of L1–2 and L1–3 obtained with different methods in the base case (without WF)

Method	CBM [MW] L_{1-2}	CBM [MW] L_{1-3}
I	89.23	8.44
II	72.56	7.22
III	63.04	15.07

5.2 CBM Evaluation of MRTS Containing sWF

As mentioned, CBM is considered to ensure secure interconnected power system with acceptable level of LOLE. With attention to stochastic nature of wind turbines, it is expected that replacing conventional units by wind turbines changes LOLE and consecutively CBM.

In order to perform the studies, a WF of 60 MW is connected to different areas of the base case. For comparison purpose, other input data in case of presence of WF is the same as the base case.

In the first case, WF is replaced by 5 conventional units of 12 MW of bus 15 in area 1. This change leads to LOLE increase of area 1 to 6.05 h/yr. With compare to the base case more power is needed to reach LOLE of area 1 to 2.40 h/yr. Table 6 show the results of LOLE compensation by the proposed methods.

Since connecting WF to area 1 causes LOLE increment, more power should be imported to this area to meet generation deficiency so that needed power of area 1 increase from 98.46 MW in the base case to 110.94 MW in the case of utilizing method I. Similarly, comparison of the results obtained by method II and III for the base case with the results of MRTS containing WF indicated that import/export power and consequently CBM increased.

Table 7 shows CBM for L_{1-2} and L_{1-3} obtained by different methods in the case of connecting the WF to the area 1. The obtained results show that with inclusion of WF to the system the CBM increased.

Table 6. Final result after optimization for MRTS with WF in area 1

Method	Area	Import/Export power [MW]	LOLE [h/year]
I	1	110.94 (import)	2.40
	2	94.97 (export)	2.04
	3	15.97 (export)	1.05
II	1	110.94 (import)	2.40
	2	107.28 (export)	2.36
	3	3.66 (export)	0.73
III	1	110.94 (import)	2.40
	2	67.66 (export)	1.51
	3	43.32(export)	2.40

Table 7. CBM between areas in different methods for MRTS with WF in area 1

Method	CBM [MW]	
	L_{1-2}	L_{1-3}
I	98.62	12.07
II	100.61	9.38
III	93.25	15.07

In order to study WF effects on CBM in case of connecting a WF of 60 MW to area 2, the capacity of conventional units of 50 MW of bus 22 are reduced to 40 MW. It causes LOLE increase of area 2 from 0.63 h/yr in the base case to 1.15 h/yr. In spite of LOLE increase, area 2 has still acceptable reliability level and can export power to area 1. The results obtained by three proposed methods are listed in Table 8. Since LOLE of area 2 has increased, this area can export lower power to support area 1. Table 9 shows CBM values provided by the methods in this case.

Table 8. Final result after optimization for MRTS with WF in area 2

Method	Area	Import/Export power [MW]	LOLE [h/year]
I	1	79.40 (import)	2.400
	2	62.25 (export)	2.367
	3	17.15 (export)	1.092
II	1	79.40 (import)	2.400
	2	59.18 (export)	2.263
	3	20.22 (export)	1.194
III	1	79.40 (import)	2.400
	2	36.38 (export)	1.717
	3	43.02(export)	2.400

Table 9. CBM between areas in different methods for MRTS with WF in area 2

Method	CBM [MW]	
	L_{1-2}	L_{1-3}
I	69.61	9.85
II	69.05	10.02
III	64.88	15.29

In the final study, WF is connected to area 3 by eliminating 3 conventional units of 20 MW from bus 1. With this change, LOLE of this area will be 2.39 h/yr. With attention to LOLT value, area 3 almost can not have supporting duty. The obtained results in this case are listed in Table 10. Table 11 lists the CBM of tie lines in this case.

As the results show, the value of import/export power in sensitive to the WF location and employing different methods the CBM optimization results different value results. Table 11 lists the CBM of tie lines.

Table 10. Final result after optimization for MRTS with WF in area 3

Method	Area	Import/Export power (MW)	LOLE (h/year)
I	1	79.40 (import)	2.40
	2	107.35 (export)	2.36
	3	27.44 (import)	0.90
II	1	79.40 (import)	2.40
	2	79.40 (export)	1.75
	3	0	2.39
III	1	79.40 (import)	2.40
	2	78.90 (export)	1.74
	3	0.5	2.40

Table 11. CBM between areas in different methods for MRTS with WF in area 3

Method	CBM [MW]	
	L_{1-2}	L_{1-3}
I	96.35	16.23
II	72.83	7.23
III	72.56	7.32

6 Conclusion

ATC is one of the important indexes in the deregulated power systems that are utilized to set future contracts. To calculate ATC between two areas in an interconnected network, TTC, TRM and CBM need to be evaluated. This chapter addresses CBM evaluation and allocation in the interconnected power system with wind power generation. CBM is a portion of transfer capability of the network that is specified in such a way that the LOLE of any area is not greater than a specified target value, typically 2.4 hrs/yr. This study emphasizes that several strategies can be considered in the CBM determination and introduces three different methods. For a given situation, one of these methods may be used either as it is or with some modifications to suit the objectives. In method I, the objective is to achieve a minimum total LOLE, method II looks for minimum total change in the LOLE of supporting areas and the third method tries to find the solution with minimum CBM. In all of the methods, the target of LOLE equal or less than a specified value is taken into account as a constraint. The CBM determination and allocation is performed simultaneously in the proposed PSO based optimization process. Finally a modified IEEE 24 bus test system is used to illustrate the results of the optimization method.

References

1. Ackerman, T.: Wind Power in Power Systems. John Wiley & Sons, Chichester (2005)
2. Karki, R., Hu, P., Billinton, R.: A simplified wind power generation model for reliability evaluation. IEEE Transactions on Energy Conversion 21(2), 533–540 (2006)
3. Fockens, S., Wijk, A.J.M., Turkenburg, W.C., Singh, C.: Reliability analysis of generating systems including intermittent sources. International Journal on Electric Power & Energy Systems 14(1), 2–8 (1992)
4. Billinton, R., Wangdee, W.: Reliability-based transmission reinforcement planning associated with large–scale wind farms. IEEE Transactions on Power Systems 22(1), 34–41 (2007)
5. Ummels, B.C., Gibescu, M., Pelgrum, E., Kling, W.L., Brand, A.J.: Impact of wind power on thermal generation unit commitment and dispatch. IEEE Transactions on Energy Conversion 22(1), 44–51 (2007)
6. American Electric Reliability Council, Available Transfer capability definitions and determination, Reference Document (1996)
7. American Electric Reliability Council, Transmission Capability Margins and Their Use in ATC Determination, White Paper (1999)
8. Ou, Y., Singh, C.: Assessment of available transfer capability and margins. IEEE Transactions on Power Systems 17(2), 463–468 (2002)
9. Shin, D.J., Kim, J.O., Kim, K.H., Singh, C.: Probabilistic approach to available transfer capability calculation. Electric Power Systems Research 77(7), 813–820 (2006)
10. Othman, M.M., Mohamed, A., Hussain, A.: Available transfer capability assessment using evolutionary programming based capacity benefit margin. International Journal of Electrical Power & Energy Systems 28(3), 166–176 (2006)
11. Park, J.B., Lee, K.S., Shin, J.R., Lee, K.Y.: A particle swarm optimization for economic dispatch with nonsmooth cost functions. IEEE Transactions on Power Systems 20(1), 34–42 (2005)
12. Zhao, B., Guo, C.X., Cao, Y.J.: A multiagent-based particle swarm optimization approach for optimal reactive power dispatch. IEEE Transactions on Power Systems 20(2), 1070–1078 (2005)
13. Masters, C.L., Mutale, J., Strbac, G., Curcic, S., Jenkins, N.: Statistical evaluation of voltages in distribution systems with embedded wind generation. In: IEE Proceeding on Generation, Transmission & Distribution, vol. 147(4), pp. 207–212 (2000)
14. Naka, S., Genji, T., Yura, T., Fukuyama, Y.: A hybrid particle swarm optimization for distribution state estimation. IEEE Transactions on Power Systems 18(1), 60–68 (2003)
15. Kennedy, J., Eberhart, R.: Particle swarm optimization. In: Proceeding IEEE International Conference on Neural Networks, pp. 1942–1948 (1995)
16. Bergh, F.V.D., Engelbrecht, A.P.: Assessment a cooperative approach to particle swarm optimization. IEEE Transactions on Evolutionary Computation 8(3), 225–239 (2004)
17. Reliability Test System Task Force of the Application of Probability Methods Subcommittee, IEEE reliability test system. IEEE Transactions on Power Apparatus and Systems PAS-98, 2047–2054 (1979)
18. Iran Meteorological Organization, `http://www.weather.ir`

Stochastic Dispatch of Power Grids with High Penetration of Wind Power Using Pareto Optimization

Ali T. Al-Awami and Mohamed A. El-Sharkawi

Abstract. Stochastic dispatch (SD) for a power system with both thermal and wind units is considered. First, the uncertainty associated with the wind power forecast is characterized. Conditional probability density functions (CPDF) of the actual wind power output given the forecast level are obtained. For each forecast level, Beta, Weibul, and Extreme Value distribution functions are compared and the best fit is selected. Then, both single-objective and multi-objective dispatch problems are investigated. In the single-objective dispatch, the objective is to minimize the expected value of the operating cost. In the multi-objective dispatch, Pareto-optimization of operating cost and emissions is studied. The sensitivity of the Pareto-optimal solution to different system conditions is examined. Multi-objective particle swarm optimization (MO-PSO) is employed to obtain the Pareto-front. Simulation results show that SD gives rise to lower average operating cost than deterministic dispatch, in which the wind schedule is identical to the forecast.

1 Introduction

In the past few years, researchers, utility companies, governmental agencies, and other stakeholders have developed the concept of the smart grid. The smart grid gives rise to a power grid that is more autonomous, more reliable, and more environmentally friendly than a traditional power grid. In addition, the smart grid can facilitate the integration of high penetration of renewable energy sources to the existing power grids [1, 2].

One of the main challenges in the operation of the smart grid is the high uncertainty associated with the power output of many renewable energy sources (RES), such as wind and solar. This high uncertainty makes the task of maintaining the generation-load balance a formidable task to system operators, especially in areas with high RES penetration [3, 4].

Ali T. Al-Awami · Mohamed A. El-Sharkawi
University of Washington, Seattle, USA

L. Wang et al. (Eds): *Wind Power Systems*, Green Energy and Technology, pp. 125–149.
springerlink.com

Several activities have been reported in the literature to address the different power system uncertainties in the economic dispatch problem, from both the market perspective [5-7] and the overall system perspective [8-12]. One of the interesting stochastic dispatch (SD) formulations that addresses the uncertainty associated with wind power is proposed in [8] for a single objective dispatch and is expanded in [9] for a multi-objective dispatch. In [8], the cost function includes the operating cost of the thermal units and the wind plants, and the imbalance cost due to the mismatch between the actual and scheduled power outputs of the wind plants. However, the formulation presented in [8, 9] does not take into consideration the wind power forecast.

In order for the wind power forecast to be considered in the SD, the stochastic nature of the wind power output given the forecast needs to be accurately characterized. A model of wind power forecast error is presented in [13, 14]. The data consists of the actual wind power output and the forecast for one year. The data is re-arranged in an ascending order according to the level of the wind forecast. The re-arranged data is divided into a number of data bins, each of which corresponds to a given forecast level. Beta distribution is used to fit the data in each bin. The resulting distribution is the conditional probability density function (CPDF) of the wind power forecast error given the wind power forecast. The set of CPDF functions is then used to generate a weighted pdf of the wind forecast error which is used to optimally size the energy storage medium for a given power system [13]. These studies were based on optimizing a single objective function.

The operation of smart grids often requires optimizing conflicting objectives such as cost, reliability, and environmental impact. Since the objectives are conflicting, the optimization of all objectives simultaneously is not achievable. In this case, Pareto optimization can be employed [15].

In this chapter, a stochastic dispatch formulation that takes into account the wind forecast level is presented for a system with wind plants and thermal units. The SD presented here builds on the work suggested in [8, 9]. First, the CPDF of the wind power output given the forecast level is modeled. In addition to the Beta distribution suggested in [13, 14], the effectiveness of Weibull and Extreme Value distribution functions is examined. The accuracy of the three distributions in characterizing the CPDF is evaluated. For each wind forecast level, the best CPDF of the three is obtained. Then, the best set of CPDFs is used in the proposed SD. The objective function of the SD is the expected value of the combined operating cost of the system given the total system load and the wind power forecast level at each wind plant.

The SD formulation is later expanded to optimize two conflicting objective functions: the combined operating cost and the thermal units' emissions. The concept of Pareto optimality is utilized. To identify the Pareto-optimal set of solutions, the enhanced multi-objective particle swarm optimization (MO-PSO) suggested in [9] is used. Several simulation runs are conducted to study the effect of different system conditions on the Pareto-optimal solutions. These system conditions include total system load, reserve cost coefficients, and penalty cost coefficients. Moreover, the dispatch results of multi-objective SD are compared with those of multi-objective DD.

2 Stochastic Dispatch of Power Systems with High Wind Penetration

This section describes the formulation of the single- and multi-objective dispatch problems. In the single-objective deterministic dispatch, the objective is to minimize the combined operating cost without taking into account the uncertainty of wind forecast. Therefore, the wind schedules are assumed to be identical to their corresponding forecasts. In single-objective stochastic dispatch, however, the intent is to minimize the expected value of the operating cost. Hence, the uncertainty of wind forecast is taken into account. These techniques are compared with the perfect schedule, which is the theoretical lower bound.

The above mentioned techniques can be expanded to include the optimization of conflicting objectives. This can be achieved by extending the search space to identify the boundary of the optimization surface, which is known as the Pareto front.

2.1 Single-Objective Deterministic Dispatch

The objective of a conventional deterministic dispatch (DD) is to minimize the combined operating cost of the power system subject to the generation-load balance constraint. In the presence of wind power, the wind power outputs are dealt with as negative loads, and perfect wind forecast is assumed. In other words, wind power schedule is taken as the wind power forecast without any consideration of the characteristics of the forecast error. Later on, if the actual wind power output of a wind plant deviates from its forecast, the mismatch is dealt with either by dispatching system reserves, limiting the wind power output, or curtailing the wind schedule [16]. As shown in [8, 9], DD can be formulated by

Minimize

$$OC_d(P_{gi}, w_i) \tag{1}$$

Subject to

$$w_i = w_{fci} \tag{2}$$

$$P_{gi}^{\min} \leq P_{gi} \leq P_{gi}^{\max} \tag{3}$$

$$\sum_{i=1}^{M} P_{gi} = L - \sum_{i=1}^{N} w_{fci} \tag{4}$$

where

OC_d	Deterministic combined operating cost of thermal units and wind plants
P_{gi}	Scheduled output of ith thermal unit
w_i	Scheduled output of ith wind plant
w_{fci}	Wind power forecast of ith wind plant

L System load including losses
M Number of thermal units
N Number of wind plants

The combined operating cost is formulated as

$$OC_d = \sum_{i=1}^{M} C_i(P_{gi}) + \sum_{i=1}^{N} C_{wi}(w_i) \tag{5}$$

$$C_i = \frac{a_i}{2} P_{gi}^2 + b_i P_{gi} + c_i \tag{6}$$

$$C_{wi} = d_i w_i \tag{7}$$

where
C_i Operating cost of ith thermal unit
C_{wi} Operating cost of ith wind plant
a_i, b_i, c_i Cost coefficients for i^{th} thermal unit
d_i Cost coefficient for i^{th} wind plant

The DD as described above does not take into account the forecast inaccuracy. In addition, some system operators apply charges to wind plants for the imbalance between their scheduled and actual wind power outputs [16, 17]. These charges reflect the cost of reserves the system operator has to secure in order to compensate for that amount of imbalance. These imbalance charges are not considered in DD. Therefore, the actual operating cost is not necessarily minimized.

2.2 Single-Objective Stochastic Dispatch

In order for the uncertainty in wind power outputs and the imbalance costs to be considered, stochastic dispatch (SD) is suggested [8, 9]. The objective of SD is to minimize the expected value of the stochastic operating cost subject to the system constraints. Single-objective SD for a system with wind and thermal units can be formulated as

Minimize

$$E[OC_s(P_{gi}, w_i)] \tag{8}$$

Subject to

$$0 \leq w_i \leq w_{ri} \tag{9}$$

$$P_{gi}^{\min} \leq P_{gi} \leq P_{gi}^{\max} \tag{10}$$

$$\sum_{i=1}^{M} P_{gi} + \sum_{i=1}^{N} w_i = L \tag{11}$$

where

$E[.]$ Expected value operator

OC_s Stochastic combined operating cost of thermal units and wind plants

w_{ri} Rated output of ith wind plant

As shown in [8, 9], the expected value of the combined operating cost including imbalance charges can be formulated as

$$E[OC_s] = \sum_{i=1}^{M} C_i(P_{gi}) + \sum_{i=1}^{N} C_{wi}(w_i) + \sum_{i=1}^{N} E[C_{pi}(W_{i,ac} - w_i)] + \sum_{i=1}^{N} E[C_{ri}(w_i - W_{i,ac})] \tag{12}$$

where

C_{pi} Imbalance cost of ith wind plant due to over-generation (under-estimation)

C_{ri} Imbalance cost of ith wind plant due to under-generation (over-estimation)

$W_{i,ac}$ Actual wind power output from i^{th} wind plant. This is a random variable since it is unknown at the time of optimization.

In (12), the expected value of the imbalance costs can be described as

$$E[C_{pi}] = E[k_{pi}(W_{i,ac} - w_i)] = k_{pi} \int_{w_i}^{w_{ri}} (w - w_i) f_W(w \mid w_{fci}) dw \tag{13}$$

$$E[C_{ri}] = E[k_{ri}(w_i - W_{i,ac})] = k_{ri} \int_{0}^{w_i} (w_i - w) f_W(w \mid w_{fci}) dw \tag{14}$$

where

k_{pi}, Penalty cost coefficient for over-generation of ith wind plant

k_{ri}, Reserve cost coefficient for under-generation of ith wind plant

$f_W(w/w_{fci})$ Conditional probability density function (CPDF) of wind power output given wind power forecast level

The commonly used second-order polynomial shown in (6) models the operating costs of the thermal units. For wind plants, the operating cost is considered linearly proportional to the power output, as shown in (7). The imbalance cost due to over-generation or under-generation of wind plants is assumed to be linearly proportional to the difference between the actual and scheduled wind powers [8, 9].

It is noteworthy that the formulation given in [8, 9] uses the pdf of the wind power output, $f_W(w)$, instead of the conditional pdf of the wind power output given the forecast, $f_W(w|w_{fci})$. Using $f_W(w)$ results in a single optimal dispatch solution for any given load with no regard to the wind power forecast. However, using $f_W(w|w_{fci})$ takes into account the wind power forecast. The process of obtaining $f_W(w|w_{fci})$ is described in details in section 1.3.

2.3 Single-Objective Perfect Scheduling

Perfect scheduling (PS) is a theoretical lower bound optimal solution that is not attainable in real systems. However, PS is considered here just for comparison purposes. With PS, the wind schedule of each wind plant and its actual wind power output are assumed to be identical. That is, $w_i = W_{i,ac}$. In this case, (1) can be used to obtain the perfect schedules after setting $w_{fci} = W_{i,ac}$.

2.4 Multi-objective Stochastic Dispatch

Stochastic dispatch can be extended to include multiple objectives. In this case, the goal is to optimize two conflicting cost functions simultaneously; the combined operating cost of thermal and wind units, and the emissions caused by the thermal units.

The economic/environmental dispatch for wind and thermal units can be formulated as

Minimize

$$[\ E[OC_s(P_{gi}, w_i)]\ ,\ S(P_{gi})\ ] \tag{15}$$

Subject to

$$0 \le w_i \le w_{ri} \tag{16}$$

$$P_{gi}^{\min} \le P_{gi} \le P_{gi}^{\max} \tag{17}$$

$$\sum_{i=1}^{M} P_{gi} + \sum_{i=1}^{N} w_i = L \tag{18}$$

where S is the emissions of the thermal units.

The first objective function is the expected value of the combined operating cost as given by (12). The environmental emissions of atmospheric pollutants can be expressed as shown in [18] as

$$S = \sum_{i=1}^{M} \left[\alpha_i + \beta_i P_{gi} + \gamma_i P_{gi}^2 + \zeta_i \exp(\lambda_i P_{gi}) \right] \tag{19}$$

where α_i, β_i, γ_i, ζ_i, and λ_i are the coefficients of the emission function of the ith thermal unit.

3 Stochastic Characterizing of Uncertainty of Wind Power Output

The methodology described in this section for characterizing the uncertainty of wind power output given the wind power forecast level follows from [13, 14]. The process of characterizing wind power uncertainty can be summarized as follows:

1. Normalize historical wind power output data:

In this work, 10-minute wind power outputs of two wind plants for 11 months are used. These wind plants are Condon and Stateline, which are partially integrated in the balancing authority of Bonneville Power Administration (BPA) of the Pacific Northwest [19].

2. Generate wind power forecast:

Several wind power forecast algorithms are reported in the literature [20], [21]. For short-term, persistence forecast has served as a benchmark with which every other forecast algorithm is compared. Persistence forecast works simply by assuming that the wind power output in the next time period is equal to the current value. Hence, persistence forecast can be formulated as given in [13], and shown in (20):

$$w_{fci}(t+k:t+k+T \mid t-T:t) = \frac{1}{T}\sum_{i=0}^{n-1} W_{i,ac}(t-i\Delta t) \tag{20}$$

$w_{fci}(t + k : t + k + T \mid t)$ is the wind forecast for the period that starts at time $t + k$ and ends at $t + k + T$ given the actual wind power output $W_{i,ac}(t - T : t)$ from time $t - T$ to t.

Fig. 1 illustrates how persistence forecast is implemented in this work. It is assumed that the forecast needs to be available 30 minutes before the start of each hour, hence $k = 30$ min. Also, the forecast is used to schedule wind power for one hour, i.e. the wind schedule cannot be changed within the hour, hence $T = 60$ min. In other words, the wind power forecast, w_{fci}, for the period from $t + 30$ to $t + 30 + 60$ is the actual wind power output, $W_{i,ac}$, averaged over the period from $t - 60$ to t. Fig. 2 shows the normalized wind power output data of Stateline for 11 months and the corresponding hour-ahead forecast.

3. Re-arrange the data based on wind forecast level:

The data corresponding to each wind plant is re-arranged based on the wind forecast level in an ascending order. Fig. 3 shows the Stateline wind plant data after re-arranging. The thick green line is the wind forecast level and the blue line is the actual wind power output. It should be noted that the x-axis is no longer time; it is just the serial numbers assigned to the data points.

4. Divide the data into bins according to wind forecast level:

The re-arranged data corresponding to each wind plant is divided according to wind forecast level into B data bins with even widths. In this work, the best results were obtained when $B = 25$. Fig. 3 shows the data corresponding to bin 7 ($w_{fci} \in [0.24, 0.28]$) and bin 19 ($w_{fci} \in [0.72, 0.76]$) of Stateline wind plant.

5. For each data bin, find the probability density functions for the actual wind power output:

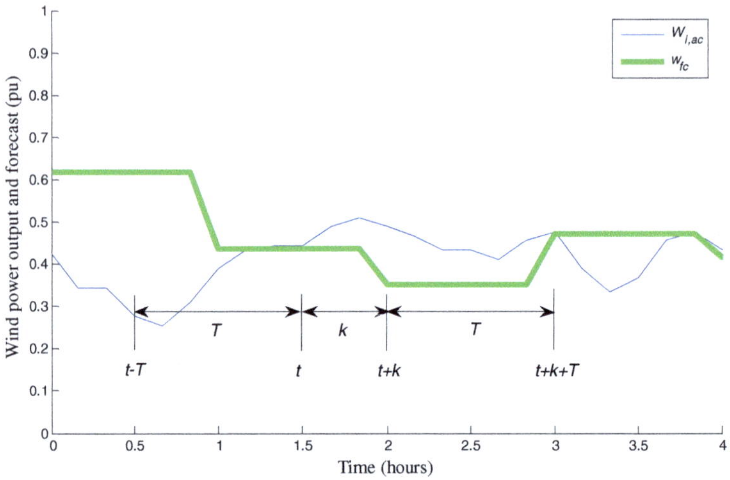

Fig. 1. Persistence forecast of wind power output

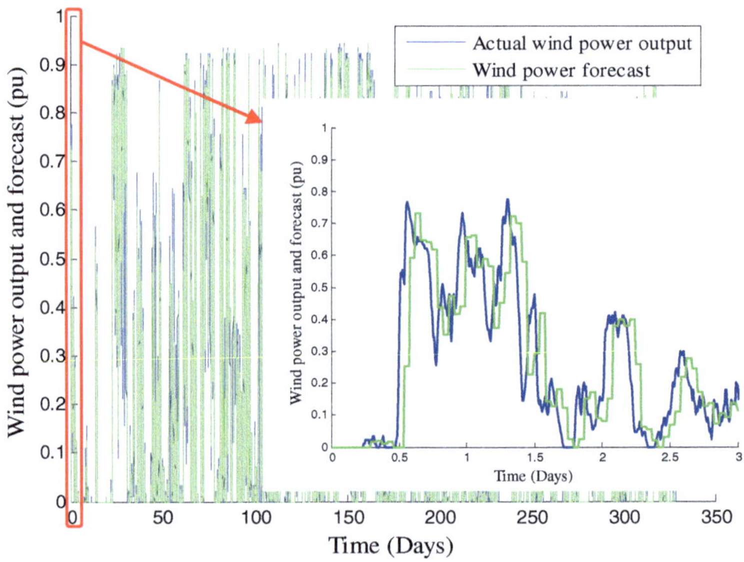

Fig. 2. Normalized wind power output and forecast data of Stateline

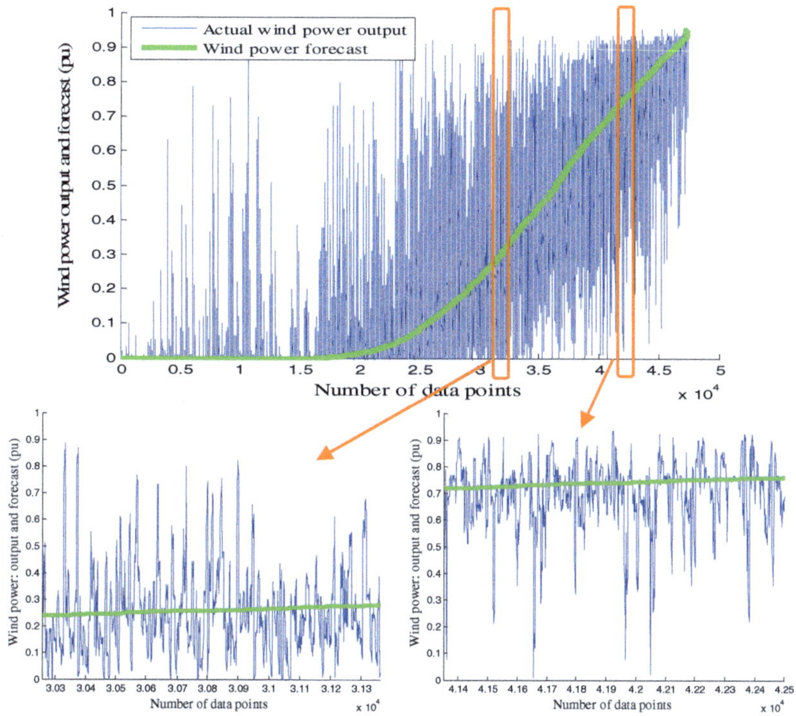

Fig. 3. Stateline wind plant data after re-arranging according to forecast. (Top). Data corresponding to Bin 7; $w_{fci} \in [0.24, 0.28]$, and Bin 19; $w_{fci} \in [0.72, 0.76]$ (Bottom).

The wind power output cannot be negative. In addition, the histogram of the wind power output of each bin is usually unsymmetrical. Therefore, assuming that the pdf follows Gaussian distribution is not a good assumption. In [13], [14], Beta distribution is considered to model the uncertainty in wind power forecast error. Our investigation using the data of Condon and Stateline wind plants shows that Beta distribution fits reasonably well when the wind power forecast is low to moderate. In many cases, however, Beta distribution is inaccurate for high wind forecast level. Therefore, we decided to consider two other distributions; Weibull and Extreme Value pdfs.

The formulae of the Weibull (f_{wbl}), Extreme Value (f_{ev}), and Beta (f_{beta}) distributions are given as

$$f_{wbl}(w_i \mid w_{fci}) = \frac{k_i}{w_i}\left(\frac{w_i}{c_i}\right)^{k_i} \exp\left(-\left(\frac{w_i}{c_i}\right)^{k_i}\right) \tag{21}$$

$$f_{wbl}(w_i \mid w_{fci}) = \frac{k_i}{w_i}\left(\frac{w_i}{c_i}\right)^{k_i} \exp\left(-\left(\frac{w_i}{c_i}\right)^{k_i}\right) \tag{22}$$

$$f_{beta}(w_i \mid w_{fci}) = \frac{w_i^{c_i-1}(1-w_i)^{k_i-1}}{\int_0^1 w_i^{c_i-1}(1-w_i)^{k_i-1}\,dw} \tag{23}$$

where c_i and k_i are pdf parameters.

Fig. 4 shows the pdf fits of the data corresponding to bins 7 and 19 of Stateline wind plant. For bin 7, both Beta and Weibull distributions outperform extreme value distribution. For bin 19, however, extreme value distribution is superior. In both cases, Weibull distribution performs better than Beta distribution. Analyzing the results of all the bins of both wind plants, we concluded that Beta pdf is only occasionally better than the other two, but with a small margin.

6. Among the three pdf distributions, select the one that best fits the bin data:

Given the actual bin data, there are several criteria that can be used to compare the performance of the three distributions. Two of them are the absolute error (*AE*) between the actual data and the data generated by the pdf fit parameters, and the squared error (*SE*) between the two sets of data:

$$AE = \sum_{b=1}^{N_b} \mid P(m_b \le W_{i,ac} \le M_b) - P(m_b \le W_{i,sim} \le M_b) \mid \tag{24}$$

$$SE = \sum_{b=1}^{N_b} [P(m_b \le W_{i,ac} \le M_b) - P(m_b \le W_{i,sim} \le M_b)]^2 \tag{25}$$

$$P(m_b \le W_{i,ac} \le M_b) = F_{W_{i,ac}}(M_b) - F_{W_{i,ac}}(m_b) \tag{26}$$

$$P(m_b \le W_{i,sim} \le M_b) = F_{W_{i,sim}}(M_b) - F_{W_{i,sim}}(m_b) \tag{27}$$

$$m_b = \frac{1}{N_b}(b-1) \tag{28}$$

$$M_b = \frac{b}{N_b} \tag{29}$$

where

$P(m_b \le W_{i,ac} \le M_b)$, The probability that the actual wind power falls between m_b and M_b

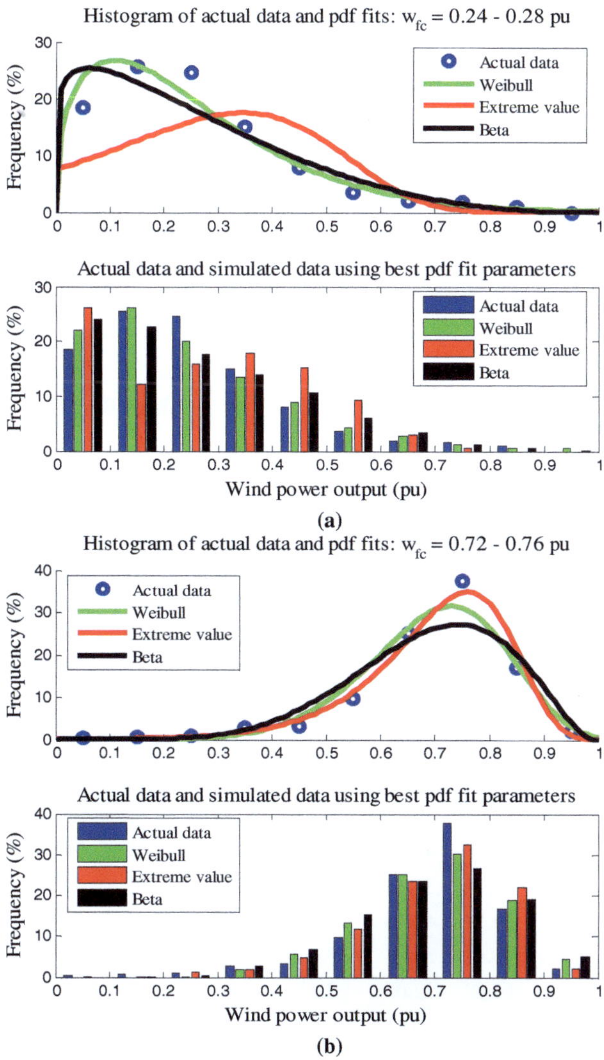

Fig. 4. The three pdf fits for the data corresponding to Stateline's: (a) Bin 7 (b) Bin 19

$P(m_b \le W_{i,sim} \le M_b)$, The probability that the simulated wind power falls between m_b and M_b

$F_{W_{i,ac}}(.)$, The cumulative distribution function of the actual wind power

$F_{W_{i,sim}}(.)$, The cumulative distribution function of the simulated wind power

m_b and M_b are lower and upper bounds, respectively, of the interval b

N_b The total number of intervals. In this study, the total number of intervals is taken as 10.

Tables 1 and 2 show the comparison results using *AE* and *SE* of the three pdf fits for all data bins of Condon and Stateline wind plants. The bold numbers of each row gives the least *AE* (or *SE*). The results show that, in almost all the cases, using either *AE* or *SE* leads to the same conclusions. In addition, these tables show that Weibull distribution usually works best at data bins corresponding to low to moderate wind forecast levels whereas Extreme Value distribution is superior at bins corresponding to high wind forecast levels. Beta distribution, however, leads to minimum *AE* or *SE* only occasionally. These conclusions are in line with those obtained from Fig. 4.

Table 1. The Absolute Errors and Squared Errors of the Three Best Distributions for the Condon Wind Plant Data

Bin No.	Absolute Error (AE)			Squared Error (SE)		
	WBL	EV	BETA	WBL	EV	BETA
1	2.17	42.91	4.40	0.93	26.66	2.61
2	10.25	42.67	8.36	4.66	24.92	4.56
3	10.78	38.59	12.92	5.59	18.96	6.56
4	16.27	46.61	23.75	8.96	25.75	12.80
5	13.57	40.93	17.85	7.52	22.77	8.85
6	19.32	59.08	18.86	10.06	27.70	9.28
7	6.26	45.65	12.69	2.45	19.84	5.71
8	13.63	50.50	24.30	6.55	21.67	11.15
9	9.29	43.00	12.88	4.30	16.67	5.80
10	12.04	37.88	11.17	5.12	14.91	4.16
11	10.72	43.76	15.55	4.73	16.86	6.18
12	15.12	35.08	19.49	6.44	13.17	8.52
13	14.68	32.65	22.33	5.93	13.15	8.21
14	15.56	30.18	20.18	5.76	12.44	8.42
15	18.68	34.37	22.34	7.26	14.60	10.42
16	12.18	14.44	7.52	4.48	5.86	2.71
17	11.52	26.42	24.76	4.18	10.32	9.30
18	27.78	10.43	24.29	10.69	4.13	9.00
19	22.53	16.98	20.21	10.81	6.82	8.83
20	21.37	17.22	12.61	8.61	7.46	5.02
21	21.71	15.40	16.77	8.81	6.37	6.19
22	26.79	17.45	19.22	12.82	9.08	9.17
23	50.83	32.23	55.59	26.20	15.92	29.37
24	29.67	25.66	17.81	15.15	12.01	7.07

Table 2. The Absolute Errors and Squared Errors of the Three Best Distributions for the Stateline Wind Plant Data

Bin No.	Absolute Error (AE)			Squared Error (SE)		
	WBL	EV	BETA	WBL	EV	BETA
1	4.74	46.82	2.48	2.63	28.98	1.28
2	8.28	45.26	11.20	3.80	24.82	4.60
3	23.31	40.54	19.21	12.29	17.29	9.46
4	21.25	42.94	21.63	10.47	18.24	10.86
5	31.48	33.57	32.58	15.84	18.26	17.94
6	16.28	50.41	28.63	8.87	22.27	13.59
7	11.43	48.82	19.00	5.78	20.22	8.31
8	20.85	43.41	25.34	10.14	18.28	13.00
9	17.18	51.15	30.46	8.29	19.72	12.77
10	17.68	40.00	19.38	7.26	16.32	8.88
11	20.34	36.73	23.42	8.91	13.97	9.88
12	12.57	41.11	17.30	5.54	16.09	7.80
13	26.82	27.77	31.71	10.12	11.14	13.77
14	11.79	22.51	13.50	4.24	9.96	5.55
15	17.70	11.75	19.87	6.59	4.54	7.45
16	19.68	12.44	15.40	6.75	5.22	6.86
17	16.51	12.33	15.11	5.84	5.63	6.48
18	28.34	12.38	38.07	11.69	5.62	15.66
19	23.64	15.37	30.32	10.79	6.71	14.14
20	20.81	9.52	24.62	8.68	3.93	10.83
21	30.58	20.58	30.48	14.81	10.52	15.56
22	27.48	19.51	38.80	13.94	9.28	21.01
23	8.04	13.18	10.49	4.09	7.31	4.94
24	34.61	30.93	36.53	21.19	18.55	23.89

4 Multi-objective Optimization

The concept of Pareto optimality is often used in trade-off analysis, where there exist multiple conflicting objectives. In this case, there is no single solution that simultaneously optimizes all objectives. Rather, there are a number of solutions that are not dominated by any other candidate solution. In Pareto optimization, a solution is said to be dominant when no objective can be further improved without degrading any of the other objectives. These non-dominated solutions form the Pareto-optimal set of solutions, or the Pareto front.

Fig. 5 illustrates the concept of Pareto optimality through an example. The goal is to minimize the two objective functions, f_1 and f_2, simultaneously. The figure shows the evaluation of several candidate solutions of the multi-objective

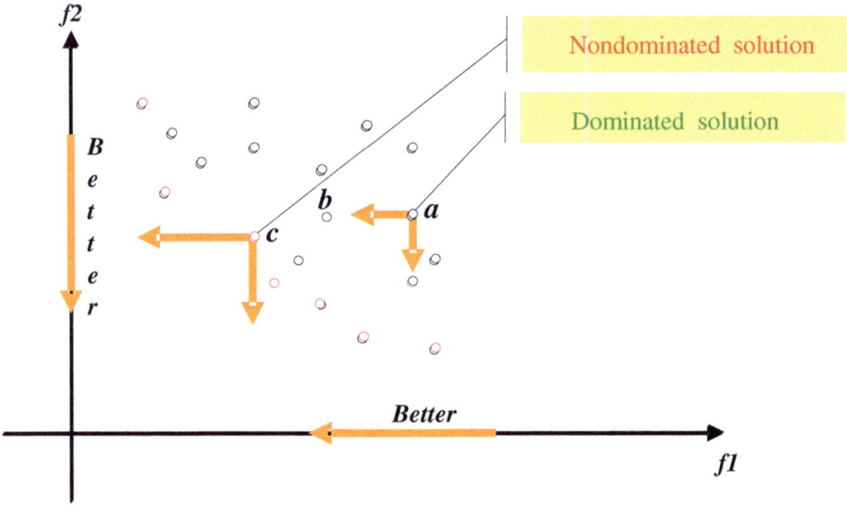

Fig. 5. Dominated and non-dominated solutions

optimization problem. Moving from *a* to *b*, *f1* can be further improved without degrading *f2*. In this case, *b* dominates *a*, and *a* is eliminated. Also, point *c* dominates *b*, thus *b* is eliminated. By continuing this process of selection and elimination, we can reach a solution that cannot be dominated. This is known as a Pareto solution. By generating enough Pareto solutions, we can identify the surface of possible optimal solutions. This surface is known as the Pareto front.

Usually, the Pareto front cannot be identified analytically [22]. Since the process of identifying the Pareto front is computationally intensive, parallel and computationally-efficient search techniques, such as the Particle Swarm Optimization (PSO), should be used.

5 Multi-Objective Particle Swarm Optimization (MO-PSO)

The identification of a Pareto front in multi-objective optimization requires extensive search of a sparse and multidimensional space. Single point search cannot be used in this environment, parallel and fast search techniques such as the PSO or evolutionary algorithms should be used.

The PSO technique is a parallel search technique that enjoys a good balance between global and local searches. It is fast, robust, and easy to code. It conducts parallel searches using a group of particles. Each particle in the group is a potential solution to the problem. PSO consists of, at each step, changing the velocity of each particle toward its personal best position as well as the group's best position. [23].

The general algorithm of a PSO is to randomly initialize all the particles with solutions in the space and then assign to each of them a direction to search and a velocity per iteration. After each iteration, all solutions are evaluated and the local and global best solutions are saved. The particles' velocities are then updated. The three components of a particle's velocity vector are the inertial component, the component pushing the particle towards its own local best, and the component pushing the particle to the swarm's global best. As the search progresses, the particles are pushed increasingly towards the global best until they converge on the solution [23].

The PSO is modified to fit into the multi-objective optimization, which is known as MO-PSO [24]. In this work, a modified MO-PSO is implemented as follows:

1. Start with n initial random positions (solutions) and n initial random velocities.
2. Evaluate the two objective functions given by (12) and (19). For each particle, if the new position dominates the old position, assign the new position as the local best solution.
3. Stack the old and new local best solutions in one vector. Test the union of the two groups for Pareto-optimality. The non-dominated solutions of the union are points on the Pareto front.
4. If the number of points on the Pareto front exceeds a certain limit, cluster neighborhood points.
5. For each particle, assign one of the Pareto-optimal solutions as a global best solution. In [24], it was suggested that the closest Pareto-optimal position to a given particle is used as its global best position. Instead of that, it was shown in [9] that choosing the global best position at random from the set of Pareto-optimal positions results in a more enhanced performance. In this paper, the modified MO-PSO presented in [9] is implemented.
6. Update the new velocities and positions.
7. Go to Step 2 unless the maximum number of iterations is reached.

6 Example

Consider a system that consists of two thermal units and two wind plants. Assume the following:

- The fuel cost coefficients of the two thermal units are $a_1 = 0.01$, $b_1 = 1.25$, $c_1 = 100$, $a_2 = 0.0125$, $b_2 = 1$, and $c_2 = 125$.
- The minimum and maximum power outputs of thermal unit 1 are $P_{g1}^{\min} = 20\,MW, P_{g1}^{\max} = 280\,MW$ and those of thermal unit 2 are $P_{g2}^{\min} = 0\,MW, P_{g2}^{\max} = 280\,MW$.
- The operating cost coefficients of the two wind plants are $d_1 = 1.0$ and $d_2 = 1.1$.
- The penalty and reserve factors are $k_{pi} = 2$ and $k_{ri} = 8$.
- The emission coefficients are $\alpha_1 = 4.091\times10^{-2}$, $\beta_1 = -5.554\times10^{-4}$, $\gamma_1 = 6.490\times 10^{-6}$, $\zeta_1 = 2\times10^{-6}$, $\lambda_1 = 2.857\times10^{-2}$, $\alpha_2 = 2.543\times10^{-2}$, $\beta_2 = -6.047\times10^{-4}$, $\gamma_2 = 5.638\times10^{-6}$, $\zeta_2 = 5\times10^{-6}$, and $\lambda_2 = 3.333\times10^{-2}$.
- The actual wind power outputs and the load profile for the system in the previous year are known.
- The wind power forecast is available 30 minutes before the start of each hour.

Published utility data such as the one on the Northwest Power Pool website for wind plants can be used [19]. For this example, 11-month data for Condon and Stateline wind plants are used. The various economic dispatch methods discussed above can now be evaluated and compared as given below.

6.1 Single Objective Optimization

The objective function to be minimized is the system operating cost. The process is carried out as follows:

1. A deterministic dispatch (DD) is implemented. At each dispatch period, the scheduled wind power of each wind plant is assumed to be equal to the plant wind forecast level, i.e. $w_i = w_{fci}$. In addition, the optimal thermal unit schedules P_{g1} and P_{g2} are calculated using (1)-(4).
2. A stochastic dispatch is carried out using (8)-(11).
3. For comparison purposes, the case of perfect scheduling (PS) is also performed. As explained earlier, PS implies no mismatch between wind schedules and actual wind power outputs.
4. The actual operating costs (AOC) of the three dispatches are compared.

The results of 11 months are shown in Table 3. These results show that SD leads to average savings of 2.13% over DD. The average savings PS achieves over DD is 9.85%. Obviously, PS provides more savings but it is not achievable in reality since perfect forecasting is unattainable.

Fig. 6 shows the load, wind forecasts, and wind schedules using SD for the month of March. It is shown that SD wind schedules are usually lower than those

Table 3. The Actual Operating Costs and Savings of SD and PS as Compared to DD

Month	DD ($/MW)	SD ($/MW)	SD Savings (%)	PS ($/MW)	PS Savings (%)
Jan	594.20	591.54	0.45	558.15	6.07
Feb	543.86	534.70	1.68	501.83	7.73
Mar	485.16	472.89	2.53	423.35	12.74
Apr	502.57	489.68	2.56	451.45	10.17
May	488.59	477.41	2.29	436.43	10.68
Jun	497.76	484.95	2.57	445.58	10.48
Jul	536.14	524.38	2.19	485.76	9.40
Aug	503.61	489.55	2.79	444.22	11.79
Sep	N/A	N/A	N/A	N/A	N/A
Oct	464.09	451.62	2.69	410.17	11.62
Nov	456.29	449.33	1.53	420.43	7.86
Dec	520.59	510.91	1.86	471.85	9.36
Average	**506.55**	**496.4**	**2.13**	**457.54**	**9.85**

of DD partially because the penalties associated with over-generation are less than those for under-generation.

The mean absolute error between the schedules and the actual power output of the DD and SD cases are shown in Table 4. It is interesting to observe that SD schedules result in higher mean absolute error than DD schedules, yet SD is cheaper, on average, than DD. This is because the objective is to minimize the operating cost regardless of the dispatch errors.

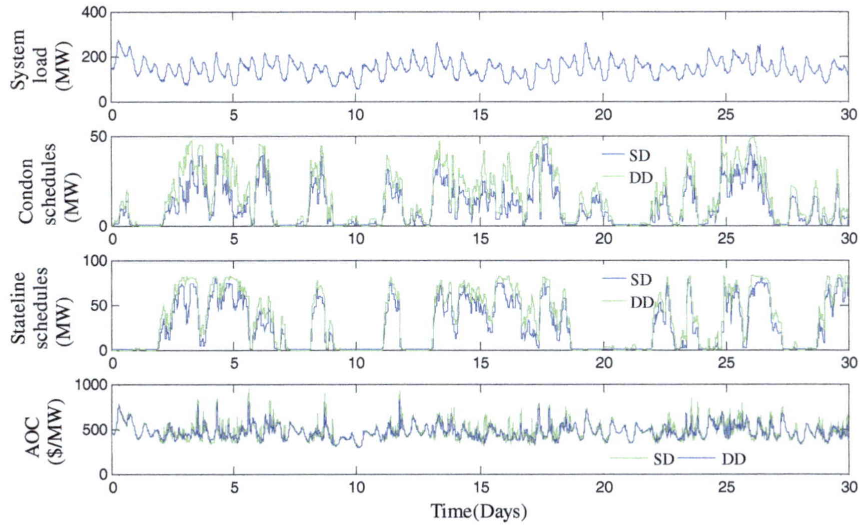

Fig. 6. Test Results for the month of March

Table 4. Mean Absolute Error between Scheduled and Actual Wind power Outputs

Wind Plant	Mean Absolute Error (%)	
	DD	SD
Condon	10.49	13.68
Stateline	7.99	11.06

6.2 Multi-objectives Optimization

The two objective functions to be minimized are the system operating cost and the emissions of the thermal units. MO-PSO is used to identify the Pareto-optimal front and the associated scheduled power outputs of the thermal and wind units. Normally, 100 particles are sufficient, and the maximum number of iterations can be limited to 1000. The particles' maximum and minimum values are the maximum and minimum power outputs of the four generating units. The velocities are limited to prevent the particles from searching out of bound. To ensure that the equality constraint given in (18) is met, a penalty term proportional to the mismatch between generation and load is added to the operating cost of (12), and another penalty term proportional to the same mismatch is added to the total emissions of (19).

Before proceeding to the results, it is worth discussing how the emissions change as a function of the outputs of the thermal units. Fig. 7 shows the emissions versus the power outputs of the two thermal units. It is apparent in the boxed region that the emission is at minimum at a certain power output, $P_{gi,opt}$. The emissions are higher at any output below or above $P_{gi,opt}$. It can be seen that $P_{g1,opt}$ = 43.0 MW and $P_{g2,opt}$ = 53.5 MW in our example.

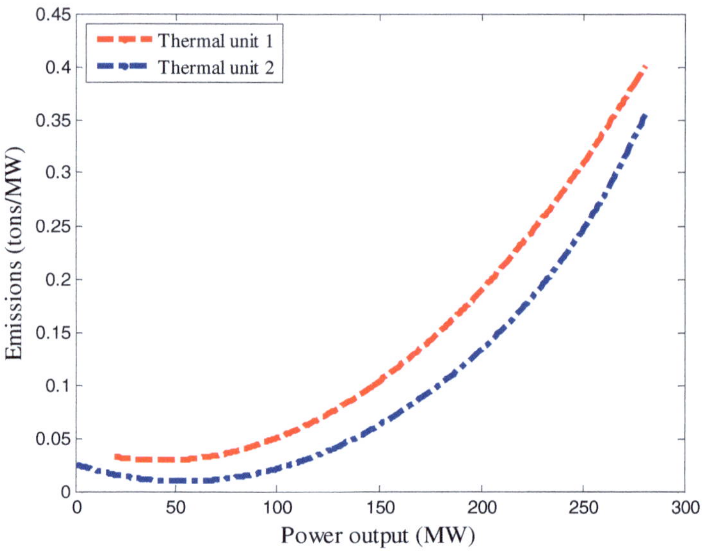

Fig. 7. Emission curves of thermal units

6.2.1 Comparison of Multi-objective SD with DD

Figs. 8 and 9 show the results for multi-objective DD and SD for one dispatch period during the month of March. Condon's and Stateline's forecasts are 36.3 and 39.2 MW, respectively. The total load is 215.3 MW. As shown in Fig. 8, the wind plant schedules are identical to their associated forecasts. Therefore, the only variables to be determined by MO-PSO are the power outputs of the two thermal units (P_{g1} and P_{g2}). Since the generation-load balance has to be maintained, as P_{g1} increases, P_{g2} has to decrease. This explains the trend of P_{g1} and P_{g2} in Fig. 8(b).

For SD, wind schedules are allowed to deviate from the forecasts. Hence, the number of variables increases to four. Fig. 9 shows the Pareto front and the four generator outputs for this case. It is shown that as the scheduled wind power increases, the combined operating cost increases and the emissions decrease.

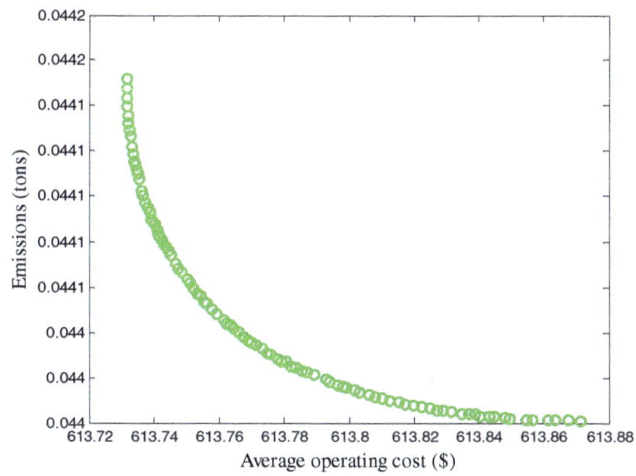

(a)

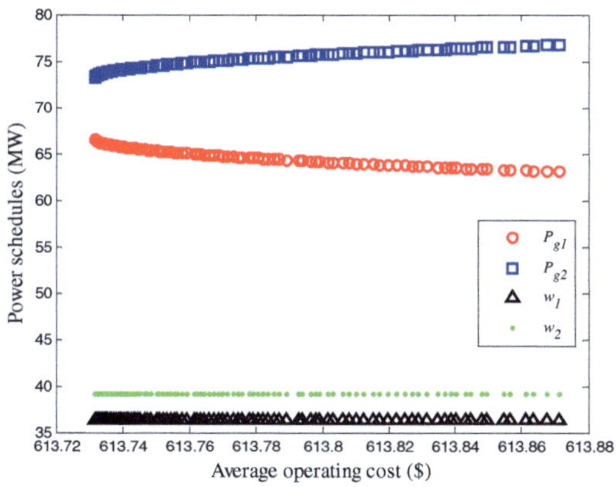

(b)

Fig. 8. Multi-objective DD simulation results: (a) Pareto front (b) Pareto-optimal power schedules

Increasing the scheduled wind power causes the imbalance charges to increase at a higher rate than the decreasing rate of the fuel cost of the thermal units. Hence, the combined operating cost increases. Furthermore, because the outputs of the thermal units slide between $P_{gi,opt}$ and $P_{gi,max}$, the emissions are directly proportional to the thermal power outputs. Therefore, increasing scheduled wind power outputs results in a decrease in thermal power outputs, hence less emissions.

6.2.2 Effect of Load Level

In order to study the effect of load level on the Pareto-optimal solutions, the load is reduced to one half of that for the base case. Fig. 10 shows the Pareto front and the four generator outputs for this case. As expected, the combined operating cost decreases with the decrease in load. The thermal power outputs are now between

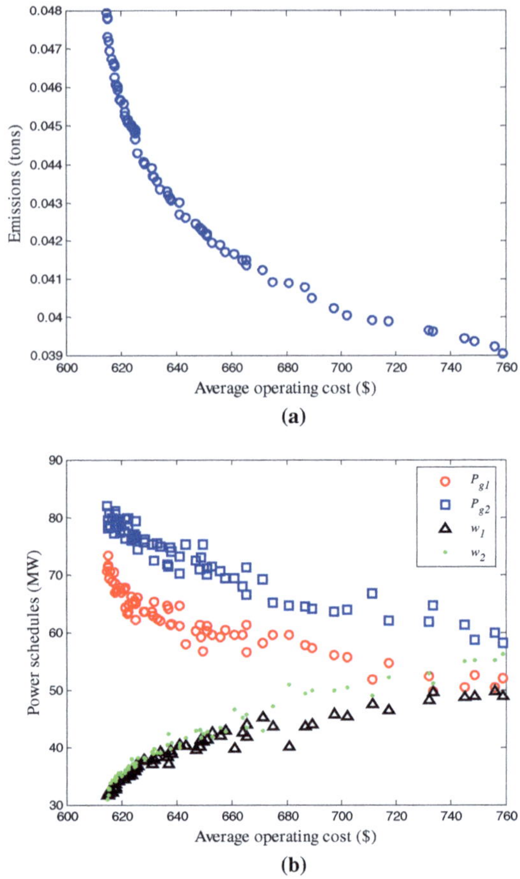

Fig. 9. Multi-objective SD: Base case (a) Pareto front (b) Pareto-optimal power schedules

$P_{gi,min}$ and $P_{gi,opt}$. Therefore, the trends of the thermal units have changed. As shown in Fig. 7, at this range of output power, the emissions are inversely proportional to the power outputs. Hence, the increase in thermal power outputs gives rise to fewer emissions. However, at low scheduled wind power outputs, fuel costs are the major components of the combined operating cost. This is exactly the opposite trend to that of the base case.

6.2.3 Effect of Reserve Cost Coefficient $\mathbf{k_{ri}}$

In order to study the effect of the reserve cost coefficient k_{ri} on the Pareto-optimal solutions, k_{ri} is decreased from 8 to 4. Fig. 11 shows the Pareto front and the four generator outputs for this case. For comparison purposes, the Pareto front of the base case is also shown in Fig. 11. As can be observed, decreasing k_{ri} decreases the combined operating cost due to the decrease in reserve costs. In addition, because the

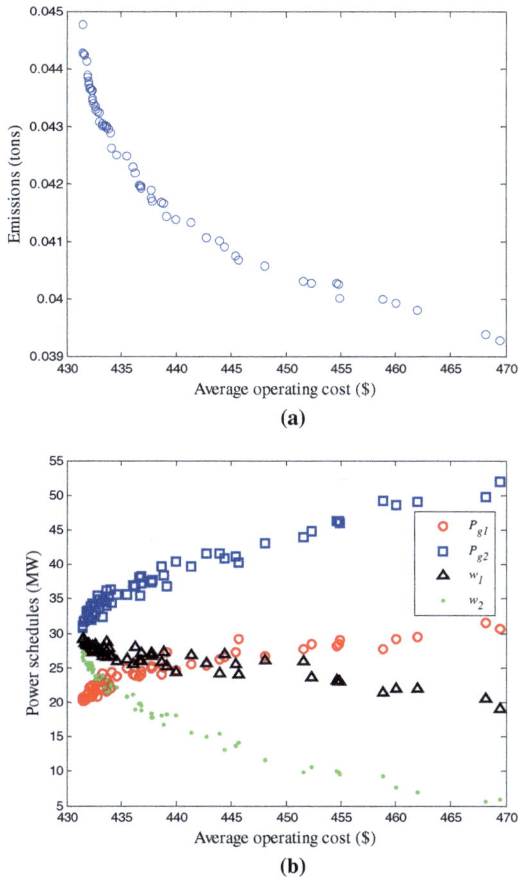

Fig. 10. Multi-objective SD: Effect of load level (a) Pareto front (b) Pareto-optimal power schedules

decrease in the reserve coefficients motivates wind power generators to increase their scheduled outputs due to the lower under-generation costs, the outputs of thermal units must decrease.

6.2.4 Effect of Penalty Cost Coefficient $\mathbf{k_{pi}}$

In order to study the effect of the penalty cost coefficient k_{pi} on the Pareto-optimal solutions, k_{pi} is decreased from 2 to 1. Fig. 12 shows the Pareto front and the four generator outputs for this case. It can be noticed that decreasing k_{pi} decreases the total operating cost due to the decrease in penalty costs. In addition, because decreasing the penalty coefficients motivates wind power generators to decrease their scheduled outputs due to the lower over-generation costs, the outputs of thermal units must increase, as indicated by comparing the generator power outputs of Figs. 9 and 12.

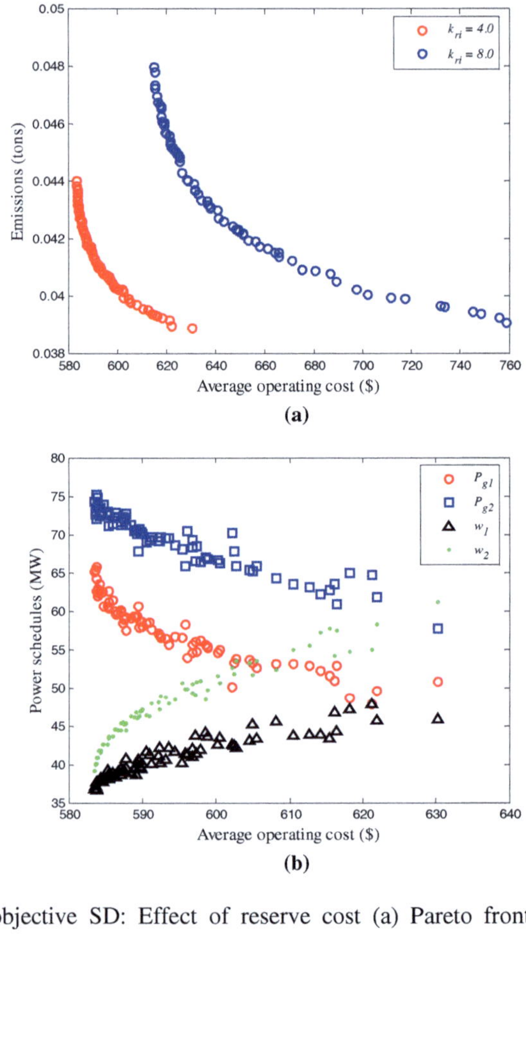

(a)

(b)

Fig. 11. Multi-objective SD: Effect of reserve cost (a) Pareto front (b) Pareto-optimal power schedules

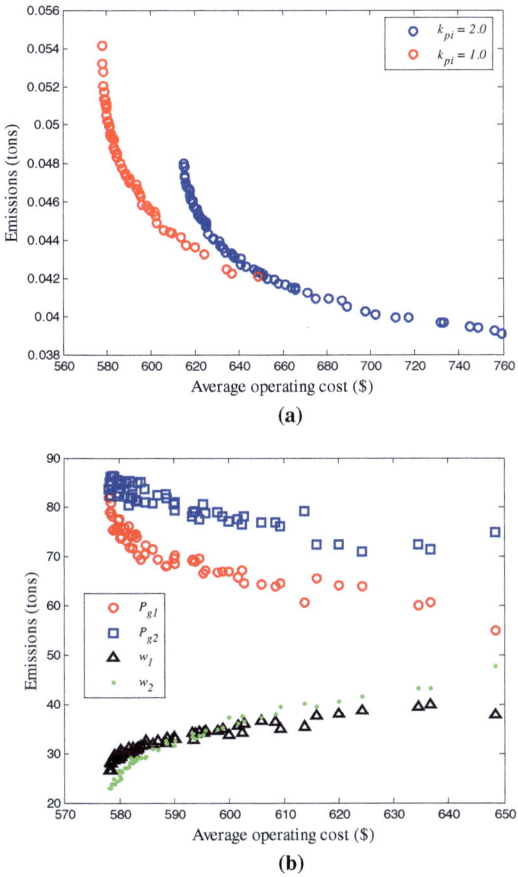

Fig. 12. Multi-objective SD: Effect of penalty cost (a) Pareto front (b) Pareto-optimal power schedules

7 Conclusions

In stochastic dispatch, the objective is to minimize the expected value of the system operating cost. The uncertainty associated with the wind power forecast is taken into account as imbalance cost that is added to the operating cost. The imbalance cost reflects the cost of securing and employing balancing reserves to maintain the system generation-load balance. The following conclusions are drawn from the example given above:

- The stochastic dispatch is, on average, cheaper than deterministic dispatch.
- The stochastic dispatch usually results in lower wind schedules than the forecast. This is partially because the imbalance cost due to under-generation is higher than that due to over-generation.

- Although SD is cheaper than DD, the average absolute deviations of the stochastic wind schedules from the actual power outputs can be higher than those of the deterministic wind schedules. The reason is that SD tends to be more conservative in scheduling wind to avoid high under-generation charges.
- Multi-objective optimization of operating cost and emissions gives an insight of how the two objectives vary with respect to the changes in schedules.

References

1. European SmartGrid Technology Platform: Vision and strategy for Europe's electricity networks of the future. EU Commission, Directorate-General for Research, Information and Communication Unit. European Commission Website (2006), `http://ec.europa.eu/research/energy/pdf/smartgrids_en.pdf` (accessted January 28, 2010)
2. U.S. Federal Energy Regulatory Commission (FERC): Smart grid policy. Docket No. PL09-4-000 (2009) FERC Website, `http://www.ferc.gov/whats-new/comm-meet/2009/071609/E-3.pdf` (accessed January 28, 2010)
3. North American Electric Reliability Corporation (NERC): Accommodating high levels of variable generation (2009) NERC Website, `http://www.nerc.com/files/IVGTF_Report_041609.pdf` (accessed January 28, 2010)
4. Departmet, U.S.: of Energy (DOE): 20% wind energy by 2030: increasing wind energy's contribution to U.S. electricity supply. National Renewable Energy Laboratory. NREL/TP-500-41869, DOE/GO-102008-2567 (2008) US DOE Website, `http://www1.eere.energy.gov/windandhydro/pdfs/41869.pdf` (accessed January 28, 2010)
5. Bathurst, G.N., Weatherill, J., Strbac, G.: Trading wind generation in short term energy markets. IEEE Trans. Power Syst. 17(3), 782–789 (2002)
6. Galloway, S., Bell, G., Burt, G., McDonald, J., Siewierski, T.: Managing the risk of trading wind energy in a competitive market. In: IEE Proc. Gener., Transm., and Distrib., vol. 153(1), pp. 106–114 (2006)
7. Pinson, P., Chevallier, C., Karniotakis, G.N.: Trading wind generation from short term probabilistic forecasts of wind power. IEEE Trans. Power Syst. 22(3), 1148–1156 (2007)
8. Hetzer, J., Yu, D.C., Bhattarai, K.: An Economic Dispatch Model Incorporating Wind Power. IEEE Trans. Energy Convers. 23(2), 603–611 (2008)
9. Al-Awami, A.T., Sortomme, E.V., El-Sharkawi, M.A.: Optimizing economic/environmental dispatch with wind and thermal units. IEEE/PES Gen. Meet. (2009)
10. El-Hawary, M.E., Mbamalu, G.A.N.: A comparison of probabilistic perturbation and deterministic based optimal power flow solutions. IEEE Trans. Power Syst. 6(3), 1099–1105 (1991)
11. Gorenstin, B.G., Campodonico, N.M., Costa, J.P., Pereira, M.V.F.: Stochastic optimization of a hydrothermal system including network constraints. IEEE Trans. Power Syst. 7(2), 791–797 (1992)

12. Contaxis, G., Vlachos, A.: Optimal power flow considering operation of wind parks and pump storage hydro units under large scale integration of renewable energy sources. In: Proc. 2000 IEEE PES Winter Meet, pp. 1745–1750 (2000)
13. Bludszuweit, H., Dominguez-Navarro, J.A., Llombart, A.: Statistical analysis of wind power forecast error. IEEE Trans. Power Syst. 23(3), 983–991 (2008)
14. Bofinger, S., Luig, A., Beyer, H.G.: Qualification of wind power forecasts. In: Proc. 2002 Global Wind Power Conf. GWPC 2002 (2002)
15. King, R.T.F., Rughooputh, H.C.S., Deb, K.: Stochastic evolutionary multiobjective environmental/economic dispatch. In: IEEE Congr. Evol. Comput., pp. 946–953 (2006)
16. Bonneville Power Administration: Wind Integration Team Work Plan (2009) BPA Website, `http://www.bpa.gov/corporate/WindPower/docs/WIT_Work_Plan_-_June_16.pdf` (accessed October 09, 2009)
17. Bonneville Power Administration: Open Access Transmission Tariff (2008) BPA Website, `http://www.transmission.bpa.gov/business/ts_tariff/documents/BPA_OATT_Filed_10_03_2008.pdf` (accessed August 21, 2009)
18. Yokoyama, R., Bae, S.H., Morita, T., Sasaki, H.: Multiobjective generation dispatch based on probability security criteria. IEEE Trans. Power Syst. 3, 317–324 (1988)
19. Wind Integration Study Group: Revised 2004 10 minute data (2006) Northwest Power Pool Website, `http://www.nwpp.org/ntac/publications.html?CommitteeID=29` (accessed March 10, 2009)
20. Sideratos, G., Hatziargyriou, N.D.: An Advanced Statistical Method for Wind Power Forecasting. IEEE Trans. Power Syst. 22(1), 258–265 (2007)
21. Damousis, I.G., Alexiadis, M.C., Theocharis, J.B., Dokopoulos, P.S.: A fuzzy model for wind speed prediction and power generation in wind parks using spatial correlation. IEEE Trans. Energy Convers. 19(2), 352–361 (2004)
22. Ngatchou, P., Zarei, A., El-Sharkawi, M.A.: Pareto Multi Objective Optimization. In: Proc. 13th Int. Conf. Intell. Syst. Appl. Power Syst., pp. 84–91 (2005)
23. Kennedy, J., Eberhart, R.: Particle swarm optimization. In: Proc. IEEE Int. Conf. Neural Netw., pp. 1942–1948 (1995)
24. Abido, M.A.: Multiobjective particle swarm for Environmental/Economic dispatch problem. In: Proc. Int. Power Engineering Conf., pp. 1385–1390 (2007)

Wind Turbine Diagnostics Based on Power Curve Using Particle Swarm Optimization

Lisa Ann Osadciw, Yanjun Yan, Xiang Ye, Glen Benson, and Eric White

Abstract. In wind energy industry, *power curve*, the plot of the generated power versus the ambient wind speed, is an important indicator of the performance and health of wind turbines. The nominal power curves differ by manufacturers and types. The actual power curve will deviate from the nominal one because of the turbulence in the incoming wind, turbine health, etc. Power curve is widely used for visual inspection and performance evaluation, but there is no et a quantified approach to use it for diagnostic purpose. We propose an inverse transformation based change detector, called *Inverse Diagnostic Curve Detector (IDCD)*, to track the variation of power curve over time for diagnostics. IDCD is adaptable to different wind turbine types. We use two example wind turbine types to illustrate the adaptation procedure. We select the Gaussian CDF (cumulative density function) in the inverse data transformation method for its fitting accuracy and one-to-one mapping property in its inversion. The dynamic fitting is optimized by particle swarm optimization (PSO) algorithm. IDCD simplifies abnormality detection with a scaler decision threshold. Some failures are predictable such as some major component failure, which causes degradation; other failures are not predictable from turbine information alone such as lightning strike, which happens suddenly and quickly. Early detection of either degradation or sudden faults is beneficial. After a deviation pattern is discovered by comparing it with historical data, the pattern can be used for prognostics to help predict the remaining useful life of a turbine and create an optimal schedule for maintenance and repair tasks.

Lisa Ann Osadciw · Yanjun Yan · Xiang Ye
Syracuse University, Department of Electrical Engineering and Computer Science, Syracuse, NY 13244, USA
e-mail: {laosadci, yayan, xye}@syr.edu

Glen Benson · Eric White
AWS Truewind, LLC, 463 New Karner Road, Albany, NY 12205, USA
e-mail: {GBenson, EWhite}@awstruewind.com

L. Wang et al. (Eds): *Wind Power Systems*, Green Energy and Technology, pp. 151–165.
springerlink.com

1 Introduction

In wind energy industry, the turbine power curve (a plot of generated power versus ambient wind speed) is an important indicator of wind turbine performance [1]. A turbine manufacturer usually provides a nominal power curve as a reference. The actual power curve will vary from this nominal curve for a variety of reasons - some inherent to the incoming wind and its characteristics such as turbulence, some due to the way the turbine actually responds to the observed wind, but some may also be caused by multiple system faults - sensor and control faults, turbine or generator faults, structural issues in the turbine main-body or generator faults, besides interferences resulting in additional air turbulence in the wind source [2]. As there is large inherent variation in the system performance expected under even normal operation, finding a means to understand and differentiate normal from abnormal behavior becomes critical in analyzing the data and focusing troubleshooting and repair efforts on fault conditions. Wind energy experts have acknowledged that power curve analysis should be site specific and some influencing factors are studied [3]. This chapter proposes an automatic and adaptive approach to analyze different turbines, and this approach integrates the multiple factors together. The power curve is often used to provide a prediction on the power generation [1] or serve as a visual tool to check the approximate performance, but there is no systematic approach to use power curve for turbine health diagnostics yet. This chapter provides a quantitative method to point to potential causes for the deviations from the nominal power curve.

On a power curve, the data points with zero wind speed but non-zero power typically indicate that the anemometer is faulty as it should provide a non-zero wind speed measurement that is clearly producing power. The data points with zero power but large wind speeds indicate a suboptimal turbine that did not work. The measured wind speed difference between turbine pairs is a good test for diagnostics [4] and for wake analysis [5]. Other than the outliers, the majority of the data points are roughly aligned around the nominal power curve. The data points on the left hand side of the nominal power curve are regarded as over-performance possibly resulting from an offset that occurred during system calibration, a faulty or degrading anemometer, or a slight difference between the system used to determine the nominal curve and the one currently being tested. The data points on the right hand side of the nominal power curve are regarded as under-performance, and there could be many reasons for it. The checking of data locations relative to the nominal power curve is not straightforward, because the nominal power curve is provided with a discrete set of wind speed, yet the actual wind speed is a real number within the full range of feasible wind speeds. A fitting of the nominal power curve is needed for such a comparison. In this paper, we propose an inverse function based approach, called *Inverse Diagnostic Curve Detector (IDCD)*, to transform the data into a linearized domain for accurate and simplified detection of the status variation. The traditional deviation method requires multiple fitting at each decision boundary, yet our IDCD requires only one fitting.

The remaining of this chapter is organized as follows. Section 2 presents the evolution of the one-segment IDCD from a two-parameter version to a four-parameter version for the first turbine type. Section 3 discusses parameters optimization using a particle swarm optimization algorithm designed for nominal power curve fitting. Section 4 extends the one-segment IDCD to two-segment IDCD to address the diagnostic problem for a second wind turbine type. Section 5 reports the numerical results on the parameters. Section 6 elaborates the power curve test procedure based on IDCD, with illustrations using real data examples in section 7. Section 8 concludes this chapter while hinting at some future problems to solve.

2 Evolution of Gaussian CDF Based IDCD for the First Turbine Type

In our earlier experimentations of state definitions [6], we compared the traditional way of decision boundary definition with three fitting functions in IDCD.

1. In the traditional way of setting up decision boundary definitions for multiple states, each boundary needs to be fitted once, which requires multiple fittings. The fitting procedure is the same, and yet it is repeated multiple times consuming calculation resources. Furthermore, once the definitions of states are changed (for instance, a 20% range is regarded too wide and hence a 15% range is used), the fitting process needs to be re-run again. What's worth mentioning is that depending on the distribution of the data, the decision boundaries should be customized accordingly. A Kaiser window is used in our modeling to capture the "big belly" pattern of the power curve data in the linearized domain [6], as shown in Figure 1. As the Kaiser parameter increases, the Kaiser window's "big belly" shape, outlined in the red line, becomes more accentuated and more tightly fits the blue data points at the beginning and ending of the plots. Although one can note the shape of the decision boundary is natural in transition, there is an over-inclusion region in the upper kink of the power curve, and the separation of states is not even, as shown in Figure 2. These limitations in the traditional decision boundary definition method motivated us to define the states in the linearized domain, and hence IDCD was proposed.
2. The first function used in IDCD is the widely accepted polynomial fitting function [6]. The polynomial function is accurate in fitting using high orders. However, when it is inverted for our state definitions, there are multiple roots. Selecting the correct root is tedious and inaccurate. The spurious roots may also lead to false decisions on the state definitions. As illustrated in Figure 3, there are too many false under-performance decisions.
3. The second function used in IDCD is the Sigmoid function [6], motivated by the seemingly rotational-symmetry of the power curve.
4. The third function used in IDCD is the Gaussian CDF (cumulative distribution function) [6]. The Gaussian CDF is similar to the Sigmoid function in nature, but it achieves better accuracy than the Sigmoid function, especially with less deviation in the upper kink region of the power curve, as shown in Figure 4. Therefore, we selected the Gaussian CDF in IDCD. The states are partitioned evenly in the power curve domain as shown in Figure 5.

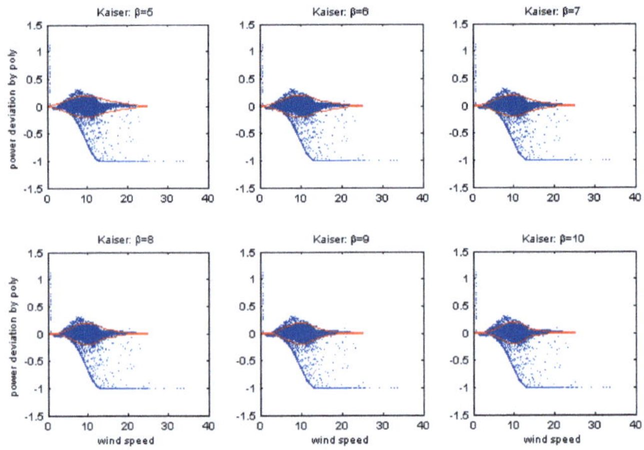

Fig. 1. Kaiser Window Modeling the Real Data Deviation

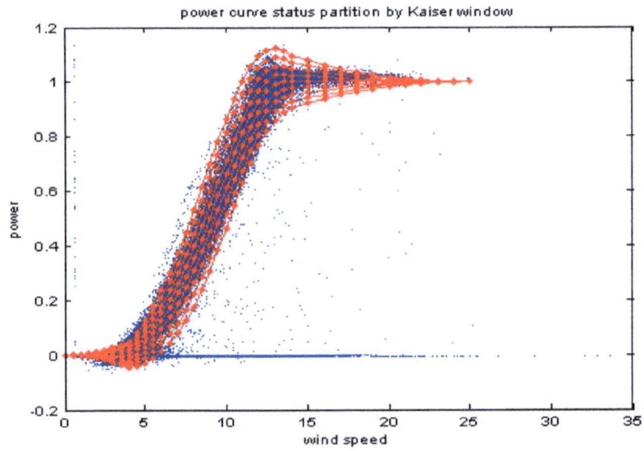

Fig. 2. A Kaiser Window Modeling for Partitioning of the states

The power curve is first normalized with unit-maximum-power. In the refinement of the Gaussian CDF based IDCD, two versions are developed.

2.1 First Version of Gaussian CDF Based IDCD

The first version of Gaussian CDF based IDCD is defined as

$$\hat{p}(w|c,a) = \int_{-\infty}^{w} \frac{1}{\sqrt{2\pi}a} e^{-\frac{(x-c)^2}{2a^2}} dx, \tag{1}$$

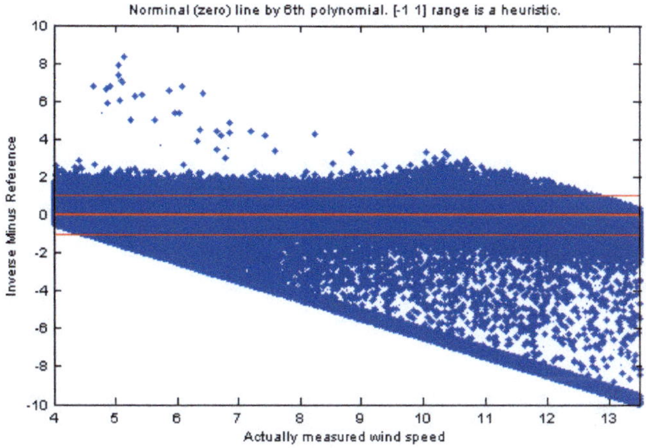

Fig. 3. Linearized Power Curve by the Polynomial Fitting. Note: Spurious roots cause too many false under-performance decisions.

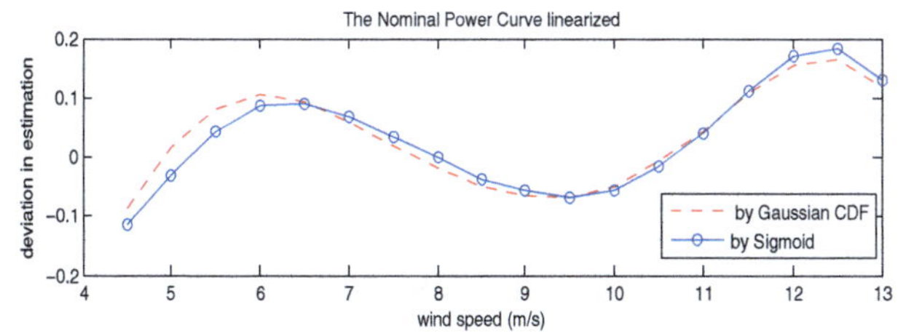

Fig. 4. Distortion of Gaussian CDF function and Sigmoid function

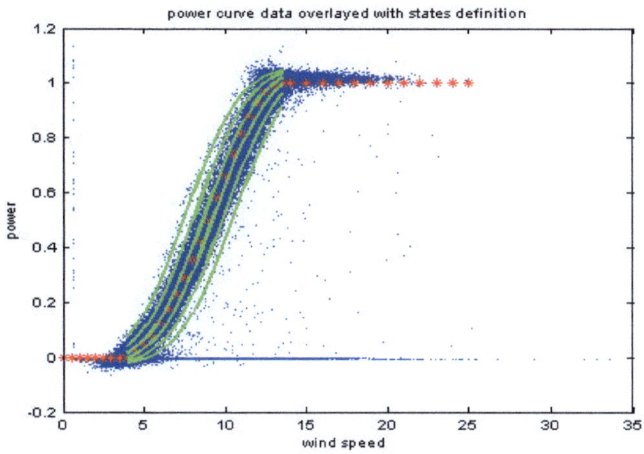

Fig. 5. Power Curve Data Overlaid with States Defined by Gaussian CDF based IDCD

where w is the wind speed, p is the generated power with $\hat{p}$ as the estimated power at that specific w, c is the mean of Gaussian CDF, and a is its standard deviation. Gaussian CDF is calculated from the error function of the normal distribution, $erf(z)$, which is available in various software packages and defined as

$$erf(z) = \frac{2}{\sqrt{\pi}} \int_0^z e^{-u^2} du. \tag{2}$$

Namely, the Gaussian CDF is calculated with new bounds by

$$\hat{p}(w|c,a) = \int_{-\infty}^{w} \frac{1}{\sqrt{2\pi}a} e^{-\frac{(x-c)^2}{2a^2}} dx = \frac{1}{2} + \int_{c}^{w} \frac{1}{\sqrt{2\pi}a} e^{-\frac{(x-c)^2}{2a^2}} dx. \tag{3}$$

Changing the bounds on the integration and substituting for the difference in the exponent gives

$$\hat{p}(w|c,a) = \frac{1}{2} + \int_{0}^{w-c} \frac{1}{\sqrt{2\pi}a} e^{-\frac{y^2}{2a^2}} dy. \quad \text{Let } y = x - c, \tag{4}$$

replacing y with u to simplify the exponential to a single variable yields

$$\hat{p}(w|c,a) = \frac{1}{2} + \int_{0}^{\frac{w-c}{\sqrt{2}a}} \frac{1}{\sqrt{2\pi}a} e^{-u^2} d(\sqrt{2}au). \quad \text{Let } u = \frac{y}{\sqrt{2}a}, \tag{5}$$

taking the derivative of the variables and simplifying the constant in the integral gives

$$\hat{p}(w|c,a) = \frac{1}{2} + \frac{1}{2} \cdot \frac{2}{\sqrt{\pi}} \int_{0}^{\frac{w-c}{\sqrt{2}a}} e^{-u^2} du. \tag{6}$$

Using the common integral given either as a table or a computer function, we can express this as

$$\hat{p}(w|c,a) = \frac{1}{2} + \frac{1}{2} erf(\frac{w-c}{\sqrt{2}a}). \tag{7}$$

2.2 Second Version of Gaussian CDF Based IDCD

As the power curve is not exactly rotational symmetric, two more parameters are introduced in the second IDCD version to increase the flexibility of the curve as

$$\hat{p}(w|c,a,s,m) = -m + s \cdot \int_{-\infty}^{w} \frac{1}{\sqrt{2\pi}a} e^{-\frac{(x-c)^2}{2a^2}} dx, \tag{8}$$

where s is the scaling factor, and m is an extra shift.

The fitting results of the two versions are compared in Figure 6, where the main differences between the two versions are in the kink regions. The second version capture the nominal working state better than the first version with less deviation.

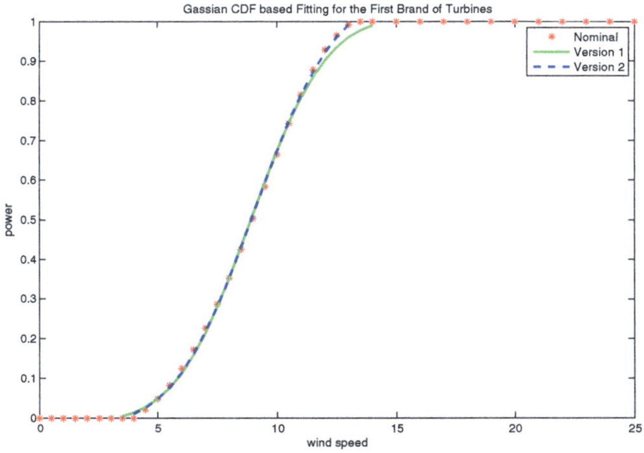

Fig. 6. Comparison of the two versions of Gaussian CDF based IDCD. The extra flexibilities by scaling and shifting further minimizes the fitting deviation.

3 Parameter Optimization in Gaussian CDF Based IDCD

Once the form of the fitting function is determined, there are multiple parameters that need to be optimized to best match the nominal power curve. As the wind turbines on the market differ greatly in all aspects as in size, processing, integrated equipment, calibration method, etc., an adaptive and efficient optimization method is needed to handle all these differences. Particle Swarm Optimization (PSO) [7] is exactly such a method to effectively search the solution space in manageable time through implicit parallelism. Comparing to other computer intelligence algorithms to deal with discrete search spaces by their original designs, such as genetic algorithm (GA), memetic algorithm (MA), or ant colony optimization (ACO), PSO is originally designed to search a continuous solution space, which suits the fitting problems.

3.1 Description of Particle Swarm Optimization (PSO)

Without losing generalizability, assume that the optimization is a minimization problem, which often describes fitness equations that minimize an error, and negating a maximization yields minimization. The function $f(\mathbf{x})$, of multivariate $\mathbf{x}$, is to be optimized,

$$\mathbf{x} = \arg minf(\mathbf{x}). \tag{9}$$

The particle in PSO is $\mathbf{x}$, and the function value $f(\mathbf{x})$ is the fitness of the solution $\mathbf{x}$. Suppose that there are N particles in the swarm with the fitness of each particle as $f(x_i)$ ($i \in [1,N]$). The particle's movement is affected by its inertia, its cognitive awareness (p_{best}, the best location that the particle has been before) and social influence (g_{best}, the best location for the entire population during any iterations). The PSO algorithm repeatedly computes the fitness and then moves the particle as follows:

1. Initialize a population of N particles. Each particle is a solution or at a location, $\mathbf{x}_i$, $i \in [1,N]$. The fitness, current particle's best location and best location for all the particles over all time ($f(x_i)$, p_{best_i}, and g_{best}), are all initialized at infinity.
2. Until the iterations, t, reach the maximum t_{max}, or some other termination condition is satisfied, do the following.
   a. Evaluate $Fitness_i = f(x_i)$.
   b. Update $p_{best_{i,\,t}}$ for each particle.
   c. Update g_{best_t} for the population.
   d. Move the particles by
   $$\mathbf{x}_{i,t+1} = \mathbf{x}_{i,t} + \mathbf{u}_{i,t+1}, \tag{10}$$
   where $\mathbf{u}_{i,t+1}$ is the velocity required to move the current particle position to the next particle position, and it is computed by
   $$\begin{aligned}\mathbf{u}_{i,t+1} =& \omega \cdot \mathbf{u}_{i,t} + \\ & c_1 \cdot r_1 \cdot (p_{best_{i,t}} - \mathbf{x}_{i,t}) + \\ & c_2 \cdot r_2 \cdot (g_{best_t} - \mathbf{x}_{i,t}),\end{aligned} \tag{11}$$
   where $\mathbf{u}_{i,t}$ is the velocity of the ith particle at time instance t, $\mathbf{u}_{i,t+1}$ is the velocity of the next step, and ω is an inertial constant, less than 1, to retain the information of previous velocity. The current particle solution, $\mathbf{x}_{i,t}$, needs to be updated. The constants, c_1 and c_2, weigh these influences. Uniform random number draws, r_1 and r_2, maintain a controlled level of randomness in the process.
   e. $t = t + 1$.
3. Finish and give the best solution of parameters that can most accurately model the power curve for the turbine.

3.2 PSO Based Parameter Optimization for Power Curve Modelling

In IDCD, the parameters in Gaussian CDF, $\mathbf{x} = \{c, a, s, m\}$, need to be optimized to fit the given set of m discrete power curve points, $\{(w_1, p_1), (w_2, p_2), \cdots, (w_m, p_m)\}$. The fitness function is the summation of the absolute fitting errors,

$$f(\mathbf{x}) = \sum_{j=1}^{m} |p_j - \hat{p}_j|, \tag{12}$$

which is to be minimized, where $\hat{p}_j$ is the estimated power from the fitting model at that specific wind speed w_j.

4 IDCD for the Second Turbine Type

IDCD is designed to linearize the curvy decision boundaries of multiple states on the performance evaluation plot (for instance, power curve, as discussed in this paper) with the need to fit only once. After the fitting of the nominal state is optimized, the changes relative to this nominal state can be easily evaluated in the linearized domain. We will compare the resulting power curves for two kinds of turbines each from a different manufacturer as an example to illustrate the generalizability of IDCD to different wind turbines. For the new type of turbine, the nominal power curve is less symmetric than the previous turbine's. As illustrated in Figure 7, if the parameters are optimized by a single Gaussian CDF, then the upper region is well fitted while the lower region can not keep up with the higher nominal values.

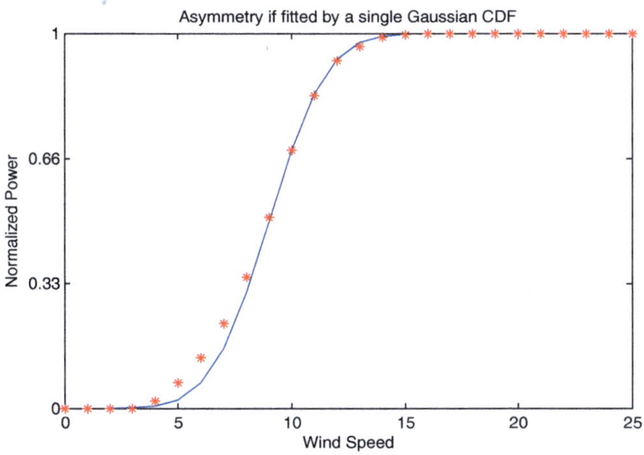

Fig. 7. Inadequacy of the single-segment IDCD on the second type of turbines

The inadequacy of the single-segment IDCD motivates us to implement the multi-segment IDCD for cases with more complicated decision boundaries. For the second type of turbines, two segments are utilized and found to be adequate, as illustrated in Figure 8. The transition point of the two segments are chosen as the middle point. With multiple segments, multiple sets of parameters need to be optimized. This makes the application of PSO instead of an exhaustive search critical, as there are many dependent parameters for a traditional approach.

5 Numerical Results on the Parameters

In summary, Gaussian CDF based IDCD expands from two parameters to four parameters to tune the fitting on the asymmetric part for relatively symmetric power curves. For turbines with more asymmetric power curves, IDCD expands from one

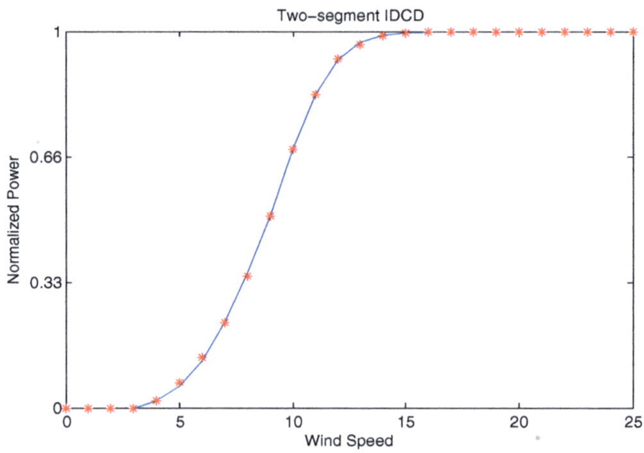

Fig. 8. Improvement by the two-segment IDCD on the second type of turbines

segment to two segments with eight parameters. These parameters are optimized to best classify the turbine states by the particle swarm optimization algorithm as outlined in section 3.2. The turbine's health is tracked using IDCD as explained from section 6. The optimized parameters for the power curve models are reported in Table 1, and the previous nominal power curve plots overlaid with estimations are plotted using these parameters.

Table 1. Parameters Optimized by PSO

Type	version		a	c	s	m
first	one-segment	two-parameter	8.9056	2.4693	NA	NA
		four-parameter	9.0579	2.6392	1.0844	0.0193
second	one-segment	two-parameter	9.0000	2.0123	NA	NA
	two-segments	left	8.7742	2.5914	0.9774	0.0126
		right	8.9551	2.0986	0.9942	-0.0044

6 Power Curve Test for Diagnostics Purpose

In a wind turbine electricity generation system, the diagnostics of potential faults is crucial to maintain and improve the efficiency of the system [8]. Data-driven approach does both sensor validation and diagnostics in an integrated way. In data-driven diagnostics and prognostics, change detection is important to detect abnormalities [9]. A change detector should be sensitive to status variation, and it should also tolerate noise and interference [10]. The earlier the degradation is detected, the sooner the degradation can be stopped through maintenance or changing the control parameters instead of harming the wind turbine. As the wind is inherently variable, but predictable, early fault detection may allow maintenance and repairs to occur during low wind periods to mitigate the impacts of failures [11].

In some cases, use of a crane for a repair is needed, which can be costly. Maximizing use of a crane once on site can be economically beneficial to the farm. Once the fault is diagnosed, then prognostics, estimation of the remaining useful life of a turbine with faulty components or the faulty system, is needed. If the faults were estimated to degrade the performance gradually, then the repair could be scheduled at the next regular maintenance trip to save the repair cost at a minimum loss of power production and overall operating efficiency [12]. Prognostics optimizes the repair scheduling and resource utilization so that the negative impact on power production is minimized [13].

The power curve test provides a set of variables that can identify key working states for the turbine, including complete shut-downs, under-performing states, abnormal default states, as well as normal working states. Based on the multiple states defined by IDCD, as illustrated in Figure 5, we keep track of the time sequence of power curve data relative to its nominal curve. Drastic change detection indicates imminent faults, and a complete shut-down when the wind speed is not nearly zero indicates apparent faults. The detailed definition of eight states relative to the linearized power curve is shown in Table 2.

Table 2. Multiple State Definition of Linearized Power Curve

State Index	Linearized Power Values ($\hat{p}$), or Raw Power Values (p), vs. Wind Speed (w)
1	$\hat{p} > 1.5$
2	$0.5 < \hat{p} < 1.5$
3	$-0.5 < \hat{p} < 0.5$
4	$-1.5 < \hat{p} < -0.5$
5	$\hat{p} < -1.5$
6	Horizontal power (p is nearly 0)
7	Vertical power ($p > 0$ when $w < 4m/s$)
8	$w < 4m/s$ and $\hat{p} < 0$

The divided sections in Figure 9 show the separation of different regions in the power curve that are either normal working states or the states with problems currently occurring or about to. For example, the normal power measurements should reside in the regions corresponding to states 2, 3 and 4. However, if the measurements lie in the horizontal region, it typically indicates complete shut-downs of the turbine. If the measurements move to the upper left region, there is usually a soft failure, such as faulty anemometer or degraded gear box, which is insufficient to stop operations yet but causing power production losses. Without prompt maintenance, the turbine may completely shut down.

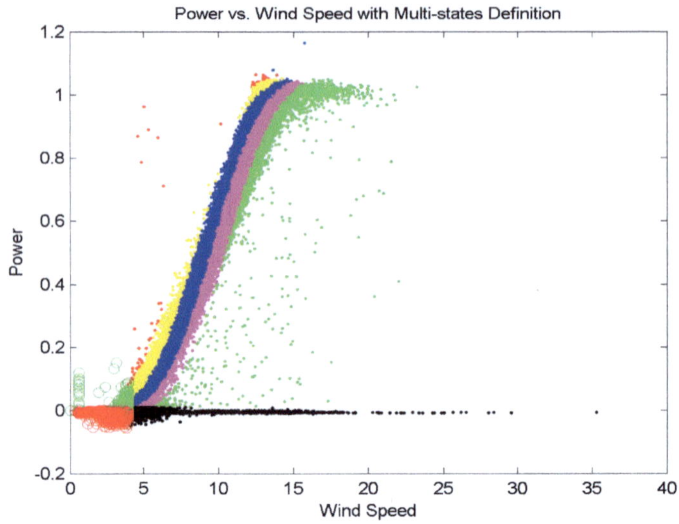

Fig. 9. State diagram for power curve vs. wind speed

7 Real Data Examples Using Power Curve Test

Data mining using the power curve test with the information of corresponding events helps associate important suboptimal states with turbine maintenance tasks. Sequencing these states will provide a more accurate prediction of maintenance needs. Our processing also tracks the changes in the percentages of states providing a clearer picture of what is happening automatically. Over time, this approach mathematically presents changes in the percentage of measurements falling in each state. Through extensive data mining of historical data and verification from turbine farm operators, some dramatic changes in specific states are discovered as strong indicators of major component failures, anemometer faults, etc.

The real data are collected by turbine SCADA (Supervisory Control And Data Acquisition) systems, where each data point of each variable is an averaged value in 10-minute interval. The averaging smoothes out the drastic temporal variations, and yet the 10-minute interval provides adequate resolution relative to the long-time operation of the turbines.

We divide the whole data set into daily data segments for the turbine of interest, with 144 samples per day. Then, we evaluate the percentage of each state, distinguished with different regions in the linearized power curve. The sharp peaks in the percentage may indicate under-performance due to a potential failure and need to be analyzed in more detail.

7.1 A Major Component Failure

One example is the detection of a major component failure. Once it happens, the turbine is forced into a complete shut-down with motionless rotor. As shown in Figure 10, state 6 (the horizontal power curve state) has a much higher occurrence rate from day 63 to day 108. It indicates that, during this period, this turbine produces no power most of the time. A major component failure is verified through the operator's monthly operational report.

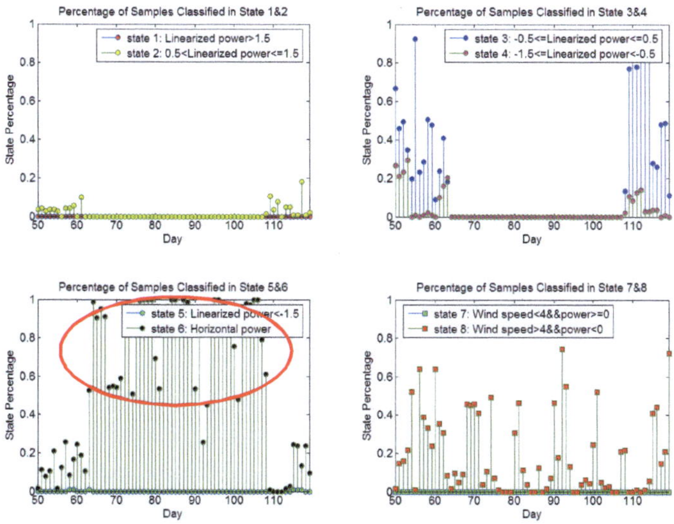

Fig. 10. Percentage of multiple states in power curve vs. wind speed used as an indicator for a major component failure

7.2 Faulty Anemometer

An anemometer measures the wind speed as seen by the turbine. The anemometer output is an important parameter on both turbine operation and maintenance. In Figure 11, the exempla turbine shows a sharp peak in the percentage of state 7 (the vertical power curve state) on day 826. It seems that, even with zero wind speed, there is abnormally large amount of power produced, which is not feasible. It implies that the anemometer does not measure the wind speed correctly. This event is also verified by the weekly wind speed difference test comparing to its neighboring turbine in [3], where day 826 falls into week 118, and abnormal event is observed in week 118. The significant difference between these two adjacent turbines on day 826 is found out to be caused by the anemometer dysfunction of this example turbine in this section, because the comparing turbine does not show such big differences when it compares with other turbines.

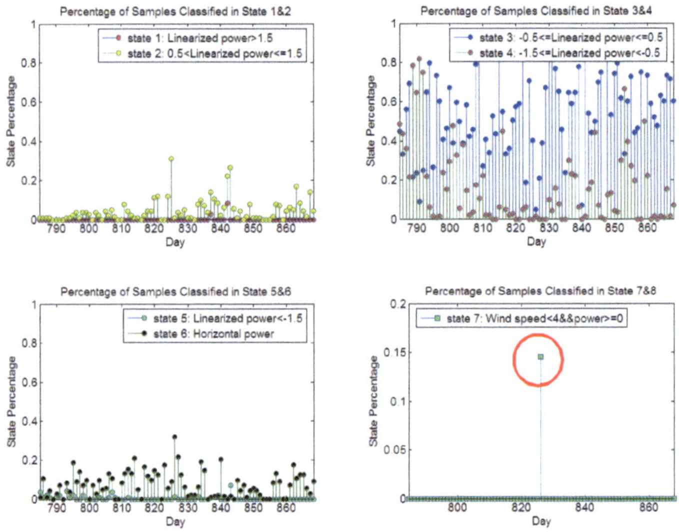

Fig. 11. Percentage of multiple states in power curve vs. wind speed used as an indicator for faulty anemometer

8 Conclusions

A power curve is a plot of generated power versus wind speed. It is a key performance evaluation tool for wind turbines. To facilitate the testing on irregularly shaped power curve, we design and propose the *Inverse Diagnostic Curve Detector* (IDCD) using Gaussian CDF (Cumulative Density Function) as the inversion function to linearize the power curve to track the turbine status. The Gaussian CDF ensures an accurate one-to-one inversion. IDCD simplifies the change detection for diagnostics, because the direct deviation detection requires multiple fitting for each state boundary definition, but IDCD calls for data fitting only once.

The nominal power curve is fitted using the particle swarm optimization (PSO) algorithm. We elaborate the evolution of Gaussian CDF based IDCD from a two-parameter version to a four-parameter version with a single segment for the first turbine type, and then from one-segment version to two-segment version for the second turbine type, to illustrate the adaptation procedure of IDCD. PSO based fitting and version adaptation makes IDCD versatile for different kinds of power curves.

We define eight states relative to the linearized power curve to track the state variation of turbines. If the turbine performance is suboptimal due to soft failures, or worse yet, if the turbine is completely shut down due to major faults, the percentage of specific states in the power curve changes dramatically. Two application examples using IDCD are provided to automatically detect a major component failure and a faulty anemometer.

If the nominal curve is unavailable, or a more customized diagnostic framework is implemented for each individual turbine, an estimation from the real data can be used to replace the nominal reference in deriving the fitting function. Besides power curves, IDCD can be also adapted for other sensor measurements. Furthermore, IDCD lays the ground for higher-level decision strategies based on multiple states such as in a Bayesian network.

References

1. Cabezon, D., Marti, I., Isidro, M.J.S., Perez, I.: Comparison of methods for power curve modelling. In: CD-Rom Proceedings of the Global WindPower 2004 Conference, Chicago, Illinois, USA (2004)
2. Robb, D.: Gearbox design for wind turbines improving but still face challenges. Windstat Newsletter 18(3) (May 2005)
3. Tindal, A., Johnson, C., LeBlanc, M., Harman, K., Rareshide, E., Graves, A.: Site-specific adjustments to wind turbine power curves. In: AWEA WINDPOWER Conference, Houston, TX, USA (2008)
4. Ye, X., Veeramachaneni, K., Yan, Y., Osadciw, L.A.: Unsupervised learning and fusion for failure detection in wind turbines. In: Proceedings of 12th International Conference on Information Fusion, Seattle,Washington, USA (July 2009)
5. Yan, Y., Kamath, G., Osadciw, L.A., Benson, G., Legac, P., Johnson, P., White, E.: Fusion for modeling wake effects on wind turbines. In: Proceedings of 12th International Conference on Information Fusion, Seattle,Washington, USA (July 2009)
6. Yan, Y., Osadciw, L.A., Benson, G., White, E.: Inverse data transformation for change detection in wind turbine diagnostics. In: Proceedings of 22nd IEEE Canadian Conference on Electrical and Computer Engineering, Delta St. John's, Newfoundland and Labrador, Canada (May 2009)
7. Kennedy, J., Eberhart, R.: Particle swarm optimization. In: Proc. IEEE Int'l. Conf. on Neural Networks (Perth, Australia)., vol. IV, pp. 1942–1948. IEEE Service Center, Piscataway (1995)
8. Blaabjerg, F., Chen, Z.: Wind energy-the world's fastest growing energy source. IEEE Power Electron. Soc. Newsl. 18(3), 15–19 (2006)
9. DePold, H.R., Gass, F.D.: The application of expert systems and neural networks to gas turbine prognostics and diagnostics. Journal of Engineering for Gas Turbines and Power 121(4), 607–612 (1999)
10. Karki, R., Billinton, R.: Cost effective wind energy utilization for reliable power supply. IEEE Trans. Energy Convers. 19(2), 435–440 (2004)
11. Ribrant, J.: Reliability performance and maintenance - a survey of failures in wind power systems. Ph.D. dissertation, XR-EE-EEK, (September 2006)
12. Burton, T., Sharpe, D., Jenkins, N., Bossanyi, E.: Wind Energy Handbook. Wiley, Chichester (2001)
13. Nilsson, J., Bertling, L.: Maintenance management of wind power systems using condition monitoring systemslife cycle cost analysis for two case studies. IEEE Transaction on Energy Conversion 22(1), 223–229 (2007)

Optimal Controller Design of a Wind Turbine with Doubly Fed Induction Generator for Small Signal Stability Enhancement

Lihui Yang, Guang Ya Yang, Zhao Xu, Zhao Yang Dong, and Yusheng Xue

Abstract. Multi-objective optimal controller design of a doubly fed induction generator (DFIG) wind turbine system using Differential Evolution (DE) is presented in this chapter. A detailed mathematical model of DFIG wind turbine with a close loop vector control system is developed. Based on this, objective functions, addressing the steady state stability and dynamic performance at different operating conditions are implemented to optimize the controller parameters of both the rotor and grid side converters. A superior ε-constraint method and method of adaptive penalties are applied to handle the multi-objective problem and the constraint with DE. Eigenvalue analysis and simulation are performed on the single machine infinite bus (SMIB) system to demonstrate the control performance of the system with the optimized controller parameters.

1 Introduction

The Doubly-fed Induction Generator (DFIG) equipped wind turbine is currently one of the most popular wind conversion systems due to its high energy

Lihui Yang · Guang Ya Yang · Zhao Xu
Centre for electric Technology, Department of Electrical Engineering,
Technical University of Denmark, DK-2800 Kongens Lyngby, Denmark

Lihui Yang
School of Electrical Engineering, Xi'an Jiao Tong University, Xi'an, 710049, China

Zhao Yang Dong
Department of Electrical Engineering, Hong Kong Polytechnic University

Yusheng Xue
State Grid Electric Power Research Institute / NARI, Nanjing, China

L. Wang et al. (Eds): *Wind Power Systems*, Green Energy and Technology, pp. 167–190.
springerlink.com

efficiency, reduced mechanical stress on the wind turbine and relatively low power rating of the connected power electronics converter (Hansen 2005). Increasing penetration level of wind power generation of DFIG type into the grid will give impact on the power system performance (Eriksen et al. 2005). As stability is a key issue for power system operations and planning, there is a genuine concern that the effect of DFIG wind turbine on power system stability needs proper investigation.

In order to achieve decoupled control of active and reactive power of DFIG, vector control strategy based on proportional-integral (PI) controllers was proposed and has been widely used in the industry (Yamamoto and Motoyoshi 1991; Pena et al. 1996; Muller et al. 2002). The decoupled control of DFIG has several different PI controllers. Suitable controller parameters are needed to achieve better control performance for system stability. However, the coordinated tuning of these controllers using the traditional trial and error method is a challenging and cumbersome task.

Recently optimization methods have been utilized in controller parameter tuning for DFIG wind turbine system. Particle Swarm Optimization (PSO) and Genetic Algorithm (GA) have been applied to optimize the controller for the rotor side converter in time domain in (Qiao et al. 2006) and (Vieira et al. 2009), respectively. The objective function is to reduce the overcurrent as well as voltage in the rotor circuit. However, as the grid side converter controller was not optimized, hence, larger oscillations of the DC-link voltage can not be avoided. PSO has been used to optimize all the five controllers in the DFIG system including both rotor and grid side controllers (Wu et al. 2007). The objective is to shift all the eigenvalues as far to the left of the left hand side of the *S*-plane. Bacteria Foraging (BF) optimization has been applied for tuning damping controller to improve the damping of the oscillatory modes of the DFIG wind turbine (Mishra et al. 2009). However, (Wu et al. 2007) and (Mishra et al. 2009) only considered single objective and single operating point. So robust damping performance as well as enough stability margin for changed operating conditions (e.g. changed wind speed) can not be obtained simultaneously.

Differential Evolution (DE), a relatively new member in the family of Evolutionary Algorithms (EAs), is a population-based method and generally considered as a parallel stochastic direct search optimizer which is very sample yet powerful (Storn and Price 1995; 1996). The main advantages of a DE are its capability in solving optimization problems which require optimization process with nonlinear and multi-modal objective functions. It employs a nonuniform crossover using parameters of child vectors to guide through the minimization process. The mutation operation with DE is conducted by arithmetical combinations which exploit the difference among randomly selected vectors, other than perturbing the genes in individuals with small probability as compared with one of the most popular EAs, Genetic Algorithm (GA). These special features make DE a precise, fast as well as robust algorithm. Therefore, DE has been attracting more and more attentions from industry applications, including the field of power system (Yang et al. 2008).

This chapter focuses on the optimizing controllers' parameters of a DFIG wind turbine to enhance its small signal stability under grid connection conditions. A comprehensive DFIG model including induction generator, two-mass drive train, pitch control, close-loop control etc. has been developed. An effective DE based multi-objective optimization method is used to get the optimal controllers' parameters of both the rotor and grid side converters, so as to obtain robust damping performance as well as enough stability margin for all the operating conditions in considerations. Small signal stability analysis and simulations are performed on a sample single machine infinite bus (SMIB) DFIG system to demonstrate the control performance of the system with the optimized controller parameters.

2 Model of Wind Turbine with DFIG

Since SMIB system is sufficient to investigate the dynamics, stability and control design of a DFIG, the DFIG wind turbine SMIB system (Hansen et al. 2004) shown in Fig. 1 is studied in this chapter. The DFIG is connected to the infinite bus through a transformer and transmission line. The DFIG utilizes a wound rotor induction generator, in which the stator windings are directly connected to the external three-phase grid and the rotor windings are fed through three-phase back-to-back bi-directional pulse width modulated (PWM) converters. The back-to-back PWM converters consist of two three-phase six-switch converters, i.e., the rotor and grid side converter, between which a DC-link capacitor is placed. In the overall control system of the DFIG wind turbine, two control levels can be distinguished: wind turbine control and DFIG control. The wind turbine control level controls the pitch angle of the wind turbine and the reference rotor speed to the DFIG control level. Two stage control strategies are used: power optimization strategy below rated wind speed and power limitation strategy above rated wind

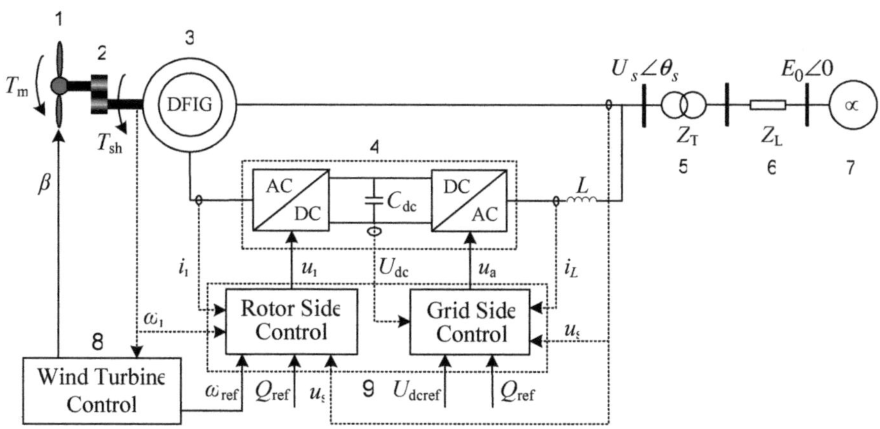

Fig. 1. Schematic diagram of a DFIG wind turbine system. 1. Wind turbine. 2. Gear box. 3. Induction generator. 4. Back-to-back PWM converters. 5. Transformer. 6. Transmission line. 7. Infinite bus. 8. Wind turbine control level. 9. DFIG control level.

speed (Hansen et al. 2004). The DFIG control level, including the rotor side and the grid side controller, aims to control the active and reactive power of the DFIG. To achieve decoupled control of active and reactive power, vector control is used for both the rotor and grid side converters. In the following subsections, a comprehensive model of a grid connected DFIG including induction generator, two-mass drive train, pitch control and close-loop control etc. will be presented. This model particularly enables small signal stability analysis of the overall system.

2.1 Generator

For power system stability studies, the generator is modeled as an equivalent voltage source based on transient impedance (Kundur 1994). The differential equations of the stator and rotor circuits of the induction generator with stator current and equivalent voltage behind transient impedance as state variables can be given in a *d-q* reference frame rotating at synchronous speed (Mei and Pal 2007) (we define this reference frame as the generator reference frame in this paper)

$$\begin{cases}\frac{1}{\omega_b}\frac{di_{ds}}{dt}=-\frac{\omega_s}{X_s'}(R_s+\frac{X_s-X_s'}{\omega_s T_0})i_{ds}+\omega_s i_{qs}+\frac{\omega_r}{X_s'}e_{ds}'-\frac{1}{X_s'T_0}e_{qs}'+\frac{\omega_s L_m}{X_s'L_r}u_{dr}-\frac{\omega_s}{X_s'}u_{ds}\\ \frac{1}{\omega_b}\frac{di_{qs}}{dt}=-\omega_s i_{ds}-\frac{\omega_s}{X_s'}(R_s+\frac{X_s-X_s'}{\omega_s T_0})i_{qs}+\frac{1}{X_s'T_0}e_{ds}'+\frac{\omega_r}{X_s'}e_{qs}'+\frac{\omega_s L_m}{X_s'L_r}u_{qr}-\frac{\omega_s}{X_s'}u_{qs}\\ \frac{1}{\omega_b}\frac{de_{ds}'}{dt}=-\frac{1}{T_0}[e_{ds}'-(X_s-X_s')i_{qs}]+(\omega_s-\omega_r)e_{qs}'-\frac{\omega_s L_m}{L_r}u_{qr}\\ \frac{1}{\omega_b}\frac{de_{qs}'}{dt}=-\frac{1}{T_0}[e_{qs}'+(X_s-X_s')i_{ds}]-(\omega_s-\omega_r)e_{ds}'+\frac{\omega_s L_m}{L_r}u_{dr}\end{cases} \quad (1)$$

where $\mathbf{i}_s=i_{ds}+ji_{qs}$ is the stator current vector; $\mathbf{e}'_s=e'_{ds}+je'_{qs}$ is the vector of equivalent voltages behind transient impedance, by defining $e'_{ds}=-\omega_s L_m\psi_{qr}/L_r$, $e'_{qs}=\omega_s L_m\psi_{dr}/L_r$; $\mathbf{u}_s=u_{ds}+ju_{qs}$ is the stator voltage vector; $\mathbf{u}_r=u_{dr}+ju_{qr}$ is the rotor voltage vector; $L_s=L_m+L_{ls}$, $L_r=L_m+L_{lr}$, $X_s=\omega_s/L_s$, $X'_s=\omega_s(L_s-L_m^2/L_r)$, $T_0=L_r/R_r$. This model adopts the generator convention meaning that stator and rotor currents are positive and negative when flowing out of the generator, respectively. Since the control design is of interest in this chapter, the full order model of the generator is necessary (Mei and Pal 2008).

2.2 Drive Train

When studying the stability of DFIG wind turbine, the two-mass model of the drive train is important due to the wind turbine shaft is relatively softer than the typical shaft used in conventional power plants (Akhmatov 2005). The equations on the two-mass model of the drive train are given by

$$\frac{d\omega_{\mathrm{r}}}{dt}=\frac{1}{2H_{\mathrm{g}}}(T_{\mathrm{sh}}-T_{\mathrm{e}}-B\omega_{\mathrm{r}}) \tag{2}$$

$$\frac{d\theta_{\mathrm{t}}}{dt}=\omega_{\mathrm{b}}(\omega_{\mathrm{t}}-\omega_{\mathrm{r}}) \tag{3}$$

$$\frac{d\omega_{\mathrm{t}}}{dt}=\frac{1}{2H_{\mathrm{t}}}(T_{\mathrm{m}}-T_{\mathrm{sh}}) \tag{4}$$

where ω_{r} and ω_{t} are the generator and wind turbine speeds, respectively. θ_{t} is the shaft twist angle. The electromagnetic torque T_{e}, the shaft torque T_{sh} and the mechanical torque T_{m}, which is the power input of the wind turbine, are

$$T_{\mathrm{e}}=L_{\mathrm{m}}(i_{\mathrm{qs}}i_{\mathrm{dr}}-i_{\mathrm{ds}}i_{\mathrm{qr}}) \tag{5}$$

$$T_{\mathrm{sh}}=K_{\mathrm{sh}}\theta_{\mathrm{t}}+D_{\mathrm{sh}}\omega_{\mathrm{b}}(\omega_{\mathrm{t}}-\omega_{\mathrm{r}}) \tag{6}$$

$$T_{\mathrm{m}}=\frac{0.5\rho\pi R^{2}C_{\mathrm{p}}(\lambda,\beta)V_{\mathrm{w}}^{3}}{\omega_{\mathrm{t}}} \tag{7}$$

where C_{p} is the power coefficient, and

$$C_{\mathrm{p}}=0.22(\frac{116}{\lambda_{\mathrm{i}}}-0.4\beta-5)e^{\frac{-12.5}{\lambda_{\mathrm{i}}}} \tag{8}$$

$$\lambda_{\mathrm{i}}=\frac{1}{1/(\lambda+0.08\beta)-0.035/(\beta^{3}+1)} \tag{9}$$

where $\lambda=\omega_{\mathrm{t}}R/V_{\mathrm{w}}$ is the blade tip speed ratio. When the generator speed is less than rated rotor speed, in order to extract the maximum power from particular wind speed, λ is tuned to the optimal value over different wind speeds by adapting the rotor speed to its reference, expressed by (Hansen et al. 2004)

$$\omega_{\mathrm{ref}}=\sqrt{\frac{T_{\mathrm{m}}}{K_{\mathrm{opt}}}} \tag{10}$$

2.3 Pitch Control

The pitch angle of the blade is controlled to optimize the power extraction of wind turbine as well as to prevent over rated power production in strong wind. The pitch servo is modeled as,

$$\frac{d\beta}{dt} = \frac{1}{T_{\beta}}(\beta_{\text{ref}} - \beta) \tag{11}$$

For the sake of simplicity, the reference of the pitch angle β_{ref} is kept zero when wind speed is below rated value, while is increased by a non-linear function at large wind speed (Hansen et. al 2004).

2.4 Rotor Side Converter

The aim of the rotor side converter is to independently control the active power (rotor speed) and reactive power at the stator terminal. The generic control scheme of the rotor side converter is illustrated in Fig. 2. In order to decouple the electromagnetic torque and the rotor excitation current, the induction generator is controlled in the stator-flux oriented reference frame, which is a synchronously rotating reference frame, with its *d*-axis oriented along the stator-flux vector position (Pena et al. 1996). The typical proportional-integral (PI) controllers are used for regulation in both rotor speed (outer) control loop and rotor current (inner) control loop. In Fig. 2, superscript φ denotes the variable is in the stator-flux oriented reference frame.

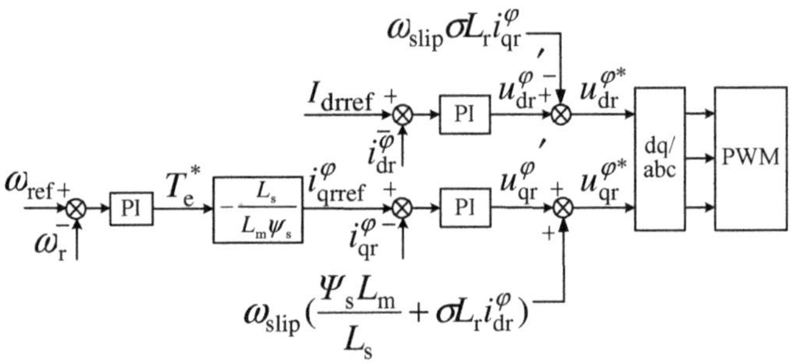

Fig. 2. Control scheme of the rotor side converter

Based on the stator-flux orientation, the stator flux can be described as $\psi_{\text{ds}}^{\varphi} = \Psi_{\text{s}}$ and $\psi_{\text{qs}}^{\varphi} = 0$ (Pena et al. 1996). According to the control scheme of the rotor side converter shown in Fig. 2, the equations regarding the rotor voltage equations can be written as

$$\begin{cases} u_{\text{dr}}^{\varphi} = -(\omega_{\text{s}} - \omega_{\text{r}})\sigma L_{\text{r}} i_{\text{qr}}^{\varphi} + K_{\text{Pir}}(I_{\text{drref}} - i_{\text{dr}}^{\varphi}) + x_{\text{idr}} \\ u_{\text{qr}}^{\varphi} = (\omega_{\text{s}} - \omega_{\text{r}})(\sigma L_{\text{r}} i_{\text{dr}}^{\varphi} + \dfrac{\Psi_{\text{s}} L_{\text{m}}}{L_{\text{s}}}) + K_{\text{Pir}}(i_{\text{qrref}}^{\varphi} - i_{\text{dr}}^{\varphi}) + x_{\text{iqr}} \end{cases} \tag{12}$$

where $\sigma = 1 - \dfrac{L_{\text{m}}^2}{L_{\text{s}} L_{\text{r}}}$ is the leakage factor.

The control equations of the rotor side converter become,

$$\begin{cases} \dot{x}_{\omega} = \dfrac{K_{\mathrm{P\omega}}}{T_{\mathrm{I\omega}}}(\omega_{\mathrm{ref}} - \omega_{\mathrm{r}}) \\ \dot{x}_{\mathrm{idr}} = \dfrac{K_{\mathrm{Pir}}}{T_{\mathrm{Iir}}}(I_{\mathrm{drref}} - i_{\mathrm{dr}}^{\varphi}) \\ i_{\mathrm{qrref}}^{\varphi} = -\dfrac{L_{\mathrm{s}}}{L_{\mathrm{m}}\Psi_{\mathrm{s}}}[K_{\mathrm{P\omega}}(\omega_{\mathrm{ref}} - \omega_{\mathrm{r}}) + x_{\omega}] \\ \dot{x}_{\mathrm{iqr}} = \dfrac{K_{\mathrm{Pir}}}{T_{\mathrm{Iir}}}(i_{\mathrm{qrref}}^{\varphi} - i_{\mathrm{qr}}^{\varphi}) \end{cases} \tag{13}$$

where $K_{\mathrm{P\omega}}$ and $T_{\mathrm{I\omega}}$ are the proportional gain and integral time constant of the rotor speed control loop, respectively. K_{Pir} and T_{Iir} are the proportional gain and integral time constant of the rotor current control loop, respectively. For the sake of simplicity, we assume the parameters are the same for *d*- and *q*-component of the rotor current control loop.

The relationship between the generator reference frame and the stator-flux oriented reference frame can be expressed as (Ledesma and Usaola 2005)

$$\begin{bmatrix} y_{\mathrm{dr}} \\ y_{\mathrm{qr}} \end{bmatrix} = \begin{bmatrix} \cos\varphi & -\sin\varphi \\ \sin\varphi & \cos\varphi \end{bmatrix} \begin{bmatrix} y_{\mathrm{dr}}^{\varphi} \\ y_{\mathrm{qr}}^{\varphi} \end{bmatrix} \tag{14}$$

where y can be current i or voltage u, $\varphi = \arctan(\dfrac{\psi_{\mathrm{qs}}}{\psi_{\mathrm{ds}}})$ is the angle between the stator-flux vector and the *d*-axis of the generator reference frame.

2.5 Grid Side Converter

The aim of the control of the grid side converter is to maintain the DC-link capacitor voltage at a set value as well as to guarantee converter operation with unity power factor. Fig. 3 shows the control scheme of the grid side converter. In order to obtain the independent control of active and reactive power flowing between the grid and the grid side converter, the converter control operates in the grid-voltage oriented reference frame, which is a synchronously rotating reference frame, with its *d*-axis oriented along the grid-voltage vector position (Pena et al. 1996). Similarly, the typical PI controllers are used for regulation in both DC-link voltage (outer) control loop and grid side inductor current (inner) control loop. In Fig. 3, ε denotes the variable is in the grid-voltage oriented reference frame.

The differential equations of the grid side converter are given by

$$\begin{cases} \dfrac{di_{\mathrm{d}L}}{dt} = \dfrac{\omega_{\mathrm{b}}}{L}(u_{\mathrm{ds}} - R_{L}i_{\mathrm{d}L} + \omega_{\mathrm{s}}Li_{\mathrm{q}L} - u_{\mathrm{da}}) \\ \dfrac{di_{\mathrm{q}L}}{dt} = \dfrac{\omega_{\mathrm{b}}}{L}(u_{\mathrm{qs}} - R_{L}i_{\mathrm{q}L} - \omega_{\mathrm{s}}Li_{\mathrm{d}L} - u_{\mathrm{qa}}) \end{cases} \tag{15}$$

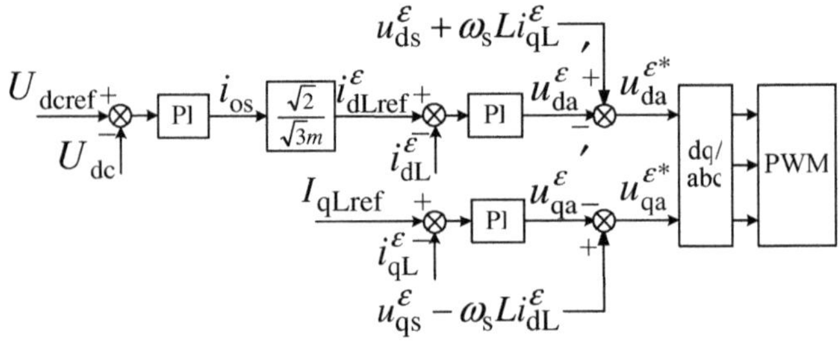

Fig. 3. Control scheme of the grid side converter

where $\mathbf{i}_L = i_{dL} + j i_{qL}$ is grid side inductor current vector, and $\mathbf{u}_a = u_{da} + j u_{qa}$ is grid side converter voltage vector.

Under the grid-voltage oriented reference frame, the grid voltage can be described as $u_{ds}^{\varepsilon} = U_s$ and $u_{qs}^{\varepsilon} = 0$. According to the control scheme of the grid side converter shown in Fig. 3, the equations regarding the voltage of grid side converter can be written as

$$\begin{cases} u_{da}^{\varepsilon} = U_s + \omega_s L i_{qL} - [K_{PiL}(i_{dLref}^{\varphi} - i_{dL}^{\varphi}) + x_{idL}] \\ u_{qa}^{\varepsilon} = -\omega_s L i_{qL} - [K_{PiL}(I_{qLref} - i_{qL}^{\varphi}) + x_{iqL}] \end{cases} \tag{16}$$

The equations regarding the controller of the grid side converter are described as

$$\begin{cases} \dot{x}_v = \dfrac{K_{Pv}}{T_{Iv}}(U_{dcref} - U_{dc}) \\ i_{qLref}^{\varphi} = \dfrac{\sqrt{2}}{\sqrt{3}m}[K_{Pv}(U_{dcref} - U_{dc}) + x_v] \\ \dot{x}_{idL} = \dfrac{K_{PiL}}{T_{IiL}}(i_{dLref}^{\varphi} - i_{dL}^{\varphi}) \\ \dot{x}_{iqL} = \dfrac{K_{PiL}}{T_{IiL}}(I_{qLef} - i_{qL}^{\varphi}) \end{cases} \tag{17}$$

where K_{Pv} and T_{Iv} are the proportional gain and integral time constant of the DC-link voltage control loop, respectively. K_{PiL} and T_{IiL} are the proportional gain and integral time constant of the grid side inductor current control loop, respectively. Similarly, the parameters are assumed to be the same for *d*- and *q*-component of the grid side inductor current control loop.

The relationship between the generator reference frame and the grid-voltage oriented reference frame can be given by (Ledesma and Usaola 2005),

$$\begin{bmatrix} y_{da} \\ y_{qa} \end{bmatrix} = \begin{bmatrix} \cos\varepsilon & -\sin\varepsilon \\ \sin\varepsilon & \cos\varepsilon \end{bmatrix} \begin{bmatrix} y_{da}^{\varepsilon} \\ y_{qa}^{\varepsilon} \end{bmatrix} \tag{18}$$

where y can be current i or voltage u, $\varepsilon = \arctan(\frac{u_{qs}}{u_{ds}})$ is the angle between the grid voltage vector and the d-axis of the generator reference frame.

2.6 DC-Link Capacitor

The equation which describes the energy balance of the DC-link capacitor can be expressed as

$$\frac{C_{dc}U_{dc}}{\omega_b}\frac{dU_{dc}}{dt} = p_a - p_r = \frac{3}{2}(u_{da}i_{dL} + u_{qa}i_{qL} - u_{dr}i_{dr} - u_{qr}i_{qr}) \tag{19}$$

where U_{dc} is DC-link voltage, p_a and p_r are powers supplied to the grid side converter and the rotor circuit, respectively.

2.7 Interfacing with Power Grid

The voltage equation describing the interface with the external system, which is the infinite bus in this chapter, can be written as

$$U_s\angle\theta_s - E_0\angle 0 = (Z_T + Z_L)(\mathbf{i}_s - \mathbf{i}_L) \tag{20}$$

where $\mathbf{E}_0$ is the voltage of the infinite bus. Z_T and Z_L are the impedance of the transformer and transmission line, respectively.

From Eqs. (1)-(20), we can obtain a set of differential equations to present the DFIG wind turbine system. They can be written in a compact form

$$\dot{\mathbf{x}} = f(\mathbf{x},\mathbf{u}) \tag{21}$$

where **x** and **u** are the vectors with respect to the state and the input variables which are defined as $\mathbf{x}=[i_{ds}\ i_{qs}\ e_{ds}\ e_{qs}\ \omega_r\ \theta_t\ \omega_t\ \beta\ x_\omega\ x_{idr}\ x_{iqr}\ x_v\ i_{dL}\ i_{qL}\ x_{idL}\ x_{iqL}\ U_{dc}]^T$, $\mathbf{u}=[I_{drref}\ I_{qLref}\ U_{dcref}\ V_w\ \beta_{ref}\ E_0]^T$.

3 DE-Based Multi-objective Optimal Control of DFIG Wind Turbine System

3.1 Differential Evolution

Originated from Darwinian natural selection theory, efforts have been dedicated to designating efficient optimization algorithms. These algorithms are commonly termed into 'evolutionary algorithms' (EA). As a member of EA, Differential Evolution (DE) was first proposed by Storn and Price at Berkeley during 1994-1996 (Storn and Price 1996; Price 1996). DE requires initial population, iterative progress and operators of mutation, recombination and selection to explore the search space. It is known for simplicity, easy implementation and capability of solving optimization problems with nonlinear and multi-modal objective functions (Price and Storn 1997). DE has been proved to be an effective and robust optimization algorithm which can be easily extended to handle different types of variables and nonlinear/nontrivial constraints (Storn and Price 1996; Lampinen and Zelinka 1999; Wong and Dong 2008). The features of DE are described in the sequel.

Encode

Binary coding scheme is commonly used in conventional EAs to represent the individuals in population, especially in GA. Binary coding uses limited number of binary digits to represent each variable in optimization. In spite of the simplicity, the major disadvantages of Binary coding lies in the limitation of ability to effectively represent variables possible of within different ranges, and the difficulty in preserving the continuum's topology, where the coding method may not map consecutive binary integers to the continuous intervals of the original variable ranges. Unlike conventional GA, DE use floating number instead of binary string to achieve better representation with higher precision.

Population Initialization

As other EA members, DE is a population based direct search algorithm. For the G-th generation, the population contains NP n-dimensional vectors $[X_{1,G}, X_{2,G}, ..., X_{NP,G}]$. If there is no prior knowledge on the problem, the first generation of population can be initialized by the equation below:

$$X_{i,1} = X^{\min} + rand(0,1) \cdot \left(X^{\max} - X^{\min}\right) \tag{22}$$

where $rand(0,1)$ is a random scalar within [0,1]. $X^{\max}$ and $X^{\min}$ represent the lower and upper bounds of variables respectively. However, if there is prior knowledge of the solution, such as a primary solution is known, this preliminary solution may be exploited in population initialization by adding probabilistic distributed deviations.

Mutation/Recombination

The main operator of DE is mutation/recombination. DE exploits the differences among individuals to probe the solution space. A recombination operator is integrated with the mutation operation to achieve the diversity of mutated individuals. The key characteristic of DE mutation/recombination lies in that it utilizing the difference among individuals to obtain a trial population which contains NP individuals. In implementation, mutation and recombination are almost finished simultaneously. And after this operation, the trial population will then be formed.

For the *i*-th individual $X_{i,G}$ in generation *G*, a typical example of DE mutation is (Storn and Price 1995),

$$X'_{i,G} = X_{r1,G} + c \cdot \left(X_{r2,G} - X_{r3,G}\right) \tag{23}$$

where $X_{r1,G}$, $X_{r2,G}$, $X_{r3,G}$ are individuals randomly selected in the current population G and $r1 \neq r2 \neq r3 \neq i$. $X'_{i,G}$ is the mutated vector which is used for recombination. c is a constant scalar normally selected in [0,1]. This procedure is further illustrated in Fig. 4.

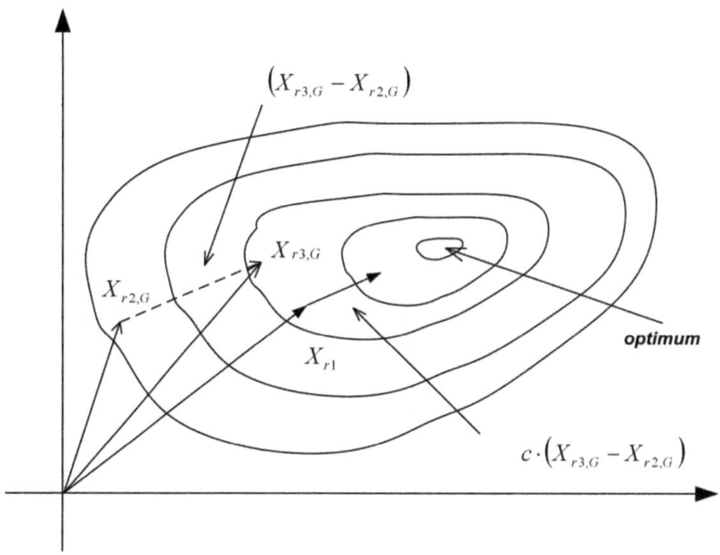

Fig. 4. Illustration of the typical mutation operation in DE

The mutation in DE is integrated with recombination operator. Recombination is a supplementary operator of DE to ensure the diversity of individuals in the trial population. For conventional DE, a probability index *CR* is introduced to control the recombination process. Also, some positions in the individual can be predefined to be compulsory to recombination operation. To demonstrate the procedure of mutation/recombination used in this chapter, the pseudo code of implementing this operation is given in Fig. 5.

Selection

The selection operator of DE is very simple. The trial population obtained from last operation will be presented to the objective function. From the trial population, if the fitness of the i-th individual $X_{i,G}^{TI}$ is better than original individual $X_{i,G}$ in the original population, then $X_{i,G}$ will be replaced by $X_{i,G}^{TI}$; otherwise $X_{i,G}$ will be retained in the running population. The whole optimization procedure of DE is shown in Fig. 6.

To obtain the i-th trial individual

1: Generate three random indices $r1$, $r2$, $r3$, $r4$ where $r1 \neq r2 \neq r3 \neq r4 \neq i$;

2: Obtain the new vector $X_{i,G}^{'}$

$$X_{i,G}^{'} = X_{\text{opt}} + c_1 \cdot \left(X_{r1,G} - X_{r2,G}\right) + c_2 \cdot \left(X_{r3,G} - X_{r4,G}\right)$$

where $c_1 = c_2 = 1$, X_{opt} is the best individual achieved so far.

3: Create a n-dimensional probability index vector PI, all the elements are randomly selected in $[0,1]$.

4: If the k-th index PI_k in PI is smaller than CR, where $k = 1,2,\ldots n$, or the position k is predefined for recombination, then

$$X_{i,G}^{TI}(k) = X_{i,G}^{'}(k),$$

else,

$$X_{i,G}^{TI}(k) = X_{i,G}(k)$$

where the $X_{i,G}^{TI}(k)$, $X_{i,G}^{'}(k)$ and $X_{i,G}(k)$ represents the k-th element of trial, mutated and the original individual respectively.

Fig. 5. Pseudo code of DE mutation/recombination process

Summery of DE

It can be seen that DE only requires few control variables for evolution and the way of generating new individuals is quite simple and straightforward. These features contribute to the efficiency and ease of implementation. In practice, DE has been recognized as a powerful and robust optimizer in many optimization issues especially nonlinear optimization problems (Price et al. 2005; Mezura-Montes et al. 2006). In this chapter, given the nonlinear and nontrivial inherence of the optimization model, the advantages of DE are exploited to optimize the controller parameters of DFIG for small signal stability issue.

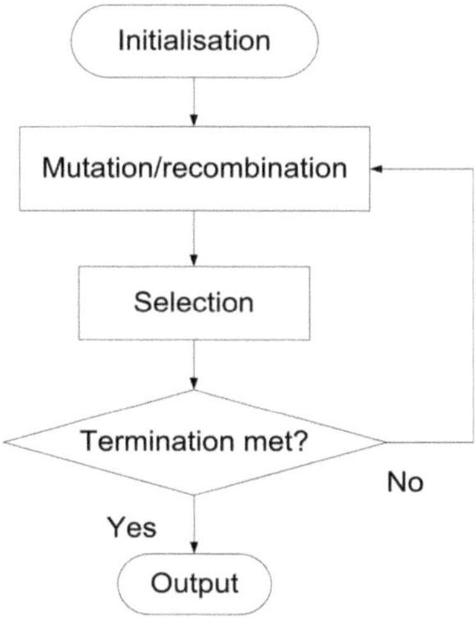

Fig. 6. The general procedure of DE optimization

3.2 *Multi-objective Optimization*

Objective Function

Considering small signal stability, the main purpose of the control system is to increase the system damping ratio as well as to guarantee enough stability margin by placing the real part of all the system eigenvalues as far to the left of the left part of the S-plane. If only the damping ratio is taken as the objective function, the eigenvalues will be limited in the wedge-shape sector as shown in Fig. 7(a),

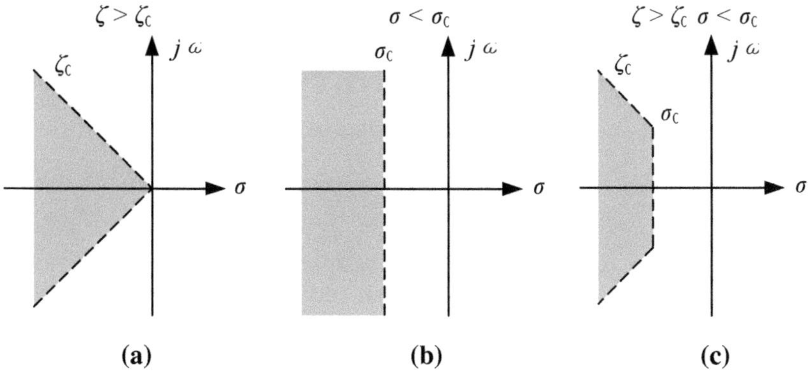

Fig. 7. Eigenvalue location regions for different objective functions. **(a)** Only the damping ratio is taken as the objective function. **(b)** Only the stability margin is taken as the objective function. **(c)** When optimized with both damping ratio and stability margin.

where the real part of some eigenvalues can not be guaranteed less than σ_0. Similarly, if only the stability margin is considered, system eigenvalues will be placed in the region to the left of dashed line as shown in Fig. 7(b), where damping ratio can not be limited larger than ζ_0. When optimized with both damping ratio and stability margin, the eigenvalues can be restricted within a D-shape area as shown in Fig. 7(c), where both robust damping performance and relative stability margin can be achieved (Abdel-Magid and Abido 2003).

Furthermore, DFIG wind turbine system works in the varying wind speed condition at most of the time. Consequently, the eigenvalues related to different operating points, which are changed along with the wind speed, should be considered during the optimal design.

Taking into account the damping ratio and stability margin at numbers of different operating conditions, the objective is formulated as follows:

Maximize,

$$F_1(X) = \min\{\zeta_{\text{ik}}\} \tag{24}$$

Minimize,

$$F_2(X) = \max\{\sigma_{\text{ik}}\} \tag{25}$$

Bound,

$$\text{X}_L \leq \text{X} \leq \text{X}_U \tag{26}$$

where X is the solution vector, ζ_{ik} and σ_{ik} represent the damping ratio and the real part of the i-th eigenvalue for the k-th operating point, respectively. It is obvious that the proposed model is actually a multi-objective optimization problem with incompatible objectives. So it is necessary to find a proper fitness function and method to solve such problem when using DE algorithm.

ε-Constraint Method

A few commonly used classical methods for handling multi-objective optimization problems have been described in (Deb 2001). The weighted-sum method is a common approach to handle the two objectives defined previously in Eqs. (24) and (25). A composite objective F can be derived by simply adding up the two objectives F_1 and F_2 with a user-supplied weight as follow:

Minimize,

$$F = w_1(-F_1) + w_2 F_2 \tag{27}$$

where w_1 and w_2 are the weighting factor assigned to denote the optimization emphasis on objectives F_1 and F_2. Note that objective F_1 is made negative to illustrate it as a maximization problem.

Although the weighted-sum approach is easy to implement, there are some major drawbacks that may affect the final solutions negatively. Due to the fact that both objectives F_1 and F_2 have different units and magnitudes, when summing them up, appropriate normalization of these objectives and weights assignments become exceptionally critical to obtain more precise solutions. Furthermore this weighted-sum approach also suffers from the difficulty in finding certain Pareto-optimal solutions in the case of non-convex objective space.

In order to alleviate the major drawbacks of the weighted-sum approach as mentioned above, a superior multi-objective optimization technique, namely the ε-constraint method, is being employed in this chapter which is capable of identifying true Pareto-optimal region regardless of whether the objective space is convex, non-convex or even discrete.

The ε-constraint method keeps one of the objectives, while restricting the rest of the objectives within user-defined values. In this application, objective F_2 is retained and the problem results in a typical nonlinear single-objective optimization problem, which can then be expressed as follows (Deb 2001):

Minimize,

$$F_2(X) \tag{28}$$

Subject to

$$G(X) = F_1(X) - \varepsilon > 0 \tag{29}$$

and bound

$$X_L \leq X \leq X_U$$

where objective F_1 becomes a soft constraint bounded by a pre-defined ε vector. Note that ε represents the lower bound of objective F_1; and ε is usually chosen in the range within the minimum and maximum values of objective F_1. The solutions obtained from the ε-constraint method are very much dependable on the values chosen for ε; thus inappropriate values used for ε can also produce inaccurate or erroneous results. In this chapter, ε can be selected as the required damping ratio of DFIG wind turbine system.

3.3 Constraint Handle Method

Using the ε-constraint method, the multi-objective optimization problem has been transferred to a constrained single-objective optimization problem. Since constraint handling is not straightforward in the algorithms of EA family, several methods have been proposed for handling constraints in EA in the past few years. They can be classified into four categories (Michalewicz and Schoenauer 1996):

- methods based on preserving feasibility of solutions;
- methods based on penalty functions;
- methods that make a clear distinction between feasible and infeasible solutions; and
- other hybrid methods.

Generally, the method based on penalty function is the most widely applied with all types of nonlinear optimization algorithms. The method is to penalize infeasible solutions, that is, try to solve an unconstrained problem using the modified fitness function.

$$\text{eval}(x) = \begin{cases} f(x) & x \in \mathcal{F} \\ f(x) + \text{penalty}(x) & x \notin \mathcal{F} \end{cases} \tag{30}$$

where penalty(x) is zero if all the solutions are in the feasible region $\mathcal{F}$. For designing the penalty function, there are several methods, which are categorized in (Michalewicz and Schoenauer 1996) as:

- method of static penalties;
- method of dynamic penalties;
- method of annealing penalties;
- method of adaptive penalties;
- death penalty method; and
- segregated genetic algorithm.

In order to design an appropriate penalty, the method of adaptive penalties is applied in this chapter. It uses a penalty function; however, the coefficient of the penalty function uses the feedback from the search process. Each individual is evaluated by (Bean and Hadj-Alouane 1993)

$$\text{eval}(x) = F_2(x) + \lambda(t)[\min\{0, G(x)\}]^2 \tag{31}$$

where $\lambda(t)$ is updated every generation t in the following way:

$$\lambda(t+1) = \begin{cases} \beta_1 \lambda(t) & b' \in \mathcal{S} - \mathcal{F} & \text{for all } t-k+1 \le i \le t \\ (1/\beta_2)\lambda(t) & b' \in \mathcal{F} & \text{for all } t-k+1 \le i \le t \\ \lambda(t) & \text{otherwise} & \end{cases} \tag{32}$$

where b' denotes the best individual, in terms of the evaluation function, in generation i; β_1, $\beta_2 > 1$, and to avoid cycling, $\beta_1 \neq \beta_2$.

This method tries to seek good solutions subject to the constraint by concurrently adjusting λ while running the DE algorithm. λ is selected as a relatively small value initially. After running the DE algorithm for a certain number of generations k, check the top solutions for these k generations. If all the best individuals in the last k generations were infeasible, that means λ is small enough. In this case, λ is increased, and then all solutions of the current generation are re-evaluated with the new λ. If all the best individuals in the last k generations were feasible, λ will be decreased for the next k generations. If there are some feasible and infeasible individuals as the best individuals in the last k generation, λ remains without change. Typically, the increasing rate β_1 is larger than the decreasing rate β_2 to allow for a fast improvement at the early stage of the algorithm.

4 Simulation and Results

The above mentioned DE based multi-objective optimization technique is applied to find out the optimal controller parameters of SMIB DFIG wind turbine system in Fig. 1. The system parameters are listed in Appendix A. The controller parameters to be optimized are X = [$K_{P\omega}$, $T_{I\omega}$, K_{Pir}, T_{Iir}, K_{Pv}, T_{Iv}, K_{PiL}, T_{IiL}], including all the controllers of both the rotor and grid side converters. To obtain robust small signal stability performance at numbers of different operating points, we consider the wind speed range from 5m/s to 12m/s, which covers sub-synchronous, synchronous and super-synchronous speed of DFIG rotor in this optimal design case. The damping ratio constraint ε, which is mentioned in Section 3.2, is selected as 0.3 in this design case. The parameters used during optimization are listed in Appendix B.

The controller parameters with and without optimization are presented in Appendix C. Table 1 shows the eigenvalues along with their damping ratios at three operating points, for both the cases when controllers are with and without optimized parameters. For ease of reference, the eigenvalues at different operating points are also portrayed in the complex S-plane as shown in Fig. 8. It can be seen that with the optimized controller parameters, the number of oscillation modes decreases. All the damping ratios increase. Most of the eigenvalues have shifted to the left in the S-plan. Only one real eigenvalue, which is relative negative, shifts to the right slightly (e.g. from -86.6 without optimization to -80.7 with optimization when V_w=8m/s). This is due to the reason that the ε-constraint method can obtain a compromised optimal solution depending on preference, which is the damping ratio in this case. Although there exists one real eigenvalue shifting to right slightly, the damping ratios are kept within the permissible limit (larger than 0.3). In the meantime, the maximum real part of all the eigenvalues is decreased from -1.12 to -4.0, which means that all the eigenvalues are shift into the region which has larger stability margin.

Table 1. Eigenvalues and damping ratios with and without optimal controller parameters

Operation points	Without optimal design		With optimal design	
	Eigenvalues	Damping	Eigenvalues	Damping
V_w=8m/s,	−50.0±j 415	0.120	−131±j367	0.336
$_r$=0.729p.u.,	−46.0±j163	0.271	−77.0±j229	0.319
sub-synchronous	−23.4±j158	0.147	−148±j198	0.599
speed	−46.1±j59.0	0.616	−24.8±j9.51	0.934
	−13.2±j10.5	0.781	−4.30±j3.71	0.757
	−1.15±j5.14	0.218	−300	
	−86.6		−246	
	−11.9		−114	
	−5.14		−80.7	
	−5.01		−12.3	
	−4.00		−5.46	
			−4.00	
V_w=11m/s,	−50.0±j 415	0.120	−128±j365	0.332
$_r$=1p.u.,	−45.7±j163	0.270	−81.7±j230	0.335
synchronous	−24.6±j159	0.153	−148±j199	0.597
speed	−45.8±j59.0	0.613	−24.9±j9.43	0.935

Table 1. *(Cont.)*

	−13.2±j10.5	0.781	−4.40±j3.74	0.763
	−1.18±j5.14	0.223	−296	
	−86.4		−246	
	−11.9		−114	
	−5.14		−80.8	
	−5.01		−12.3	
	−4.00		−5.25	
			−4.00	
V_w=12m/s,	−49.9±j 415	0.120	−128±j365	0.330
$_r$=1.09p.u.,	−45.5±j163	0.269	−83.6±j229	0.343
super-	−25.1±j159	0.155	−148±j199	0.596
synchronous	−45.8±j58.9	0.614	−24.9±j9.41	0.935
speed	−13.1±j10.5	0.781	−4.45±j3.75	0.764
	−1.18±j5.14	0.225	−294	
	−86.3		−248	
	−11.9		−114	
	−5.15		−80.8	
	−5.01		−12.3	
	−4.00		−5.18	
			−4.00	

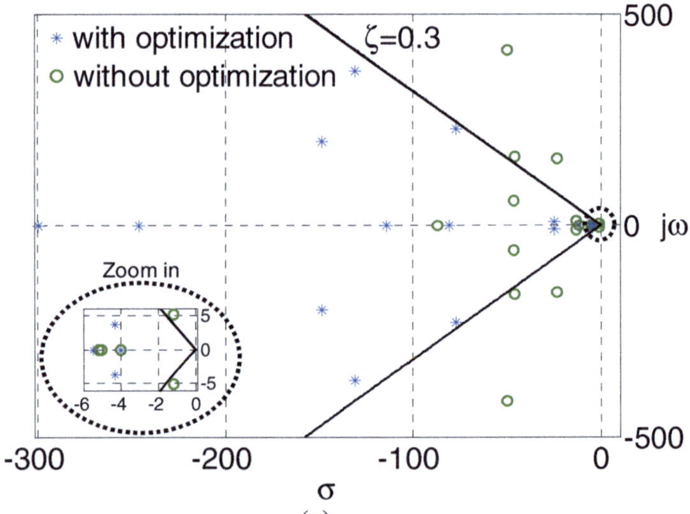

(a)

Fig. 8. Eigenvalues associated with and without optimal controller parameters. **(a)** V_w=8m/s, sub-synchronous speed. **(b)** V_w=11m/s, synchronous speed **(c)** V_w=12m/s, super-synchronous speed.

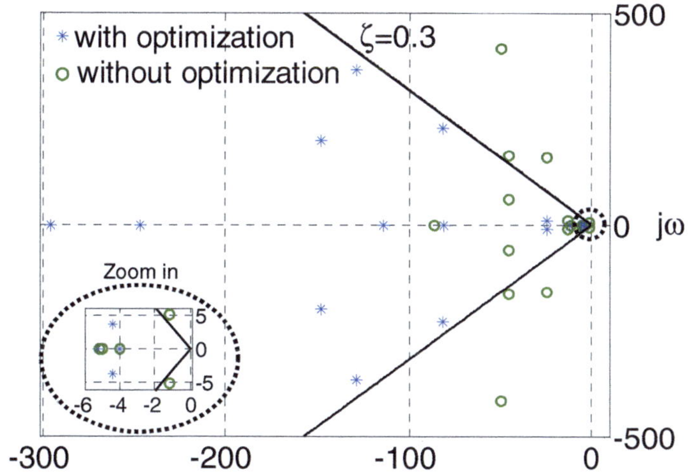

(b)

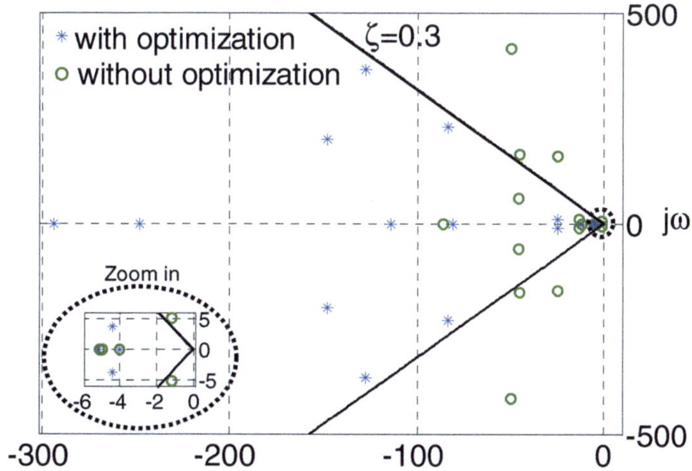

(c)

Fig. 8. (*continued*)

In order to verify the effectiveness of the optimal controller design, dynamic simulations are carried out in Matlab/Simulink environment to observe the response of the DFIG wind turbine system under small perturbation. The system is subjected to small disturbance by a small increase of 4.3% (from 11.5m/s to 12m/s) in the wind speed at the 5th second. The dynamic responses of the rotor speed, electrical torque, output active power, reactive power, terminal voltage and DC-link voltage with and without optimized controller parameters are shown in Figs. 9 (a)-(f), respectively. It can be seen that, with optimized controller parameters, the dynamic performance of the studied DFIG wind turbine system is well improved where the oscillation after the small-disturbance is well damped.

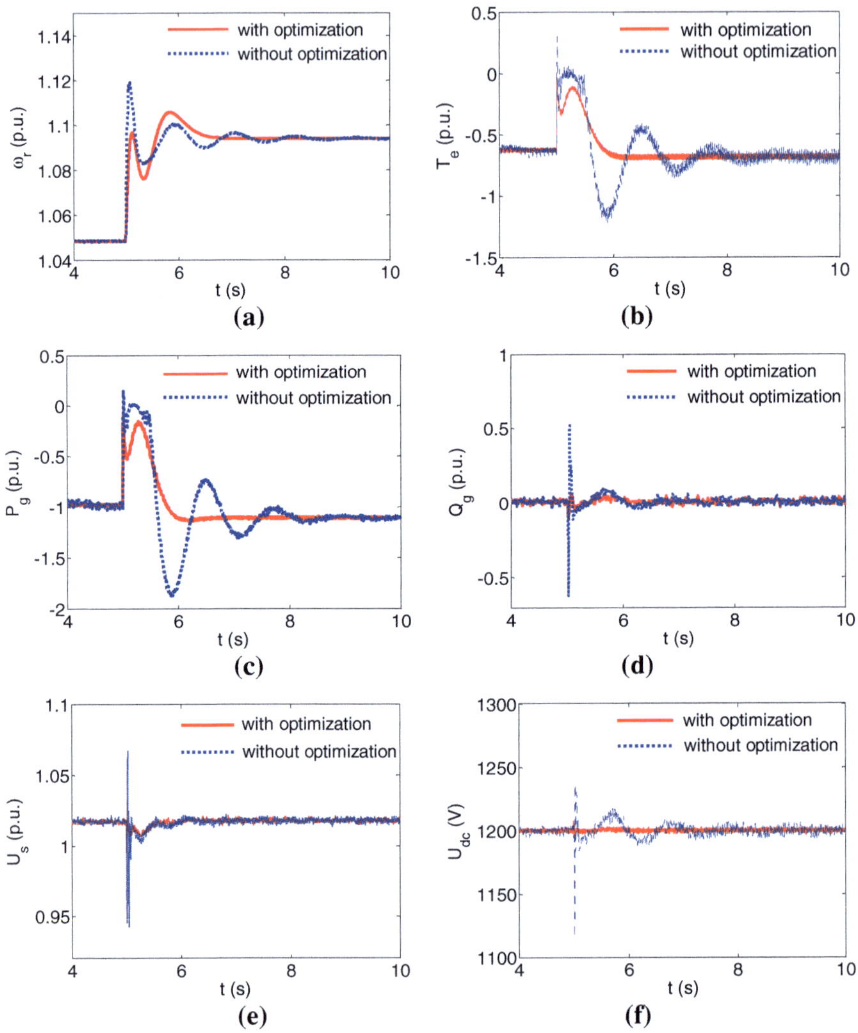

Fig. 9. Responses of DFIG wind turbine under small increase of wind speed. **(a)** Rotor speed. **(b)** Electrical torque. **(c)** Output active power. **(d)** Output reactive power. **(e)** Terminal voltage. **(f)** DC-link voltage.

5 Summary

Optimal controller design is necessary for the coordinated tuning of the controllers' parameters of the doubly fed induction generator (DFIG) wind turbine system. A differential evolution (DE) based multi-objective method has been proposed to optimize the paramters of the controllers. In this optimization, multi-objective including the damping ratios and real parts of eigenvalues at a number of operation points, are considered so that both robust damping performance and relative stability margin for changed operating conditions can be achieved. The superior ε-constraint method is applied to deal with the proposed multi-objective problem. It retains the stability margin as the objective, while restricts the damping ratio within a pre-defined value. Using adaptive penalties method, the appreciate penalty function has been designed so as to handle constraints with DE.

Taking DFIG SMIB system as a sample system, the detailed system model including induction generator, two-mass drive train, pitch control, close-loop vector control etc. is presented. Based on this model, the optimal controller parameters of both the rotor and grid side converters are obtained using the effective DE based multi-objective optimization method. Eigenvalue analysis and simulation results show that damping performance and the stability margin are well improved simultaneously with the optimized controller parameters.

References

Abdel-Magid, Y.L., Abido, M.A.: Optimal multiobjective design of robust power system stabilizers using genetic algorithms. IEEE Transactions on Power Systems 18(3), 1125–1132 (2003)

Akhmatov, V.: Induction Generators for Wind Power. Multi-science Publishing Co. Ltd., Brentwood (2005)

Bean, J.C., Hadj-Alouane, A.B.: A dual genetic algorithm for bounded integer programs. Technical Report TR 92-53. Ann Arbor, MI: University of Michigan, Department of Industrial and Operations Engineering (1993)

Deb, K.: Multi-objective optimization using evolutionary algorithms, 1st edn. John Wiley & Sons, New York (2001)

Eriksen, P.B., Ackermann, T., Abildgaard, H., Smith, P., Winter, W., Rodriguez Garcia, J.M.: System operation with high wind penetration. IEEE Power and Energy Magazine 3(6), 65–74 (2005)

Hansen, A.D., Sørensen, P., Iov, F., Blaabjerg, F.: Control of variable speed wind turbines with doubly-fed induction generators. Wind Engineering 28(4), 411–434 (2004)

Hansen, A.D.: Generators and power electronics for wind turbines. In: Ackermann, T. (ed.) Wind Power in Power systems. John Wiley&Sons, Ltd., New York (2005)

Kundur, P.: Power System Stability and Control. McGrawHill, New York (1994)

Lampinen, J., Zelinka, I.: Mechanical engineering design optimization by differential evolution. In: Corne, D., Dorigo, M., Glover, F., et al. (eds.) New ideas in optimization. McGraw-Hill Ltd., UK (1999)

Ledesma, P., Usaola, J.: Doubly fed induction generator model for transient stability analysis. IEEE Transactions on Energy Conversion 20(2), 388–397 (2005)

Mei, F., Pal, B.C.: Modal analysis of grid-connected doubly fed induction generators. IEEE Transactions on Energy Conversion 22(3), 728–736 (2007)

Mei, F., Pal, B.C.: Modelling of Doubly-fed induction generator for power system stability study. In: Power and Energy Society General Meeting, pp. 1–8 (2008)

Mezura-Montes, E., Velazquez-Reyes, J., Coello, C.A.C.: Comparing differential evolution models for global optimization. In: Genetic and Evolutionary Computation Conference, GECCO 2006 (2006)

Michalewicz, Z., Schoenauer, M.: Evolutionary algorithms for constrained parameter optimization problems. Evolutionary Computation 4(1), 1–32 (1996)

Mishra, Y., Mishra, S., Tripathy, M., Senroy, N., Dong, Z.Y.: Improving stability of a DFIG based wind power system with tuned damping controller. IEEE Transactions on Energy Conversion 24(3), 650–660 (2009)

Muller, S., Deicke, M., De Doncker, R.W.: Doubly fed induction generator system for wind turbines. IEEE Industry Applications Magazine 8(3), 26–33 (2002)

Pena, R., Clare, J.C., Asher, G.M.: Doubly fed induction generator using back-to-back PWM converters and its application to variable speed wind-energy generation. IEE Proceedings on Electric Power Applications 143(3), 231–241 (1996)

Price, K.V.: Differential evolution: a fast and simple numerical optimizer. In: Annual Conference of the North American Fuzzy Information Processing Society - NAFIPS, Berkeley, CA, USA, pp. 524–527 (1996)

Price, K., Storn, R.: Differential evolution. Dr. Dobb's Journal 22(4), 18–20 (1997)

Price, K.V., Storn, R.M., Lampinen, J.A.: Differential evolution: a practical approach to global optimization. Springer, Berlin (2005)

Qiao, W., Venayagamoorthy, G.K., Harley, R.G.: Design of optimal PI controllers for doubly fed induction generators driven by wind turbines using particle swarm optimization. In: Proceeding of International Joint Conference on Neural Network, Canada, pp. 1982–1987 (2006)

Storn, R., Price, K.: Differential evolution- a simple and efficient adaptive scheme for global optimization over continuous spaces. Berkeley, CA, ICSI Technical Report TR-95-012 (1995)

Storn, R., Price, K.: Minimizing the real functions of the ICEC 1996 contest by differential evolution. In: Proceedings of IEEE International Conference on Evolutionary Computation, pp. 842–844 (1996)

Vieira, J.P.A., Nunes, M.V.A., Bezerra, U.H., do Nascimento, A.C.: Designing optimal controllers for doubly fed induction generators using genetic algorithm. IET Generation, Transmission & Distribution 3(5), 472–484 (2009)

Wong, K.P., Dong, Z.Y.: Differential Evolution, an Alternative Approach to Evolutionary Algorithm. In: Lee, K.Y., El-Sharkawi, M.A. (eds.) Modern Heuristic Optimization Techniques: Theory and Applications to Power Systems. Wiley, New York (2008) (invited)

Wu, F., Zhang, X.P., Godfrey, K., Ju, P.: Small signal stability analysis and optimal control of a wind turbine with doubly fed induction generator. IET Generation, Transmission & Distribution 1(5), 751–760 (2007)

Yamamoto, M., Motoyoshi, O.: Active and reactive power control for doubly-fed wound rotor induction generator. IEEE Transactions on Power Electronics 6(4), 624–629 (1991)

Yang, G.Y., Dong, Z.Y., Wong, K.P.: A modified differential evolution algorithm with fitness sharing for power system planning. IEEE Transactions on Power System 23(2), 514–522 (2008)

Appendix A: Parameters of DFIG Wind Turbine System

Description	Value	Description	Value
base power:	S_{base}=1.5MW	base frequency:	ω_{base}=314rad/s
rated power of DFIG:	S_{rated}=1.5MW	rated rotor speed:	ω_{rated}=314rad/s
stator frequency:	ω_s=1p.u.	stator voltage:	U_s=575V
stator resistance:	R_s=0.00706p.u.	leakage inductance of stator:	L_{ls}=0.171p.u.
rotor resistance:	R_r=0.005p.u.	leakage inductance of rotor:	L_{lr}=0.156p.u.
mutual inductance:	L_m=3.5p.u.	DC-link capacitor:	C_{dc}=0.06F
inertia constant of generator:	H_g=0.5s	friction coefficient of generator:	B=0.01p.u.
inertia constant of wind turbine:	H_t=3s	shaft stiffness coefficient of wind turbine:	K_{sh}=0.5
damping coefficient of wind turbine:	D_{sh}=0.01	optimal constant of wind turbine:	K_{opt}=0.579
time constant of the pitch servo:	T_β=0.25s	rated wind speed:	V_{wrated}=12m/s
resistance of grid side incuctor:	R_L=0.003p.u.	inductance of grid side inductor:	L=0.3p.u.
DC-link voltage reference:	U_{dcref}=1200V	voltage of the infinite bus:	E_0=25kV
resistance of the transformer and transmission line	R_{TL}=4Ω	inductance of the transformer and the transmission line:	L_{TL}=0.085H

Appendix B: Parameters Used for DE Base Multi-objective Optimization

- DE parameters:

1. population size *NP*: 50;
2. maximum number of generations: 150;
3. crossover constant *CR*: 0.5;
4. weighting factor *F*: 0.8

- Additional parameters:

5. Lower bounds for controller parameters:
   X_L= [1, 0.01, 0.01, 0.001, 0.01, 0.001, 0.01, 0.001].
6. Upper bounds for controller parameters:
   X_U= [200, 1, 10, 1, 1, 1, 1, 1].
7. ε-constraint parameter: $\varepsilon=\zeta_0=0.3$.
8. parameters for adaptive penalty method:
   $\beta_1=4$, $\beta_2=2.8$, $\lambda_0=0.1$, $k=16$.

Appendix C: Controller Parameters with and without Optimal Design

- Without optimization:
  $K_{P\omega}$=25, $T_{I\omega}$=0.15, K_{Pv}=1, T_{Iv}=0.1, K_{Pir}=0.1, T_{Iir}=0.003, K_{PiL}=0.1, T_{IiL}=0.2.
- With optimization:
  $K_{P\omega}$=22.67, $T_{I\omega}$=0.5, K_{Pv}=0.83, T_{Iv}=0.067, K_{Pir}=0.33, T_{Iir}=0.0046, K_{PiL}=0.39, T_{IiL}=0.016.

Eigenvalue Analysis of a DFIG Based Wind Power System under Different Modes of Operations

Y. Mishra, S. Mishra, Fangxing Li, and Z. Y. Dong

Abstract. This chapter discussed the various mode of operation of the Doubly Fed Induction Generator (DFIG) based wind farm system. The impact of a auxiliary damping controller on the different modes of operation for the DFIG based wind generation system is investigated. The co-ordinated tuning of the damping controller to enhance the damping of the oscillatory modes using Bacteria Foraging (BF) technique is presented. The results from eigenvalue analysis are presented to elucidate the effectiveness of the tuned damping controller in the DFIG system under Super/Sub-synchronous speed of operation. The robustness issue of the damping controller is also investigated.

1 Introduction

Increasing power generation from renewable sources such as wind would help reduce carbon emissions, hence minimize the effect on global warming. Increasing steps have been taken by the various utilities/states across the world to achieve the above mentioned goal. Most of the states in USA has Renewable Portfolios Standard (a state policy aiming at obtaining certain percentage of the their power form

Y. Mishra
The University of Tennessee, Knoxville
e-mail: ymishra@utk.edu

S. Mishra
Indian Institute of Technology Delhi, India
e-mail: sukumar@ee.iitd.ac.in

Fangxing Li
The University of Tennessee, Knoxville
e-mail: fli6@utk.edu

Z.Y. Dong
Department of Electrical Engineering, Hong Kong Polytechnic University, Hong Kong
e-mail: eezydong@polyu.edu.hk

L. Wang et al. (Eds): *Wind Power Systems*, Green Energy and Technology, pp. 191–213.
springerlink.com

renewable energy sources by certain date) ranging from 10-20% of total capacity by 2020 [1]. This increasing penetration of renewable sources of energy, in particular wind energy conversion systems (WECS), in the conventional power system has put tremendous challenge to the power system operators/planners, who have to ensure the reliable and secure grid operation. As power generation from WECS is significantly increasing, it is of paramount importance to study the effect of wind integrated power systems on overall system stability.

The Doubly Fed Induction Generator (DFIG) has been popular among various other techniques of wind power generation, because of its higher energy transfer capability, low investment and flexible control [2]. DFIG is different from the conventional induction generator in a way that it employs a series voltage-source converter to feed the wound rotor. The feedback converters consist of a Rotor Side Converter (RSC) and a Grid Side Converter (GSC). The control capability of these converters give DFIG an additional advantage of flexible control and stability over other induction generators.

The decoupled control of DFIG has controllers to track various reference variables namely P_{ref}, V_{sref}, V_{dcref} and q_{cref}. These controllers are required to maintain maximum power tracking, stator terminal voltage, DC voltage level and GSC reactive power level respectively. The co-ordinated tuning of these controllers by hit and trial method is a cumbersome job. The co-ordinated tuning using particle swarm optimization (PSO) has been proposed [14, 18]. However, the damping of low frequency oscillatory modes were not given due importance. Moreover, the operation of DFIG under various operating conditions is not emphasized in [18]. The effect of optimized controllers on different mode of operation is also lacking in [18].

The impact of wind generation on the oscillatory modes is presented in [8, 3, 17, 4]. The auxiliary control loop for oscillation damping that adjusts the active power command to damp the inter area oscillation is proposed in [8, 3]. Moreover, a power system stabilizer using a speed deviation is proposed in [4]. It is reported that the presence of the PSS in the DFIG system improves the damping of the oscillations in the network. Nevertheless, it is very important to optimize the controller parameters of the PSS to achieve the best performance. However, the co-ordinated tuning of these controllers is not presented. Moreover, it is necessary to study the impact of these damping controllers under Super/Sub-synchronous mode of operation.

In this chapter, the auxiliary signal derived from ω_r is added to the rotor phase angle control to enhance the low frequency damping of the system. This simple PI controller is called damping controller. Moreover, all the DFIG controllers for tracking P_{ref}, V_{sref}, V_{dcref} and q_{cref} are implemented in this paper. Hence, the co-ordinated effect of these controllers on the system damping is examined. The effectiveness of damping controller under Super/Sub-synchronous modes of operation is also investigated. The issue of robustness in the performance of the damping control is discussed.

Wind speed is seldom constant and hence proper tuning of its control parameters is necessary for stable and reliable operation of wind generators. For many generation companies investing in wind power generation, wind speed is among the most important factors in determining the investment decisions. The impact of wind speed can significantly change the output of the wind generator over short time, this again cause significant challenges for operation and scheduling of the system especially with respect to stability of the system. For example, if wind gust happens during low demand period when other generators had been scheduled to supply a limited amount of load at a region, sudden increase of wind speed may results into 20-50% increase of total generation in that region because of the sudden increase of wind generations. In fact, for many power companies & system operators, this is an open problem/difficulty in their control of wind generation. The proposed tunning method in this chapter provides a very useful input to help power companies/operators in wind power generation operations, control and management. The proposed tuning method can be a key approach to ensure stable operation.

The contributions of this chapter are: (i) to study the impact of tuned damping controller on the electromechanical modes, (ii) to study its impact under Super/Sub-synchronous mode of operation, (iii) to propose the optimally tuned damping controller which is effective under variable operating condition. This paper is structured as follows : section 2 presents the modeling of the DFIG system. The detailed control methodology is discussed in section 3 with special emphasis on damping controller. Section 4 describes the bacteria foraging algorithm for the optimization of the controllers parameters. Section 5 discusses simulation and results followed by conclusions in section 6.

2 Modeling of DFIG

This section deals with the modeling of the DFIG based WT generation system. The grid connected single machine infinite bus system is as shown in Fig. 1. The stator and rotor voltages of the doubly excited DFIG are supplied by the grid and the power converters respectively. Simulation of the realistic response of the DFIG system requires the modeling of the controllers in addition to the main electrical and mechanical components. The components considered include, (i) turbine, (ii) drive train, (iii) generator and (iv) converter system.

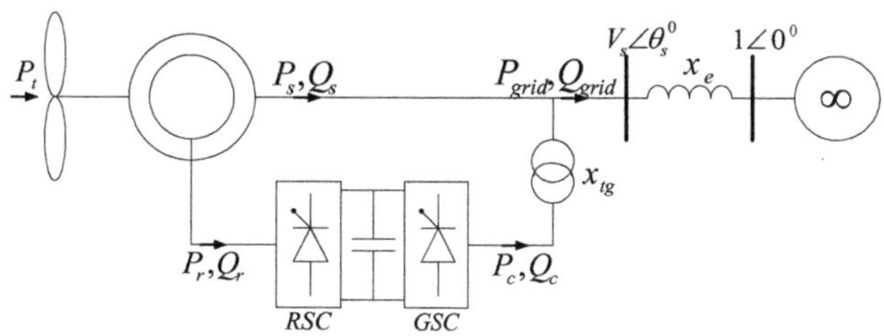

Fig. 1. A DFIG system

2.1 Turbine

The turbine in DFIG system is the combination of blades and hub. Its function is to convert the kinetic energy of the wind into the mechanical energy, which is available for the generator. In general the detailed models of the turbine are used for the purpose of design and mechanical testing only. The stability studies done in this paper do not require detailed modeling of the wind turbine blades and hence it is neglected in this paper. The mechanical power input to the WT is considered as constant, i.e. the wind speed and the blade pitch angle do not change during the period of study.

2.2 Drive Train

In stability studies, when the response of a system subjected to any disturbance is analyzed, the drive train system should be modeled as a series of rigid disks connected via massless shafts. A single or lumped mass model is used for the small signal analysis for the conventional synchronous generator as the drive train behaves as a single equivalent mass. This is because of the much greater mechanical stiffness than the equivalent "electrical stiffness" [15]. However, the presence of gearbox in the DFIG system, makes the shaft more slender resulting in a mechanical stiffness of the same order as that of equivalent "electrical stiffness". Thus, there is no mode for which the drive train behaves as a single mass. Hence, multi-mass drive train model must be considered for the stability studies of DFIG system.

In this work, the two mass drive train model is considered and the dynamics can be expressed by the differential equations below [7],

$$2H_t \frac{d\omega_t}{dt} = T_m - T_{sh} \tag{1}$$

$$\frac{1}{\omega_{elb}} \frac{d\theta_{tw}}{dt} = \omega_t - \omega_r \tag{2}$$

$$2H_g \frac{d\omega_r}{dt} = T_{sh} - T_e \tag{3}$$

where, the electrical (T_e) and shaft (T_{sh}) torque are given by $T_e = \frac{P_s}{\omega_s}$ and $T_{sh} = K_{sh}\theta_{tw}$ respectively. The wind torque, T_m, is considered constant in this work [6]. The expression for the wind torque is given by:

$$T_m = \frac{0.5\rho\pi R^2 C_p V_w^3}{\omega_t} \tag{4}$$

where the power coefficient, C_p, is defined as:

$$C_p = \frac{1}{2}\left(\frac{RC_f}{\lambda} - 0.022\beta - 2\right) e^{-0.255RC_f/\lambda} \tag{5}$$

And the blade tip speed ratio, λ, is defined as:

$$\lambda = \frac{\omega_t R}{V_w} \tag{6}$$

where,

θ_{tw} shaft twist angle (rad);
ρ air density;
R wind blade radius;
V_w wind speed;
C_f blade design constant coefficient;
β blade pitch angle;
λ blade tip speed ratio;
C_p power coefficient;
H_t inertia constant of turbine (s);
H_g inertia constant of the generator (s);
ω_t wind Turbine angular speed (p.u.);
ω_r generator angular speed (p.u.);
ω_s synchronous speed (p.u.);
ω_{elb} electrical base speed (rad/s);
K_{sh} shaft stiffness (p.u./el.rad);

At low wind speed, a wind turbine generates no power at all, because the airflow contains too less energy. The electrical output of the generator at different wind speeds is shown in Fig. 2. The cut-in speed is 3-5 m/s and the nominal power can be extracted between 11m/s to 16m/s. The aerodynamical efficiency of the rotor depends on the tip speed ratio (blade speed divided by the wind speed). Tip speed ratio is between 6 and 8 for maximum aerodynamic efficiency. This ratio can not be changed in the case of fixed wind speed turbines and that is the reason why variable speed turbines have higher aerodynamical efficiency.

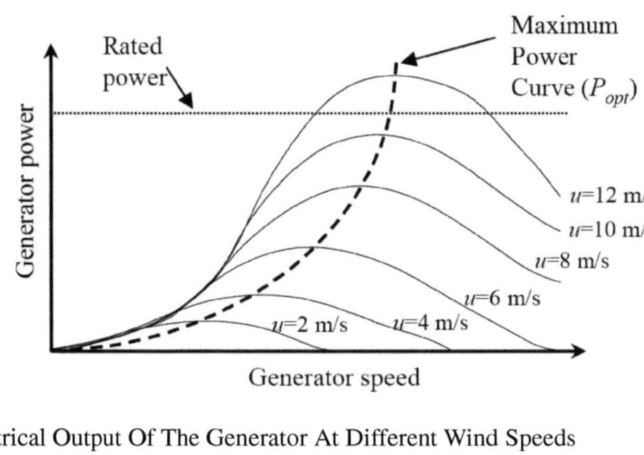

Fig. 2. Electrical Output Of The Generator At Different Wind Speeds

2.3 Generator

The most common way of representing DFIG for the purpose of simulation and control is in terms of direct and quadrature axes (dq axes) quantities, which form a reference frame that rotates synchronously with the stator flux vector [6]. The various variables are defined as: $e'_{qs} = K_{mrr}\omega_s\psi_{dr}$, $e'_{ds} = -K_{mrr}\omega_s\psi_{qr}$, $L'_s = L_{ss} - \frac{L_m^2}{L_{rr}}$, $T_r = L_{rr}/R_r$, $K_{mrr} = L_m/L_{rr}$, and $\omega_e = \omega_{elb}\omega_s$.

For balanced and unsaturated conditions, the corresponding p.u. DFIG model can be expressed as [6],

$$\frac{\omega_s L'_s}{\omega_e}\frac{di_{qs}}{dt} = -R_1 i_{qs} + \omega_s L'_s i_{ds} + \frac{\omega_r}{\omega_s}e'_{qs} - \frac{1}{T_r\omega_s}e'_{ds} - v_{qs} + K_{mrr}v_{qr} \tag{7}$$

$$\frac{\omega_s L'_s}{\omega_e}\frac{di_{ds}}{dt} = -R_1 i_{ds} - \omega_s L'_s i_{qs} + \frac{\omega_r}{\omega_s}e'_{ds} + \frac{1}{T_r\omega_s}e'_{qs} - v_{ds} + K_{mrr}v_{dr} \tag{8}$$

$$\frac{1}{\omega_e}\frac{de'_{qs}}{dt} = R_2 i_{ds} - \frac{1}{T_r\omega_s}e'_{qs} + \left(1 - \frac{\omega_r}{\omega_s}\right)e'_{ds} - K_{mrr}v_{dr} \tag{9}$$

$$\frac{1}{\omega_e}\frac{de'_{ds}}{dt} = -R_2 i_{qs} - \frac{1}{T_r\omega_s}e'_{ds} - \left(1 - \frac{\omega_r}{\omega_s}\right)e'_{qs} + K_{mrr}v_{qr} \tag{10}$$

where

$$R_2 = K_{mrr}^2 R_r$$
$$R_1 = R_s + R_2$$

where the parameters used above are explained below;

e'_{ds}, e'_{qs} d and q axis voltages behind-transient reactance (p.u.);
ψ_{dr}, ψ_{qr} d and q axis rotor fluxes (p.u.);
i_{ds}, i_{qs} d and q axis stator currents (p.u.);
v_{ds}, v_{qs} d and q axis stator voltages (p.u.);
i_{dr}, i_{qr} d and q axis rotor currents (p.u.);
i_{dg}, i_{qg} d and q axis currents of the Grid side Converter (p.u.);
v_{dg}, v_{qg} d and q axis voltages of the Grid side Converter (p.u.);
v_{dr}, v_{qr} d and q axis rotor voltages (p.u.);
v_{dc}, i_{dc} voltage and current of DC capacitor (p.u.);

2.4 Converter Model

The converter model in DFIG system comprises of two pulse width modulation invertors connected back to back via a dc link. The rotor side converter (RSC) is a controlled voltage source since it injects an AC voltage at slip frequency to the rotor. The grid side converter (GSC) acts as a controlled current source since it injects an AC current at grid frequency to the grid and maintains the dc link voltage constant.

The power balance equation for the converter model can be written as:

$$P_r = P_g + P_{dc} \tag{11}$$

where P_r, P_g, P_{dc} are the active power at RSC, GSC and DC link respectively, which can be expressed as,

$$P_r = v_{dr}i_{dr} + v_{qr}i_{qr} \tag{12}$$

$$P_g = v_{dg}i_{dg} + v_{qg}i_{qg} \tag{13}$$

$$P_{dc} = v_{dc}i_{dc} = Cv_{dc}\frac{dv_{dc}}{dt} \tag{14}$$

The details of converter controllers are discussed in the later section.

3 Controllers for DFIG

This section describes the controllers used for the DFIG system. As mentioned above, there are two back to back converters hence we need to control these two converter sides. Primarily, these controller are known as RSC and GSC controllers. This section also introduces a new auxiliary control signal which is added to the angle control in the RSC to enhance the damping. This is known as damping control.

3.1 RSC Controllers

The phasor diagram in Fig. 3 describes the control scheme (based on FMAC), for the RSC controller.

The magnitude of the e_{ig}, internally generated voltage vector in the stator, depends on the magnitude of the rotor flux vector, ψ_r. This flux can be controlled by

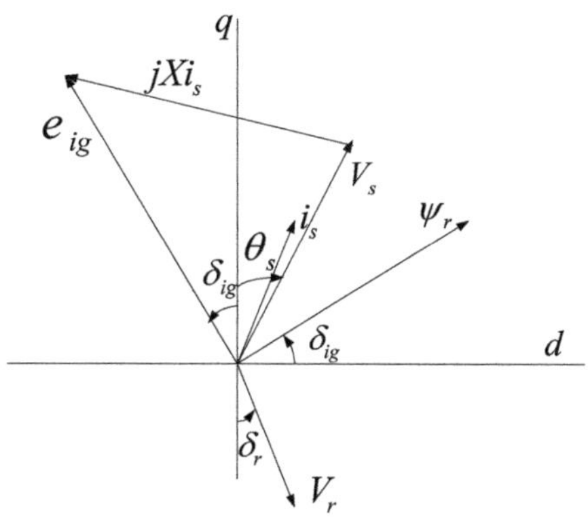

Fig. 3. Phasor diagram illustrating the operation of DFIG system [4]

V_r, the rotor voltage. The angle δ_{ig}, between the voltage vectors e_{ig} and V_s (stator terminal voltage and hence q-axis of the reference frame), is determined by the power output of the DFIG. Since vector e_{ig} is orthogonal to ψ_r, the angle between d-axis and ψ_r is also given by δ_{ig}. The adjustment of the magnitude of the rotor voltage vector, $|V_r|$ and its phase angle, δ_r, is employed for the control of terminal voltage and electrical power respectively [4].

The configuration of the feedback controllers for the DFIG system is as shown in the Fig. 4. The RSC controller is as shown in the Fig. 4(a). One part aims at controlling the active power so as to track the P_{ref} while the second part is to maintain the terminal voltage.

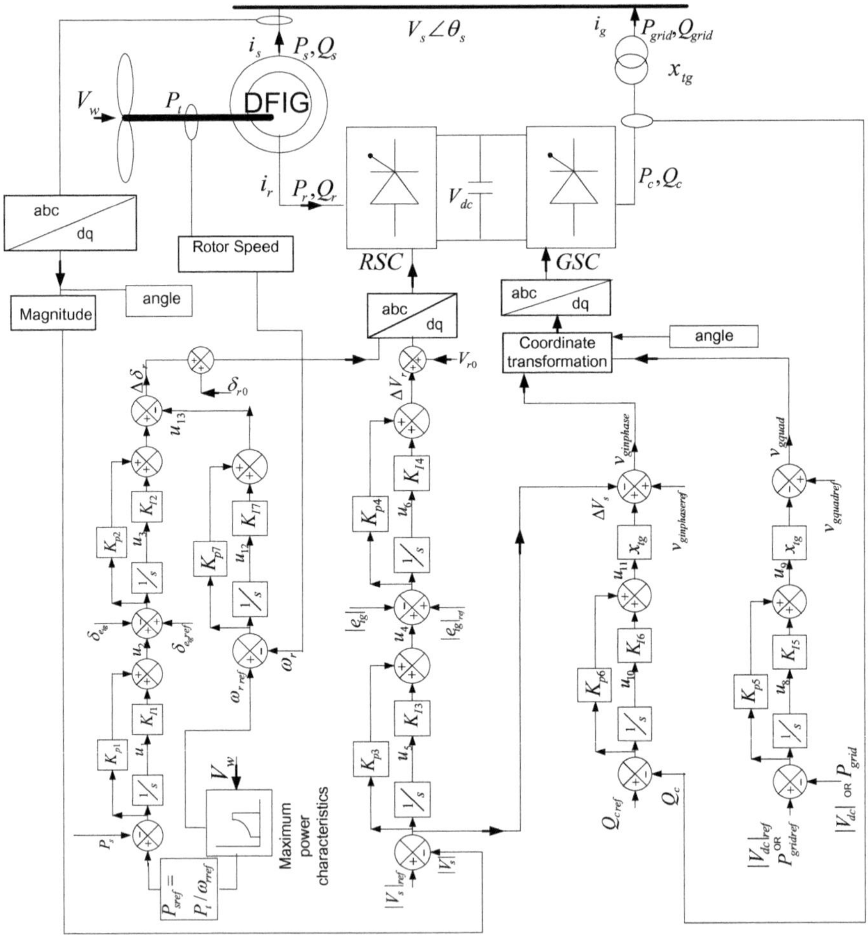

Fig. 4. Control scheme for the DFIG system (a) RSC, (b) GSC, (c) Damping controller

The P_{ref} is determined by the wind turbine power speed characteristic ($C_p - \lambda$ curve) for maximum power extraction [15]. Under normal operating condition, the active power set-point, P_{ref} for the RSC is defined by the maximum power tracking point, which is a function of optimal generation speed. Mathematically, the above mentioned concept can be expressed by the set of differential equations as below,

$$\frac{du_1}{dt} = P_{ref} - P \tag{15}$$

$$u_2 = K_{p1}(P_{ref} - P) + K_{I1}u_1 \tag{16}$$

$$\frac{du_3}{dt} = (\delta_{e_{ig}ref} + u_2 - \delta_{e_{ig}}) \tag{17}$$

$$u_4 = K_{p3}(V_{sref} - V_s) + K_{I3}u_5 \tag{18}$$

$$\frac{du_5}{dt} = V_{sref} - V_s \tag{19}$$

$$\frac{du_6}{dt} = (\left|e_{ig}\right|_{ref} + u_4 - \left|e_{ig}\right|) \tag{20}$$

$$\Delta\left|V_r\right| = K_{p4}(\left|e_{ig}\right|_{ref} + u_4 - \left|e_{ig}\right|) + K_{I4}u_6 \tag{21}$$

$$\frac{du_{12}}{dt} = \omega_{rref} - \omega_r \tag{22}$$

$$u_{13} = K_{p7}(\omega_{rref} - \omega_r) + K_{I7}u_{12} \tag{23}$$

$$\Delta\delta_r = (K_{p2}(\delta_{e_{ig}ref} + u_2 - \delta_{e_{ig}}) + K_{I2}u_3) - u_{13} \tag{24}$$

where K_{pi} and K_{Ii} are the proportional and integral gain constant respectively for i^{th} PI controller. The internal generated voltage vector, e_{ig} is, $\left|e_{ig}\right| = \sqrt{e'^2_{ds} + e'^2_{qs}}$ and the angle is defined as $\delta_{e_{ig}} = \tan^{-1}\left(e'_{ds}/e'_{qs}\right)$. The controller variables in Fig. 4(a) are added to their corresponding reference values to obtain updated values, which can be expressed as $|V_r| = |V_r|_{ref} + \Delta|V_r|$ and $\delta_r = \delta_{r\_ref} + \Delta\delta_r$ respectively.

3.2 GSC Controllers

The GSC controller scheme is represented in Fig. 4(b). The reference signal for the dc voltage, V_{dcref}, is set to a constant value independent of the wind speed. And V_{dc} is regulated by the following equation:

$$\frac{dV_{dc}}{dt} = \frac{1}{V_{dc}C}\left(v_{dr}i_{dr} + v_{qr}i_{qr} - v_{dg}i_{dg} - v_{qg}i_{qg}\right) \tag{25}$$

$$\frac{du_8}{dt} = V_{dcref} - V_{dc} \tag{26}$$

$$u_9 = K_{P5}(V_{dcref} - V_{dc}) + K_{I5}u_8 \tag{27}$$

The reactive power set point, q_{cref}, is set to zero, to reduce the GSC power rating. This implies that GSC only exchanges active power with the grid and hence the reactive power transmission to the grid by DFIG is only through the stator.

$$\frac{du_{10}}{dt} = q_{cref} - q_c \tag{28}$$

$$u_{11} = K_{P6}(q_{cref} - q_c) + K_{I6}u_{10} \tag{29}$$

The inphase and quadrature component of the GSC voltage is modified by

$$v_{ginphase} = v_{ginphaseref} + u_{11}x_{tg} - (V_{sref} - V_s) \tag{30}$$

$$v_{gquad} = v_{gquadref} - u_9 x_{tg} \tag{31}$$

where $v_{ginphaseref} = V_s + i_{cgquadref} * x_{tg}$ and $v_{gquad} = i_{cginphaseref} * x_{tg}$. And x_{tg} is the 3-winding transformer reactance between GSC and the stator terminal. $i_{cginphaseref}$ and $i_{cgquadref}$ are the inphase and quadrature component of GSC current to the stator terminal voltage defined as $i_{cginphaseref} = P_r/V_s$ and $i_{cgquadref} = (v_{ds} * i_{qg} - v_{qs} * i_{dg})/V_s$. The corresponding GSC control scheme is implemented in this chapter.

3.3 Damping Controller

Damping controller is employed in the RSC by (20) as shown in the Fig. 4(c). The auxiliary signal u_{13} is added to the angle control of the RSC controller to enhance the damping of low frequency angular oscillations. The auxiliary signal helps in increasing the damping torque by controlling the angular position of the rotor flux vector with respect to the stator flux vector.

Thus, in summary, the state equations of the DFIG are (1-7), while RSC and GSC controller state equations are (15), (17), (19), (20), (25), (26) and (28). The damping controller state equation is (22). Hence, there are total 15 states of the DFIG system including the damping controller.

4 Bacteria Foraging for the Optimal Control Of DFIG System

The idea of BF is based on the fact that, natural selection tends to eliminate animals with poor foraging strategies and favor the propagation of genes of those animals that have successful foraging strategies since they are more likely to enjoy reproductive success. After many generations, poor foraging strategies are either eliminated or reshaped into good ones. The E. coli bacteria that are present in our intestines also undergo a foraging strategy. The control of these bacteria is basically governed by four processes namely Chemotaxis, Swarming, Reproduction, Elimination and Dispersal [13].

a) *Chemotaxis*: This process is achieved through swimming and tumbling. Depending upon the rotation of the flagella in each bacterium it decides whether it should move in a predefined direction (swimming) or an altogether different direction (tumbling), in the entire lifetime of the bacterium. To represent a tumble, a unit length random direction, say $\phi(j)$, is generated; this will be used to define the direction of movement after a tumble. In particular

$$\theta^i(j+1,k,l) = \theta^i(j,k,l) + Cl(i)\phi(j) \tag{32}$$

Where $\theta^i(j,k,l)$ represents the ith bacterium at j^{th} chemotactic k^{th} reproductive and l^{th} elimination and dispersal step. $Cl(i)$ is the size of the step taken in the random direction specified by the tumble (run length unit).

b) *Swarming*: During the process of reaching towards the best food location it is always desired that the bacterium which has searched optimum path should try to attract other bacteria so that they reach the desired place quickly. Swarming makes the bacteria congregate into groups and hence move as concentric patterns of groups with high bacterial density. Mathematically, Swarming can be represented by

$$\begin{aligned} J_{cc}(\theta,P(j,k,l)) &= \sum_{i=1}^{S} J^i_{cc}\left(\theta,\theta^i(j,k,l)\right) \\ &= \sum_{i=1}^{S}\left[-d_{attract}\exp\left(-\omega_{attract}\sum_{m=1}^{p}(\theta_m-\theta^i_m)^2\right)\right] \\ &+ \sum_{i=1}^{S}\left[h_{repelent}\exp\left(-\omega_{repelent}\sum_{m=1}^{p}(\theta_m-\theta^i_m)^2\right)\right] \end{aligned} \tag{33}$$

where $J_{cc}(\theta,P(j,k,l))$ is the cost function value to be added to the actual cost function to be minimized to present a time varying cost function. 'S' is the total number of bacteria. 'p' is the number of parameters to be optimized which are present in each bacterium. $d_{attract}$, $\omega_{attract}$, $h_{repelent}$ and $\omega_{repelent}$ are different coefficients that are to be chosen judiciously. $d_{attract}$ is the depth of the attractant released by the cell and sets the magnitude of secretion of attractant by a cell. $\omega_{attract}$ is the width of the attractant signal and determines the chemical cohesion signal diffusion (smaller value makes it diffuse more). Whereas, $h_{repelent}$ is the height of the repellant effect and $\omega_{repelent}$ is the measure of the width of the repellant which controls the tendency to repel other cells.

The magnitude of $d_{attract}$ and $h_{repelent}$ should be same [10]. It is so chosen such that there is no penalty added to the cost function when the bacterial population converges, i.e. J_{cc} of (33) will be 0. Their numerical value should be decided based on the required variation in the magnitude of the actual cost function J to obtain a satisfactory result. The value of $\omega_{attract}$ and $\omega_{repelent}$ should be such that if the Euclidian distance between bacteria is large, the penalty J_{cc} is large.

c) *Reproduction*: The weakest bacteria die and the healthiest bacteria splits into two, which are placed in the same location. This makes the population of bacteria constant. Instead of taking the average value of all the chemotactic cost functions, the minimum value is selected for deciding the health of the bacteria [16].

Mathematically, for particular k^{th} and l^{th}, the health of the ith bacteria would be given by, $J^i_{health} = \min_{j\varepsilon\{1,2...N_c\}} \{J_{sw}(i,j,k,l)\}$, where $J_{sw} = J + J_{cc}$.

d) *Elimination and Dispersal*: It is possible that in the local environment the life of a population of bacteria changes either gradually (e.g., via consumption of nutrients) or suddenly due to some other influence. Events can occur such that all the bacteria in a region are killed or a group is dispersed into a new part of the environment. They have the effect of possibly destroying the chemotactic progress, but they also have the effect of assisting in chemotaxis, since dispersal may place bacteria near good food sources. From a broad perspective, elimination and dispersal are parts of the population-level long-distance motile behavior. It helps in reducing the behavior of stagnation, (i.e. being trapped in a premature solution point or local optima) often seen in such parallel search algorithms. This section is based on the work in [9]. The detailed mathematical derivations as well as theoretical aspect of this new concept are presented in [13, 16, 9, 5, 11].

In this chapter, optimization using Bacterial Foraging (BF) scheme is carried out to find the optimal controller parameters of the DFIG system. The algorithm is presented in the flow-chart as shown in Fig. 5. The algorithm is discussed here in brief. It consists of two major steps, i.e. Initialization followed by the Iterative process.

- The following variables are initialized
  1. Number of bacteria (S) to be used in the search.
  2. Number of parameters (p) to be optimized.
  3. Swimming length N_s.
  4. N_c the number of iteration in a chemotactic loop. ($N_c > N_s$).
  5. N_{re} the no of reproduction.
  6. N_{ed} the no of elimination and dispersal events.
  7. P_{ed} the probability of elimination and dispersal.
  8. Location of each bacterium $P(p,S,1)$ i.e. random numbers on [0-1].
  9. The values of $d_{attract}$,$\omega_{attract}$, $h_{repelent}$ and $\omega_{repelent}$.
- Iterative process. This consists of various steps in the bacterial foraging like chemotactic, swarming, reproduction, elimination and dispersal (initially, $j = k = l = 0$). This algorithm results in updating automatically 'P'.
  1. Elimination-dispersal loop: $l = l + 1$
  2. Reproduction loop: $k = k + 1$
  3. Chemotactic loop: $j = j + 1$
     a. For $i = 1,2,...,S$, calculate cost function value for each bacterium i as follows. Compute the value of cost function $J(i,j,k,l)$. Let J_{sw}be the location of bacterium corresponding to the global minimum cost function out of all the generations and chemotactic loops till that point (i.e., add on the cell-to-cell attractant effect for swarming behavior). It is expressed as:

$$J_{sw}(i,j,k,l) = J(i,j,k,l) + J_{cc}(\theta^i(j,k,l), P(j,k,l))$$

• Let $J_{last} = J_{sw}(i,j,k,l)$ to save this value since a better cost can be obtained in a run.
• For loop ends.

b. For $i = 1,2,,S$ take the tumbling/swimming decision.
• Tumble: Generate a random vector with each element $m = 1,2,...,p$, a random number on [0,1].
• Move: let

$$\theta^i(j+1,k,l) = \theta^i(j,k,l) + C(i)\frac{\Delta(i)}{\sqrt{\Delta^T(i)\Delta(i)}}$$

Fixed step size in the direction of tumble for bacterium i is considered.
• Compute $J(i,j+1,k,l)$ and then let

$$J_{sw}(i,j+1,k,l) = J(i,j+1,k,l) + J_{cc}(\theta^i(j+1,k,l), P(j+1,k,l))$$

.
• Swim : i)Let $m = 0$; (set counter for the swim length)
ii) While $m < N_s$ (have not climbed down too long)
• Let $m = m+1$
• If $J_{sw}(i,j+1,k,l) < J_{last}$ (if doing better), let $J_{last} = J_{sw}(i,j+1,k,l)$ and use the value of $\theta^i(j+1,k,l)$ to compute the new $J(i,j+1,k,l)$. • Else, let $m = Ns$. This is the end of the while statement.

c. Go to next bacterium $(i+1)$ if $i \neq S$ (i.e. go to b) to process the next bacterium.

4. If $j < N_c$, go to step 3. In this case, continue chemotaxis since the life of the bacteria is not over.
5. Reproduction.
   a. For the given k and l, and for each $i = 1,2,..S$, let

$$J^i_{health} = \min_{j\in\{1\cdots N_c\}} \{J_{sw}(i,j,k,l)\}$$

   be the health of the bacterium i. Sort bacteria in order of ascending cost J_{health} (higher cost means lower health).
   b. The $S_r = S/2$ bacteria with highest J_{health} values die and other S_r bacteria with the best value split (and the copies that are made are placed at the same location as their parent).
6. If $k < N_{re}$ go to 2, in this case, as the number of specified reproduction steps are not completed, hence the next generation in the chemotactic loop starts.
7. Elimination-dispersal: For $i = 1,2,...,S$, with probability P_{ed}, eliminate and disperse each bacterium (this keeps the number of bacteria in the population constant). To do this, the eliminated bacterium is replaced by simply dispersing one to a random location on the optimization domain.

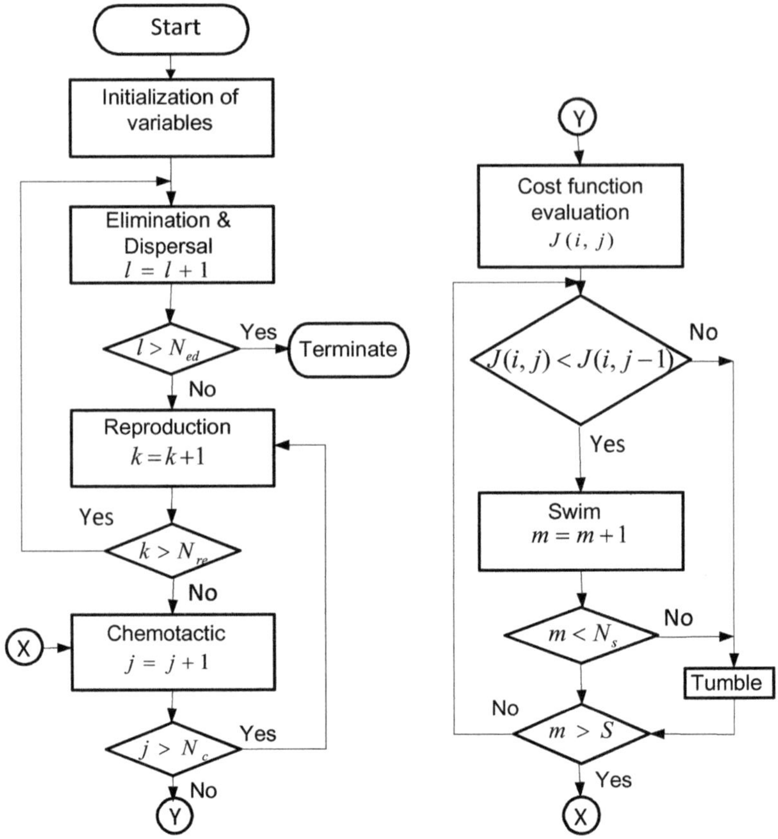

Fig. 5. Flow Chart Summarizing The Bacteria Foraging Algorithm For The Optimization Of Controller Parameters

5 Simulation and Results

The above mentioned optimization technique is applied to a SMIB DFIG system. The DFIG system with controllers can be represented by the set of Differential and Algebraic Equations (DAEs) as

$$\begin{aligned} \dot{x} &= f(x,y,u) \\ 0 &= g(x,y,u) \end{aligned} \tag{34}$$

where x, y and u are the vectors of DFIG state, algebraic and control variables respectively. The state vector is defined by

$$\begin{aligned} lx = [&i_{qs}, i_{ds}, e'_{qs}, e'_{ds}, \omega_r, \theta_{wt}, \ldots \\ &\ldots \omega_t, v_{dc}, u_1, u_3, u_5, u_6, u_8, u_{10}, u_{12}]^T \end{aligned} \tag{35}$$

Linearizing the above DAE about an operating point (x_0, y_0, u_0) (which is obtained by the load flow at a particular wind speed), the system matrix A_{sys} can be calculated as below,

$$\Delta \dot{x} = A_{sys} \Delta x \tag{36}$$

The parameters of the DFIG system is given in the Appendix.

5.1 Objective Function

The parameters of DFIG controllers are selected so as to minimize the following objective function,

$$J = 1/(\min_{\forall i} \zeta_i) \tag{37}$$

where ζ_i is the damping ratio of the ith eigenvalue of the system. This objective function makes sure that the minimum damped eigenvalue is heavily damped and the system small signal stability is ensured.

5.2 Performance of DFIG Based WT System under Different Modes of Operation

It is important to emphasize the steady-state operation of the DFIG based WT system under different modes of operation. Based on the average wind speed in the particular area, the WT (blade radius) can be designed to operate DFIG near the synchronous speed. The near synchronous speed would be desired by the manufacturer so as to extract maximum power from the stator and hence put less burden on rotor convertors. This would help the rotor current to not exceed RSC and GSC thermal rating in the event of wind speed variation. Hence, this would enhance the life cycle of the convertors. From the operation point of view, the additional rotor power at increased wind speed would give some reserve which can be added to the grid.

With the given WT rating (in appendix), DFIG operates at Super-synchronous mode when the wind speed (V_w) is more than 8 m/sec and at Sub-synchronous mode at lower wind speeds. Table 1 shows different modes of operation of the given DFIG WT system. At Super-synchronous mode, the power is supplied by stator as well as rotor, whereas rotor absorbs power at Sub-synchronous mode of operation. The positive value of the power at RSC and GSC is as shown in the Fig 1.

5.3 Need of Damping Controller

The eigenvalues of the WT system without any control at wind speed of 8 m/sec is shown in Table 2. The system looks stable with well damped eigenvalues. The first mode is stator or electrical mode and the second is electromechanical mode, which can be identified by looking at the participation factors. As electrical state (e_{ds}) and mechanical state (ω_r) participates in the second mode, hence this mode

Table 1. Different Modes of Operation of DFIG based WT system

Mode of Operation (Wind Speed)	Power from Stator (P_s)	Power from Rotor (P_r)	Total Power ($P_t = P_r + P_s$)
Normal Mode (8 m/sec)	1.1562	-0.024	1.1322
Super-synchronous (10 m/sec)	1.8017	0.4096	2.2113
Sub-synchronous (6 m/sec)	0.6565	-0.1789	0.4776

Table 2. Selected eigenvalues of the WT system without any controllers at Wind Speed of 8 m/sec

Mod No.	Eigenvalue λ	Freq (Hz)	Damp (%)	Participation factor (%)
1	-38.22±j504.77	80.33	7.55	$i_{qs}(49.2\%), i_{ds}(48.2\%), e'_{ds}(1.2\%)$
2	-4.80±j53.41	8.5	8.95	$e'_{ds}(47.6\%), \omega_r(47.9\%), i_{ds}(1.9\%)$
3	-0.26±j3.36	0.53	7.9	$\theta_{tw}(48.03\%), \omega_t(49.7\%), e'_{qs}(1.96\%)$

is electromechanical mode. The stator mode has the lowest damping ratio but its frequency is high and hence out of the range of interest. The low frequency mode, i.e mechanical mode (0.53 Hz) is well damped. However, the application of controllers (necessary to enhance the performance of the DFIG system) affects its damping as shown in Table 3, 4 and 5.

Table 3. Selected eigenvalues of the WT system. The controllers are optimized at the rated wind speed of 8 m/sec

	Without Damping Controller					With Damping Controller				
Mode of Operation (Wind Speed)	Mod No.	Eigenvalue λ	Freq (Hz)	Damp (%)	Participation factors	Mod No.	Eigenvalue λ	Freq (Hz)	Damp (%)	Participation factors
Normal operation (8 m/sec)	1	-42.11±j509	81.1	8.22	i_{qs}, i_{ds}	1	-42.05±j509	81.1	8.22	i_{qs}, i_{ds}
	2	-15.03±j64.6	10.2	22.6	e'_{ds}, ω_r	2	-13.93±j64.9	10.3	20.9	e'_{ds}, ω_r
	3	**-0.98±j3.53**	**0.56**	**26.9**	θ_{tw}, e'_{qs}	**3**	**-0.97±j3.34**	**0.53**	**27.8**	ω_t, θ_{tw}
	4	-0.014±j0.04	0.006	31.8	u_1, u_8	4	-0.714±j0.319	0.05	91.2	u_{12}, u_8
Super-Synchronous operation (10 m/sec)	1	-26.93±j469	74.7	5.7	i_{qs}, i_{ds}	1	-27.50±j468	74.6	5.85	i_{qs}, i_{ds}
	2	-48.4±j261	41.56	18.2	e'_{ds}, e'_{qs}, i_{ds}	2	-45.73±j261	41.65	17.2	e'_{ds}, e'_{qs}, i_{ds}
	3	**0.61±j11.1**	**1.76**	**-5.4**	θ_{tw}, e'_{qs}	**3**	**-1.35±j12.82**	**2.04**	**10.52**	$\omega_r, \theta_{tw}, e'_{qs}$
	4	-0.46±j0.259	0.04	87.1	V_{dc}, u_8	4	-0.41±j1.82	0.29	21.8	ω_t, u_8
	5	-0.12±j0.04	0.007	93.7	u_3, u_1	5	-0.54±j0.016	0.002	99.9	u_6, u_8
Sub-Synchronous operation (6 m/sec)	1	-50.33±j590	93.9	8.4	i_{ds}, i_{qs}	1	-49.21±j590	93.9	8.3	i_{qs}, i_{ds}
	2	**3.82±j219**	**34.9**	**-1.7**	e'_{qs}, e'_{ds}	**2**	**4.82±j219**	**34.8**	**-2.2**	e'_{ds}, e'_{qs}
	3	**-0.60±j11.1**	**1.76**	**5.4**	θ_{tw}, e'_{qs}	**3**	**-2.82±j12.7**	**2.03**	**21.5**	$\omega_r, \theta_{tw}, e'_{qs}$
	4	-0.18±j0.06	0.01	94.2	u_1, u_5	4	-0.46±j1.90	0.30	23.4	ω_t, u_8

Table 4. Selected eigenvalues of the WT system. The controllers are optimized at super-synchronous wind speed of 8.5 m/sec

	Without Damping Controller					With Damping Controller				
Mode of Operation (Wind Speed)	Mod No.	Eigenvalue λ	Freq (Hz)	Damp (%)	Participation factors	Mod No.	Eigenvalue λ	Freq (Hz)	Damp (%)	Participation factors
Normal operation (8 m/sec)	1	-39.50±j507	80.7	7.76	i_{qs}, i_{ds}	1	-39.06±j507	80.7	7.6	i_{qs}, i_{ds}
	2	-5.09±j53.55	8.52	9.4	e'_{ds}, ω_r	2	-4.87±j54.2	8.62	8.95	ω_r, e'_{ds}
	3	**-0.34±j3.67**	**0.58**	**9.42**	$\theta_{tw}, \omega_t, e'_{qs}$	**3**	**-0.41±j3.57**	**0.56**	**11.4**	ω_t, θ_{tw}
	4	-0.61±j4.94	0.78	12.3	u_8, V_{dc}	4	-0.61±j4.94	0.78	12.3	u_8, V_{dc}
	5	-0.02±j0.07	0.01	36.6	u_1, ω_t	5	-1.15±j1.00	0.16	75.4	u_6, u_{12}
Super-Synchronous operation (10 m/sec)	1	-30.69±j463	73.7	6.6	i_{qs}, i_{ds}	1	-38.85±j461	73.4	8.3	i_{qs}, i_{ds}
	2	-38.06±j149	23.8	24.64	$e'_{ds}, e'_{qs}, \omega_r$	2	-11.92±j164	26.1	7.25	$e'_{ds}, e'_{qs}, \omega_r$
	3	**1.56±j10.42**	**1.65**	**-14.8**	θ_{tw}, e'_{qs}	**3**	-	-	-	-
	4	-0.63±j5.06	0.80	12.3	V_{dc}, u_8	4	-0.54±j4.99	0.79	10.9	$V_{dc} u_8$
	5	-0.32±j0.574	0.09	49.3	ω_t, u_1, u_6	5	-0.86±j2.69	0.42	30.6	ω_t, θ_{tw}
Sub-Synchronous operation (6 m/sec)	1	-45.48±j557	88.8	8.12	i_{qs}, i_{ds}	1	-37.03±j557	88.8	6.6	i_{qs}, i_{ds}
	2	**2.21±j136**	**21.7**	**-1.6**	$e'_{qs}, e'_{ds}, \omega_r$	**2**	**14.97±j135**	**21.5**	**-11.6**	$e'_{qs}, e'_{ds}, \omega_r$
	3	**-1.61±j10.47**	**1.66**	**15.2**	θ_{tw}, e'_{qs}	**3**	-	-	-	-
	4	-0.56±j4.80	0.76	11.6	V_{dc}, u_8	4	-0.65±j4.87	0.77	13.3	V_{dc}, u_8
	5	-0.25±j0.198	0.03	78.7	ω_t, u_1, u_6	5	-0.85±j2.98	0.47	27.4	$\omega_t, \theta_{tw}, u_{12}$

Table 5. Selected eigenvalues of the WT system. The controllers are optimized at sub-synchronous wind speed of 7 m/sec.

	Without Damping Controller					With Damping Controller				
Mode of Operation (Wind Speed)	Mod No.	Eigenvalue λ	Freq (Hz)	Damp (%)	Participation factors	Mod No.	Eigenvalue λ	Freq (Hz)	Damp (%)	Participation factors
Normal operation (8 m/sec)	1	-39.13±j504	80.3	7.73	i_{qs}, i_{ds}	1	-39.08±j504	80.3	7.72	i_{qs}, i_{ds}
	2	-7.01±j56.14	8.93	12.4	ω_r, e'_{ds}	2	-6.16±j56.97	9.05	10.7	ω_r, e'_{ds}
	3	**-0.45±j3.38**	**0.53**	**13.3**	$\theta_{tw}, \omega_t, e'_{qs}$	**3**	**-0.43±j3.38**	**0.53**	**12.8**	ω_t, θ_{tw}
	4	-0.005±j0.03	0.005	14.3	u_1, u_3	4	-	-	-	-
Super-Synchronous operation (10 m/sec)	1	-37.14±j499	79.4	7.4	i_{qs}, i_{ds}	1	-37.53±j499	79.4	7.5	i_{qs}, i_{ds}
	2	-15.37±j140	22.4	10.8	$e'_{ds}, e'_{qs}, \omega_r$	2	-9.06±j141	22.5	6.3	$e'_{ds}, e'_{qs}, \omega_r$
	3	**0.86±j10.4**	**1.65**	**-8.2**	θ_{tw}, e'_{qs}	**3**	**-3.79±j9.56**	**1.52**	**36.8**	$\omega_r, \theta_{tw}, e'_{qs}$
	4	-0.02±j0.417	0.06	6.03	u_1, u_3, u_6	4	-0.58±j1.16	0.18	44.8	u_8, ω_t
Sub-Synchronous operation (6 m/sec)	1	-41.83±j505	80.5	8.2	i_{qs}, i_{ds}	1	-40.54±j506	80.5	7.9	i_{qs}, i_{ds}
	2	**-7.5±j146**	**23.3**	**5.1**	$e'_{qs}, e'_{ds}, \omega_r$	**2**	**-2.44±j145**	**23.2**	**1.6**	$e'_{qs}, e'_{ds}, \omega_r$
	3	**-1.65±j10.51**	**1.67**	**15.5**	θ_{tw}, e'_{qs}	**3**	**-9.51±j7.96**	**1.26**	**76.65**	$\omega_r, \theta_{tw}, e'_{qs}$
	4	-0.13±j0.23	0.03	51.2	u_1, u_5	4	-1.04±j1.17	0.18	66.2	ω_t, u_{15}

Initially the P_{ref}, V_{sref}, V_{dcref} and q_{cref} are implemented in the DFIG system. The parameters of these four controllers are optimized using BF algorithm for a wind speed of 8 m/sec (Normal mode of operation). It is observed from Table 3 that Mode# 3 is unstable at higher wind speed (Super-synchronous mode of operation). With the implementation of the optimized damping controller (using BF and keeping all other controller parameter as constant), the system is stable. The further detailed discussion can be found in [12].

However, the closer look at the eigenvalues gives a new dimension to the whole problem. The damping controller enhances the damping at Normal and Super-synchronous mode of operation, but at lower wind speeds (Sub-synchronous operation), mode# 2 becomes unstable. This mode remains unstable even after the implementation of damping controller as shown in Table 3. This requires further investigation as to how this mode is excited and what can be done to make the system stable.

The Wind speed is seldom constant and hence there is a need for a robust damping controller which can perform well under various operating conditions. Therefore, to further investigate the impact of damping controller at different operating conditions, the DFIG system is studied for Super-synchronous and Sub-synchronous modes of operation.

5.4 Optimal Tuning of DFIG Controllers

The robust performance of DFIG controllers is desired at all modes (Normal and Super/Sub-synchronous) to ensure the stable operation of the wind turbine under stochastically varying wind speed. Hence, it is necessary to find the optimal parameters of all the controllers (including damping controller) for the stable operation under changing wind speeds. Usually, the wind speed varies in the range of 6-14 m/sec. However, these extreme wind speeds are rare and hence the speed of 8 m/sec is selected for the near synchronous or normal mode of operation (it depends on turbine manufacturer and the average wind speed selection). For the optimal performance, controllers should be optimized at the speed near to the rated/normal speed. Therefore, the DFIG controllers are optimized at 3 different wind speeds (V_w), i.e. 8 m/sec (normal), 8.5 m/sec (Super-synchronous) and 7 m/sec (Sub-synchronous). This is shown in Table 3, 4 and 5.

When the controllers are optimized for V_w=8 m/sec, the system is stable for Normal (V_w=8 m/sec) and Super-synchronous (V_w>8 m/sec) mode of operation. However, at Sub-synchronous (V_w<8 m/sec) mode of operation, mode# 2 becomes unstable as shown in Table 3. It is interesting to observe that the damping controller still works well at this mode of operation, as mode# 3 is stable with damping controller. However, mode# 2 has become unstable because the participation of the mechanical state (ω_r) has been reduced and that of electrical states (e'_{ds}, e'_{qs}) have increased at the Sub-synchronous mode of operation. It can be seen from the Table 3. This suggests that the system is not small signal stable at Sub-synchronous mode of operation, if the controllers are optimized at near synchronous (normal) wind speed.

In the next step, the controller are optimized at wind speed (V_w) of 8.5 m/sec. Again the system is unstable at Sub-synchronous mode of operation. This can be illustrated from Table 4. As shown in the Table 4, the optimization at Super-synchronous speed would help in eliminating the electromechanical mode (mode# 3). But on the downside, the system has an additional mode, mode# 4, associated with state V_{dc}, voltage across the DC link capacitor. This can have detrimental effect on the system stability if not properly damped. Moreover, it affects the convertor rating and therefore the oscillations of this kind should be damped if the controllers are optimized for this operating condition. Mode# 2 is still unstable at lower wind speeds (Sub-synchronous operation).

Lastly, the controllers are optimized at lower wind speeds i.e. V_w=7 m/sec, and the eigenvalues of the system with the different controllers are shown in Table 5. The system is stable for all the operating wind speeds. Therefore, DFIG controllers can be made robust if optimized at any Sub-synchronous speed. Again, the speed depends on the rating of the DFIG WT. For example, if the installed WT has near synchronous operation at V_w=8 m/sec, then the controllers can be designed at any Sub-synchronous speed (V_w< 8 m/sec). This would ensure the stable performance of DFIG system across the wide range of wind speed from Sub-synchronous (V_w< 8 m/sec) to Super-synchronous (V_w> 8 m/sec) mode of operation.

5.5 *Super/Sub-Synchronous Mode of Operation*

It is interesting to observe the change in the frequency of oscillations of different eigenmodes under Super/Sub-synchronous operation of DFIG. The eigenvalues of the DFIG based WT system is observed under Super/Sub-synchronous mode of operation. It is observed that the frequency of Mode# 2 has increased from $\sim 10Hz$ (Normal operation) to $\sim 40Hz$ (Super/Sub-synchronous operation) when the controllers are optimized at 8 m/sec. The participation of electrical state variables (e'_{ds}, e'_{qs}) in Mode# 2 has been increased under Super/Sub-synchronous operation. This can be easily verified by the participation factor analysis as shown in Table 3. Moreover, when controllers are optimized for the wind speed of either 8.5 or 7 m/sec, the frequency of Mode# 2 changes to $\sim 20Hz$ for Super/Sub-synchronous operation as in Table 4and 5. Therefore, Mode# 2 has increased participation of electrical states (e'_{ds}, e'_{qs}) than the mechanical state (ω_r). Nevertheless, the change in the typical electromechanical mode (Mode# 3) is hardly noticed. The efficacy of the auxiliary controller can be easily identified from the Tables showing the comparison of the eigenvalues with and without the damping controller. It is observed that the damping controller had stabilized the system by improving the damping of the electromechanical mode# 3.

6 Conclusion

The damping controller gives promising results in damping out the low frequency oscillations and hence improves the system stability of the grid connected DFIG system. The tuning of the controllers is emphasized to ensure the stable operation under variable wind speed. The Super-synchronous, Normal and Sub-synchronous mode of operation of DFIG based WT is thoroughly investigated. It is observed that when the tuning of the controllers is done at any Sub-synchronous speed, the system is stable for all modes of operations. The change in the frequency of the electromechanical modes under Super/Sub-synchronous operation is presented. This study would help in understanding the interaction of the oscillatory modes of DFIG based WT with other components of power systems. Further, it would also help in proper tuning of the DFIG controllers to enhance the system small-signal stability.

With the increasing penetration of DFIG based wind farms into the grid, it is important to study the implications of large scale DFIG systems on grid stability. As this study is based on SMIB DFIG system, conclusions of this paper should not be extended to multi-machine DFIG based WT system. Nevertheless, this paper does provide a good initial study of the DFIG system with controller. Computations with multi-machine DFIG system as well as the effect of the DFIGs on the conventional synchronous machines will be required to confirm the obtained results and determine if its possible to quantify the impact of DFIGs on power system stability.

References

1. US department of Energy, Eere state activities and partnerships (June 2007)
2. Eriksen, P.B., Ackermann, T., Abildgaard, H., Smith, P., Winter, W., Rodriguez Garcia, J.M.: System operation with high wind penetration. IEEE power energy management 3(6), 65–74 (2005) (English)
3. Fan, L., Miao, Z., Osborn, D.: Impact of doubly fed wind turbine generation on inter-area oscillation damping. In: Proc. IEEE PES General Meeting, Pittsburg (July 2008) (English)
4. Hughes, F.M., Lara, O.A., Jenkins, N., Strbac, G.: A power system stabilizer for dfig-based wind generation. IEEE Transactions on Power Systems 21(2), 763–772 (2006) (English)
5. Hunjan, M., Venayagamoorthy, G.K.: Adaptive power system stabilizers using artificial immune system. In: IEEE symposium on artificial life, pp. 440–447 (April 2007) (English)
6. Mei, F., Pal, B.C.: Modeling snd small signal analysis of a grid connected doubly fed induction generator. In: Proc. of IEEE PES General Meeting, San Fransisco, pp. 358–367 (2005) (English)
7. Mei, F., Pal, B.C.: Modal analysis of grid connected doubly fed induction generator. IEEE Transactions on Energy Conversion 22(3), 728–736 (2007) (English)
8. Miao, Z., Fan, L., Osborn, D., Yuvarajan, S.: Control of dfig based wind generation to improve inter area oscillation damping. In: Proc. IEEE PES General Meeting (July 2008) (English)
9. Mishra, S.: A hybrid least square-fuzzy bacteria foraging strategy for harmonic estimation. IEEE Trans. Evolutionary Computation 9(1), 61–73 (2005) (English)
10. Mishra, S.: Hybrid least-square adaptive bacteria foraging strategy for harmonic estimation. IEE Proc.-Gener. Transm. Distrib. 152(3), 379–389 (2005) (English)
11. Mishra, S., Tripathy, M., Nanda, J.: Multimachine power system stabilizer design by rule based bacteria foraging. Electrical Power system research 77(12), 1595–1607 (2006) (English)

12. Mishra, Y., Mishra, S., Tripathy, M., Senroy, N., Dong, Z.Y.: Improving stability of a dfig based wind power system with tuned damping controller. Accepted to appear on IEEE trans. Energy Conv. (2008-09) (English)
13. Passino, K.M.: Biomimicry of bacterial foraging for distributed optimization and control. IEEE Control System Magazine, 52–67 (June 2002) (English)
14. Qiao, W., Venayagamoorthy, G.K., Harley, R.G.: Design of optimal pi controllers for doubly fed induction generators driven by wind turbines using particle swarm optimization. In: Int. Joint conf. on neural Networks, Canada, July 2006, pp. 1982–1987 (2006) (English)
15. Salman, S.K., Teo, A.L.J.: Windmill modeling consideration and factors influencing the stability of a grid-connected wind power based embedded generator. IEEE Transactions on Power Systems 18(2), 793–802 (2003) (English)
16. Tripathy, M., Mishra, S.: Bacteria foraging-based solution to optimize both real power loss and voltage stability limit. IEEE Transactions on Power Systems 22(1), 240–248 (2007) (English)
17. Vowles, D.J., Samasinghe, C., Gibbard, M.J., Ancell, G.: Effect of wind generation on small signal stability- a new zealand example. In: Proc. IEEE PES General Meeting, Pittsburg (July 2008) (English)
18. Wu, F., Zhang, X.P., Godfrey, K., Ju, P.: Small signal stability analysis and optimal control of a wind turbine with doubly fed induction generator. IET on Generation, transmimssion and Distribution 1(5), 751–760 (2007) (English)

Appendix

Parameters of the SMIB DFIG System (p.u.)

H_t=4; H_g=0.4; X_m=4; L_m=4;x_{tg}=0.55; C=0.01; x_e=0.06;
L_{ss}=4.04; L_{rr}=4.0602; R_s=(X_m/800); R_r=1.1*R_s;

Parameters Used for the Optimization (BF Algorithm)

S=4; Cl=0.07; $d_{attract}$=1.9; $\omega_{attract}$=0.1; $\omega_{repelant}$=10;
$h_{repelant}$=$d_{attract}$; N_{ed}=10; N_{re}=100; N_c=4; N_s=3;

Optimized Controller Parameters for Different Speeds

a) ***8.0 m/sec:=*** K_{p1}=0.4963; K_{I1}=0.1050;
K_{p2}=0.4949 ; K_{I2}=0.4470; K_{p3}=0.4800 ; K_{I3}=0.3843; K_{p4}=0.4969 ; K_{I4}=0.4405; K_{p5}=0.4688 ; K_{I5}=0.2490; K_{p6}=0.24757 ; K_{I6}=0.1794; K_{p7}=1.6 ; K_{I7}=19.74;
b) ***8.5 m/sec:=*** K_{p1}=0.4340 ; K_{I1}=0.2613;
K_{p2}=0.4952 ; K_{I2}=0.4632; K_{p3}=0.0225 ; K_{I3}=0.1713; K_{p4}=0.0062 ; K_{I4}=0.1590; K_{p5}=0.0116 ; K_{I5}=0.2360; K_{p6}=0.2495 ; K_{I6}=0.1238; K_{p7}=16.1174 ; K_{I7}=42.3049;
c) ***7 m/sec:=*** K_{p1}=0.0492 ; K_{I1}=0.1223;
K_{p2}=0.0017 ; K_{I2}=0.1998; K_{p3}=0.0229 ; K_{I3}=0.1495; K_{p4}=0.4936 ; K_{I4}=0.1928; K_{p5}=0.1593 ; K_{I5}=0.3519; K_{p6}=0.4469 ; K_{I6}=0.4938; K_{p7}=1.6 ; K_{I7}=2.3049;

An ANN-Based Power System Emergency Control Scheme in the Presence of High Wind Power Penetration

Bevrani H. and Tikdari A.G.

Re-evaluation of emergency control and protection schemes for distribution and transmission networks are one of the main problems posed by wind turbines in power systems. Change of operational conditions and dynamic characteristics influence the requirements to control and protection parameters.

Introducing a significant wind power into power systems leads to new undesirable oscillations. The local and inter-modal oscillations during large disturbances can cause frequency and voltage relays to measure a quantity at a location that is different to the actual underlying system voltage and frequency gradient. From an operational point of view, this issue is important for those networks that use the protective voltage and frequency relays to re-evaluate their tuning strategies.

In this chapter, an overview of the key issues in the use of high wind power penetration in power system emergency control is presented. The impact of wind power fluctuation on system frequency, voltage and frequency gradient is analyzed, the need for the revising of tuning strategies for frequency protective relays, automatic under-frequency load shedding (UFLS) and under-voltage load shedding (UFLS) relays are also emphasized.

In the present chapter, necessity of considering both system frequency and voltage indices to design an effective power system emergency control plan is shown. Then, an intelligent artificial neural network (ANN) based emergency control scheme considering the dynamic impacts of wind turbines is proposed. In the developed algorithm, following an event, the related contingency is determined by an appropriate ANN using the online measured tie-line powers. A suitable set of voltage sensitivity indices based on a comprehensive voltage stability analysis in the presence of the wind turbines is proposed. Another intelligent ANN is used to

Bevrani H. · Tikdari A.G.
Department of Electrical and Computer Engineering,
University of Kurdistan, Sanandaj, PO Box 416, Kurdistan, Iran

L. Wang et al. (Eds): *Wind Power Systems*, Green Energy and Technology, pp. 215–254.
springerlink.com

examine the stability margin by estimating the system power-voltage (P-V) curves. Finally, the system frequency gradient, voltage sensitivity indices and stability information are properly used by an effective load shedding algorithm. The proposed emergency control scheme and discussions are supplemented by computer nonlinear simulations on the IEEE 9-bus test system.

1 Introduction

To prevent power system blackout following a severe contingency, the emergency control actions are needed. There are few reports on the role of distributed wind turbines in emergency conditions. Frequency and voltage are more frequent decision tools in the emergency control strategies. Interconnection of wind turbines into power system significantly affects the frequency and voltage behavior following the contingencies [2, 7, 11, 12, 16, 21, 23, 36, 42]. Therefore, emergency control schemes may need a revision in the presence of a high penetration wind turbines.

Studying of wind generation impacts on the parameters that used in the emergency control actions make them more useful for the future of the power system that are moving to the systems with high wind power penetration [32].

Annually, many contingencies occur in the real-world power systems, but only some of them lead to the blackout [43]. Identifying the severe contingencies and making adequate analysis help the power system organizer to adjust the emergency control parameters, effectively.

Serious load generation imbalance which is usually the result of a severe contingency may lead the system to the cascading failures and blackout. Engaging the spinning reserve and starting up the non-spinning hydro generators and also, generation re-dispatching may be used to compensate the load generation unbalance [25, 43]. However, even in a power system these services exist, they may be not fast enough to supply the loads rapidly following a large disturbance. Load shedding is a well-known emergency control scheme used to curtail a part of system load in an acceptable time duration. The load shedding algorithms are implemented to shed the loads before loss of remained generation [2, 27].

A variety of studies are recommended to analyze the protection-based penetration limits with consideration of the wind turbine capacity, location and technology. The studies aid in determining mitigation strategies to increase the protection based penetration limit. The loss of coordination, de-sensitization, nuisance fuse blowing, bidirectional relay requirements and overvoltage, should be studied in order to arrive at the penetration limits of wind power in an existing distribution system [2].

The effect of adding wind power units to distribution feeder can produce blind zones for protection devices or upset the coordination between two (or more) protective devices and should be studied carefully [7]. In normal operation, protection devices are coordinated such that the primary protection operates before the back-up can take action. Interconnecting distributed wind turbines increases the short circuit level. Depending on the original protection coordination settings along with the size, location and type of the units, uncoordinated situations may be found.

In these situations, the backup operates before the primary, which results in nuisance tripping to some of the loads.

In response to the existing control and coordination challenges, this chapter presents a new intelligent based emergency control scheme that could be more useful for the actual power systems and will be suitable for the power system in the presence of high wind power penetration. Artificial neural networks are effectively used to identify severe faults (contingencies) and estimate the system security level by predicting the post-fault stability margin. Finally, the result is used to run an appropriate load shedding algorithm.

This chapter is organized as follows: in Section 2, the impacts of the wind turbines on the system voltage, frequency, and the rate of frequency change are illustrated. The using of initial frequency gradient in the presence of wind turbines is emphasized. It is shown that simultaneous using of voltage and frequency is required to design an effective load shedding scheme.

Contingency analysis is presented in Section 3. The contingency ranking methods are briefly introduced and some points are suggested to construct a more suitable analysis technique in emergency conditions. The continuation power flow (CPF) which is a common tool to derive the P-V curves is reviewed, and a simple method to obtain the system P-V curves for the suddenly load increasing, corresponds to the emergency conditions, is presented. The P-V curves derived by this method are carefully studied in the power systems with wind turbines for a wide range of load/generation outage.

An overall view of the developed intelligent based emergency control scheme is given in Section 4. In Section 5, a new load shedding algorithm based on the system voltage and frequency is proposed, and its advantages are clarified (in comparison of conventional load shedding plans). To achieve a robust and fast load shedding scheme, appropriate ANNs are suggested for predicting the contingencies and estimating the P-V curves. The proposed analysis and synthesis methodologies are supplemented by adequate nonlinear simulation on the IEEE 9-bus test system.

2 Wind Power Penetration

Increasing the penetration of wind turbine generators (WTGs) in the power system may affects the security/stability limits, frequency, voltage and dynamic behavior of a power system [11, 12, 16, 21, 36, 42]. The WTGs commonly use the induction generators to convert the wind energy into electrical energy [16, 21]. The induction generators are reactive power consumer. Therefore, the voltage of system would be affected in the presence of wind turbines especially in the case of fixed-speed type of WTGs [21]. The wind turbines impacts on the power system frequency and voltage have been studied in many research works [2, 7, 11, 12, 16, 21, 23, 32, 42]. Power system frequency response model in the presence of high WTG penetration, frequency control issue, a survey and some new perspectives are addressed in [2, 7]. In [21], the effects of the doubly-fed induction generator (DFIG) and induction generator (IG) type of WTGs on the voltage transient behavior are explained and the disadvantages of the IG type are shown. Frequency nadir in the presence of different type of the WTGs has been compared and it is shown in [12, 16]. The load ability of various types of WTGs is compared in [42], and it is shown that the DFIG has larger load ability than IGs.

2.1 Test System

For the sake of dynamic simulations and to describe/examine the proposed methodology, the IEEE nine-bus power system is considered as a test system. A single line diagram for the test system is shown in Fig. 1. As shown, two wind farms are added in buses 5 and 9. Simulation data and system parameters are given in Appendix (Tables 3 to 6).

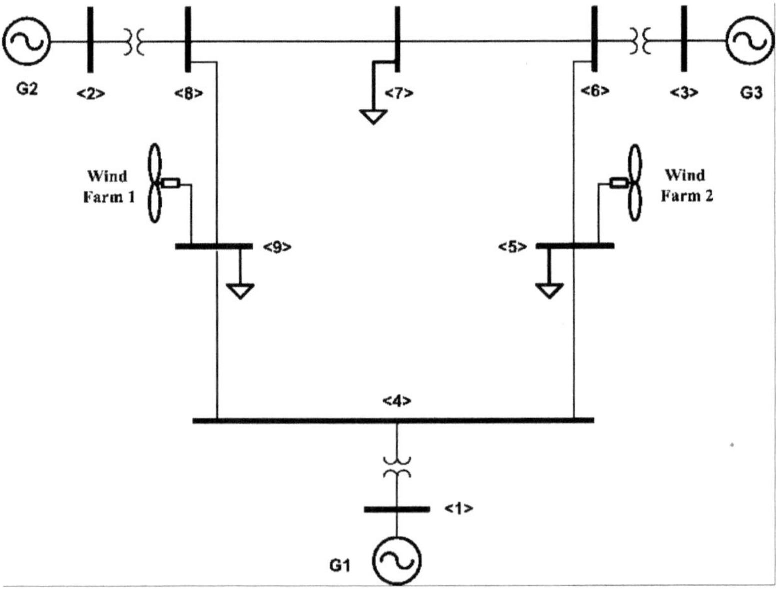

Fig. 1. Nine-bus test system including two wind farms

2.2 Impacts on Voltage Profile

Some reports have addressed the impacts of various WTGs technologies on the voltage deviation following a contingency event, and have analyzed their influences on the transient voltage stability [21, 23].

Fig. 2 shows the voltage response at bus 5 of the test system after tripping of generator G2 (the largest generator) for the following cases: without wind turbine, with 10% DFIG penetration, with 10% IG type penetration, and with 10% IG type wind turbine compensated with a static compensator (STATCOM). As shown, the voltage deviation is significantly affected by integration of wind turbines into the power system. All of them are unstable cases. To protect system against blackout

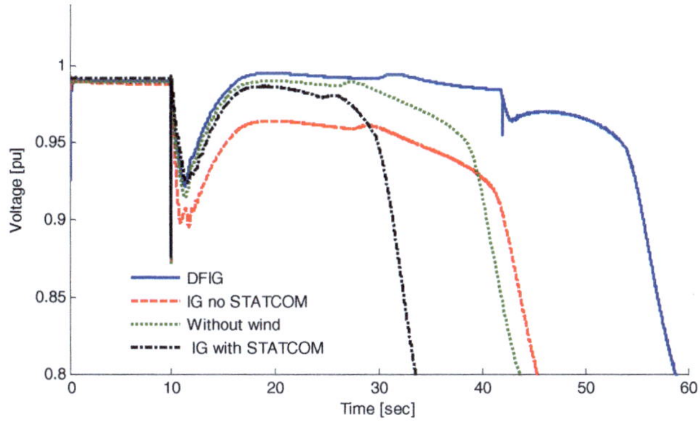

(a)

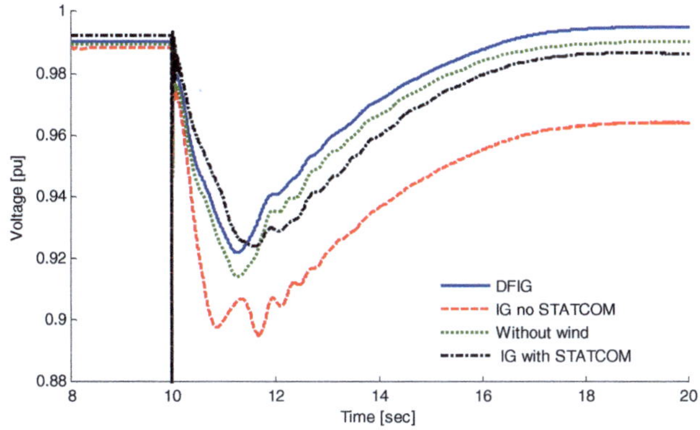

(b)

Fig. 2. Voltage deviation at bus 5 following loss of G2; figure (b) shows a zoomed view around 10s

in these situations, using a load shedding scheme that should be applied at the first few seconds is necessary. Fig. 2 b shows the zoomed view of Fig. 2 a around 10s.

The performed simulation illustrates that the DFIG and IG plus STATCOM represent a better post-contingency performance. But the system performance for the IG type without any reactive support is worse than them under the same test scenario. Absorbing of large amount of grid reactive power by IG wind turbines in the voltage dip cases, can be considered as a reason to present the mentioned behavior [20, 42]. The STATCOM can compensate the reactive power absorption and removes the voltage decline problem. The DFIG type is capable to regulate the power factor by either consuming or producing reactive power [16].

The P-V plot can be considered as a suitable tool to analyze the voltage stability of a power system in the presence of wind turbines. Fig. 3 illustrates the effects of the different wind turbine technologies on the P-V curve for bus 5 of the test system. All of the curves are derived by a slow rate of load admittance change at bus 5. As shown in

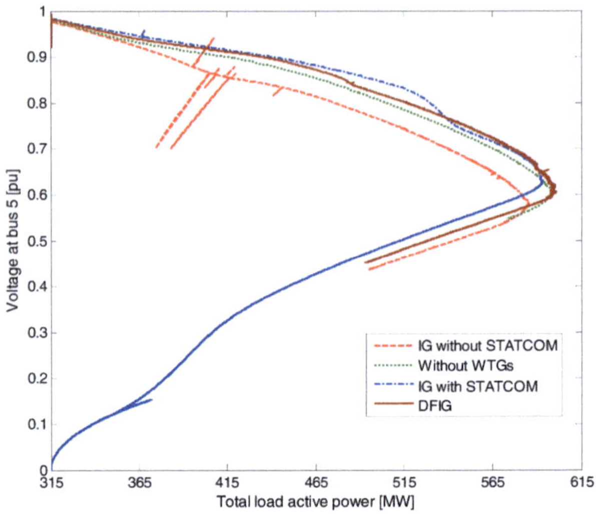

Fig. 3. Comparing the P-V curves in the presence of different type of wind turbine technologies and without WTGs

this figure, the DFIG type of WTG improves the P-V curve and represents a better stability margin for the system. On the other hand, when the IG type is used without any reactive power compensation device, the stability margin decreased, significantly. Since, the STATCOM can compensate the reactive power, in the case of using the IG type equipped with STATCOM, the voltage has been improved.

However, by increasing the power demand, the operating point reaches a point that STATCOM is not able to compensate the reactive power demand for the induction generator more, and the reactive power will be absorbed from the network. Therefore, the P-V curve cannot reach its expected nose and the stability margin to be less than the expected value.

When a generator/line outage takes place, the voltage starts to deviate as shown in Fig. 2. In Section 1.4.2, it will be illustrated that the P-V characteristic is significantly affected by changing the network topology. Therefore, to determine the new power system operating point following a contingency (on the related P-V curve), the post-contingency P-V curve for the test system (considering the network topology changes) to be needed.

2.3 Impacts on System Frequency

Following a large generation loss disturbance, the system frequency may drop quickly if the remaining generation no longer matches the load demand [2, 3]. Some parameters such as power system reserve and inertia constant are influenced by interconnecting the WTGs on the power system. Therefore, the frequency deviation will be affected in the presence of the WTGs [2, 3, 7, 16]. Fig. 4a shows the frequency response of the IEEE 9-bus test system, following loss of G2 at 10s. This simulation is done for the four cases: without WTG, with DFIG type, with IG type, and with IG type equipped by STATCOM. The initial rate of frequency change for some test scenarios is shown in Fig. 4b. The simulation results show that immediately following a contingency, the frequency behavior of the system in the presence of IG type of WTGs is better than DFIGs. The frequency decline and initial rate of frequency change in the presence of IGs is smaller than the DFIGs case. Because of their structure, IGs add more inertial response to the power system than DFIGs. It is the reason of different frequency behavior illustrated in the Fig. 4.

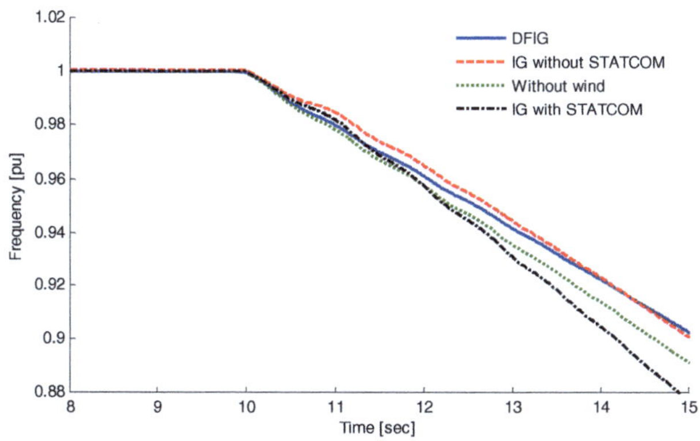

(a)

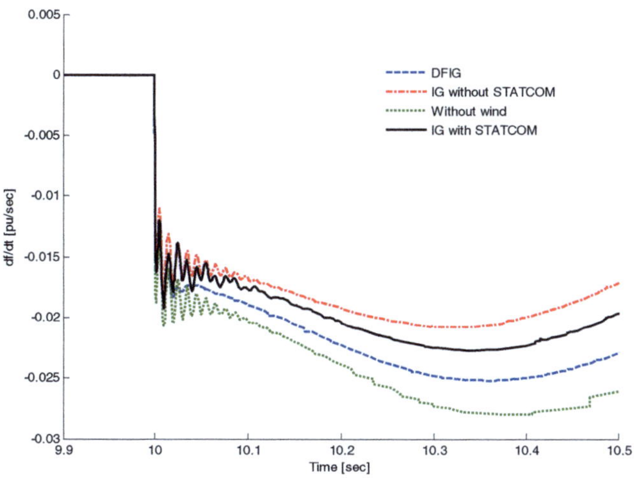

(b)

Fig. 4. System response following G2 outage with and without WTGs (DFIG, IG, and IG equipped with STATCOM); a) frequency deviation, b) rate of frequency change

The initial rate of frequency change is in use as an additional control variable in the recent introduced emergency control algorithms [2, 3, 4, 5, 6, 7, 13]. The initial frequency gradient is proportional to the amount of power imbalance [2]. Therefore, it can be used to determine the amount of load to be shed at the first step of load shedding algorithms. Fig. 2 illustrates that the post contingency voltage behavior in the presence of IG type WTG without STATCOM is the worst case. But, the magnitude of the frequency gradient in this case is the smallest one (Fig. 4b).

Fig. 5 illustrates the simulation results for G1 outage test scenario. The above phenomenon is also exists in this case. The magnitude of frequency gradient in the case of IG with STATCOM is larger than the other cases, while voltage behavior in this case is the best one. On the other hand, voltage deviation in the case of IG without STATCOM is the worst case, while it cannot be justified from the frequency gradient.

This phenomenon indicates that using initial rate of frequency change for the power system emergency control in the presence of wind turbines needs to be revised. Since, in the presence of WTGs, the undesirable oscillations are added to the frequency deviation, the measuring of frequency gradient introduces another difficulty to achieve this variable in emergency control strategies. This issue encourages power design engineers to use $\Delta f/\Delta t$ instead of df/dt [3, 7]. Furthermore, as it is studied in the Section 1.5, voltage and frequency behavior does not address the same results about contingency conditions. This phenomenon is encourages us to re-evaluate the emergency control schemes for the future of the power systems which are integrated with high wind power penetration.

3 Contingency Analysis

A contingency can be created by a three phase fault in a transmission line, a bus bar fault near a generator, etc. These faults are generally eliminated during an acceptable clearing time. But, the consequence protection actions may trip a generator, or curtail a line.

In many cases, generation loss and line outage have a serious impact on the power system performance, and even stability. A major load generation imbalance, line overload and voltage problem may lead the system to a blackout. Identifying the severe contingencies in a power system, performing comprehensive analysis to predict the post contingency conditions and preparing appropriate preventive actions are necessary to design an effective emergency control scheme. This chapter is focused on the major contingencies, and presents a fast, flexible and accurate new algorithm for power system control in emergency conditions.

Contingency screening is a suitable method to rank all contingencies based on their severity and post contingency effects on the power system operating conditions. There are many reports that introduce different algorithms and indices to rank the contingencies [1, 8, 10, 22, 34, 40]. In a contingency analysis, all of possible outages should be simulated and the post contingency conditions should be evaluated [10]. In practice, insist of nonlinear simulation, the contingency analysis problem is generally done in a static environment by considering the steady state operating conditions.

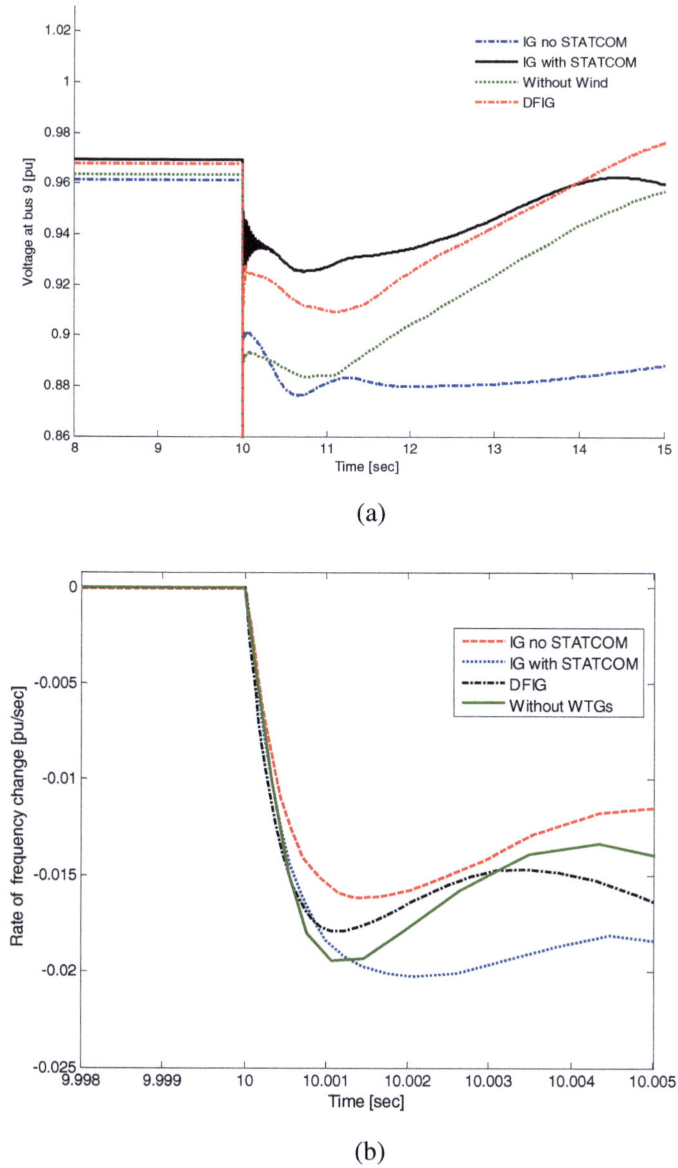

Fig. 5. System response following G1 outage with and without WTGs (DFIG, IG, and IG equipped with STATCOM); a) voltage deviation, b) initial rate of frequency change

The indices that used to measure the severity of the contingencies are divided into two general groups: the first group is based on the distance to collapse parameter or the magnitude of the stability margins. The second group of indices is based on the steady state parameter violations such as voltage violation and tie-line over loading.

The stability margins could be easily calculated by using P-V curves, Q-V characteristic [1, 22], or studying on the eigen-values of the Jacobin matrix [9]. Ref. [16] introduces some criteria to calculate the stability margin and also performs a direction for load variation at different buses to determine the minimum stability margin or minimum load-ability of system.

Some performance indices are used in above second group. Two typical indices are defined as follows:

$$PI_v = \sum_{\substack{i=1,\\ V^i>V_{max}}}^{N} \left|\frac{V_{max}-V^i}{V_{max}}\right| + \sum_{\substack{i=1,\\ V^i<V_{min}}}^{N} \left|\frac{V_{min}-V^i}{V_{min}}\right| \tag{1}$$

$$PI_p = \sum_{\substack{j=1\\ |P^i|>P^i_{limit}}}^{L} \left|\frac{P^i}{P^i_{limit}}\right| \tag{2}$$

Here, PI_v and PI_p are the performance indices based on the voltage violations and line overloading, respectively. The V^i, P^i, V_{min}, V_{max}, and P^i_{limit} are post-contingency voltage magnitude at bus i, post-contingency of active power line i, the allowable voltage magnitude limits, and the maximum transmission capacity of line i, respectively. Eq. (1) shows that, wherever the voltage deviations in a contingency violate the specified limits, the performance index of that contingency to be increased. This concept is also used for line over loading in Eq. (2) [10, 34, 40].

Some methods have been introduced for the online and fast contingency ranking [17, 40]. These methods could be more useful in uncertain conditions; such as in a restructured environment, in the presence of high penetration of renewable sources [18], and in a condition that the operating point of the system is unpredictable. The ANNs, fuzzy logic, and other artificial intelligence methods have been widely used in contingency ranking for different purposes such as online ranking [17], and combination of different performance indices [19].

In the proposed emergency control strategy in the present chapter, the contingency ranking step is used. Following, some important points are expanded using a comprehensive simulation on the test system to achieve a more efficient contingency ranking algorithm which is used in the proposed emergency control scheme.

In an emergency condition caused by a serious fault such as the generation outage, a load generation imbalance will be created. The frequency decline and droop characteristic of generator units in a power system lead to increase the unit outputs. It should be considered that in a generation loss case, all remained generator units initially response to the event.

Usually, in steady state analysis and classical AC/DC power flow computation following a generation loss, only the output of the slack generator will be changed. But according to the reality, dividing the output of a tripped generator (in a pre-contingency condition) between the remained units via specific weights on their droop characteristics is more reasonable. A simple method to determine a new operating point following a fault is given in [41].

As mentioned above, a contingency analysis methodology without considering a slack unit could be more useful. Here, also to determine new operating point after a generator outage, the amount of generation loss is divided between all operating units. Then, the power flow calculations will be done, and the performance indices are extracted. Although, this method is not so accurate, it gives an acceptable approximation.

Some nonlinear simulations for a wide range of contingencies in a dynamic environment are performed. Fig. 6 shows the results of the contingency analysis on the IEEE 9-bus test system. Figs. 6a and 6b show the performance indices for the generator outage cases. These figures show the risk of losing of each generator in the nine-bus test system is relatively high. Since, the given power system is small and each generator supplies a considerable portion of load, the above results seems to be reasonable. Figs. 6c and 6d show the performance indices for the line outage cases. For the present example, it is denoted that the loss of lines 7-8, 8-9 and 9-4 show more undesirable post-contingency conditions.

Figs. 7, 8, and 9 confirm the above results, in time-domain responses. Fig. 7 shows the voltage of the weakest bus (bus 9), following various contingencies. Considering Fig. 7 and Figs. 6b and 6d illustrate that the outage of G1 and line 9-4 are significantly degrade the overall system performance. Fig. 6b denotes that the loss of G2 and G3 does not show a serious impact on the bus voltage.

As shown in Fig. 7, the voltage instability starts about 25 seconds after outage of G2 and G3. While, Fig 8 shows the system frequency rapidly falls, immediately after the same event. Therefore, for the present test scenario, system frequency can be considered as a more suitable (than voltage) index to detect the system

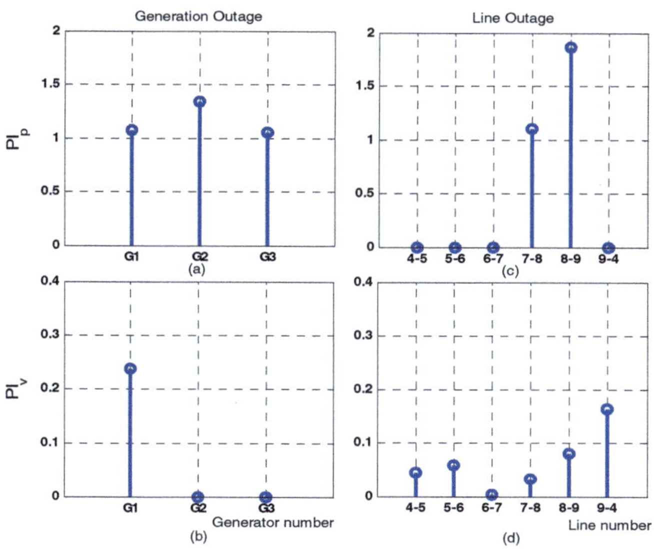

Fig. 6. Performance indices in different contingencies

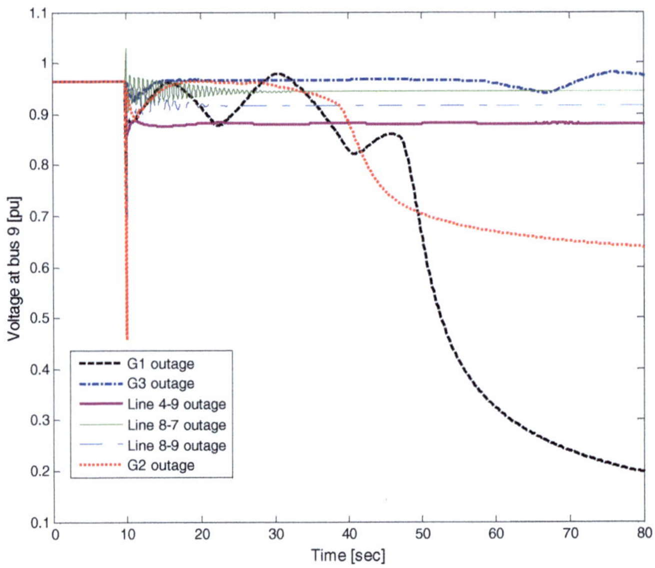

Fig. 7. Weakest bus voltage in different contingencies

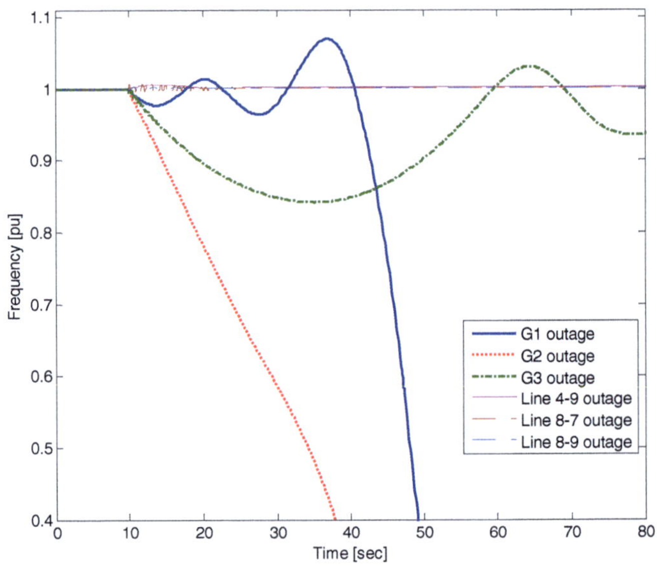

Fig. 8. Frequency deviation in different contingencies

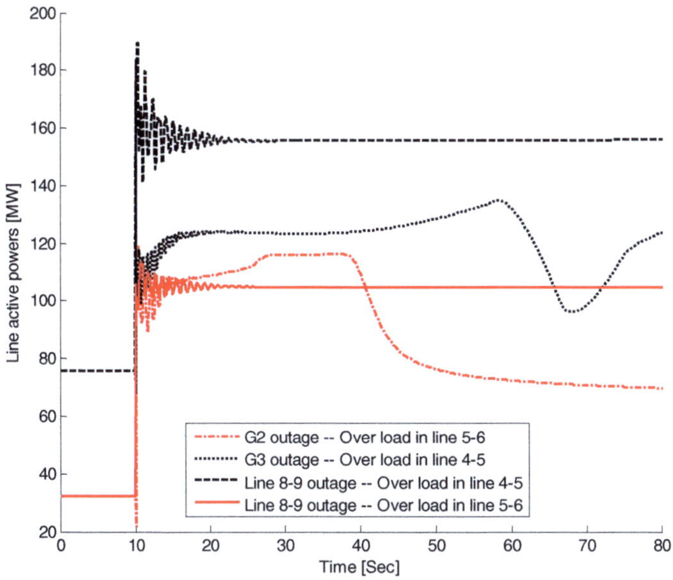

Fig. 9. Line over-loading in different contingencies

instability in an emergency condition. However, it is possible to see an inverse behavior during another test scenario. For example, considering Figs. 7 and 8 shows that diminishing of voltage is more considerable than frequency, when G1 is tripped.

Fig. 9 shows the system line powers, following a range of possible contingencies. This figure shows those line powers which are exceeded from the specified limits. It is shown that the outage of line 8-9 increases the overload on lines 4-5 and 5-6, significantly. This event may lead to lose these lines and to start a cascading failure.

In summary, the performed nonlinear simulations clearly indicate that individual monitoring of voltage and frequency may not determine the rate of severity for a contingency at the starting time of an event, securely. Both performance indices are needed to measure the severity rate of a contingency (see also G3 outage scenario in Figs. 7 and 8). It will be shown later that how the mentioned indices could be used for the sake of contingencies ranking.

4 Developed ANN-Based Emergency Control Scheme: An Overall View

4.1 Proposed Intelligent Control Framework

In the present chapter, a new intelligent emergency control is proposed. The developed emergency control scheme is summarized in Fig. 10. Following an event, the first artificial neural network (ANN-1) labels the contingency by a specific number. If the predicted contingency is a sever one, it triggers the designed load shedding algorithm which is described in Section 1.5. For this purpose, an other trained neural network (ANN-2) uses the specified contingency number together with the measured tie-line active/reactive powers to estimate the system P-V curve. The amount of load should be shed is immediately computed using an estimated P-V curve. The main steps of the proposed algorithm are as follows:

1. At the first step, all contingencies should be identified and a special binary number to be allotted to each contingency. The sever contingencies have different binary numbers but for other contingencies an identical number is used. A special number is also considered for normal operation of the system. The contingency numbering process is done by the ANN-1.
2. The ANN-1 should be properly trained using the tie-lines active and reactive powers. The training data could be generated through a static analysis, and are used for a wide range of load patterns and various generation dispatching schemes. These scenarios are repeated for different contingencies.
3. The ANN-2 should be trained to predict the P-V curve. The inputs of ANN-2 are the severe contingency's number estimated in the previous step, and the tie-line active and reactive powers before the related event. The ANN-2 outputs are the coefficients of a 4th degree polynomial function that estimate the P-V curve. It should be mentioned that ANN-2 is designed to predict the P-V curve for severe contingencies, only. Therefore, generation of training data is not so time consuming. The training data are generated for the various load patterns, generation dispatching scenarios, and the wind speeds. Finally the P-V curve is extracted and fitted by a 4th degree polynomial function, which is described in Section 1.4.4. The updated weights after training of ANNs are saved, and then will be used to implement ANN-2 for the online prediction.
4. The estimated PV curve in the previous step is used to determine the amount of load that should be shed to move the operating point into a desirable region. The distance between operating point and the bounds of desirable region determines the cumulative amount of load which is needed in load shedding. The operating point on the PV curve is calculated by crossing the normal power (overall power system load before the contingency event e.g., 315 MW for the present test system) with PV curve. The desirable point is a point that the weakest bus voltage is upper than the allowable magnitude or the stability margin to be larger than the specified value (see Section 1.5.4).
5. A new load shedding algorithm (Section 5) is implemented to shed amount of load in the assigned steps based on simultaneous using of voltage and frequency indices. The cumulative amount of loads that should be shed is limited by the determined value in the previous step. When the cumulative amount of load shedding reaches to this limit the load shedding is interrupted for 10 seconds. After this relatively long delay, if more load shedding is needed, the load shedding will be continued. This delay also exists in the Florida Reliability Coordinating Council (FRCC) regional UFLS implementation schedule [37]. But it is noteworthy that, in the FRCC schedule, the 10 seconds delay is

used when the cumulative loads reaches to a predetermined value (about 41 percent of total the load).

6. To assign the first step of load shedding, the initial rate of frequency change is used to determine the amount of load that should be shed. For example if the magnitude of frequency gradient is larger than 1.5 Hz/sec, the amount of load that should be shed in the first step to be fixed at 15% (instead of 9%) of total power system load.

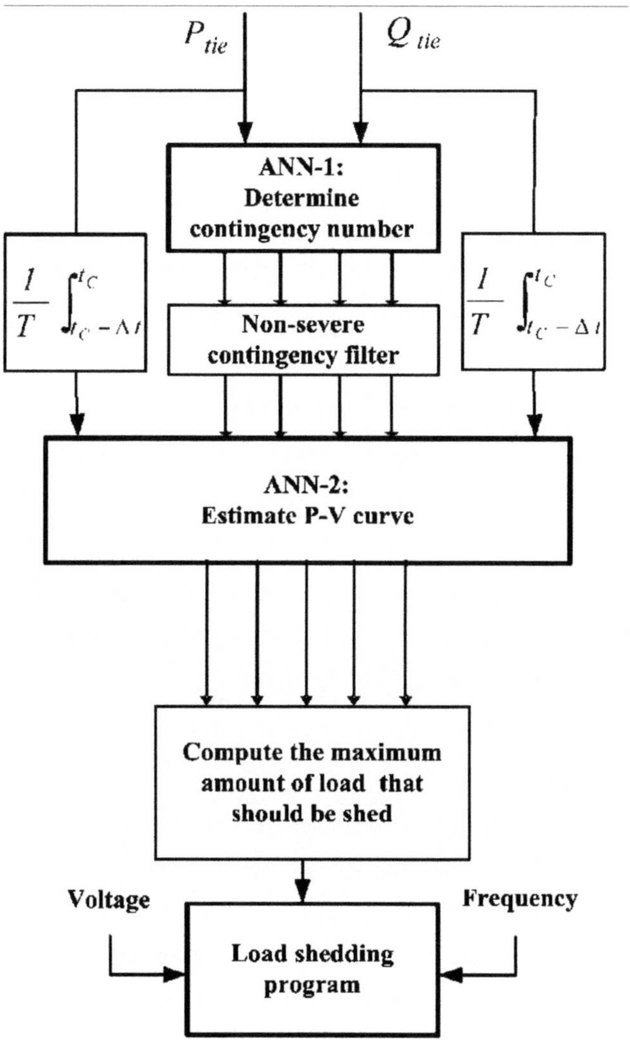

Fig. 10. The proposed intelligent power system emergency control scheme

4.2 Voltage Stability Margin and P-V Curve

In the proposed power system emergency control plan, the on-line predicting of P-V curves is used to determine the system stability margin. A P-V curve is generally obtained from the static studies, and for application in dynamic/real-time studies, it should be revised. On the other hand, a large penetration of wind turbines certainly affects the system P-V curves, which may not be clearly observed in a well-known pure static analysis.

Voltage Stability and Conventional Power Flow

Voltage stability is the ability of a power system to maintain/restore the voltages of all buses in a stable region after a disturbance [37]. The P-V curve is a suitable tool to estimate the stability margin of a power system around its operating point in the steady state. By increasing the load power connected to a bus, the bus voltage decreases and leads to change the voltages of other buses. Therefore, the system operating point moves, until reaches a point that no more power injection is possible. This is the bifurcation point and usually is used to determine the maximum load-ability of a system [20].

The behavior of the active power injection and bus voltage for two bus system can be calculated from the following equation [37].

$$V^4 + (2P.tan\emptyset - 1)V^2 + P^2.sec^2\emptyset = 0 \tag{3}$$

where V, P and $\emptyset$ are load bus voltage, active power absorbed by the load, and the angle between the active and reactive parts of load, respectively.

The critical point of this equation which shows the maximum load ability of the power system can be easily computed. For a power system with more than two buses, deriving a relationship between the active power injection and the bus bar voltages, and also tracing the P-V curve is not as simple as derived in (3). The reason is in complexity of the nonlinear power system equations that should be solved by appropriate recursive methods. Therefore, a power flow program should be run to achieve each point of P-V curve. By increasing the load power and near to the bifurcation point because of singularity of the Jacobin matrix in these points, the power flow equations cannot be solved.

To overcome above problem, the continuation power flow (CPF) method can be used via an appropriate predictor-corrector technique [36, 37]. Adding rows and columns to the Jacobin matrix, and predicting initial conditions for each point of P-V curve are helpful to solve the equations. In Fig. 11, the P-V curves of the IEEE 9-bus test system produced by the CPF method are shown. As illustrated, the maximum loadability of the given system is approximately limited to 720 MW.

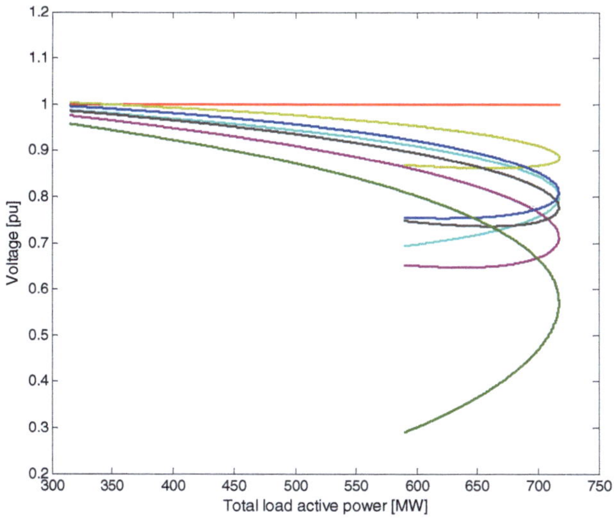

Fig. 11. The produced P-V curves by the static conventional CPF for the test system

With considering various important issues such as the generator Q-limits [26, 24], different bus load patterns and generator dispatching [31], the CPF method to be more useful. The same technique can be also used to analysis the line outages [15].

P-V Curves for Sudden Load Changes and Line/Generator Outage

As mentioned, the CPF method is known as a powerful tool for static voltage stability analysis. In this method, it is assumed that the rate of load change is slow, and all points on the P-V curve are in their steady state values. It is also assumed that all generators in the system have enough time to trace the load changes on the produced P-V curve. Since, the load variations are time dependent and the ability of power system to trace the load variations depends on the turbine governor droop characteristic, total power system reserve, the generator Q-limits, and the generator ramp rates. In other hand, an emergency condition following a sudden load change, generator trip or line outage is generally occurred in a short time. Therefore, using of P-V curve in the CPF techniques for an emergency control issue needs a revision [36].

Dynamic continuation power flow (DCPF) can be considered as a solution to plot the P-V curve in an emergency condition. In the DCPF, each point of a P-V curve is also known as a system equilibrium point. While, in emergency control strategies, such as the presented one in this chapter, the real-time voltage magnitude (not equilibrium point) is used to determine the voltage stability margin. To achieve a more suitable P-V curve for the emergency control studies, it is needed to incorporate the load variation time in the P-V curve producing algorithm.

In the conventional CPF, a continuation parameter is defined to lead the operating point in the P-V curve tracing. This parameter can be moved between the state variables. Therefore, it is possible for the P-V trajectory to turn back after reaching to the bifurcation point. However, to implement the CPF in a time domain simulation, it is needed to use one continuation parameter only, that varies through tracing path, before and after the bifurcation point. To find a suitable parameter, consider the bellow equations [20, 36]:

$$S_L = P_L + jQ_L = V_L \times {I_L}^* = \frac{|V_L|^2}{{Z_L}^*} = {Y_L}^* \times |V_L|^2 \tag{4}$$

$$V_L = \frac{Z_L}{Z_L + Z_t} \times E_t \tag{5}$$

where S_L, Y_L, V_L, Z_L, Z_t, and E_t are load power, load admittance, equivalent load voltage, load impedance, transmission line impedance and bus voltage, respectively. Usually, the increasing of load power is the conclusion of adding more consumer devices to the power system. The loads as parallel admittances and adding more loads at a bus increase the equivalent admittance of the loads at the same bus. The voltage of the load bus V_L is decreased by increasing the load admittance Y_L (5). As it is shown in (4), Y_L and V_L determine the total power injection. At the beginning of an overload event, the increasing of Y_L is more considerable than decreasing of V_L; and in result the amount of power injection to be increased. But by adding more loads, the P-V curve reaches a point that the reduction of voltage square overcomes the increasing of the equivalent admittance and the power injection start to decrease. Based on above description the load admittance increases on the whole P-V curve trajectory. Therefore, it could be a suitable continuation parameter for the implementation of CPF in the time domain simulation. Using the Z-constant load model [20] and decreasing the load impedance during the simulation time with a proper rate, and measuring the load active power and its voltage, a full trajectory for the P-V curve will be produced. It should be noted that to solve the system differential equations, the simulator/solver software must be equipped with appropriate variable stepped-size techniques.

In comparison of conventional P-V curve producing methods, the above mentioned methodology in the time domain simulation environment contains following advantages:

1. By changing the rate of load admittance variation in the P-V curve tracing scenario, different P-V curves for different tasks such as short term and long term studies could be produced.
2. Many constraints, such as generator Q-limits, ramp-rates, droop-characteristics, and other limits could be easily included.
3. Various load patterns and generation dispatching scenarios could be considered.

4. Using the proposed method, it is possible to obtain the system P-V curves for sudden load changes in an emergency condition, while they cannot be achieved by the conventional CPF algorithms. It will be easily done by modeling the sudden load changes with a sleep slop ramp function.
5. The proposed method can be used to derive a P-V curve for the power system with WTGs by considering their dynamic models. It may be useful to explore unknown behaviors of the power system in the presence of wind turbines, and analyze their effects on the voltage stability margin [36].

It should be mentioned that using this analysis technique may not suitable for the fast dynamic cases such as fast contingency ranking. In the present emergency control scheme, this algorithm is only used for the severe contingencies. Fig. 12 compares the P-V curves derived by conventional CPF and above mentioned algorithm. As shown, the stability margin of a power system following a sudden load change is considerably less than one achieved by the steady state analysis. Therefore, to use the P-V curves in the emergency control strategies, a revision (like as explained above) is needed.

A comparison between the derived P-V curves using two different load rate changes is given in Fig. 13. The applied load variation patterns by increasing the load admittance are shown in Fig. 14. It can be seen that the stability margin for the rapid load change is less than the slow load change. These figures also show that the present method is capable to drive P-V curve for the rapid load change cases.

The P-V curves for the cases that a line or a generator is out of service or has been tripped are also important. When the system reaches its stable state, the tracing P-V curve scenario will be start. In order to plot P-V curve for the severe contingencies that

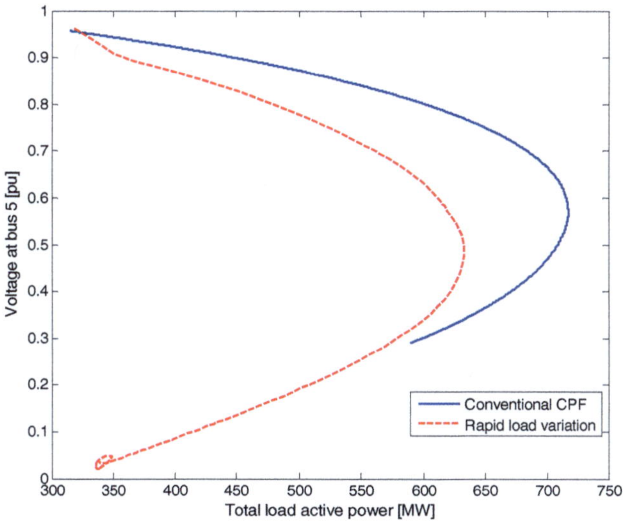

Fig. 12. P-V curves for the conventional CPF, and rapid load variation

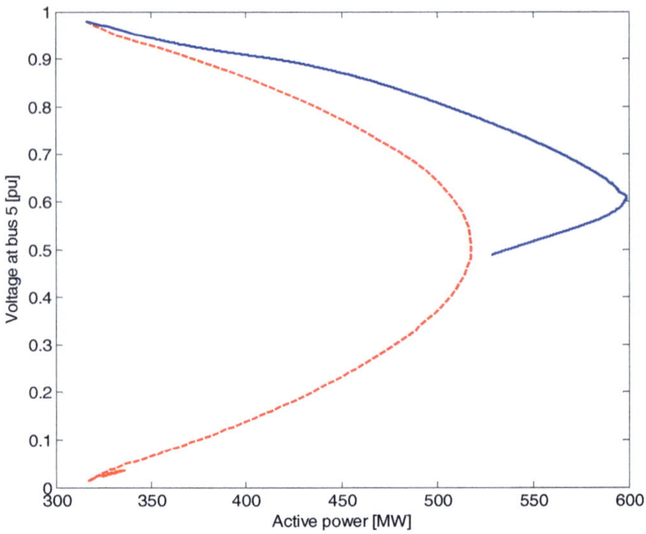

Fig. 13. P-V curve for the rapid (dashed) and slow (solid) load variation patterns (shown in Fig. 14). Only load at bus 5 has been changed

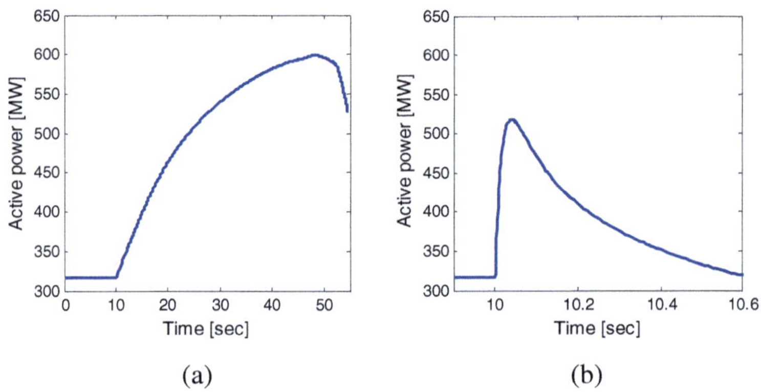

Fig. 14. Applied load variation patterns; (a) slow change, (b) rapid change

system never finds a stable state, an auxiliary technique should be used to lead the system into a stable condition. For example, one may curtails some load blocks to find a steady state instead of increasing the generator outputs.

The P-V curves for the normal (base case) condition and two different scenarios (line 8-9 and G1 outages) are shown in Fig. 15. These curves are achieved by immediate increasing load at bus 5. It is shown that the G1 outage is more unsecure than other test scenarios. The system stability margin is significantly reduced, when G1 is tripped.

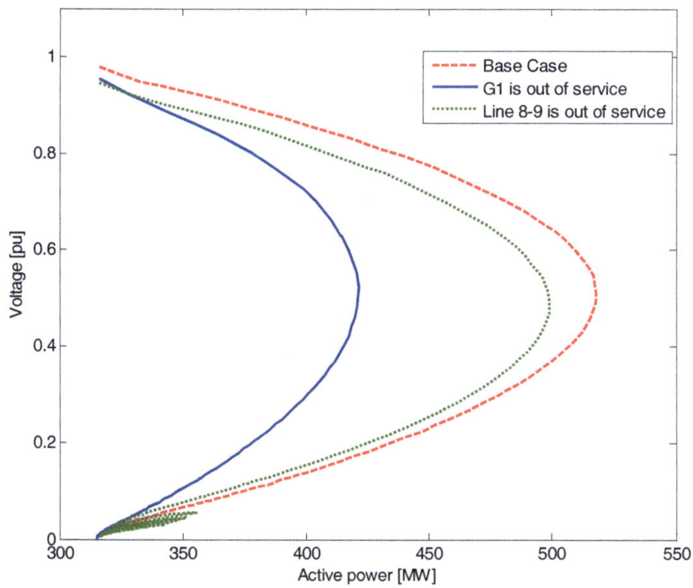

Fig. 15. P-V curves for two severe contingencies at the weakest bus

4.3 ANN-1 to Determine Contingency Number

In an emergency condition, an appropriate control decision should be immediately made. Therefore, using recursive algorithms to find optimal solutions are not implementable for this kind of situations. On the other hand, direct using of some tools such as deriving P-V curve to approach optimal solutions are usually time-consuming. However, these tools could be applicable, if one uses the fast estimation/prediction techniques. Because of predictability and immediate responsibility, the ANN can be considered as a powerful alternative to do the mentioned task.

A contingency number is an arbitrary code that determine which generator/line has been tripped. In order to obtain the contingency numbers, one may use remote terminal unit (RTU) signals in a supervisory control and data acquisition (SCADA) center. This method is too slow to use in an emergency condition. That is why, the measurements form the existing fast devises such as phase measurement units (PMU) are used to feed the performed neural network to provide a contingency number which actually shows the fault location.

Using P and Q by an ANN to predict the contingency number is suggested in this section. As shown in Fig. 10, if a severe contingency occurs in one of power system regions, the assigned ANN (ANN-1) is able to predict the contingency number by measuring the tie-line powers. This method could be also replaced by the existing methods that use the system frequency gradient as an input variable.

Selection of Input Variable

Selecting the features that used as input variables for the ANN-2 plays an important role in the efficiency of the proposed intelligent framework. The ANN-2 should be enough fast to response in a few milliseconds following a contingency event. Since, the majority of the variables oscillate in the contingency duration, using the measured values in a real-time measurement and making appropriate decisions based on results is difficult.

An experimental study (under specific assumptions) on the tie-line active powers measurement following loss of generation shows some interesting results. Fig. 16 shows the tie-line active powers following loss of G1.

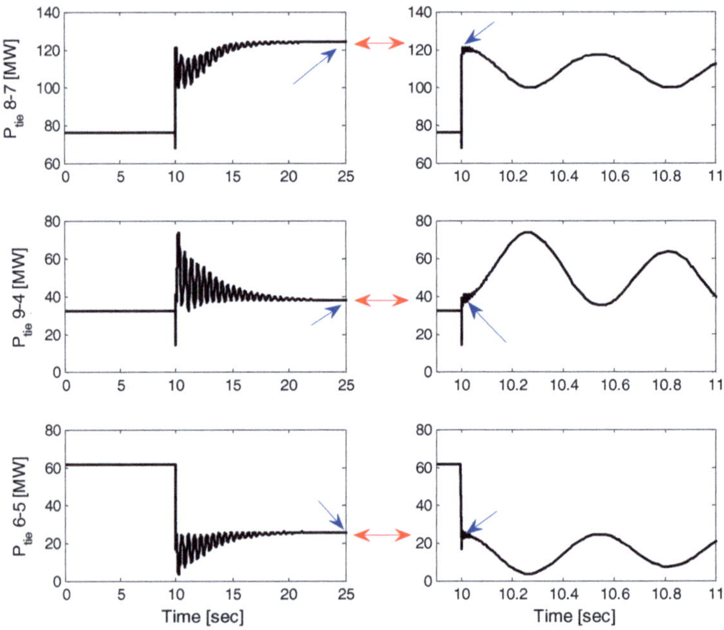

Fig. 16. Tie-line active powers following G1 loss; right-side figures show the zoomed views around 10s

As illustrated, the values of line active powers during a few milliseconds after the contingency (right-side column in the figure) are equal to their steady state values that could be derived even from the steady state power flow programs. Fig. 17 shows this phenomenon in the presence of DFIG type WTGs. In the emergency control strategies it is important to find a picture from the post contingency conditions, and analyze it to make a suitable decision and effective control action. Based on the above study the tie-line powers are used as inputs for ANN-1 to determine the contingency at the beginning of a serious fault.

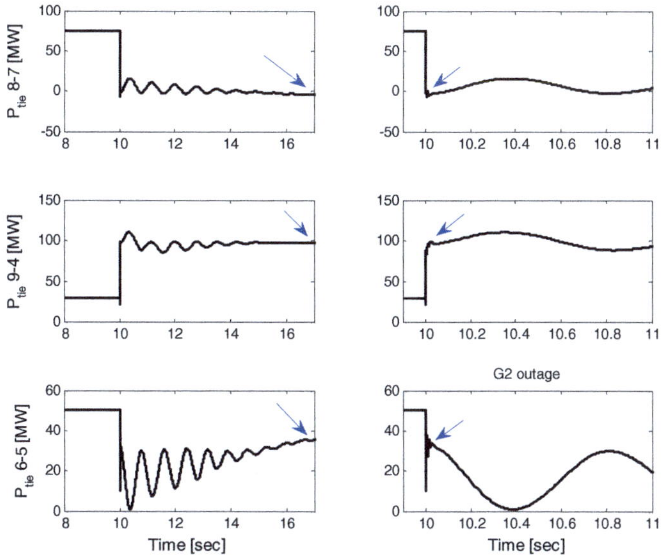

Fig. 17. Tie-line active powers in the presence of DFIG following G2 loss; right-side figures show the zoomed views around 10s

Identify the Contingency Number

As shown in Fig. 15, the P-V curves completely depends on the power system topology. The line/generator outage affects the P-V curves, significantly. In the other hand, the performance of an ANN is considerably depended on the features that used as ANN inputs [30]. The main target of ANN-2 is to estimate the power system P-V curve following a serious event to determine the maximum amount of load for shedding, in order to restore the power system operating point into a point with an acceptable stability margin.

The performed study shows that using the tie-line active and reactive powers as inputs are not adequate to predict the P-V curve. Extracting suitable new features and adding them to the inputs array make ANN-2 more efficient. The new features must concern the power system topology.

The inputs of the first ANN (ANN-1) are the tie-lines active and reactive powers and the output is a binary code representing the number of a severe contingency. For each severe contingency, a specific number is allotted and for all other (non-severe) ones an identical number is allotted. Fig. 18 shows the outputs of ANN-1 following loss of G3. The number that allotted for this contingency is 3, and the number that allotted to the base (normal) case is 7. As shown, the output of the power system before the contingency is 0111 and after loss of G3 is 0011. It should be mentioned that appropriate thresholds are applied to the outputs of the ANN-1, such that the binary digits bellow 0.5 are considered as 0, and above 0.5 are fixed at 1.

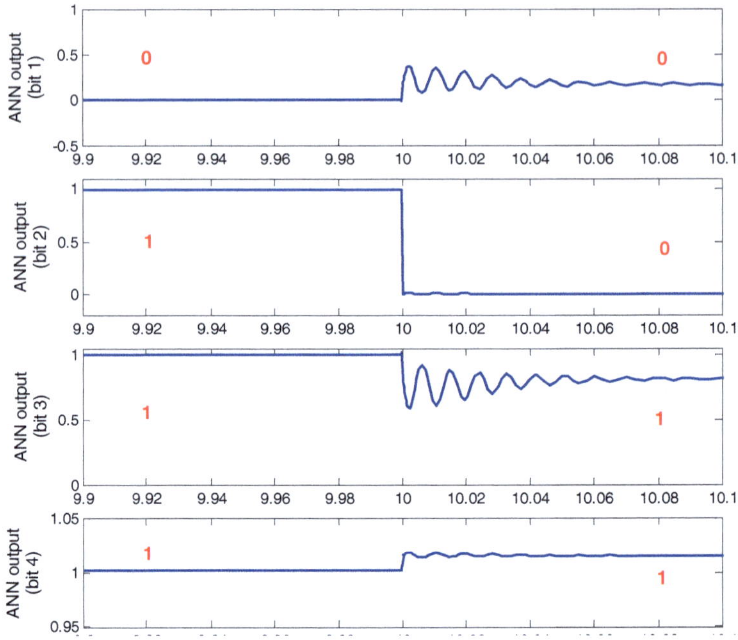

Fig. 18. The output of ANN exactly at the time of G3 outage. The target output for this case is the binary digits 0011 (3).

4 ANN-2 to Estimate the P-V Curve

The ANN-2 target is estimating of system P-V curve for post-contingency topology in a few milliseconds following a serious event. The method used for this purpose should satisfy some characteristics. It must be enough fast, and should consider system topology and various uncertainties due to variable nature of load, generation and wind speed.

The Thevenin's equivalent circuit for an existing real power system is known as a solution to formulate power system topology [33]. However, on-line finding of an appropriate Thevenin's equivalent system concerning the variable nature of load, complexity of real system, various uncertainties and practical constraints, specifically in the presence of wind turbines is difficult.

Considering the real-time prediction property and learning ability of neural networks, the ANN-2 is designed to solve above problem. In order to taken account the system post fault topology, the contingency information is used as the ANN-2 input.

As has already mentioned, the contingencies are numbered and the numbers can be identified in a few milliseconds after the contingency event by ANN-1, or an alternative method. For the example at hand, contingencies are numbered as four digits binary codes.

Furthermore, the tie-line active and reactive powers are used to formulate the load/generation and wind speed changes. As shown in Fig. 10, the average of these powers among Δt (here, 5 seconds) before the contingency is supplied to the ANN-2. This input can be formulated as follows.

$$P_{ave} = \frac{1}{T}\int_{tc-5}^{tc} P(t)dt \ , \ Q_{ave} = \frac{1}{T}\int_{tc-5}^{tc} Q(t)dt \tag{6}$$

where, t_C is the time of fault (contingency time), P (Q) is the tie-line active (reactive) power, and T is the time period (here, $T = \Delta t = 5s$).

The output layer of ANN-2 includes five neurons, corresponding to the coefficients of a four-degree polynomial function to estimate the P-V curve in the following form,

$$P = a_1V^4 + a_2V^3 + a_3V^2 + a_4V^1 + a_5 \tag{7}$$

where a_1 to a_5 to be determined by the outputs of ANN-2. The data for above P-V curves is produced by the algorithm described in Section 1.4.2. The developed P-V curve estimation method is applied to the test system.

A comparison between original and estimated P-V curves (using the given 4th order polynomial in (7)) for three tests (normal operation, G2 outage, and line 8-9 outage) are shown in Fig. 19. The obtained polynomial coefficients (7) are given in Table 1.

Table 1. 4th order polynomial coefficients

		a_1	a_2	a_3	a_4	a_5
Case1:	Target	0.9349	-2.7315	-0.3821	2.0858	0.0118
Normal operation	Predicted	0.9237	-2.6442	-0.4827	2.1199	0.0088
Case2:	Target	0.2960	-1.4384	0.0129	1.0492	0.0048
G1 outage	Predicted	0.2347	-1.2797	-0.1056	1.0721	0.0021
Case3:	Target	-0.3862	0.3249	-2.7286	2.7086	-0.0694
Line 8-9 outage	Predicted	-0.5644	0.6790	-2.9786	2.7757	-0.0759

4.5 ANN-1 and ANN-2 Configurations

A three layer back-propagation neural network is used for both ANN-1 and ANN-2. The activation functions for the hidden layers are in form of tangent-sigmoid function, and the output activation functions are linear.

The number of neurons in hidden layer is chosen via try and error method, and for the scaling of neural network's inputs/outputs, *min-max* method is applied. The parameters of applied min-max function to the training data are saved and will be used for the actual test data. Fig. 20 shows a general configuration of ANN-1 and ANN-2 for the online applications.

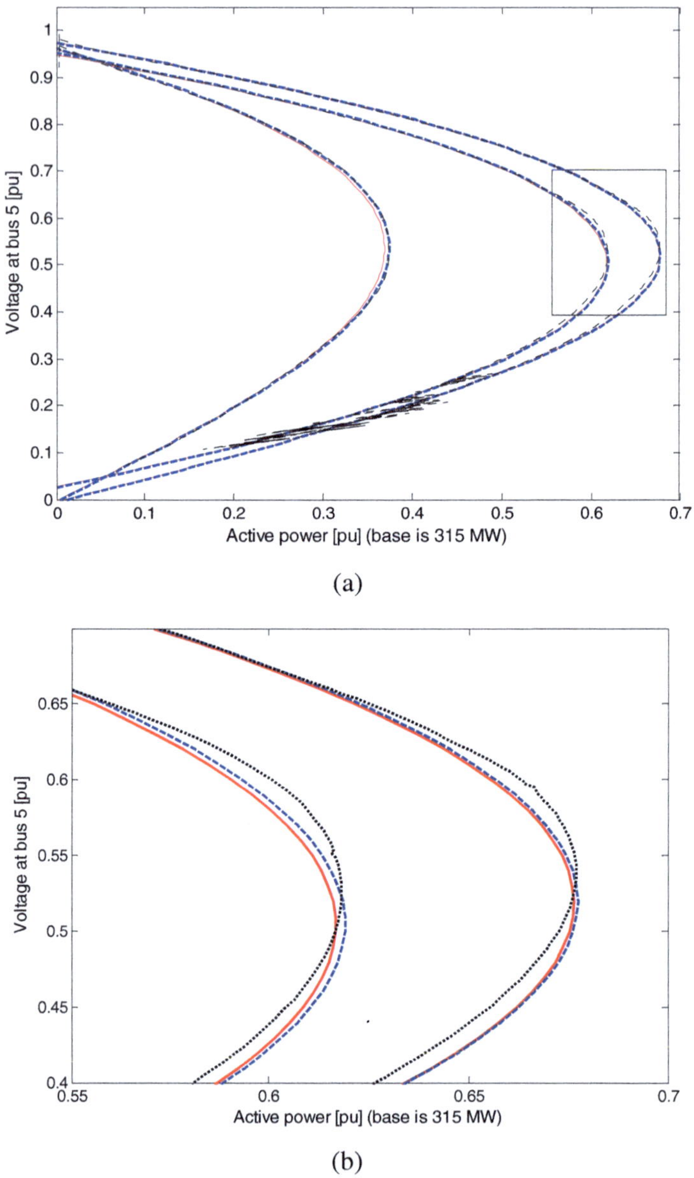

Fig. 19. P-V curves; ANN-2 output (solid), Original (dotted), and fitting by polynomial (dashed). Figure (b) shows a zoomed view around 0.6 pu.

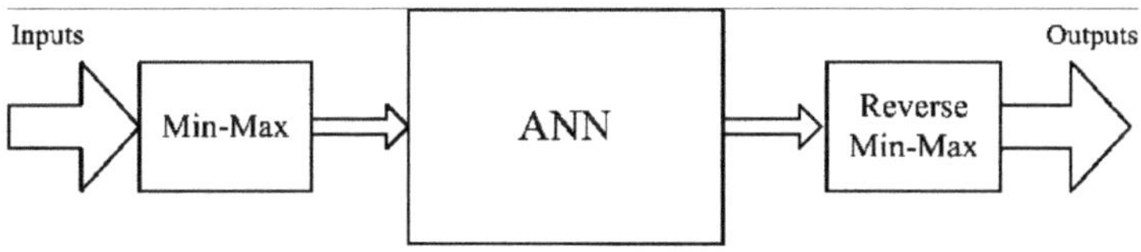

Fig. 20. ANN configuration for the online application; the min-max and reverse min-max parameters come from training data

In the min-max method, each element of the input/output vector is mapped into interval [-1, 1], based on the training data samples corresponding to the following variable.

$$y_i^t = (y_{max} - y_{min})\frac{x_i^t - x_{min}^i}{x_{max}^i - x_{min}^i} + y_{min} \tag{8}$$

where, x_i^t is the i-th element of the t-th sample vector of the training data, y_i^t is the i-th element of the t-th sample vector of the training data mapped into the interval [-1, 1], and $x_{max}^i(x_{min}^i)$ is the maximum (minimum) value of the training vector that contains i-th input/output element (here, $y_{min} = -1$, and $y_{max} = 1$).

Therefore, for each input and output vector elements there are a minimum and a maximum values derived from training data. These values should be saved and used beside the trained neural network for predicting purposes.

The used tangent sigmoid activation function is as follows:

$$f(x) = \frac{2}{1 - e^{-x}} - 1 \tag{9}$$

For learning of neural network, the corresponding P-V curves for both normal operation and major contingencies (which are identified in the contingency ranking step) should be obtained. The proposed ANN-2 has 10 linear neurons in its first layer, corresponding to the number elements in the input vector (3 tie-line active powers, 3 tie-line reactive powers, and 4 digits for the contingency number). The number of neurons in form of sigmoid function (8) is 15, and the number of output neurons is fixed at 5; corresponding to the coefficients of estimating polynomial (7).

5 A New Load Shedding Algorithm

5.1 Load Shedding

Load shedding (LS) is one of emergency control actions to protect the system following a major fault which seriously deviates the system frequency and/or voltage. The LS curtails amount of load in the power system until the available generation could supply the remind loads. If the power system is unable to supply its active and reactive load demands, the under-frequency and under-voltage conditions will be intense. Many algorithms for UFLS and UVLS have been proposed

[2, 3, 6, 13, 15, 28, 35, 29]. The number of LS steps, amount of load that should be shed in each step, the delay between the stages, and the location of shed load are the important objects that should be determined in an LS algorithm.

To prevent the post-load shedding problems and over loading, the location bus for LS will be determined based on the load importance, cost, and distance to the contingency location. The most LS schemas proposed so far used voltage and frequency parameters, separately. The underfrequency and undervoltage relays are working in the power system without any coordination. Following, some examples are given to illustrate the necessity of considering both of voltage and frequency indices to achieve a more effective/adaptive and comprehensive LS scheme.

5.2 Simultaneous Using of Voltage and Frequency

In this section, it is shown that considering both voltage and frequency indices are needed to achieve a more effective emergency control plan [36]. Fig. 21 shows the voltage and frequency deviations for two different LS scenarios following the same contingency. In these tests, G1 is tripped at 10s. In scenario 1, only 9% of total system active power is curtailed, while in scenario 2, in addition to 9% active power, 9% of total reactive power is also discarded. Both scenarios shed the load when the frequency falls below 59.7 Hz as used in some existing LS standard such as FRCC standard [14]. Considering the frequency and voltage behavior in the performed two scenarios, some important points are achieved.

A majority of published research on the UFLS issue considered the active part of load only; while by considering the reactive power part, the frequency decline will be affected as shown in Fig. 21b. Furthermore, in the actual power system, the loads contain both active and reactive parts. Figs. 21a and 21b do not show which scenario is more effective. Fig. 21a illustrates a better performance for scenario 2, while Fig. 21b shows an inverse result. This simulations show that by individual monitoring/using of frequency and voltage there is no guaranty to achieve an effective LS strategy.

Recall Figs. 2, 4 and, 5 and consider the frequency and voltage deviation for the case that IG with STATCOM is interconnected to the power system. Fig. 2b shows that the post contingency voltage behavior of the system in the case of IG with STATCOM is much better than IG without STATCOM, while Fig. 4 shows an inverse result for the system frequency response. Fig. 5 also shows that voltage and frequency may behave in the opposite directions.

From above descriptions and the performed simulation results, it is realized that considering just one of frequency and voltage indices cannot lead to an effective/optimal LS plan, especially when the reactive power is incorporated into studies. The WTGs generally use induction generators that consume the reactive power. The FACTS devices are usually used to compensate the reactive power for these cases may influence the amount of frequency decline.

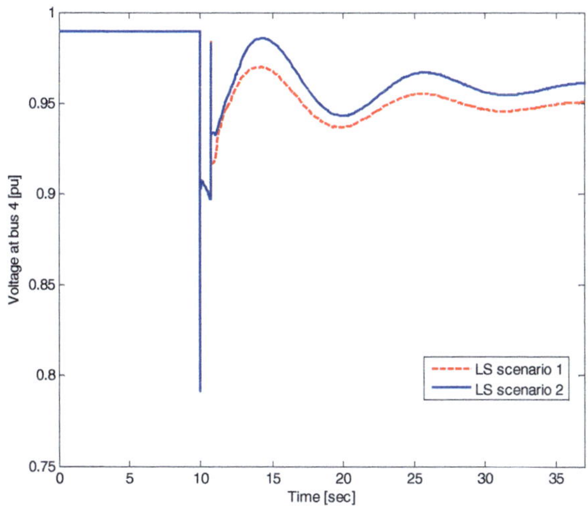

(a)

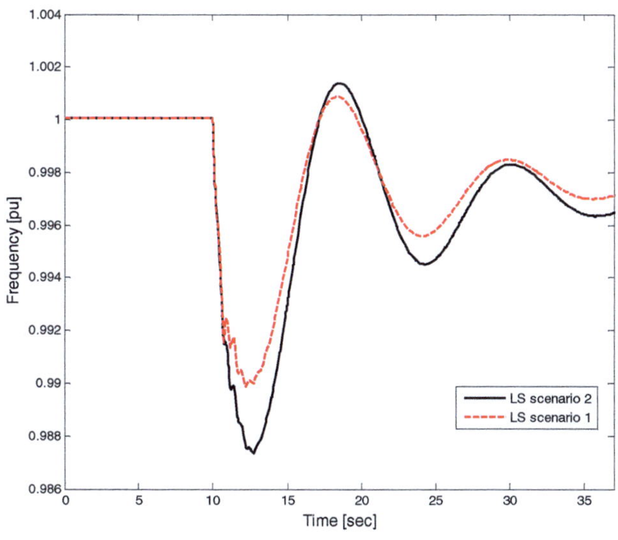

(b)

Fig. 21. System response for two different LS scenarios; a) voltage deviation, b) frequency deviation

5.3 Proposed LS Algorithm

The coordination between conventional UFLS and UVLS as separate algorithms is very difficult and even may be impossible. Therefore, both voltage and frequency should be used in the same LS program, simultaneously.

Following a contingency event, the power system operating point deviates from its stable pre-contingency state. If the contingency is not severe enough, the system may converge to another stable point. Otherwise, the system states move to unstable region. As mentioned, for the emergency control purposes, voltage and frequency are two suitable observable variables that could illustrate the state of system following an event. Fig. 22 shows the trajectory of the system states in the voltage-frequency plane, following a contingency with two scenarios: unstable trajectory, and stabilized trajectory using an LS plan. The states that used in this trajectory are Δf and ΔV, as follows.

$$S = \widehat{\Delta f} + j\widehat{\Delta V} \tag{10}$$

and,

$$\widehat{\Delta f} = \frac{\Delta f}{f_0}, \qquad \widehat{\Delta V} = \frac{\Delta V}{V_0} \tag{11}$$

where, f_0 and V_0 are the frequency and voltage before contingency. To design a new LS algorithm based on the above state variables, some threshold boundaries should be defined instead of threshold values that are used in the conventional LS algorithms.

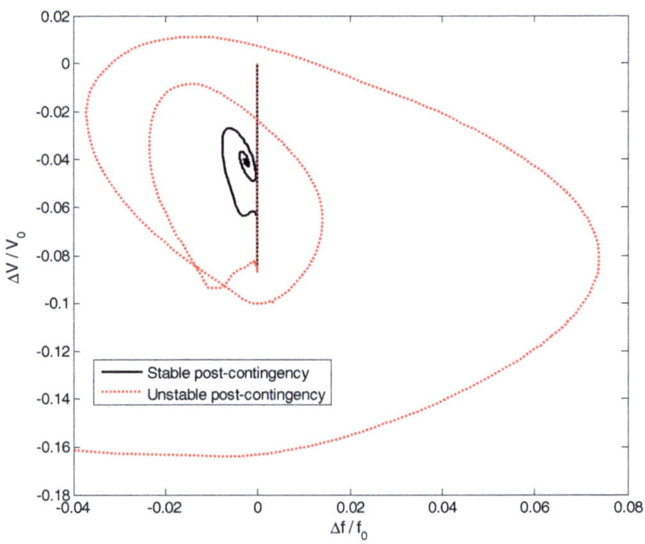

Fig. 22. Phase trajectory for stable and unstable post-contingency scenarios

One may suggest the circular boundaries. But since the threshold movement sizes in two directions (the UFLS and the UVLS) are not the same, the elliptical boundaries are much better. Therefore, assuming

$$x^t = \widehat{\Delta f} = \frac{f_t - f_0}{f_0}, \; y^t = \widehat{\Delta V} = \frac{V_t - V_0}{V_0} \tag{12}$$

and according to Fig. 23, the control load shedding (CLS) switches can be defined as follows:

$$CLS_i = \left(\frac{x^t}{a_i}\right)^2 + \left(\frac{y^t}{b_i}\right)^2 \tag{13}$$

Where, the x^t and y^t are new variable states. The f_t and V_t are the frequency and voltage at time *t*. The a_i and b_i parameters can be computed as shown in Table 2, and the threshold can be fixed in a point satisfying CLS_i>1. As shown in Table 2, the overall framework of the present LS scheme is close to the LS schedule introduced by FRCC [14], and it is rewritten in Appendix (Table 7).

Each step is determined by an ellipse and when the phase trajectory reaches to each ellipse, the corresponding LS step will be triggered. Fig. 23 shows the LS steps on the phase trajectory plan in the case of losing G2, the largest generator in the test system. The time delay between steps should be added to the LS algorithm. The voltage and frequency may need to pass through a low pass filter before entering into the algorithm. Existing practical constraints should be also, considered in the proposed scheme.

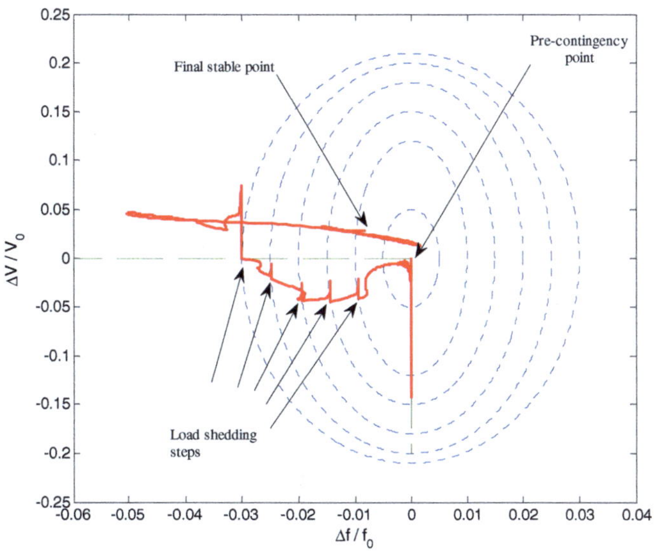

Fig. 23. New LS scheme

Table 2. New LS schedule

LS Steps	a_i	b_i	Time delay [sec]	LS (%)
A	0.3/60	0.05	0.28	9
B	0.6/60	0.12	0.28	7
C	0.9/60	0.15	0.28	7
D	1.2/60	0.18	0.28	6
E	1.5/60	0.2	0.28	5
F	1.8/60	0.21	0.28	7
....			....	

The initial conditions at time of each LS step are an important factor to obtain a more effective protection plan. For example, shedding the load in the conditions that voltages of the buses are above their pre-contingency values may result more bad circumstance.

That is why, in the proposed algorithm, the loads are allowed to be shed only at the third quarter of the phase trajectory plane. In order to prevent shedding of load in the voltage dip conditions (e.g., when a short circuit is not eliminated and the frequency is decreased while the voltage is normal), a specific margin will be considered. Based on this margin, the loads are allowed to shed only when the angle of *S* in (4) is between $(\pi+\alpha_1)$ and $(\frac{3\pi}{2}-\alpha_2)$.

The phase trajectory of the presented LS algorithm that contains time delay and dead-margin is shown in Fig. 24. Because of time delay, the LS steps are not exactly fired on the elliptical borders. Fig. 25 shows the system voltage response at bus 4, following loss of G2. The new LS scheme is compared with the conventional UFLS. It is shown that the new LS scheme is more efficient to prevent over load shedding and over voltage conditions after running the algorithm.

5.4 Determine Amount of Cumulative Load That Should Be Shed

The ANN-2 output determines the coefficients of the polynomial function (7) that estimate the P-V curve of the power system. In the present chapter, this curve is used to predict the amount of load that should be shed to stabilize the system after a severe contingency. Here, it is explained that how the P-V curve is applicable for the mentioned purpose.

The P-V curve is simply estimated using a polynomial function and the designed neural network. The horizontal axis of the P-V curve which is obtained from the ANN-2 output is in per-unit at the base of total power system load, and only the amount of load that is added to the power system is considered. As has mentioned, to produce a P-V curve following a severe contingency, it may be needed to curtail some loads manually to lead the power system operating point to a stable point before starting the scenario. By considering these points, the P-axis of the P-V curve to be re-scaled as follows.

$$P_{new} = P_{old} \times P_{base} - P_c + P_{base} \tag{14}$$

where, P_{old}, P_{base}, P_c and, P_{new} are per-unit values of P-axis, base case total active power, manually curtailed active load, and re-scaled P-axis, respectively.

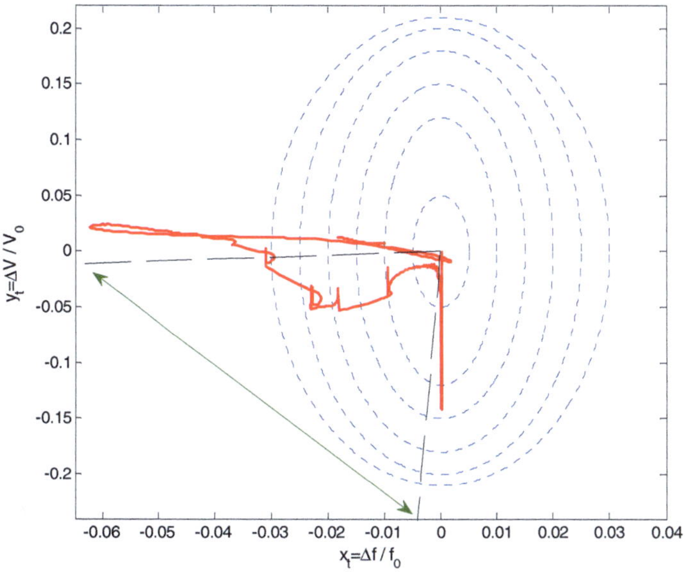

Fig. 24. The proposed LS scheme considering time delay and the permitted shedding region

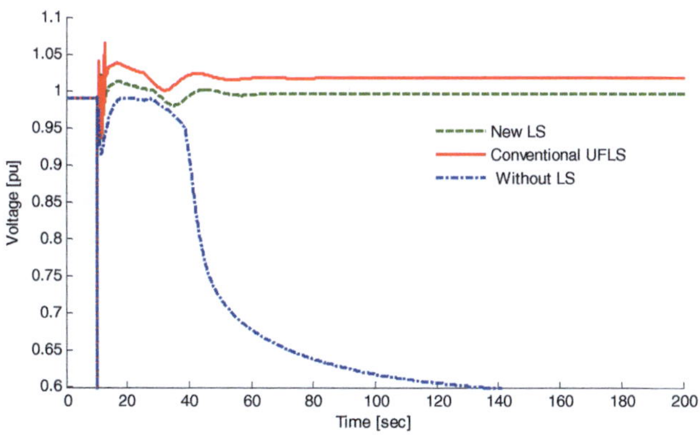

Fig. 25. Bus 4 voltage using different LS schemes following loss of G2

Using the produced P-V curve, the amount of load that should be shed could be calculated by means of one of following methods: 1) shed amount of load to restore the power system bus voltages upper than a minimum permissive voltage value (for example, 0.95 pu), 2) shed amount of load to achieve a minimum permissive power system voltage stability margin.

As shown in Fig. 26, the stability margin is defined as distance between the power system operating critical points. The critical point is the saddle node of the P-V curve, or it can be defined as a point that its corresponding voltage value reaches to a permissive value (for example, 0.8 pu [29]). Fig. 26 shows the P-V curve for the case of G2 outage. To provide this curve, 125 MW has been manually curtailed after making the G2 out of service. This curve is predicted by ANN-2, and re-scaled by means of equation (1.14). For the sake of comparison, the PV curve for the base case (normal operation) is also shown in Fig. 26. In the present example, the base case active load is 315 MW. By intercrossing the vertical line of 315 MW with the produced PV curve, the post-contingency power system operating point to be estimated. By considering this point and using one of above described methods, the amount of load that should be shed will be calculated. For example, as shown in Fig.26, to restore the bus voltages to 0.95 pu, it is needed to shed amount of 96 MW system load.

The LS steps and the amount of shaded load in each step are illustrated in Fig. 27. As shown in this figure, the cumulative load that should be shed is fixed to 96MW. Fig. 28 shows the system voltage and frequency deviation following application of the proposed emergency control strategy. In the performed simulations, all wind turbines are DFIG type, which supply ten percent of total system load.

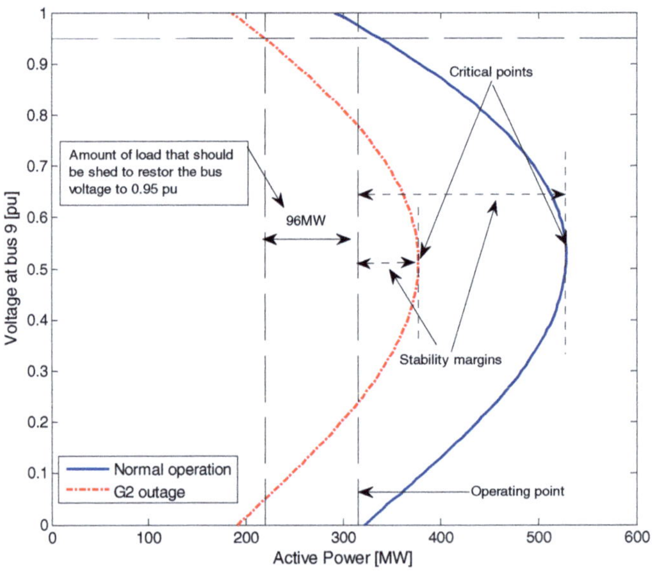

Fig. 26. Determine amount of load that should be shed after loss of G2

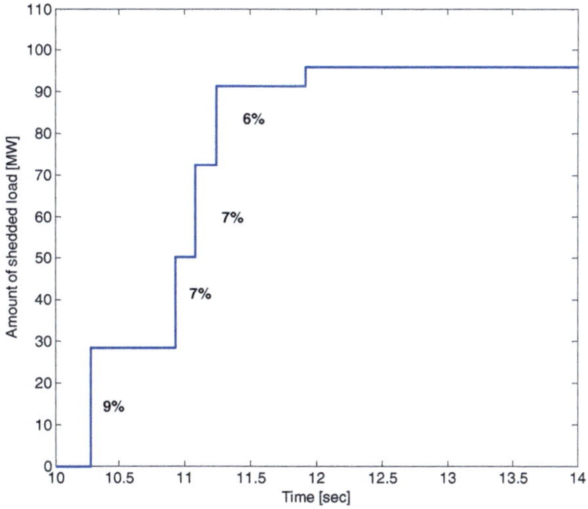

Fig. 27. The LS steps and amount of load that should be shed in each step in the case of G2 outage

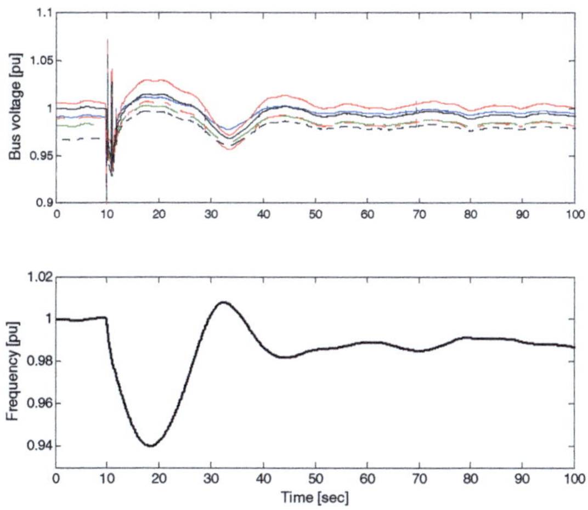

Fig. 28. System voltage and frequency deviation, following loss of G2

6 Summary

This chapter presents a new intelligent based power system emergency control scheme in the presence of high wind power penetration. The impacts of wind power fluctuation on the system frequency, voltage and frequency gradient are analyzed, and an overview of the key issues in the use of wind power penetration in power system emergency control is addressed.

The necessity of considering both system frequency and voltage indices to design an effective load shedding algorithm is shown. Then, an intelligent artificial neural network (ANN) based emergency control scheme considering the dynamic impacts of wind turbines is proposed.

The ANNs are effectively used to identify severe contingencies and to estimate the system security level by predicting the post-fault stability margin. An ANN based methodology to estimate the post-contingency power-voltage (P-V) curve is introduced, and finally, the results are used to run an optimal load shedding algorithm. The proposed emergency control scheme and discussions are supplemented by computer nonlinear simulations on the IEEE 9-bus test system.

References

1. Alexander, J.F., Renuka, G., Jayabharath, R.D.: New Power Sensitivity Method of Ranking Branch Outage Contingencies for Voltage Collapse. IEEE Transactions on Power Systems 17(2), 265–270 (2002)
2. Bevrani, H.: Robust Power System Frequency Control. Springer, New York (2009)
3. Bevrani, H., Ledwich, G., Ford, J.J.: On the Use of df/dt in Power System Emergency Control. In: Proceedings 2009 IEEE Power Systems Conference & Exposition, Seattle, Washington, USA (2009)
4. Bevrani, H., Ledwich, G., Dong, Z.Y., Ford, J.J.: Regional Frequency Response Analysis under Normal and Emergency Conditions. Electric Power Systems Research 79, 837–845 (2009)
5. Bevrani, H., Ledwich, G., Ford, J.J., Dong, Z.Y.: On Power System Frequency Control in Emergency Conditions. Journal of Electrical Engineering & Technology 3(4), 499–508 (2008)
6. Bevrani, H., Hiyama, T.: On Load-Frequency Regulation with Time Delays: Design and Real Time Implementation. IEEE Transactions on Energy Conversion 24(1), 292–300 (2009)
7. Bevrani, H., Ghosh, A., Ledwich, G.: Renewable Energy Resources and Frequency regulation: Survey and New Perspectives. Will be submitted to IET Renewable Power Generation (2009)
8. Bijwe, P.R., Nanda, J., Puttabuddhi, K.L.: Ranking of line outages in an AC-DC system causing overload and voltage problems. IEE Proceedings-C 138(3), 207–211 (1991)
9. Claudio, A.C., Nadarajah, M., Federico, M., John, R.: Linear Performance Indices to Predict Oscillatory Stability Problems in Power Systems. IEEE Transactions on Power Systems 19(2), 1104–1114 (2004)
10. Ejebe, G.C., Wollenberg, B.F.: Automatic Contingency Selection. IEEE Transactions on Power Apparatus and Systems PAS-98(1), 97–109 (1979)
11. El-Saadawi, M.M., Kaddah, S.S., Osman, M.G., Abdel-Wahab, M.N.: Impact of wind farms on contingent power system voltage stability. In: 12th International Middle-East Power System Conference, pp. 637–644 (2008)

12. Erlich, I., Rensch, K., Shewarega, F.: Impact of large wind power generation on frequency stability. In: Proc. of Power Engineering Society General Meeting (2006) (CD ROM)
13. Ford, J.J., Bevrani, H., Ledwich, G.: Adaptive Load Shedding and Regional Protection. International Journal of Electrical Power and Energy Systems 31, 611–618 (2009)
14. FRCC Automatic Underfrequency Load Shedding Program, PRC-006-FRCC-01 (2009), `https://www.frcc.com/`
15. Fu, X., Wang, X.: Load Shedding Scheme Ensuring Voltage Stability. In: Power Engineering Society General Meeting IEEE, pp. 1–6 (2007)
16. Gillian, L., Alan, M., Mark, O.M.: Frequency Control and Wind Turbine Technologies. IEEE Transactions on Power Systems 20(4), 1905–1913 (2005)
17. Gu, X., Canizares, C.A.: Fast prediction of load ability margins using neural networks to approximate security boundaries of power systems. IET Gener. Transm. Distrib., 466–475 (2007)
18. IEEE PES, power and energy magazine 7(2), March/April Issue (2009)
19. Jadid, S., Jalilzadeh, S.: Application of Neural Network for Contingency Ranking Based on Combination of Severity Indices. In: Proceedings of World Academy of Science, Engineering and Technology, vol. 5 (2005)
20. Kundur, P.: Power System Stability and Control. McGraw-Hill, New York (1994)
21. Marcus, V.A.N., Ja, P.L., Hans, H.Z., Ubiratan, H.B., Rogério, G.A.: Influence of the Variable-Speed Wind Generators in Transient Stability Margin of the Conventional Generators Integrated in Electrical Grids. IEEE Transactions on Energy Conversion 19(4), 692–701 (2004)
22. Moura, R.D., Prada, R.B.: Contingency screening and ranking method for voltage stability assessment. IEE Proc.-Gener. Transm. Distrib. 152(6), 891–898 (2005)
23. Mukhtiar, S., Ambrish, C.: Power Maximization and Voltage Sag/Swell Ride- through Capability of PMSG based Variable Speed Wind Energy Conversion System. In: Annual Conference of IEEE on Industrial Electronics, vol. 34, pp. 2206–2211 (2008)
24. Naoto, Y., Hua-Qiang, L., Hiroshi, S.: A Predictor/Corrector Scheme for Obtaining Q-Limit Points for Power Flow Studies. IEEE Transactions on Power Systems 20(1), 130–137 (2005)
25. Oscar, E.M.: A Spinning Reserve, Load Shedding, and Economic Dispatch Solution by Bender's Decomposition. IEEE Transactions on Power Systems 20(1), 384–388 (2005)
26. Pengcheng, Z., Gareth, T., Malcolm: A Novel Q-Limit Guided Continuation Power Flow Method. In: Power and Energy Society General Meeting - Conversion and Delivery of Electrical Energy, July 20-24, pp. 1–7 (2008)
27. Power Systems Relaying Committee, IEEE Guide for the Application of Protective Relays Used for Abnormal Frequency Load Shedding and Restoration. IEEE Std C37.117™, pp. c1–c43 (2007)
28. Faranda, R., Pievatolo, A., Tironi, E.: Load Shedding: A New Proposal. IEEE Transactions on Power Systems 22(4), 2086–2093 (2007)
29. Mark, S.H., Keith, A.H., Robert, A.J., Lee, Y.T.: Slope-Permissive Under-Voltage Load Shed Relay for Delayed Voltage Recovery Mitigation. IEEE Transactions on Power 23(3), 1211–1216 (2008)
30. Simon, H.: Neural Network a Comprehensive foundation. Prentice hall international, Inc., Englewood Cliffs (1999)
31. Shao-Hua, L., Hsiao-Dong, C.: Continuation Power Flow with Multiple Load Variation and Generation Re-Dispatch Patterns. In: Proc of Power Engineering Society General Meeting (2006) CD ROM

32. Smith, J.C.: Winds of change: Issues in utility wind integration. IEEE Power Energy Mag. 3(6), 20–25 (2005)
33. Shu-Jen, S.T., Kim-Hoi, W.: Adaptive Under-voltage Load Shedding Relay Design Using Thevenin Equivalent Estimation. In: Power and Energy Society General Meeting, pp. 1–8, July 20-24 (2008)
34. Tarlochan, S.S., Lan, C.: Contingency Screening for Steady-State Security Analysis By Using FFT and Artificial Neural Networks. WEE Transactions on Power Systems 15(1), 421–426 (2000)
35. Thelma, S.P.F., Lenzi, J.R., Miguel, A.M.: Load Shedding Strategies Using Optimal Load Flow with Relaxation of Restrictions. IEEE Transactions on Power Systems 23(2), 712–718 (2008)
36. Tikdari, A.: Load Shedding in the Presence of Renewable Energy Sources in a Restructured Power System Environment, Master Thesis, University of Kurdistan (2009)
37. Venkataramana, A.: Computational Techniques for Voltage Stability Assessment and Control. Springer, Heidelberg (2006)
38. Venkataramana, A., Colin, C.: The Continuation Power Flow a Tool for Steady State Voltage Stability Analysis. Transactions on Power Systems 7(1), 416–423 (1992)
39. Vladimir, V.T.: Under-frequency Load Shedding Based on the Magnitude of the Disturbance Estimation. IEEE Transactions on Power Systems 21(3), 1260–1266 (2006)
40. Vidya, S.S.V., Nutakki, D.R.: Contingency Screening through Optimizing Hopfield Neural Networks Canadian Conference on Electrical and Computer Engineering, vol. 1, pp. 199–204 (1993)
41. Yeu, R.H., Sauer, P.W.: Post-Contingency Equilibrium Analysis Techniques for Power Systems. In: Annual North American Power Symposium, vol. 37, pp. 429–433 (2005)
42. Yongning, C., Yanhua, L., Weisheng, W., Huizhu, D.: Voltage Stability Analysis of Wind Farm Integration into Transmission Network. In: International Conference on Power System Technology, pp. 1–7 (2006)
43. Yuri, V.M., Viktor, I.R., Vladimir, A.S., Nikolai, I.V.: Blackout Prevention in the United States, Europe, and Russia. IEEE Transactions on Energy Conversion 14(3), 749–753 (1999)

Appendix

- Nine-bus test system simulation data

Table 3. Branch data

From bus	To bus	R (pu)	X (pu)	B (pu)	Limit (MW)
1	4	0	0.0576	0	200
4	5	0.017	0.092	0.158	200
5	6	0.039	0.17	0.358	100
3	6	0	0.0586	0	250
6	7	0.0119	0.1008	0.209	100
7	8	0.0085	0.072	0.149	200
8	2	0	0.0625	0	200
8	9	0.032	0.161	0.306	200
9	4	0.01	0.085	0.176	200

Table 4. Generator data

Generator	G1	G2	G3
Nominal Power [MVA]	128	247.5	192
Type	Hydro	Steam	Steam
Speed [rpm]	180	3600	3600
V_{L-L} [KV]	16.5	18.3	13.8
X_d	0.146	0.8958	1.3125
X'_d	0.0608	0.1198	0.1813
X''_d	0.205	0.155	0.22
X_q	0.0969	0.8645	1.2578
X'_q	0.0969	0.1969	0.25
X''_q	0.221	0.143	0.292
X_L	0.0336	0.0521	0.0742
T'_d	8.96	6	5.89
T''_d	0.02	0.02	0.02
T'_{qo}	0.00002	0.535	0.6
T''_{q0}	0.02	0.02	0.02
R_S	2.8544e-3	2.8544e-3	2.8544e-3
H (on 100MW)	23.64	3.01	6.4

Reactance values are in pu on 100 MVA base, and all generators are equipped with governor and PSS

Table 5. Load data

Load	Load A	Load B	Load C
Bus NO.	9	5	7
Active power [MW]	125	90	100
Reactive power [MVAR]	50	30	35
Number of blocks	6	4	5

Table 6. Exciter data

Generator	Low pass filter	Regulator		Exciter		Output limits	
	Tr	Ka	Ta	Ke	Te	Efmin	Efmax
G1	20e-3	200	0.001	1	0	0	7
G2	20e-3	200	0.001	1	0	0	12.3
G3	20e-3	200	0.001	1	0	0	12.3

- FRCC Scheme

Table 7. FRCC regional UFLS implementation schedule

UFLS step	Frequency (Hz)	Time delay (s)	Amount of load (% of member system)	Cumulative amount of load (%)
A	59.7	0.28	9	9
B	59.4	0.28	7	16
C	59.1	0.28	7	23
D	58.8	0.28	6	29
E	58.5	0.28	5	34
F	58.2	0.28	7	41
L	59.4	10	5	46
M	59.7	12	5	51
N	59.1	8	5	56

Intelligent Control of Power Electronic Systems for Wind Turbines

Bharat Singh, S.N. Singh, and Elias Kyriakides

Abbreviations

ANFIS	Adaptive neuro-fuzzy inference system
ANN	Artificial neural network
CI	Computational intelligence
DFIG	Doubly-fed induction generator
DPC	Direct power control
EC	Evolutionary computation
FIS	Fuzzy inference system
FL	Fuzzy logic
FLC	Fuzzy logic controller
FRT	Fault-ride-through
GCR	Grid connection requirement
GSC	Grid-side converter
LSE	Least-square estimation
MPPT	Maximum power point tracking

Bharat Singh · S.N. Singh
Department of Electrical Engineering
Indian Institute of Technology Kanpur
Kanpur 208016, India
e-mail: brajpurohit@gmail.com, snsingh@iitk.ac.in

Elias Kyriakides
Department of Electrical and Computer Engineering
KIOS Research Center for Intelligent Systems and Networks
University of Cyprus
Nicosia 1678, Cyprus
e-mail: elias@ucy.ac.cy

L. Wang et al. (Eds): *Wind Power Systems*, Green Energy and Technology, pp. 255–295.
springerlink.com

PCC	Point of common coupling
PES	Power-electronic systems
PGSC	Parallel grid-side converter
PI	Proportional-integral
PID	Proportional-integral-derivative
PMSG	Permanent magnet synchronous generator
PWM	Pulse width modulation
RMSE	Root mean square error
RSC	Rotor-side converter
SCIG	Squirrel-cage induction generator
SG	Synchronous generator
SGSC	Series grid-side converter
UA	Unified architecture
VSC	Voltage source converter
WPG	Wind power generation
WRIG	Wound-rotor induction generator
WT	Wind turbine

Abstract. Electric power generation from wind is becoming a major contributing energy source in the power systems around the world. Modern variable-speed wind turbines (WTs) systems that process power through power-electronic systems (PESs) have found better acceptance and have captured most of the market share. PES technologies enhance the controllability of WTs substantially. The PES employed in the wind power generation (WPG) system can effectively face the challenges of grid connection requirements (GCRs). Computational intelligence (CI) techniques, such as fuzzy logic (FL), artificial neural network (ANN), evolutionary computation (EC), etc. are recently proposed and utilized for the control of power electronics systems. Overall, the dynamic performance of a wind turbine system can be substantially improved by the intelligent control of the PESs that are used in WPG systems.

In this chapter, a computational strategy directed more towards intelligent behavior is employed as a tool for fast, accurate, and efficient control of PES used in double fed induction generator (DFIG) based wind power generation. The conventional proportional-integral (PI) controller is replaced with a nonlinear adaptive neuro-fuzzy inference system (ANFIS) based controller. The fundamental concepts of CI based techniques like ANN, fuzzy logic, hybrid methods, and evolutionary programming are briefly described. The design and procedure for selection of parameters and training of ANFIS are described. A unified architecture (UA) of the DFIG and its control strategies is also addressed. The performance of the conventional PI and ANFIS based controllers is compared using simulation results on a detailed power system test model having wind farms.

1 Introduction

Electric power generation from wind energy is becoming a major contributing energy source in the power systems around the world. A significant emphasis is placed on the cost-effective utilization of this energy resource to simultaneously achieve a quality and reliable power supply. In terms of operational and security aspects, wind power integration in the power grid has evolved into one of the most important concerns due to its variable nature and the increased penetration level; both of these aspects have a significant influence on the operation and control of power systems. New electricity grid codes are being set up by several countries to specify the relevant requirements to integrate the wind power generation into the existing electric power system [1]. Under the new grid codes, it is required for the wind farms to fulfill the same requirements as the conventional power plants and should remain connected to actively support the grid during the abnormal conditions [1], [2].

Traditionally, wind power generation systems used variable-pitch, constant speed wind turbines (WTs) that were coupled to the squirrel-cage induction generators (SCIGs) and fed power directly to the utility grids. Modern variable-speed WT systems that process power through power-electronic systems (PES) have found better acceptance and have captured most of the market share. PES technologies enhance the controllability of WTs substantially. The PES employed in the wind power generation (WPG) system can effectively face the challenges of grid connection requirements (GCRs) and in turn, improve substantially the integration and operation of wind farms connected to the grid.

Computational intelligence (CI) techniques, such as fuzzy logic (FL), artificial neural network (ANN), and evolutionary computation (EC), are recently promoted for the control of power electronics systems. Overall, the dynamic performance of a wind turbine system can be substantially improved by the introduction of CI based techniques for the intelligent control of the PES that are used in WPG systems. Hence, the objectives of reliable and efficient wind power integration in the power system can be effectively realized.

In this work, a computational strategy directed more towards intelligent behavior is employed as a tool for controlling PES employed in wind power generation. The conventional proportional-integral (PI) controller is replaced with a nonlinear adaptive neuro-fuzzy inference system (ANFIS) based controller, that is applied for fast, accurate, and efficient control of PES used in WPG systems. The design and procedure for selection of parameters and training of ANFIS are described. The performance of the conventional PI and ANFIS based controllers is compared using simulation results on a test system.

This chapter covers basic concepts and applications of computational intelligence methodologies applied to WPG systems. Section 2 covers the fundamental concepts of CI based techniques like ANN, fuzzy logic, hybrid methods, and evolutionary programming. The basic theory and mathematical formulation of the ANFIS algorithm and its implementation are presented in Section 3, whereas Section 4 briefly introduces the various elements of wind energy conversion systems (WECSs) and the main generator system topologies used for WPG systems.

Section 5 briefly covers the formulation and design of the conventional vector control for grid connected doubly-fed induction generator (DFIG) system. Section 6 presents a unified architecture (UA) of the DFIG and its control strategies. Section 7 presents the steps for designing the ANFIS based controller for DFIG and UA based WTs. In Section 8, a detailed power system model having wind farms connected to the power system grid is developed. The performance of the conventional PI and the ANFIS based controllers is tested and simulation results are presented. Section 9 concludes the chapter.

2 Computational Intelligence

The goal of computational intelligence (CI) is to enable computer or data processing devices to take decisions or suggest actions based on some kind of "intelligent thinking" similar to human beings. A system with embedded CI is often defined as an "intelligent" system, which has "learning," "self-organizing," or "self-adapting" capabilities. Recent trends and advances in the field of CI have stimulated the development of various CI based systems for various applications in the area of electric power systems. Its application has penetrated deeply in PES and motion control, and appears very promising. For example, CI applications in the area of electric drives are presented in depth in [3]-[5]. Some of the main computational intelligence based techniques include ANNs, FL, EC and hybrid methods. These techniques are finding increasing scope and applications in the area of electric power systems.

Artificial Neural Network

The artificial neural network, often called neural network, is the most generic form of computational intelligence for emulating the human thinking process compared to the rule-based expert system and fuzzy logic. ANNs are universal function approximators and are capable of closely approximating complex mappings, which can be extended to include the modeling of complex, nonlinear systems. Unlike some other types of models, an ANN model can be formed directly by using the input-output data of the unknown system, without the need for any prior model structure. It is very important to note that, in contrast to linear theories, a general ANN model is not assumed to be linear. A supervised neural network can learn the non-linear input–output function of a system in the learning phase by observing a set of input–output examples.

One of the most widely used supervised ANNs uses the back-propagation training algorithm. In the training phase, the inputs and outputs of the ANN are used to obtain the ANN architecture (e.g., weights, if the number of layers, number of nodes, biases, and activation functions are fixed) by back-propagating the error (desired output minus actual output).

Fuzzy Logic

Fuzzy logic (FL) basically maps an input space to an output space, primarily through a list of *if-then* statements, called rules. All rules are evaluated in parallel, and the order of the rules is not important. The rules, themselves, are useful, because they refer to the variables and the adjectives that describe those variables. A fuzzy logic controller (FLC) utilizes fuzzy logic to convert the linguistic control strategy based on expert knowledge into an automatic control strategy. Fuzzy logic systems are universal function approximators as well. In order to use fuzzy logic for control purposes, one has to add a front-end "fuzzifier" and a rear-end "defuzzifier" to the usual input-output data set. A simple fuzzy logic controller contains four components: rule base, fuzzifier, inference engine, and defuzzifier. Once the rules have been established, it can be considered as a nonlinear mapping from the input to the output. These rules may directly originate from experts, but if the experts are not available, these can also be obtained by the appropriate processing (e.g., clustering) of available input-output data. It is important to note that the system can be nonlinear, and the nonlinearity is incorporated into the fuzzy logic system.

Hybrid Methods

It was shown in recent years that computational intelligent techniques are complementary methodologies in the design and implementation of intelligent systems. Each approach has its own merits and demerits. To take advantage of their strengths and to eliminate their drawbacks, several hybrid methods have been proposed [3]-[6]. In many cases, these hybrid methods have proven to be more effective in designing intelligent control systems. The combination of ANNs and fuzzy logic can be realized in three different ways, resulting in systems with different characteristics such as [6]:

1. *Neuro-fuzzy system*: It provides the fuzzy system with automatic tuning systems using ANN as a tool. This chapter presents an application of this hybrid system to wind power generation systems.
2. *Fuzzy-neural network*: It retains the functions of ANNs with fuzzification of some of their elements. For instance, fuzzy logic can be used to determine the learning steps of the ANN structure.
3. *Fuzzy-neural hybrid system*: It utilizes both fuzzy logic and ANNs in a system to perform separate tasks for a decoupled subsystem. The architecture of the system depends on the particular application.

Evolutionary Computation

In recent years, various evolutionary computation (EC) methodologies have been proposed to solve problems of common engineering applications. Applications often involve automatic learning of nonlinear mappings that govern the behavior of control systems, as well as parallel search strategies for solving multi-objective optimization problems. These algorithms have been particularly appealing in the scientific communities since these allow autonomous adaptation/optimization without human intervention. These strategies are based on the biological evolution according to Darwinian concept. There are various approaches to the evolutionary optimization algorithms including the concept of evolution, genetic programming and genetic algorithms. These algorithms are similar in their basic concepts of evolution and differ mainly in their approach to the parameter representation. The evolutionary optimization algorithms operate by representing the optimization parameters via a gene-like structure and subsequently utilizing the basic mechanisms of Darwinian natural selection to find a population of superior parameters [6].

In the majority of computational intelligence (CI) research work in electric power systems, the fuzzy logic and ANNs are used, where an existing PI or proportional-integral-derivative (PID) controller is simply replaced by a CI based controller. Although, this is an important application and the CI based controllers can lead to improved performance, enhanced tuning and adaptive capabilities, there are further possibilities for a much wider range of CI based applications such as controllers, observers, optimizers, estimators, etc. [3]-[5]. CI based techniques have been applied to a variety of problems associated with wind power generation (most of them employ FL based controllers) [7]-[11]. FLC suffers from some serious drawbacks presented in the next section and hence, a hybrid system of FL and ANN is suggested for use in the wind power generation systems.

3 Nonlinear Adaptive Neuro-fuzzy Inference System

Fuzzy logic applications for the control of power electronics systems (PES) have been increased in the past few years due to the non-linearities and unavailability of precise models associated with PES, something that makes them well suited for FL control [7], [12], [13]. The FLC of a given system is capable of embedding, in the control strategy, the qualitative knowledge and experience of an operator or field engineer about the process. In spite of its practical success, FL has been criticized for its limitations, such as the lack of a formal design methodology, and the difficulty in predicting stability and robustness of FL controlled systems. In reality, these aspects have been improved considerably in the past few years, as the heuristic approaches commonly used in FLC design have gradually been replaced by methods that are more formal.

Further, in certain systems, it is difficult or, sometimes, impossible to define crisp rules for control. In such situations, fuzzy rules can be used to control the system dynamics. By using FLC, it is possible to explain why a particular value appeared at the output of a fuzzy system. However, sometimes, it is very difficult to define the fuzzy rules and the process of tuning the fuzzy system parameters requires a long time, specifically, when the number of fuzzy rules is large. ANN based controllers have also been utilized, as these require a small computational time after training. However, to choose their optimal structure and their parameter values and to minimize the training set are some of the issues to be addressed in the ANN applications. A major drawback of ANN is that, although they are designed to map any non-linear mapping, it is considered as a black box, without providing any general information.

Utilizing the advantages of both the ANN and FL, researchers have tried to combine these two to make a hybrid system to perform a required non-linear mapping of input-output relation by adjusting their parameters in order to improve their performance. Hybrid systems, particularly neuro-fuzzy techniques, have advanced rapidly in recent years and their potential for application in PES seems enormous. In a neuro-fuzzy system, there is no need of prior information regarding the membership function (MF) and rule base. These can be obtained by appropriate tuning of ANN. The state-of-the-art intelligent control is at the level of adaptive neuro-fuzzy controllers and it is generally termed as an adaptive neuro-fuzzy inference system (ANFIS).

In this work, in order to overcome the limitations of conventional control techniques for the complex PES based WPG systems, an ANFIS controller is used. Based on the ability of an ANFIS to learn from training data, it is possible to create an ANFIS structure from an extremely limited mathematical representation of the system. The ANFIS architecture can identify the near-optimal MFs of FLC for achieving desired input-output mappings. The basic concepts of fuzzy inference system (FIS) are briefly presented in the rest of this section along with the basic theory, mathematical formulation, and implementation of the ANFIS algorithm.

3.1 Fuzzy Inference Systems

Fuzzy inference systems are also known as fuzzy rule based systems, fuzzy models, fuzzy associative memories, or fuzzy logic controllers (when used as controllers). A fuzzy inference is a method that interprets the values in the input vector and assigns values to the output vector based on a set of rules. The FIS maps a crisp set of input variables into a fuzzy set by using MFs and based on these fuzzy input sets, according to the predefined logic, output is assigned. The process of fuzzy inference involves all of the processes that are described above. Basically, a FIS consists of five functional blocks as shown in Fig. 1 [14].

- A rule base, containing a number of fuzzy if-then rules.
- A data base, which defines the MFs of the fuzzy sets used in the fuzzy rules.
- A decision-making unit, which performs the inference operations on the rules.
- A fuzzification interface, which transforms the crisp inputs into linguistic values.
- A defuzzification interface, which transforms the fuzzy results into a crisp output.

Generally, the rule base and data base are jointly referred to as the *knowledge base*.

There are various types of FIS, but three main types are the Mamdani-type [15], the Sugneo-type [16] and the Tsukamoto-type [17] FIS. Mamdani's fuzzy inference method, which expects the output MFs to be fuzzy sets, is the most commonly used fuzzy methodology. Mamdani's method was among the first control systems built using fuzzy set theory. After the aggregation process, there is a fuzzy set for each output variable that needs defuzzification, which is done by finding the centroid of a two-dimensional aggregate output function.

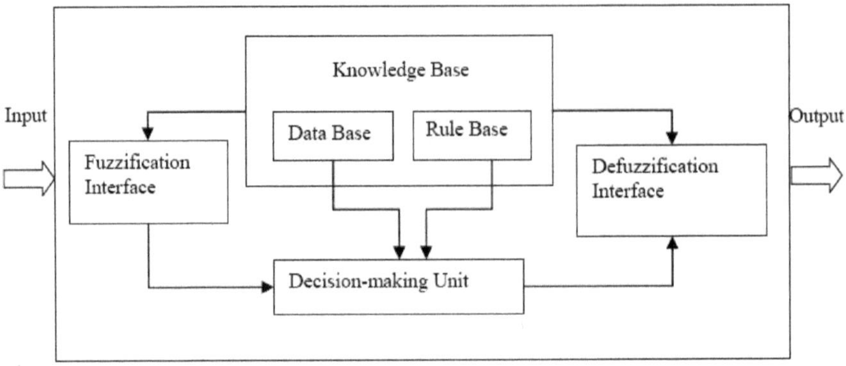

Fig. 1. Block diagram of fuzzy inference system [14]

3.2 Sugeno-type Fuzzy Inference System

In this work, the Sugeno or Takagi-Sugeno-Kang method of fuzzy inference has been used. The Sugeno method was first introduced in 1985 [16]. It is similar to the Mamdani method in many aspects. The first two parts of the fuzzy inference process, fuzzifying the inputs and applying the fuzzy operator, are exactly the same. The difference is that unlike the Mamdani method, in the Sugeno method the output MFs are only constants or have linear relations to the inputs. With a constant output MF, this method is known as the zero-order Sugeno method, whereas with a linear relation, it is known as the first-order Sugeno method.

A typical rule in a Sugeno fuzzy model has the following form:

If Input-1 = x, and Input-2 = y, then, Output $z = ax + by + c$

For a zero-order Sugeno model, the output level z is a constant ($a = b = 0$). The output level z_i of each rule is weighted by the firing strength w_i of the rule. For example, for an AND rule with Input-1 = x, and Input-2 = y, the firing strength is

$$w_i = \text{AND}\,(F_1(x), F_2(y))$$

where, $F_1(.)$ and $F_2(.)$ are the MFs for Inputs 1 and 2.

The final output of the system is the weighted average of the outputs of all the rules, computed as

$$\text{Final output} = \frac{\sum_{i=1}^{N} w_i z_i}{\sum_{i=1}^{N} w_i} = \sum_{i=1}^{N} \overline{w}_i z_i, \qquad \overline{w}_i = \frac{w_i}{\sum_{i=1}^{N} w_i} \tag{1}$$

A Sugeno rule operates as shown in Fig. 2.

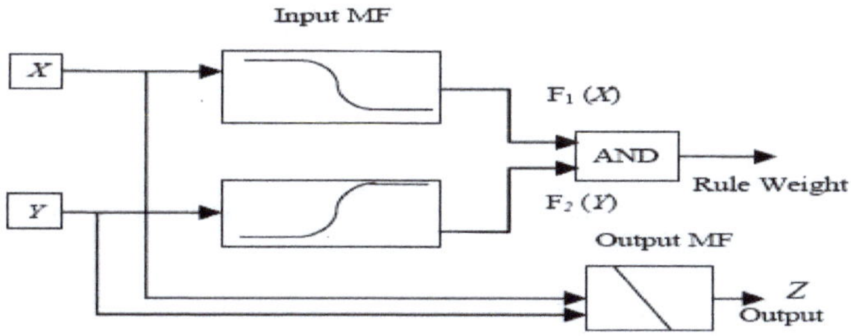

Fig. 2. First order Sugeno-type inference system

It is a great advantage of the Sugeno-type FIS, that it avoids the use of a time consuming defuzzification. Since, it is a more compact and computationally efficient representation than the Mamdani system, the Sugeno system lends itself to the use of adaptive techniques for constructing fuzzy models. These adaptive techniques can be used to customize the MFs so that the fuzzy system accurately models the data. Some of the advantages of the Sugeno-type method are that it is computationally efficient, it works well with linear techniques (e.g., PID control), it works well with optimization and adaptive techniques, it has guaranteed continuity of the output surface, and it is well-suited to mathematical analysis.

3.3 Adaptive Neuro-Fuzzy Inference System

The basic structure of FIS described above, is a model that maps input characteristics to input MFs, input MFs to rules, rules to a set of output characteristics, output characteristics to output MFs, and the output MFs to a single-valued output or a decision associated with the output. In both Mamdani and Sugeno-type of inference systems, when used for data modeling, MFs and rule structure are essentially predetermined by the human interpretation of the characteristics of the variables of the data model.

The shape of the MFs depends on the values of the parameters. Instead of just looking at the data to choose the MF parameters, by using ANFIS MF, the parameters can be chosen automatically, requiring a minimum human intervention for tuning. The neuro-adaptive learning techniques provide a method for the fuzzy modeling procedure to learn information about a data set, in order to compute the MF parameters that best allow the associated FIS to track the given input/output data. This learning method works similar to the ANN. In an adaptive neuro-fuzzy inference technique, using a given input/output data set, a FIS is constructed, whose MF parameters are tuned (adjusted) using a back-propagation algorithm (gradient descent learning method), and in general, this technique is slow. However, it is possible to use a hybrid-learning rule where the gradient method is combined with the least square error techniques to quickly identify the parameters [18]. Fig. 3 shows the basic structure of the ANFIS algorithm for a Sugeno-type FIS. The various layers shown in Fig. 3 are explained below [14].

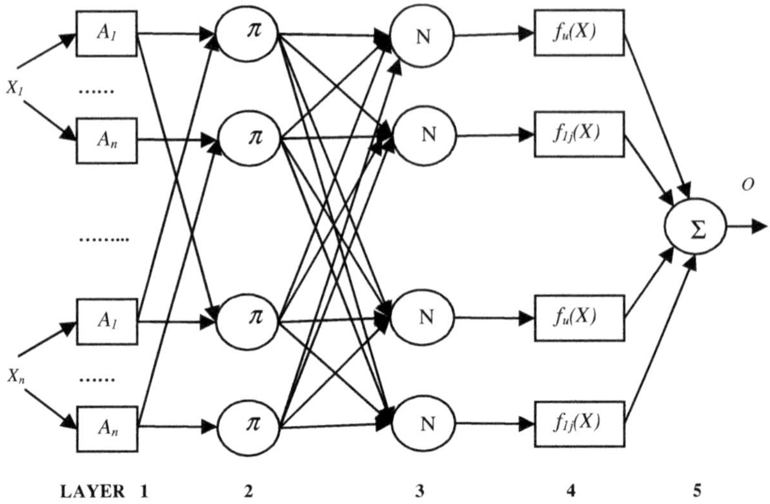

Fig. 3. Typical ANFIS structure

Layer 1 (Membership layer)

Every node i, in this layer, is an adaptive node with a node function

$$O_i^1 = \mu_{A_i}(x) \tag{2}$$

where, x is the input to node i, and A_i is the linguistic label (small, large, etc.) associated with this node function. In other words, o_i^1 is the MF of A_i and it specifies the degree to which the given x satisfies the quantifier A_i.

Usually, $\mu_{Ai}(x)$ is selected to be bell shaped with maximum value equal to 1, and minimum value equal to 0, such as

$$\mu_{A_i}(x) = \frac{1}{1 + \left[\left(\frac{x - c_i}{a_i}\right)^2\right]^{b_i}} \tag{3}$$

where, $\{a_i , b_i , c_i\}$ is the parameter set. As the values of these parameters change during learning, the bell-shaped functions vary accordingly, thus exhibiting various forms of MFs on linguistic label A_i. In fact, any piecewise differentiable function, such as commonly used trapezoidal or triangular-shaped MF, is also qualified candidates for node functions in this layer. Parameters in this layer are referred to as *premise parameters*.

Layer 2 (Fuzzy AND layer)

Every node in this layer is a fixed node, labeled π, which multiplies the incoming signals, for example, $w_i = \mu_{A_i}(x) \times \mu_{A_j}(y)$. Each node output represents the firing strength of a rule. In fact, other T-norm operators that perform a generalized AND operation can be used as the node function in this layer.

Layer 3 (Normalizing layer)

Every node in this layer is a fixed node, labeled *N*. The ith node calculates the ratio of the ith rule's firing strength to the sum of all rule's firing strengths, as given below.

$$\overline{w_i} = \frac{w_i}{\sum_{\forall j} w_j} \tag{4}$$

Outputs of this layer are known as normalized firing strengths.

Layer 4 (Partial output layer)

Every node *i* in this layer is an adaptive node with a node function

$$O_i^4 = \overline{w_i} f_i = \overline{w_i}(p_i x + q_i y + r_i) \tag{5}$$

where, $\overline{w_i}$ is the output of Layer-3, and $\{p_i , q_i , r_i\}$is the parameter set. The parameters in this layer are referred to as *consequent parameters*.

Layer 5 (Output layer)

The single node in this layer is a circle node labeled *Σ* that computes the overall output as the summation of all incoming signals, i.e.,

$$O_i^5 =_i overall\,output = \sum_i \overline{w_i} f = \frac{\Sigma_i w_i f_i}{\Sigma_i w_i} \tag{6}$$

The adjustment of modifiable parameters is a two-step process. First, information are propagated forward in the network until Layer-4, where the parameters are identified by a least-square estimation (LSE) method. Then, the parameters in Layer-2 are modified using gradient descent. The only user specified information is the number of MFs in the universe of discourse for each input and output as training information. ANFIS uses back-propagation learning to learn the parameters related to MFs and LSE to determine the consequent parameters. Every step in the learning procedure includes two parts. The input patterns are propagated, and the optimal consequent parameters are estimated by an iterative LSE. The premise parameters are assumed fixed for the current cycle through the training set. The pattern is propagated again, and in this epoch (iteration), back-propagation is used to modify the premise parameters, while the consequent parameters remain fixed. The application of this hybrid learning rule will lead to a decrease in the dimension of the search space in the steepest gradient descent method, but it will also substantially reduce the convergence time.

4 Wind Energy Conversion Systems (WECS)

The development in wind power generation systems has been steady for the last 25 years and four to five types of wind turbine (WT) system exist. The main components of a WT system, including the turbine rotor, gearbox, generator, transformer, and possible PES, are illustrated in Fig. 4.

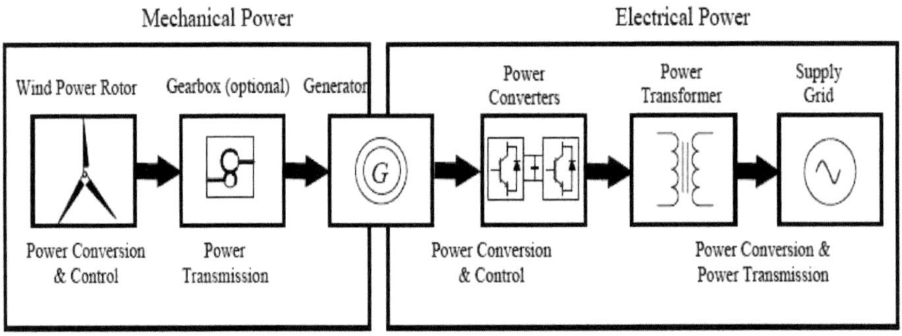

Fig. 4. Main components of a wind turbine system

The turbine rotor converts the fluctuating wind energy into mechanical energy, which is converted into electrical power through the generator, and then fed into the grid through transformers and transmission lines. The WT captures the power from the wind by means of aerodynamically designed blades and converts it to rotating mechanical power. The most efficient way to convert the low-speed, high-torque power to electrical power is to use a gearbox and a generator with standard speed. The gearbox converts the low speed of the turbine rotor to the high speed of the generator. The connection of wind power generators to the grid is possible at low voltage, medium voltage, high voltage, and even at the extra high voltage level since the transmittable power of an electricity system usually increases with increasing the voltage level. While most of the wind farms are nowadays connected to the medium voltage system, large offshore wind farms are connected to the high and extra high voltage levels. For modern WT systems, each generator has its own transformer to raise the voltage. The transformer is normally located close to the WT systems to avoid long low-voltage cables [19].

4.1 Generator Systems for Wind Turbines

The main generator systems for WECS are broadly classified in four categories. However, WTs can operate either at fixed speed (actually within a speed range of about 1%) or at variable speeds.

Fixed-Speed Wind Turbines

For fixed-speed WT systems as shown in Fig. 5, the squirrel cage induction generator (SCIG) is directly connected to the grid, with a soft-starter and a capacitor bank to reduce the reactive power demand from the wind generators. The fixed rotor speed is set by the frequency of the supply grid, the gear ratio, and the generator design. Since the speed is almost fixed to the grid frequency, the turbulence of the wind will result in power variations, and thus, will affect the power quality of the grid, which in turn causes a varying amount of reactive power drawn from the grid. The advantage of a fixed-speed WT is that it is relatively simple, robust and reliable, and therefore, the investment cost is relatively low.

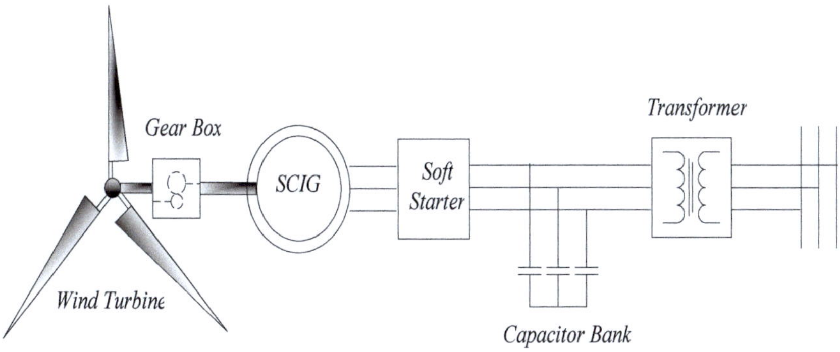

Fig. 5. Schematic of fixed-speed SCIG wind turbine system

Fixed-speed WTs are designed to achieve maximum efficiency at one particular wind speed. In order to increase power production, often, these are provided with two fixed speeds. This is accomplished by using two generators with different ratings and pole pairs, or it can be a single generator with two windings having different ratings and pole pairs (one is used at low wind speeds and the other at medium and high wind speeds). This increases the aerodynamic capture as well as reduces the magnetizing losses of the generator at low wind speeds. This system (one or two-speed) is known as the *conventional* concept used by many Danish manufacturers in the 1980s and 1990s [20].

Variable-Speed Wind Turbines

In variable-speed WTs, the generators are controlled and connected to the grid via a power electronic system (PES), which makes it possible to control the rotor speed. The power fluctuations caused by wind variations can be more or less absorbed by changing the rotor speed and thus, power variations, originating from the wind conversion and the drive train, can be reduced. Variable-speed WT systems are designed to achieve maximum aerodynamic efficiency over a wide range of wind speeds by employing the maximum power point tracking (MPPT) algorithm. Another advantage is that variable-speed WTs also allow the grid voltage to be controlled, as the reactive power generation can be varied. The drawbacks of variable speed wind turbines are that the built-in PES is sensitive to the voltage dips caused by faults and/or switching and that these PESs are typically expensive. However, a variable speed generating system can also give major savings by increased efficiency, better power quality, reduced mechanical stress and lighter foundations in offshore applications i.e., limiting the overall cost increase.

The introduction of the variable-speed wind turbine system gives choice to select appropriate WECS configuration with various combinations for generator types and PES. Mainly for variable-speed wind systems, SCIG or wound rotor induction generator (WRIG), synchronous generator (SG) or permanent magnet synchronous generator (PMSG), and doubly fed induction generator (DFIG), as shown in Fig. 6, are used. Fig. 6 (a) shows the limited variable-speed (typically 2-5% of rated speed) wind turbine system having variable WRIG rotor resistance which is controlled by PES, known as OptiSlip™ (used by Vestas). This system still needs a soft starter and reactive power compensation. Drawbacks of this method are that the energy is unnecessarily dissipated in the external rotor resistances and also it is not possible to control the reactive power.

For variable-speed wind systems with limited slip (±30%), the DFIG systems can be an interesting solution, as shown in Fig. 6 (b). The major advantage of the DFIG, which has made it popular, is that the PES handles only a fraction (20–30%) of the total system power. Another advantage in using DFIGs is the ability to transfer maximum power over a wide speed range in both sub- and super-synchronous speeds. According to [21] the DFIG concept was the most successful variable speed concept with more than 55% market share in 2004.

The WECS having full-scale power converters between the generator and the grid corresponds to the full variable speed range WTs and gives the added technical performance, as shown in Fig. 6 (c). Usually, a back-to-back voltage source converter is used to achieve full control of the active and reactive powers. The power converter to the grid enables the system to control active and reactive powers very fast. However, the major disadvantage is its complexity and expensive system with more sensitive electronic parts. With the permanent magnet, high power densities can be achieved in a less space. A notable advantage of this system is the absence of gearbox and its low acoustic noise. The main limitation for PMSG is the high cost of the materials for the magnet. Also, generators with permanent magnet excitation have poor power factors and can be compensated by inverter technology.

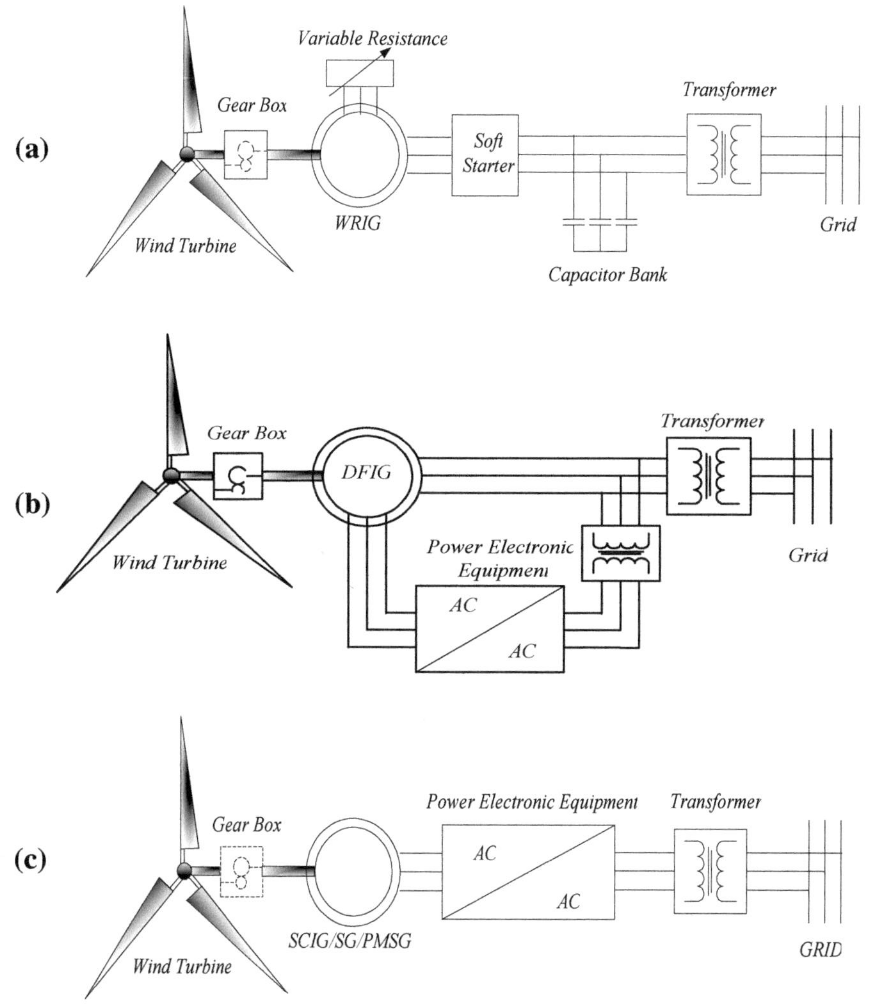

Fig. 6. Schematic of variable-speed wind turbine with (a) WRIG, (b) DFIG and (c) SCIG/SG/PMSG, connected to the grid

4.2 Power Electronic Solutions for Wind Power Generation

Variable-speed wind turbine systems use PES as interface to the electric grid. Since, the WT operates at variable rotational speed, the electric frequency of the generator varies and must, therefore, be decoupled from the frequency of the grid. This can be achieved by using PES. During recent years, different converter topologies have been investigated such as diode bridge rectifier, back-to-back converters, multilevel converters, tandem converters, matrix converters, and resonant converters. In a fixed-speed WT system where wind power generator may be directly connected to the grid, thyristors are used as soft-starters and for switching of capacitor banks. The working principles of different topologies including their advantages and disadvantages are given in [21]. In this chapter, only the most popularly used configuration of PES for wind power generation i.e., the back-to-back converter is described in detail.

5 DFIG with Back-to-Back PWM Converter

The DFIG is a wound-rotor induction generator having three-phase windings on the rotor and stator. The stator is directly connected to the grid, and the rotor power is fed by variable frequency bidirectional PES, as shown in Fig. 6 (c). The PES consists of two widely used back-to-back PWM voltage-fed current-regulated converters (VSC), as shown in Fig. 7, namely, the rotor-side converter (RSC) and grid-side converter (GSC), which are controlled independently. The RSC is used to convert the rotor-frequency power to dc power and then feedback to the ac utility grid using the GSC, which converts dc power to ac power at the grid frequency. DFIG can be operated as a generator as well as a motor in both sub-synchronous and super-synchronous speeds utilizing the RSC control appropriately. Only generating modes at sub-synchronous and super-synchronous speeds are of interest for wind power generation.

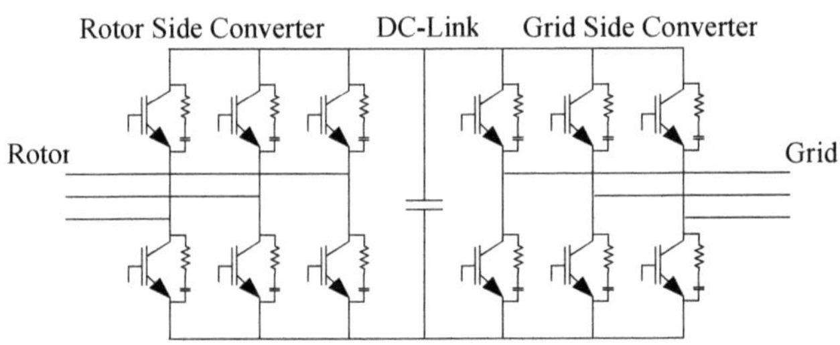

Fig. 7. Back-to-back connected power converter bridge

PWM techniques have been used to decrease the harmonic distortion and to increase controllability of the system, as well as to improve the dynamic performance. Due to the popularity of this scheme, several control techniques of two-level back-to-back PWM-VSC are described in the literature [3]-[5], [22]. The main drawback of back-to-back PWM-VSC is the high cost and the reduced lifetime due to the presence of a dc-link capacitor. Another important drawback is the switching losses. The high switching speed to the grid may also require extra electromagnetic interference filters.

The vector control method [23]-[27] is very extensively used for DFIG systems. The objective of the vector control method for the GSC is to keep the dc-link voltage constant regardless of the magnitude and direction of the rotor power, while keeping sinusoidal grid currents. It may also be responsible for controlling reactive power flow between the grid and the GSC. The vector control method for the RSC ensures a decoupled control of stator-side active and reactive power exchanges from the grid. PWM modulation techniques or alternatively space vector modulation (SVM) techniques can be used in order to achieve a better modulation index [27].

Other control schemes aided by a rotor speed encoder may give excellent tracking results. However, these encoders are expensive and the performance of such schemes depends on the computational accuracy of the stator flux and the accuracy of the rotor position information derived from the position encoder. Alignment of the position sensor is, however, difficult in a doubly-fed wound rotor machine [28]. Position sensorless vector control methods have been proposed by several researchers in the recent past [29]-[30]. The direct power control (DPC) methods [28], [31] which are advanced closed loop controllers using hysteresis approach, are based on the measurement of active and reactive powers on the grid side. These controllers directly trigger a sequence of voltage vectors in the RSC, based on power errors and on position of the rotor flux. These methods are inherently position independent and do not depend on machine parameters like stator/rotor resistance. In the following sub-sections, the vector control of grid connected DFIG is briefly presented.

5.1 Equivalent Circuit of DFIG

The equivalent circuit of the DFIG, with inclusion of the magnetizing losses, can be seen in Fig. 8 [32]. If the DFIG is delta-connected, the machine can still be represented by the equivalent star (Y) representation.

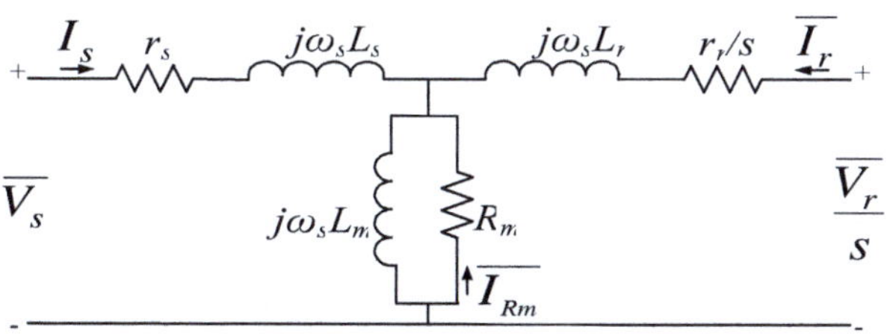

Fig. 8. Equivalent circuit of doubly-fed induction generator

If the rotor voltage, $\overline{V_r}$, in Fig. 8, is short circuited, the equivalent circuit for the DFIG becomes the equivalent circuit for an ordinary SCIG. On the stator side, r_s and $j\omega_s L_s$ are the resistance and leakage reactance per phase of the stator winding. On the rotor side, r_r and $j\omega_s L_r$ are the resistance and leakage reactance per phase, respectively, of the rotor winding. The mutual reactance is $j\omega_s L_M$. When the rotor rotates at angular velociy of ω_r electrical rad/s, the rotor resistance, r_r, is modified as r_r/s where, s is the slip and equals (1- ω_r/ω_s). The RSC injects balanced three-phase voltages (V_{ra}, V_{rb}, V_{rc}) at slip frequency, $s\omega_s$. Because Fig. 8 is represented based on the stator-side frequency, ω_s, the voltage phasor

$\overline{V_r} = V_r \angle \delta$ has also to be divided by the slip, *s*, resulting in the equivalent rotor voltage $(V_r \angle \delta)/s$. Applying Kirchhoff's voltage law to the equivalent circuit shown in Fig. 8, one can obtain

$$\overline{V_s} = r_s \overline{I_s} + j\omega_s L_s \overline{I_s} + j\omega_s L_m (\overline{I_s} + \overline{I_r} + \overline{I_{Rm}}) \tag{7}$$

$$\frac{\overline{V_r}}{s} = \frac{r_r}{s} \overline{I_r} + j\omega_s L_r \overline{I_r} + j\omega_s L_m (\overline{I_s} + \overline{I_r} + \overline{I_{Rm}}) \tag{8}$$

$$0 = R_m \overline{I_{Rm}} + j\omega_s L_m (\overline{I_s} + \overline{I_r} + \overline{I_{Rm}}) \tag{9}$$

where, $\overline{V_s}$ is the stator complex voltage per phase; $\overline{I_{Rm}}$ is the core loss current. The air-gap flux, stator flux and rotor flux are defined as

$$\overline{\Psi_m} = L_m (\overline{I_s} + \overline{I_r} + \overline{I_{Rm}}) \tag{10}$$

$$\overline{\Psi_s} = L_s \overline{I_s} + L_m (\overline{I_s} + \overline{I_r} + \overline{I_{Rm}}) = L_s \overline{I_s} + \overline{\Psi_m} \tag{11}$$

$$\overline{\Psi_r} = L_r \overline{I_r} + L_m (\overline{I_s} + \overline{I_r} + \overline{I_{Rm}}) = L_r \overline{I_r} + \overline{\Psi_m} \tag{12}$$

5.2 Back-to-Back Converter Configuration

The back-to-back converter consists of two voltage source converters (VSCs) for RSC and GSC, as shown in Fig. 7. Between the two converters, a dc-link capacitor is placed, as an energy storage, in order to keep the small voltage variations (or ripple), in the dc-link voltage. Standard three-phase bridge topology is employed for the converters. With a PWM converter in the rotor circuit, the rotor currents can be controlled in a desired phase, frequency and magnitude. This enables bidirectional flow of active power between the rotor and grid and the system can operate in sub-synchronous and super-synchronous speeds. The dc-link capacitor acts as a source of reactive power and it is possible to supply the magnetizing current, partially or fully, from the rotor side. The stator side power factor can thus, be controlled. Using vector control techniques for controlling the RSC, the stator side active and reactive powers can be controlled independently and hence, fast dynamic performance can be achieved.

Unlike the RSC, the GSC operates at the grid frequency. Flow of active and reactive powers are controlled by adjusting the phase and amplitude of the inverter terminal voltage with respect to the grid voltage. Active power can flow either to the grid or to the rotor circuit depending on the mode of operation. Since the inverter operates at a high frequency, usually between 1 kHz to 5 kHz, the harmonics in the input current are greatly reduced.

It should be noted that since the slip range is limited, the dc-link voltage is lower in this case compared to stator side control. A transformer is, therefore, necessary to match the voltage levels between the grid and the dc-side of the GSC. This arrangement presents enormous flexibility in terms of control of active and reactive powers in variable speed applications. In the following sections, the model of the dc-link is presented and also the control schemes of RSC and GSC are discussed.

5.3 DC-Link Model

The losses in the converter can be considered small and thereby, it can be neglected. The rate of change of energy in the dc-link capacitor is dependent on the power delivered to the grid through GSC, P_g and the power delivered to the rotor circuit of the DFIG, P_r.

$$\frac{d}{dt}E_{dc} = \frac{1}{2}C_{dc}\frac{d}{dt}V_{dc}^{\ 2} = -P_g - P_r \tag{13}$$

where, E_{dc} is energy stored in the dc-link capacitor, C_{dc}. V_{dc} is the dc-link voltage. This means that the dc-link voltage will vary as

$$C_{dc}V_{dc}\frac{d}{dt}V_{dc} = -P_g - P_r \tag{14}$$

which means that, if $P_g = -P_r$, the dc-link voltage will be constant.

5.4 Rotor-Side Converter (RSC) Control

To exploit the advantages of variable speed operation, the tracking of optimum torque-speed curve is essential. Speed can be adjusted to the desired value by controlling torque or active power. For easy control, both stator and rotor quantities are transformed to a reference frame that rotates at an angular frequency identical to the stator magneto-motive force. At steady state, the reference frame speed equals the synchronous speed. In stator flux-oriented control, the active power or torque can be controlled by the *q*-axis rotor current and reactive power can be controlled by controlling the *d*-axis rotor current. The control scheme can be explained as follows.

This scheme makes use of the stator flux angle, which is determined dynamically to map the stator and rotor quantities into the new reference frame. The stator flux angle (ρ_s) is the angle between the stator flux linkage phasor and the stationary *d*-axis (assuming that all stator and rotor currents are calculated in stationary reference frame).

$$\rho_s = \tan^{-1}\left(\frac{\Psi_{qs}}{\Psi_{ds}}\right) \tag{15}$$

where, Ψ_{ds} and Ψ_{qs} are the *d-q* axis stator flux linkage components in the stationary reference frame. It can be shown that the choice of the stator flux-oriented reference frame results in a decoupled control of stator side active and reactive powers. The stator flux linkages expressed in the new reference frame are

$$\Psi_{ds}^{\Psi s} = L_s i_{ds}^{\Psi s} + L_m i_{dr}^{\Psi s} \text{ and } \Psi_{qs}^{\Psi s} = L_s i_{qs}^{\Psi s} + L_m i_{qr}^{\Psi s} \tag{16}$$

$$\text{where } L_s = L_{ls} + L_m \tag{17}$$

Since, the x-axis of the new reference frame is aligned with the stator flux linkage vector, $\psi_{sq}^{\psi s} = 0$. Thus,

$$i_{qs}^{\Psi s} = -\frac{L_m}{L_s} i_{qr}^{\Psi s} \text{ and } i_{ds}^{\Psi s} = \frac{L_m}{L_s}(i_{ms} - i_{dr}^{\Psi s}) \tag{18}$$

where the stator magnetizing current is

$$i_{ms} = \frac{\psi_s^{\psi s}}{L_m} = \frac{\psi_{ds}^{\psi s} + j\psi_{qs}^{\psi s}}{L_m} = \frac{\psi_{ds}^{\psi s}}{L_m} \tag{19}$$

The stator side active and reactive powers are given by

$$P_s = \frac{3}{2}\mathrm{Re}(V_s^{\psi s} i_s^{\psi s*}) = \frac{3}{2}(V_{ds}^{\psi s} i_{ds}^{\psi s} + V_{qs}^{\psi s} i_{qs}^{\psi s}) = -\frac{3}{2}|\overline{V_s}|\frac{L_m}{L_s} i_{qr}^{\psi s} \tag{20}$$

$$Q_s = \frac{3}{2}\mathrm{Im}(V_s^{\psi s} i_s^{\psi s*}) = \frac{3}{2}(V_{qs}^{\psi s} i_{ds}^{\psi s} - V_{ds}^{\psi s} i_{qs}^{\psi s}) = \frac{3}{2}|\overline{V_s}|\frac{L_m}{L_s}(i_{ms} - i_{dr}^{\psi s}) \tag{21}$$

where in the stator flux-oriented reference frame, $V_{ds}^{\psi s} = 0$, $V_{qs}^{\psi s} = |\overline{V_s}|$.

Thus, the variations in rotor currents will also reflect in the variation of stator side currents $i_{ds}^{\psi s}$, $i_{qs}^{\psi s}$ and hence, in the stator side active and reactive powers as well. This principle has been used in the control of stator active and reactive powers. The control scheme uses a conventional PI controller to obtain the reference value for $i_{qs}^{\psi s}$ from active power error, i.e., the difference between desired and actual values of active power. Similarly, a PI controller can be tuned to get the reference value for $i_{ds}^{\psi s}$ from the reactive power error. Since, the objective is to capture the maximum available energy in the wind, the active power reference is made equal to the maximum available WT power. This can be done by implementing a maximum power point tracking (MPPT) algorithm in the outer control loop [25].

The reactive power reference value is derived from the active power reference and the desired value of the power factor. Usually, the reactive power reference is made equal to zero, in order to operate the DFIG at unity power factor.

Control of Rotor-Side Converter Currents

The inner loops of the controllers can be derived as follows. The equations of the DFIG rotor variables in the stator flux reference frame can be written as

$$V_{dr}^{\psi s} = r_r i_{dr}^{\psi s} + \frac{d}{dt}\Psi_{dr}^{\psi s} - (\omega - \omega_r)\Psi_{qr}^{\psi s} \tag{22}$$

$$V_{qr}^{\psi s} = r_r i_{qr}^{\psi s} + \frac{d}{dt}\Psi_{qr}^{\psi s} + (\omega - \omega_r)\Psi_{dr}^{\psi s} \tag{23}$$

$$\Psi_{dr}^{\psi s} = (L_{lr} + L_m) i_{dr}^{\psi s} + L_m i_{ds}^{\psi s} \tag{24}$$

$$\Psi_{qr}^{\psi s} = (L_{ls} + L_m) i_{qr}^{\psi s} + L_m i_{qs}^{\psi s} \tag{25}$$

By substituting fluxes with currents in Eqs. (22) and (23), we get

$$V_{dr}^{\psi s} = r_r i_{dr}^{\Psi s} + L_{rr}\sigma \frac{d}{dt} i_{dr}^{\Psi s} - s\omega_s L_{rr}\sigma\, i_{qr}^{\psi s} \tag{26}$$

$$V_{qr}^{\psi s} = r_r i_{qr}^{\psi s} + Lrr\sigma \frac{d}{dt} i_{qr}^{\Psi s} + s\omega_s (L_{rr}\sigma\, i_{dr}^{\psi s} + L_{rr}(1-\sigma) i_{ms}) \tag{27}$$

$$\text{where } \sigma = L_{rr} - \frac{L_m^{\ 2}}{L_{ss}} \tag{28}$$

Equations (26) and (27) indicate that the dynamics of direct and quadrature components of the machine rotor current are coupled. However, these can be decoupled if they are expressed in terms of auxiliary variables (V_{dr}' and V_{qr}') as

$$V_{dr}' = V_{dr}^{\Psi s} + s\omega_s L_{rr}\sigma\, \mathrm{i}_{qr}^{\Psi s} \tag{29}$$

$$V_{qr}' = V_{qr}^{\Psi s} - s\omega_s (L_{rr}\sigma\, \mathrm{i}_{dr}^{\Psi s} + L_{rr}(1-\sigma) i_{ms}) \tag{30}$$

By substituting Eqs. (29) and (30) in Eqs. (26) and (27), we get

$$V_{dr}' = r_r i_{dr}^{\Psi s} + L_{rr}\sigma \frac{d}{dt} i_{dr}^{\Psi s} \tag{31}$$

$$V_{qr}' = r_r i_{qr}^{\psi s} + L_{rr}\sigma \frac{d}{dt} i_{qr}^{\Psi s} \tag{32}$$

Then, a PI controller can be designed such that

$$V_{dr}' = \left(K_{p2} + \frac{K_{i2}}{s}\right)\left((i_{dr}^{\Psi s})^* - i_{dr}^{\Psi s}\right) \tag{33}$$

$$V_{qr}' = \left(K_{p2} + \frac{K_{i2}}{s}\right)\left((i_{qr}^{\Psi s})^* - i_{qr}^{\Psi s}\right) \tag{34}$$

The complete block diagram of the RSC is shown in Fig. 9.

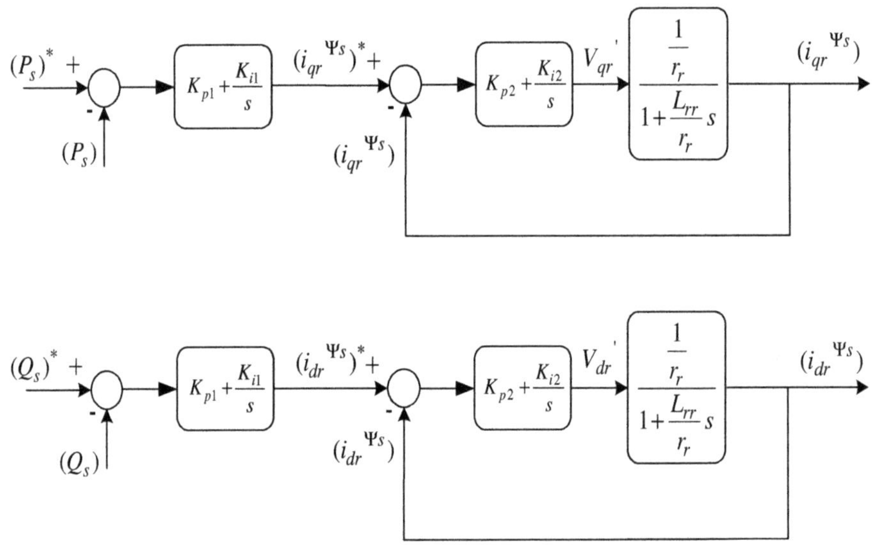

Fig. 9. Rotor-side converter control scheme

5.5 Grid-Side Converter (GSC) Control

The objective of the GSC is to keep the dc-link voltage constant irrespective of the direction of rotor power flow. Decoupled controls of active and reactive powers flowing between rotor and grid are performed by using supply voltage vector oriented control. In such a scheme, the d-axis current (i_d) is controlled to keep the dc-link voltage constant and the q-axis current (i_q) is used to obtain the desired value of reactive power flow between the GSC and the point of common coupling (PCC). The scheme makes use of the supply voltage angle determined dynamically to map the supply voltage, the converter terminal voltage and the phase currents onto the new reference frame. First, the supply voltage angle (θ_s) has to be determined. By definition, the supply voltage angle is

$$\theta_s = \tan^{-1}\left(\frac{V_{qs}}{V_{ds}}\right) \tag{35}$$

where, V_{ds} and V_{qs} are the *d*-*q* axis stator voltages in the stationary reference frame. The active axis (*d*-axis) is aligned with the supply voltage phasor. Thus, $V_{qs}{}^{Vs} = 0$. Hence, the active and reactive powers between the GSC and the grid are

$$P_g \approx \frac{3}{2} V_{ds}{}^{Vs} i_{dg}{}^{Vs} \text{ and } Q_g \approx -\frac{3}{2} V_{ds}{}^{Vs} i_{qg}{}^{Vs} \tag{36-37}$$

The energy balance in the dc-link capacitor is governed by

$$\frac{1}{2} C_{dc} \frac{d}{dt} V_{dc}{}^2 = -P_r - P_g = -P_r - \frac{3}{2} V_{ds}{}^{Vs} i_{dg}{}^{Vs} \tag{38}$$

From Eq. (38), it can be seen that the dc-link capacitor voltage can be controlled through the direct-axis component of the GSC. The control scheme uses a PI controller to get the *d*-axis current reference from the dc-link capacitor voltage error. The *q*-axis current reference can be obtained from the grid voltage error. Usually, no reactive power to the grid is supported from the GSC. So, the *q*-axis current reference can be set to zero. But advanced GCR requires active support from wind power generation to utility grid, hence, reactive power can be demanded from GSC.

Control of Grid-Side Converter Currents

The equations of the grid side converter in stator voltage reference frame are given below (as shown in Fig. 10)

$$V_{ds}{}^{Vs} = R_f i_{dg}{}^{Vs} + L_f \frac{d}{dt} i_{dg}{}^{Vs} - \omega_s L_f i_{qg}{}^{Vs} + V_{dg}{}^{Vs} = |V^s| \tag{39}$$

$$V_{qs}{}^{Vs} = R_f i_{qg}{}^{Vs} + L_f \frac{d}{dt} i_{qg}{}^{Vs} - \omega_s L_f i_{dg}{}^{Vs} + V_{qg}{}^{Vs} = 0 \tag{40}$$

where, R_f and L_f are the grid filter resistance and inductance, respectively; $i_{dg}{}^{Vs}$ and $i_{qg}{}^{Vs}$ are the *d*-*q* axis grid filter currents in grid voltage reference frame; $V_{dg}{}^{Vs}$ and $V_{qg}{}^{Vs}$ are the *d*-*q* axis GSC voltages at the ac terminals in grid voltage reference frame.

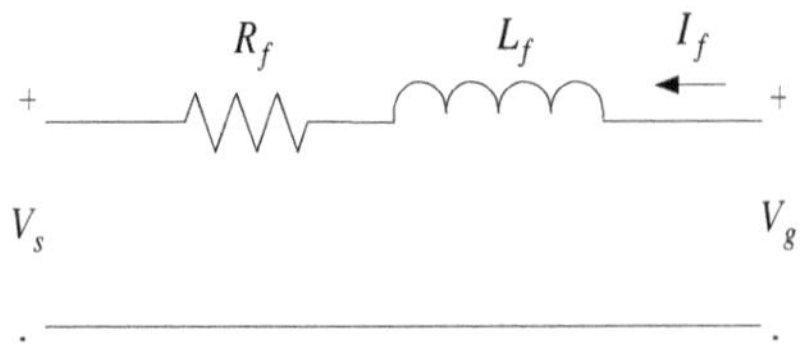

Fig. 10. Grid-side converter filter model

Eqs. (39) and (40) indicate that the dynamics of direct and quadrature components of the GSC current are coupled. However, these can be decoupled if these are expressed in terms of auxiliary variables (V_{dg}' and V_{qg}') as

$$V_{dg}' = V_{dg}^{Vs} + \omega_s L_f \, i_{qg}^{Vs} - V_{dg} \text{ and } V_{qg}' = -\omega_s L_f \, i_{dg}^{Vs} - V_{qg} \tag{41-42}$$

By substituting Eqs. (41) and (42) in Eqs. (39) and (40), we have

$$V_{dg}' = R_f i_{dg}^{Vs} + L_f \frac{d}{dt} i_{dg}^{Vs} \text{ and } V_{qg}' = R_f i_{qg}^{Vs} + L_f \frac{d}{dt} i_{qg}^{Vs} \tag{43-44}$$

Then, a PI controller can be designed such that

$$V_{dg}' = \left(K_{p4} + \frac{K_{i4}}{s} \right) \left((i_{dg}^{Vs})^* - i_{dg}^{Vs} \right) \tag{45}$$

$$V_{qg}' = \left(K_{p4} + \frac{K_{i4}}{s} \right) \left((i_{qg}^{Vs})^* - i_{qg}^{Vs} \right) \tag{46}$$

The complete block diagram of the GSC is shown in Fig. 11.

6 Unified Architecture of DFIG

In this section, a modified DFIG architecture, termed as unified architecture (UA), is presented to eliminate the shortcomings of the conventional DFIG system to meet the advanced GCR. The strength of the conventional DFIG system in power processing, namely, the direct connection of the stator windings to the PCC, turns out to be a source of weakness in regards to tolerating PCC voltage disturbances [32]-[36].

As the penetration of large scale wind power generation (WPG) into the electric power grid continues to increase, the response of wind power generation system to grid disturbances is an important issue. New electricity grid codes require the WPG to have fault-ride-through (FRT) capabilities during the disturbances [1], [2]. Many authors have studied the performance of the DFIG system through time domain simulation studies [32]-[35]. Time-domain studies offer a direct appreciation of the dynamic behavior in terms of visual clarity. However, these are not

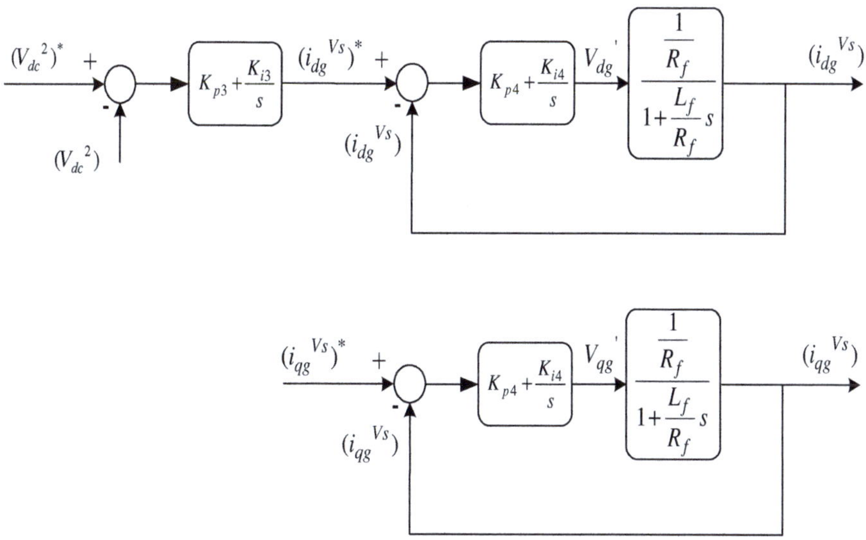

Fig. 11. Grid-side converter control scheme

able to identify and quantify the cause and nature of interactions and problems. This complementary information can be obtained with eigenvalue analysis. In [37], by simulations, it was found that the flux is influenced both by load changes and stator power supply variations. Several authors have investigated the stability of the DFIG analytically, showing that the dynamics of the DFIG have poorly damped eigenvalues (poles) corresponding to natural frequency near the line frequency, and, also, that the system is unstable for certain operating conditions, at least for a stator flux oriented system [38]-[39]. These poorly damped poles influence the rotor current dynamics through the back electromotive force.

In the literature, different methods to modify the DFIG system in order to accomplish voltage sag ride-through and to damp the flux oscillations during abnormal conditions have been proposed by different authors. A possible solution is to limit the high current in the rotor to protect the converter and to provide a bypass for this current via a set of resistors connected to the rotor windings, known as crowbar [34], during the faults without disconnecting the WPG from the grid. During the period of operation of the crowbar, the control action of the converter is trimmed to shut down and allow the DFIG to operate as a conventional slip-ring induction machine. Even though, the current magnitude is brought into the limits, the oscillations may still be present in the voltages/currents. An extra (third) converter that substitutes the star-point of the stator winding may be used [33]. Extra converter method actively damp the flux oscillations via using an extra converter that is connected in series with the stator windings to provide an extra degree of freedom [32], [36]. Even though this architecture actively damps out the stator flux oscillations, the inverter is idle during normal operating conditions. The extra converter involves extra cost and losses. Reactive power support can be provided from extra series converter during normal operation [40].

A detailed schematic diagram of grid connected DFIG with an additional series grid-side converter (SGSC) is shown in Fig. 12. This architecture is a unified architecture that has three bidirectional VSCs (i.e., RSC, parallel grid-side converter (PGSC), and SGSC) sharing the common dc-link. The dc-terminals of the RSC are connected to the PGSC and the ac-terminals of PGSC are connected to PCC via transformer. The three-phase terminals of the SGSC are connected in series with the three-phase terminals of the stator circuit and the line. Thus, the stator power also passes through the SGSC providing the active mechanism for effective grid disturbance ride through.

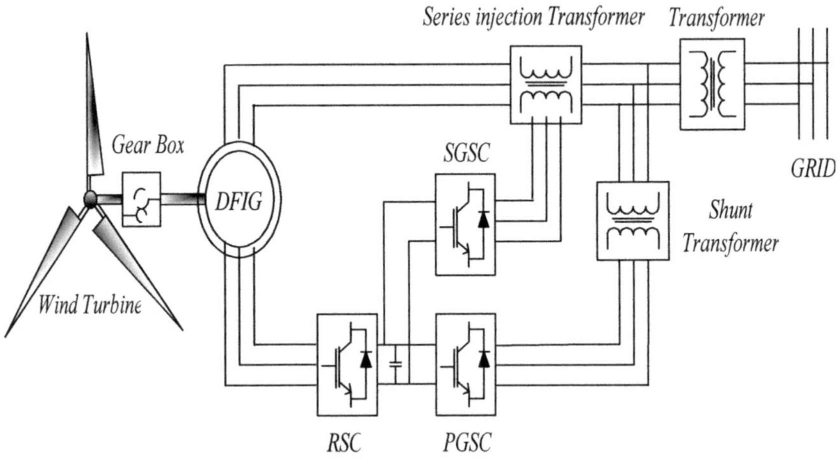

Fig. 12. Schematic of unified architecture of DFIG (with additional series grid side converter)

The only difference between conventional DFIG and UA is an extra SGSC. The main objective of the third converter, i.e., the SGSC, is to control the stator flux during abnormal conditions. During normal operating conditions, it can be either inactive or it can be used to inject reactive power into the grid.

6.1 Unified Architecture Dynamic Model

The equivalent circuit of the UA with the SGSC connected at the neutral point of the stator of the DFIG is shown in Fig. 13. The rotor circuit elements are referred to the stator side. The stator voltage is equal to the sum of the PCC voltage and the series injected voltage. The equations of the DFIG can be modified as

$$\overline{V_f} + \overline{V_{se}} = r_s \overline{I_s} + j\omega_s \overline{\Psi_s} \tag{47}$$

$$\frac{\overline{V_r}}{s} = \frac{r_r}{s}\overline{I_r} + j\omega_s \overline{\Psi_r} \tag{48}$$

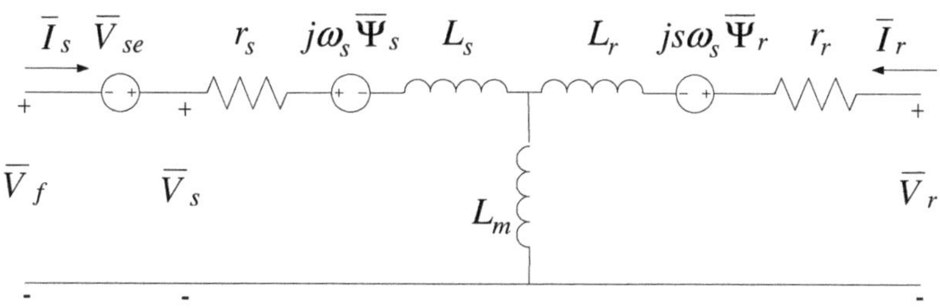

Fig. 13. Equivalent circuit of unified architecture

6.2 Control Structure

The main objective of the various controls for the UA includes regulation of stator active and reactive powers, dc-link voltage control, stator flux magnitude, and SGSC reactive power injection. During normal grid conditions, the RSC and PGSC of the UA are controlled in a conventional manner as described in section 1.5. A high bandwidth current regulator on the RSC and stator flux aligned field orientation allows decoupled control of active and reactive powers at the stator terminals. Likewise, a current regulator for the PGSC aligned with the farm collection transformer voltage enables control of the dc-link voltage and net reactive power regardless of the direction of rotor power.

6.3 SGSC Control

The main objective of the SGSC control is to regulate the stator flux of DFIG during abnormal conditions. During normal operating conditions, SGSC can be idle or can inject reactive power to the electric grid. In the following sections, both cases are considered.

SGSC Control without Reactive Power Injection

In this case, SGSC is made inactive during normal operating conditions and the inverter is maintained at zero voltage vector switch state to eliminate the switching losses. During abnormal condition, using sag detection logic, SGSC is made active and the resultant stator voltage is controlled in order to transfer the total stator flux to a new level without oscillations.

The sag detection logic can be derived from Eq. (47) as

$$\frac{d}{dt}\overline{\Psi_s} = \overline{V_f} + \overline{V_{se}} - r_s\overline{I_s} - j\omega_s\overline{\Psi_s} = 0 \quad \text{(in steady state)} \tag{49}$$

In steady state, the series injected voltage, $\overline{V}_{se}$, and the derivative of the stator flux of DFIG ($d\psi_s/dt$) are zero. This results in a balance between the applied stator voltage, $\overline{V}_s$, the sum of stator flux voltage, $j\omega_s\psi_s$, and the ohmic voltage drop $r_s\overline{I_s}$. During abnormal conditions, this equilibrium is disturbed and the derivative of the stator flux is not zero.

$$|\overline{V}_f - r_s\overline{I}_s - j\omega_s\overline{\Psi}_s| > 0 \text{ or } \varepsilon \tag{50}$$

where, ε is some threshold value.

The threshold value serves as the criterion for activating the stator flux controller during abnormal conditions. As long as this quantity is zero (or less than some threshold value), the SGSC is in zero voltage state and when it crosses the threshold, the SGSC controller is activated. The SGSC control scheme uses a proportional controller to obtain the series voltage, $\overline{V}_{se}$, from the stator flux error. A complete block diagram of SGSC control scheme, including sag detection logic, is shown in Fig. 14. As there is no exchange of active or reactive power to the grid from SGSC, no decoupling control is required. Hence, the control can be done in any arbitrary reference frame. The command for the SGSC controller (i.e., ψ_s^{Vf}) is generated from the measured PCC voltage and stator current in the synchronous frame as

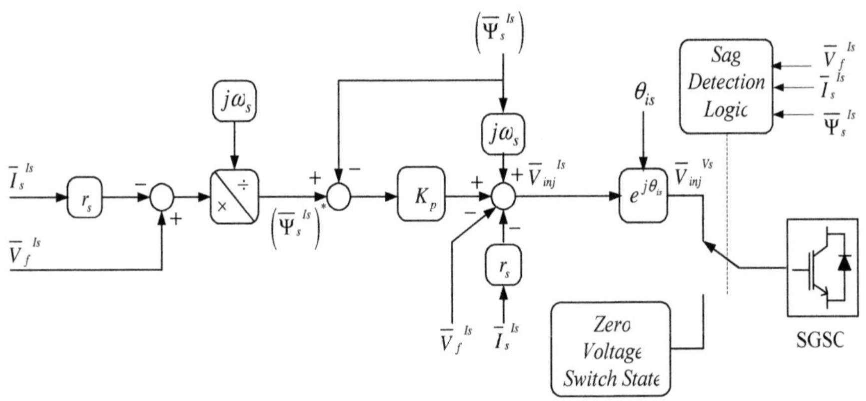

Fig. 14. SGSC control scheme without reactive power injection

$$\overline{\Psi_s}^{*} = \frac{\overline{V_f} - r_s\overline{I_s}}{j\omega_e} \tag{51}$$

At abnormal conditions, the PCC voltage, $\overline{V_f}$, in Eq. (51) is the reduced PCC voltage. This total stator flux command is chosen to synchronize the flux level in the DFIG such that it will be in equilibrium with the new wind farm collector voltage during the sag. After the sag, the SGSC is again switched back to zero voltage state.

SGSC Control with Reactive Power Injection

The SGSC can be used to control the stator flux during abnormal conditions and inject reactive power during healthy conditions. It is assumed that no active power is injected or absorbed by the SGSC. Hence, the series voltage component that is in phase with the stator current is zero and the series voltage component that is perpendicular to the stator current is proportional to the reactive power to be injected. In this case, the stator current reference frame has been chosen to get the decoupling between active and reactive powers injected by the SGSC. The active and reactive powers injected by the SGSC are given by

$$P_{se} = \frac{3}{2}\mathrm{Re}(\overline{V}_{se}^{Is}\overline{I}_s^{Is*}) = \frac{3}{2}(V_{dse}^{Is}i_{ds}^{Is} + V_{qse}^{Is}i_{qs}^{Is}) \tag{52}$$

$$Q_{se} = \frac{3}{2}\mathrm{Im}(\overline{V}_{se}^{Is}\overline{I}_s^{Is*}) = \frac{3}{2}(V_{qse}^{Is}i_{ds}^{Is} - V_{dse}^{Is}i_{qs}^{Is}) \tag{53}$$

In the stator current reference frame, $i_{ds} = |\overline{I_s}|$ and $i_{qs} = 0$. Thus, Eqs. (52) and (53) become

$$P_{se} = \frac{3}{2}V_{dse}^{Is}|\overline{I_s}|\ and\ Q_{se} = \frac{3}{2}V_{qse}^{Is}|\overline{I_s}| \tag{54-55}$$

In order to make P_{se} equal to zero, V_{dse}^{Is} can be made equal to zero or in order to ensure the zero active power exchange of SGSC, a PI controller can be used. The vector control scheme uses a PI controller to obtain the value of V_{qse}^{Is} from the reactive power error as shown in Fig. 15.

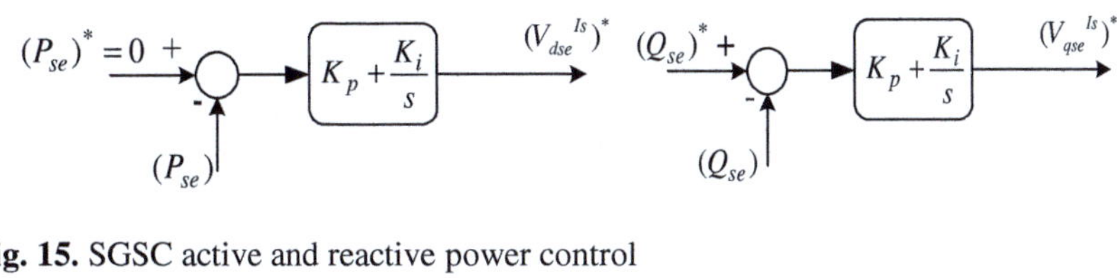

Fig. 15. SGSC active and reactive power control

The series injected voltage, $\overline{V}_{se}^{Is}$ (= $V_{dse}^{Is} + j\,V_{qse}^{Is}$), measured wind farm collector voltage, $\overline{V}_f^{Is}$, and the measured stator current $\overline{I}_s^{Is}$, are used to derive the stator flux command as

$$\left(\overline{\Psi}_s^{Is}\right)^* = \frac{\overline{V}_f^{Is} + \left(\overline{V}_{se}^{Is}\right)^* - r_s \overline{I}_s^{Is}}{j\omega_e} \tag{56}$$

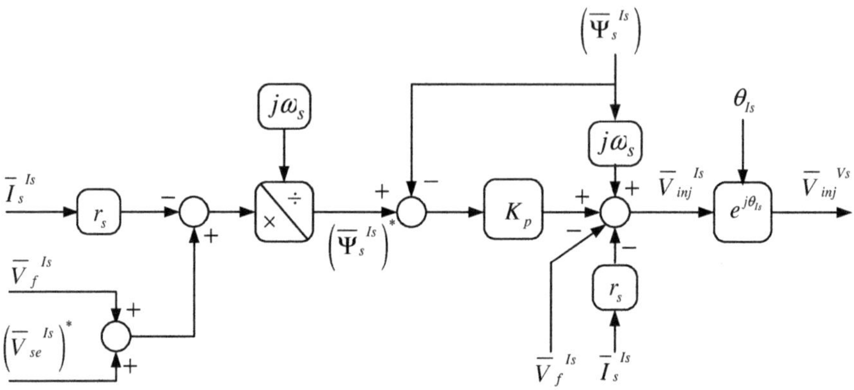

Fig. 16. SGSC control scheme with reactive power injection

The complete block diagram of the control scheme is shown in Fig. 16. In order to inject reactive power, a certain magnitude of voltage in quadrature to stator current, has to be injected into the line. Typically, it is desirable to operate the DFIG close to unity stator flux magnitude (p.u.), to use the complete capacity of the DFIG at the point of peak torque production which normally occurs at the maximum rotor speed. During lightly loaded conditions, the magnitude of the series injected voltage will be larger and causes more deviation from the nominal stator voltage.

7 Design of ANFIS Controllers

Classical PI and PID controllers that are used in conventional PES interfaced variable-speed WPG systems are mainly tuned using specific methods. Several methods provide initial values of the controller parameters. The most commonly used methods are based on the Ziegler-Nichols approach. However, these methods can be time consuming and fixed controllers cannot necessarily provide acceptable dynamic performance over the complete operating range of the WPG system. Performance will degrade mainly because of factors such as machine non-linearities and parameter variations. Adaptive controllers can be used to overcome these problems. Alternatively, performance-index based optimal control techniques can be adopted, but these may suffer from convergence related problems.

The purpose of using a computational intelligence based controller is to reduce the tuning efforts for improved response and to remove the shortcomings of conventional controllers. The design of the ANFIS controller is shown in Fig. 17. There are various possibilities to obtain the training data from the classical PI controlled transient simulation of the WPG system. The ANFIS controller is trained with the input and output data obtained from the transient simulations of the conventional PI controller with a wide range of operating conditions. The ANFIS controller acts like the conventional PI controller without the need to design and tune for different operating conditions repeatedly. The fuzzy logic toolbox in MATLAB™ has been used for designing and testing the ANFIS controllers [41]. The major steps are as follows.

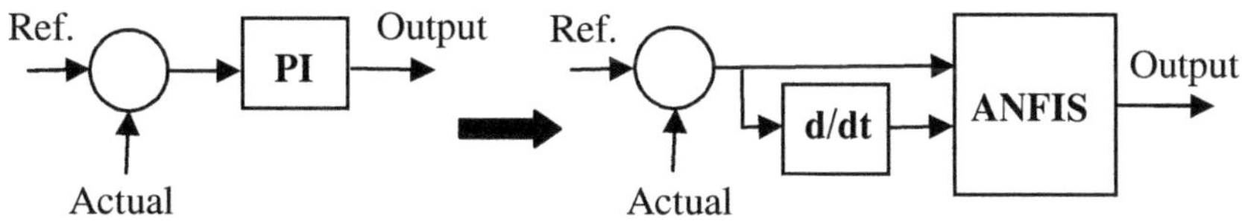

Fig. 17. Design of ANFIS controllers

- The data required for training and testing the ANFIS is generated by designing and testing conventional PI controllers for different operating conditions utilizing the vector control technique as discussed in sections 1.5 and 1.6 for DFIG and UA, respectively. These operating conditions are:
  - Various wind speed profiles
  - Varying ramp increase in wind speed
  - Varying step increase in wind speed
  - Performance during a single-phase fault at different loading conditions
  - Performance during a three-phase fault at different loading conditions
  - Performance during various voltage sag conditions at different loading conditions
- After obtaining the data, the ANFIS structure gets generated by using the grid-partitioning method. For input and output, the Gaussian- and linear-shaped MFs have been considered in this work, respectively, as suggested in [17]. Analysis can be extended for different types of MFs and results can be compared. It may be possible that any other MF may perform better than the MF chosen during the above process, but the main objective of this study is to show the effectiveness and application of CI based techniques to the WPG system.
- After generating the ANFIS structure, it is trained using a hybrid optimum method (LSE combined with back-propagation) with a root mean square error (RMSE) tolerance of 0.0001 p.u. for 500 epochs. While training the ANFIS, the center and spread of the Gaussian MFs are changed by using the back-propagation algorithm [14]. The overall output of the system is determined using Sugeno-type defuzzification, by combining the outputs of all the rules. In arriving at the ANFIS controlled WPG system, a simultaneous automated tuning procedure is adopted. However, a sequential tuning procedure is

possible. Simultaneous tuning is more preferred and popular for hybrid systems.

- Conventional PI controllers of RSC currents are replaced with the trained ANFIS based controllers. It is also possible to replace the remaining active and reactive powers conventional PI controllers of RSC of DFIG/UA, as shown in Fig. 9, by an ANFIS controller and results can be obtained by extending the same analysis.
- The performance of the ANFIS controllers has been shown for a test case having grid connected wind farm with transmission line and results are compared with the conventional PI controllers.

8 Simulation Results and Discussion

The proposed ANFIS based controllers have been implemented for a wind farm connected to the grid. The wind farm consists of six, 1.5 MW DFIG/UA WTs connected to a 25 kV line exporting power to 120 kV grid through a 25 km, 25 kV line, as shown in Fig. 18. ANFIS controllers replace the classical PI controllers that are used for the control of RSC currents of the DFIG and the UA. The dynamic performance of the ANFIS based controllers for three-phase fault at t = 20s for 150 ms duration at 25 kV line (f = 60 Hz) is performed and compared with classical PI controllers. The fault is introduced at the line at a distance 10 km from the wind farm as shown in Fig. 18. The system parameters have been given in Appendix. The simulation is carried out using MATLAB/Simulink^{TM} and the fuzzy logic toolbox [41].

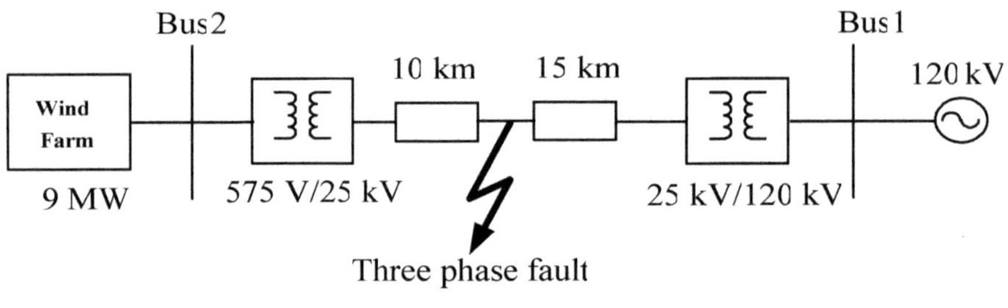

Fig. 18. Schematic of the simulated system

8.1 Simulation Results for DFIG

The impact of a three-phase fault on the stator flux, stator and rotor currents, dc-link voltage, and active and reactive powers of the stator of the DFIG is presented in Figs. 19 to 24, respectively. The conventional DFIG has more difficulty to deal with asymmetrical faults, as it experiences a large amount of negative sequence components. It is seen from the simulation results that the stator and rotor currents exceed 2.0 p.u., which is unsafe for the converters.

The *d* and *q* axis stator currents have oscillations and their magnitudes are not within the safe limits, as shown in Figs. 19 and 20. The dynamic response of ANFIS based controllers for stator current settles in 0.05 s earlier as compared to the conventional PI controllers. The peak overshoot is reduced considerably with ANFIS based controllers.

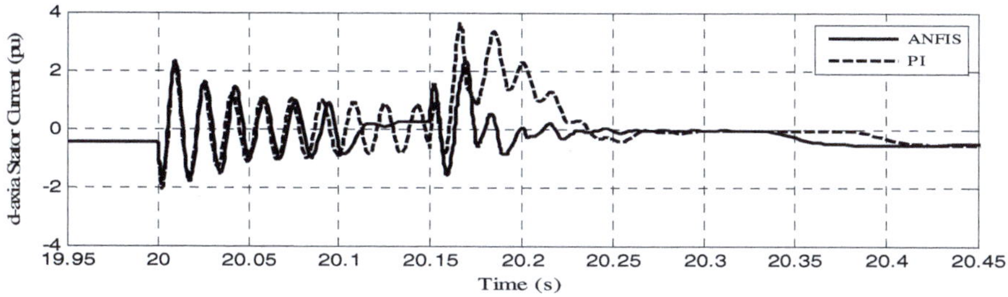

Fig. 19. d-axis stator current of DFIG during three-phase fault

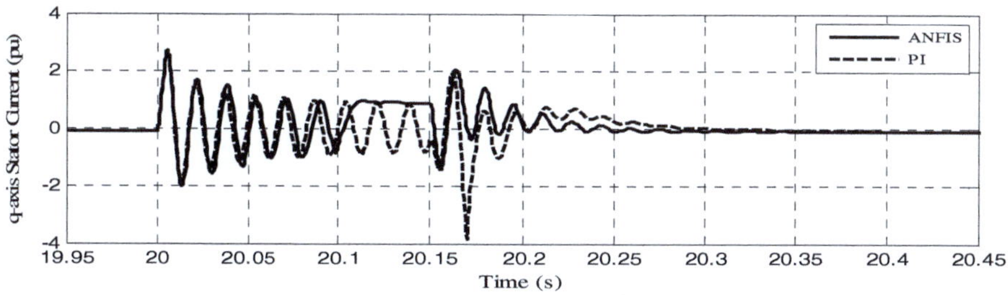

Fig. 20. q-axis stator current of DFIG during three-phase fault

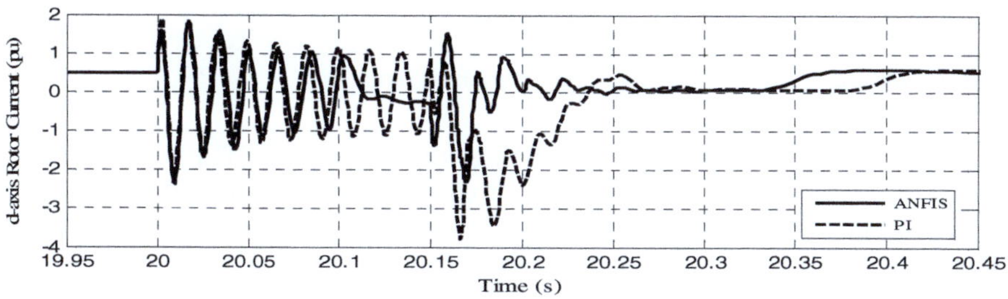

Fig. 21. d-axis rotor current of DFIG during three-phase fault

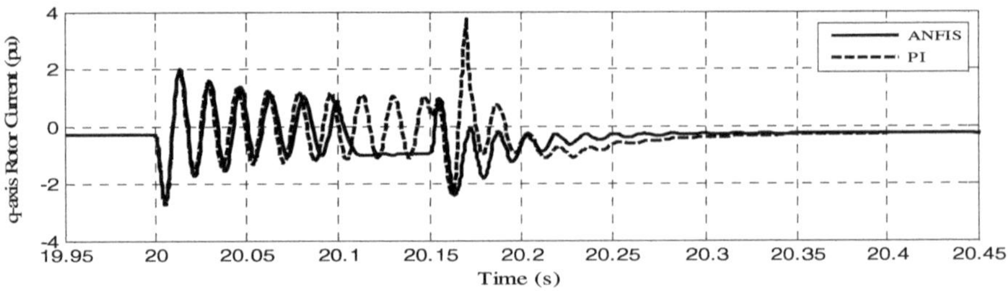

Fig. 22. q-axis rotor current of DFIG during three-phase fault

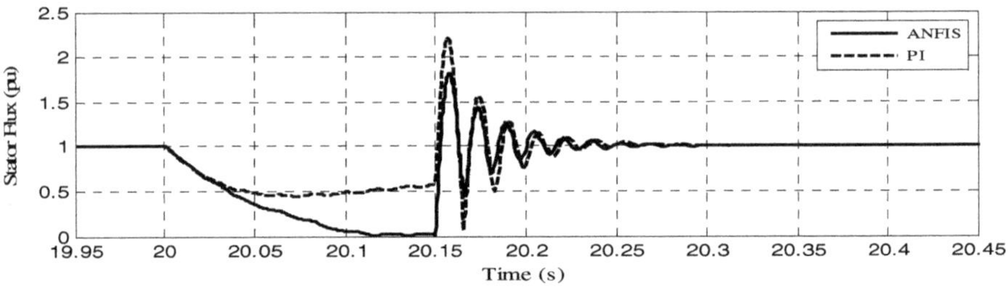

Fig. 23. Stator flux of DFIG during three-phase fault

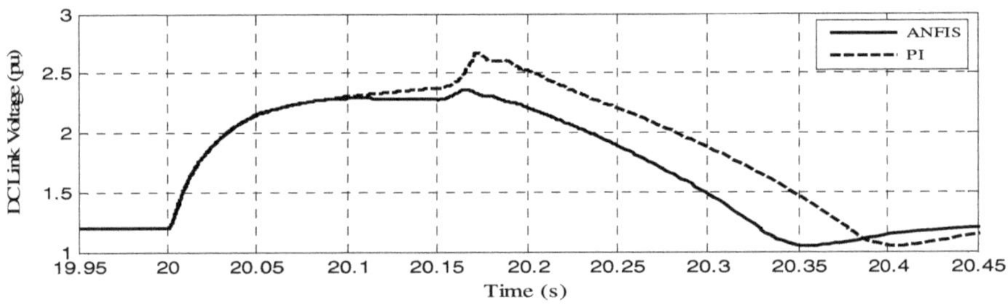

Fig. 24. DC-link voltage of DFIG during three-phase fault

The *d* and *q* axis rotor current responses during and after fault conditions are shown in Figs. 21 and 22, respectively. It can be seen from these figures that the rotor current magnitude is not within the safe limits and has severe oscillations. Moreover, with ANFIS based controllers, the rotor current settles earlier than the conventional PI controllers. The peak overshoot is also reduced considerably with ANFIS based controllers as compared to conventional PI controllers.

The stator flux, as shown in Fig. 23, goes to zero during the fault condition and oscillates after the fault clearance. With ANFIS based controllers as compared to PI controllers, the peak overshoot is reduced considerably. The settling time is almost the same for both controllers. The dc-link voltage, as shown in Fig. 24, largely deviates from its set value. With ANFIS based controllers, the rise in the dc-link voltage is only 2.3 p.u. whereas it is 2.7 p.u. in the case of PI controllers. The normalized stator active and reactive power outputs of the DFIG for a three-phase fault are shown in Figs. 25 and 26, respectively. ANFIS based controllers perform better than conventional PI controllers in terms of settling time and peak overshoot. The settling time for ANFIS based controllers is 0.05 s less as compared to PI controllers.

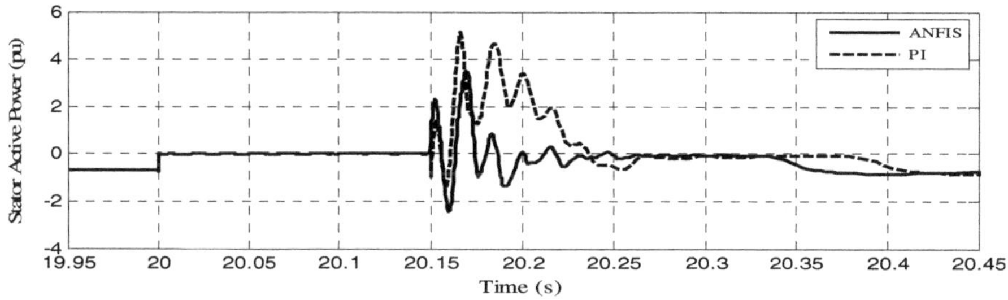

Fig. 25. Stator active power of DFIG during three-phase fault

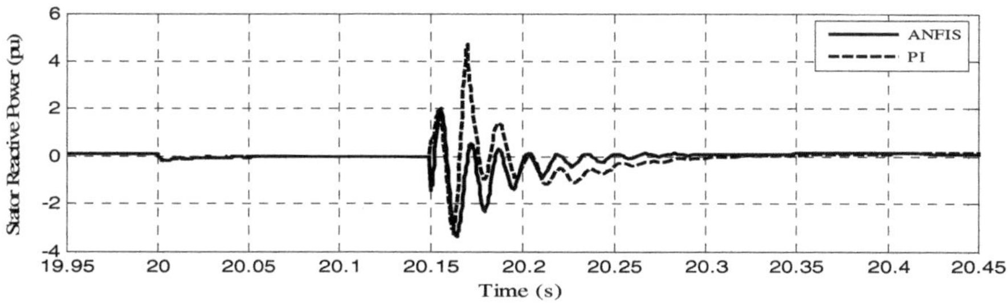

Fig. 26. Stator reactive power of DFIG during three-phase fault

8.2 Simulation Results for Unified Architecture

The impact of a three-phase fault on the stator flux, stator and rotor currents, dc-link voltage, and active and reactive powers of stator for UA is presented in Figs. 27 to 34. It is seen that the stator flux becomes smooth and its oscillations are well damped. Variations in dc-link voltage are reduced significantly. Stator and rotor currents behave smoothly and remain within their limits. The peak overshoots of the stator and rotor currents are reduced drastically, and showing the safe and reliable UA operation as compared to the conventional DFIG. The oscillations are also well damped out by injecting series voltage from the SGSC.

The d and q axis stator and rotor currents have reduced oscillations and the magnitudes are within safe limits, as shown in Figs. 27 and 28 and in Figs. 29 and 30, respectively. The dynamic response of ANFIS based controllers for stator and rotor currents settles 0.025 s earlier as compared to conventional PI controllers. The peak overshoot of d-axis stator and rotor currents is reduced considerably

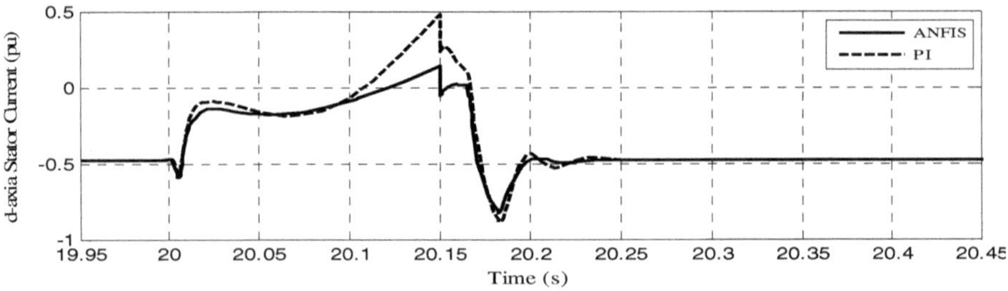

Fig. 27. *d*-axis stator current of UA during three-phase fault

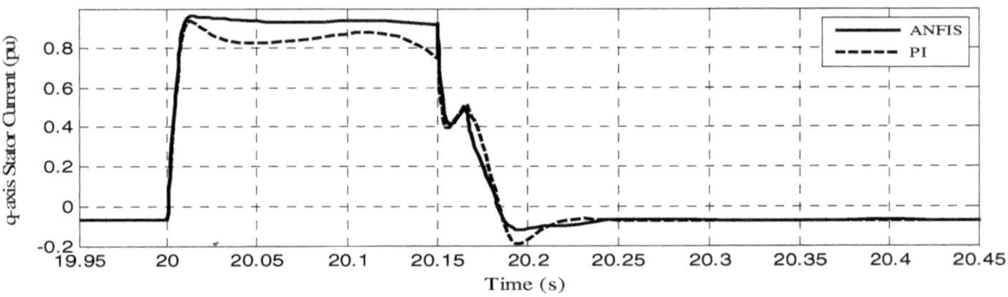

Fig. 28. q-axis stator current of UA during three-phase fault

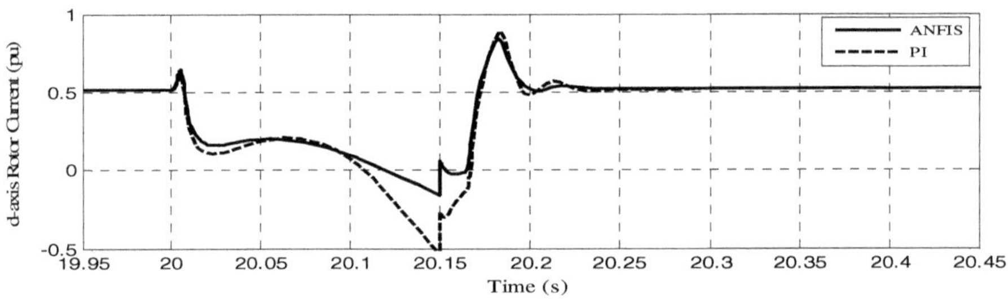

Fig. 29. d-axis rotor current of UA during three-phase fault

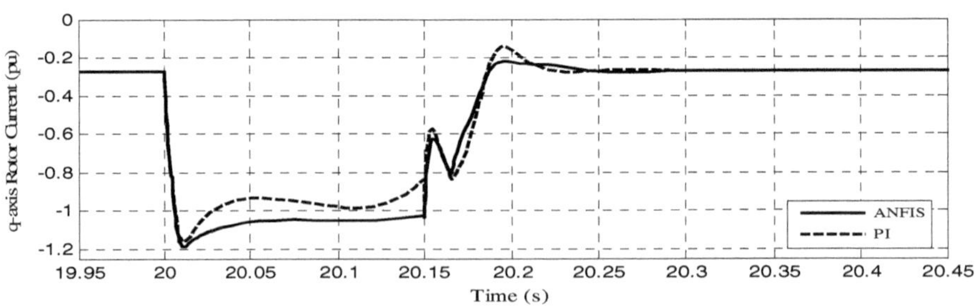

Fig. 30. q-axis rotor current of UA during three-phase fault

with ANFIS based controllers but for q-axis stator current, it is higher compared to conventional PI controllers.

For UA architecture, it can be observed that the stator flux becomes smooth and its oscillations are damped out, as compared to the DFIG system. The stator flux is controlled smoothly even when the PCC voltage falls to 10% of its nominal value during the fault condition. ANFIS based controllers perform better compared to PI controllers, as shown in Fig. 31, for peak overshoot and settling time. The active and reactive powers developed by UA have no severe fluctuations and no large peak overshoots (as shown in Figs.32 and 33) and hence, the stress on the mechanical system will be smaller as compared to the DFIG system. The rise in dc-link voltage is reduced for both types of controllers (as shown in Fig. 34). ANFIS based controllers perform better compared to PI controllers for UA in terms of peak overshoot and settling time.

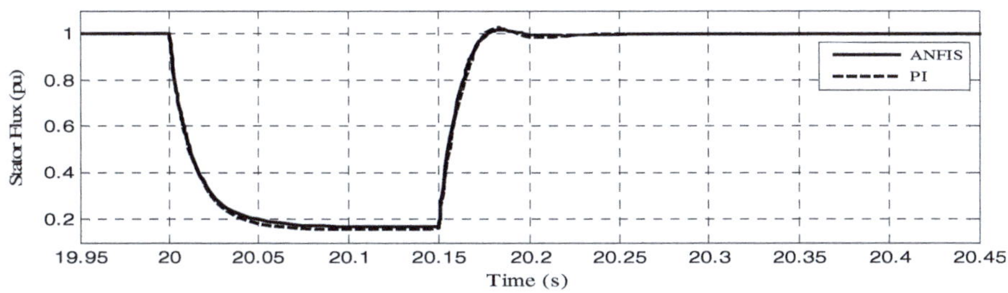

Fig. 31. Stator flux of UA during three-phase fault

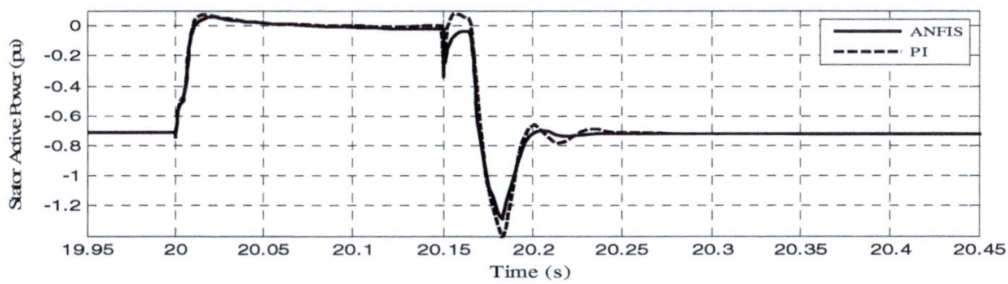

Fig. 32. Stator active power of UA during three-phase fault

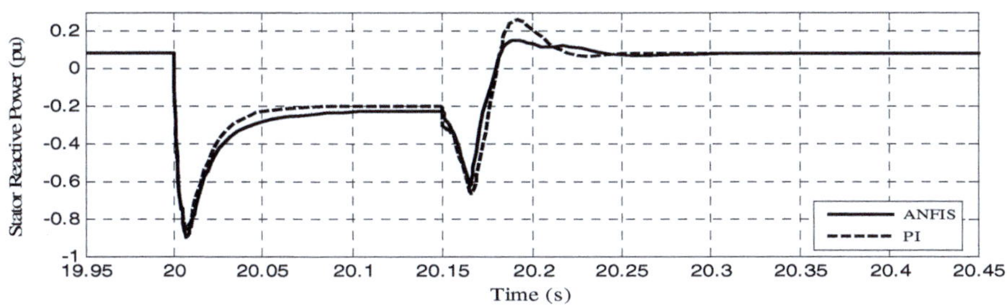

Fig. 33. Stator reactive power of UA during three-phase fault

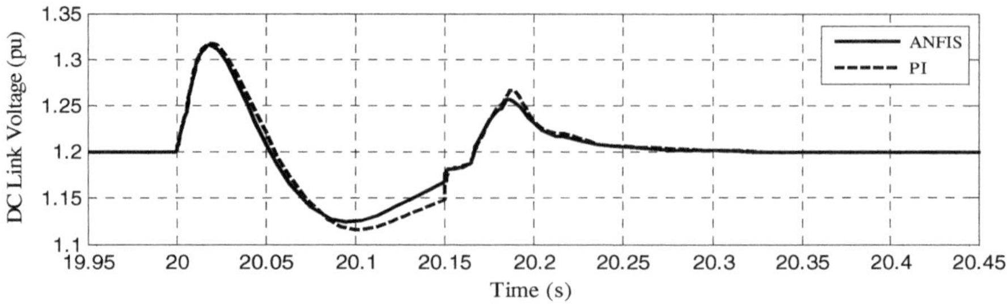

Fig. 34. DC-link voltage of UA during three-phase fault

8.3 Discussion

The control structure of the conventional PI controllers is simple due to the linear nature which results in simple digital implementations. However, it is extremely beneficial to implement the computation intelligence (CI) based controller, which is, though, more complex, non linear control surfaces, but results in effective and efficient dynamic performance. By appropriate tuning, it is possible to obtain better dynamic characteristics under all operating conditions. CI controller response does not exhibit the under-damped characteristics which are seen in the response of the conventional PI controller. It should be noted that the tuning effort required with CI based controllers is significantly less than that required for the conventional PI controllers. ANFIS based controllers easily overcome most of the shortcomings of conventional controllers as discussed earlier.

9 Concluding Remarks

Renewable energy resources are growing in electric power generation and the emphasis is given to the cost effective utilization of these energy resources. Wind power is one of the most popular renewable energy sources as it combines a number of economic and technical advantages. With increased penetration of wind power generation in the electric power system, the efficient, stable, economical, and secure operation of power systems is becoming a major concern. This chapter introduces basics of the electrical aspects involved with the modern variable-speed wind generation systems that are equipped with power-electronic systems (PES). The control techniques using conventional and computational intelligence methods for wind power generation systems have been described.

In this chapter, a nonlinear adaptive neuro-fuzzy inference system (ANFIS) is proposed to control the rotor side converter (RSC) of conventional and unified architecture (UA) of a doubly-fed induction generator (DFIG). The proposed ANFIS controllers, to make the DFIG more suitable for highly varying operating conditions and parameter sensitiveness, have been used in place of the conventional proportional-integral (PI) controllers of the RSC. The ANFIS have been trained with the input and output data of the conventional PI controller for different operating conditions. The training of ANFIS controllers has been done by combining the back-propagation gradient descent learning algorithm to choose the parameters related to membership functions and the least-squares estimation to determine the consequent parameters. In order to evaluate the performance of ANFIS, the transient simulation for a three-phase fault is carried out on conventional and UA of the DFIG and results are compared with classical PI controllers. The results show that, with ANFIS, the settling time is reduced considerably, peak overshoot of values are limited, and oscillations are damped out quickly as compared to the conventional PI controllers.

The vast potential of CI based techniques has yet to be explored for PES and wind power generation applications. Hybrid CI techniques, particularly neuro-fuzzy techniques, have enormous potential for application in PES. The area is vast, and the authors provided a discussion on the subjects which are the most relevant to the power electronic systems used in the wind energy conversion system.

References

1. Jauch, C., Matevosyan, J., Ackermann, T., Bolik, S.: International comparison of requirements for connection of wind turbines to power systems. Wind Energy 8(3), 295–306 (2005)
2. Erlich, I., Winter, W., Dittrich, A.: Advanced grid requirements for the integration of wind turbines into the German transmission system. In: IEEE PES General Meeting, Montreal, Canada (June 2006)
3. Vas, P.: Artificial-intelligence-based electrical machines and drives. Oxford University Press, NewYork (1999)
4. Bose, B.K.: Modern power electronics and ac drives. Prentice Hall PTR, New Jersey (2001)
5. Dote, Y., Hoft, R.G.: Intelligent Control: Power Electronic Systems. Oxford University Press, NewYork (1998)
6. Zilouchian, A., Jamshidi, M. (eds.): Intelligent control systems using soft computing methodologies. CRC Press, Boca Raton (2001)
7. Simoes, M.G., Bose, B.K., Spiegel, R.J.: Design and performance evaluation of a fuzzy-logic-based variable-speed wind generation system. IEEE Transactions on Industry Applications 33(4), 956–965 (1997)
8. Simoes, M.G., Bose, B.K., Spiegel, R.J.: Fuzzy-logic-based intelligent control of a variable-speed cage machine wind generation system. IEEE Transactions on Industry Applications 12(1), 87–95 (1997)
9. Hillowala, R.M., Sharaf, A.M.: A rule base fuzzy logic controller for a PWM inverter in a stand alone wind energy conversion scheme. IEEE Transactions on Industry Applications 32(1), 57–65 (1996)
10. Chen, Z., Gomez, S.A., McCormick, M.: A fuzzy logic controlled power electronic system for variable speed wind generation system. In: Eighth International conference on power electronics and variable speed drives, pp. 114–119 (2000)
11. Soloumah, H.M., Kar, N.C.: Fuzzy logic based vector control of a doubly-fed induction generator for wind power application. Wind Engineering 30(3), 201–224 (2006)
12. Sousa, G., Bose, B.K.: Fuzzy logic applications to power electronics and drives-an overview. In: Proceedings of IECON 1995, November 1995, pp. 57–62 (1995)

13. Bose, B.K.: Neural network applications in power electronics and motor drives-an introduction and perspective. IEEE Transactions on Industrial Electronics 54(1), 14–33 (2007)
14. Jang, J.S.R.: ANFIS: Adaptive-network based fuzzy inference system. IEEE Transactions on System Man, Cybernetics 23(2), 665–685 (1993)
15. Mamdani, E.H.: Applications of fuzzy algorithm for simple dynamic plant. Proceedings of IEEE 121(12), 1585–1588 (1974)
16. Sugeno, M.: Industrial applications of fuzzy control. Elsevier Science Pub. Co., Amsterdam (1985)
17. Tsukamoto, Y.: An approach to fuzzy resoning method. In: Gupta, M.M., Saridis, G.N., Gaines, B.R. (eds.) Fuzzy Automata and Decision Processes, pp. 89–102. North-Holland, NY (1977)
18. Jang, J.S.R.: Fuzzy modeling using generalized neural networks and kalman filter algorithm. In: Proceedings of the Ninth National Conference on Artificial Intelligence (AAAI 1991), July 1991, pp. 762–767 (1991)
19. Blaabjerg, F., Chen, Z.: Power electronics for modern wind turbines, Morgan & Claypool Publishers, USA (2006)
20. Ackermann, T. (ed.): Wind power in power system. John Wiley & Sons, Ltd., England (2005)
21. Hansen, L.H., et al.: Conceptual survey of generators and power electronics for wind turbines. RISO National Laboratory, Roskilde, Denmark (December 2001)
22. Homes, D.G., Lipo, T.A.: Pulse Width Modulation for Power Converters: Principles and Practice. IEEE Press, Los Alamitos (2003)
23. Leonhard, W.: Control of Electrical Drives. Springer, Heidelberg (1985)
24. Tang, Y., Xu, L.: Flexible active and reactive power control strategy for a variable speed constant frequency generating system. IEEE Transactions on Power Electronics 10(4), 472–478 (1995)
25. Pena, R., Clare, J.C., Asher, G.M.: Doubly fed induction generator using back-to-back PWM converters and its application to variable-speed wind-energy generation. IEE Proceedings of Electric Power Applications 143, 231–241 (1996)
26. Chowdhury, B.H., Chellapilla, S.: Double-fed induction generator control for variable speed wind power generation. Electrical Power System Research 76, 786–800 (2006)
27. Rabelo, B., Hofmann, W.: Optimal active and reactive power control with the doubly-fed induction generator in the MW-class wind-turbines. In: Proceedings of International Conference on Power Electronics and Drives Systems, Denpasar, Indonesia, October 22–25, pp. 53–58 (2001)
28. Datta, R., Ranganathan, V.T.: Direct power control of grid-connected wound rotor induction machine without rotor position sensors. IEEE Transactions on Power Electronics 16, 390–399 (2001)
29. Xu, L., Cheng, W.: Torque and reactive power control of a doubly fed induction machine by position sensorless scheme. IEEE Transactions on Industrial Applications 31, 636–642 (1995)
30. Morel, L., Godfroid, H., Mirzaian, A., Kauffmann, J.M.: Double-fed induction machine: converter optimization and field oriented control without position sensor. IEE Proceedings of Electric Power Applications 145, 360–368 (1998)
31. Zhi, D., Xu, L.: Direct power control of DFIG with constant switching frequency and improved transient performance. IEEE Transactions on Energy Conversion 22(1), 110–118 (2007)

32. Rajpurohit, B.S.: Reactive power capability and performance analysis of grid connected unified DFIG for wind power application. Ph.D. Thesis, IIT Kanpur, India (2009)
33. Petersson, A.: Analysis, modeling and control of doubly-fed induction generators for wind turbines. Ph.D. Thesis, Chalmers University of Technology, Goteborg, Sweden (2005)
34. Morren, J., Haan, S.W.H.: Ride through of wind turbines with doubly-fed induction generator during a voltage dip. IEEE Transactions on Energy Conversion 20(2), 435–441 (2005)
35. Flannery, P., Venkataramanan, G.: A grid fault tolerant doubly fed induction generator wind turbine via series connected grid side converter. In: WINDPOWER 2006, Pittsburgh, USA, June 4-7 (2006)
36. Flannery, P., Venkataramanan, G.: A unified architecture for doubly-fed induction generator wind turbines using a parallel grid-side rectifier and series grid side converter. In: Power Conversion Conference-Nagoya 2007, April 2007, pp. 1442–1449 (2007)
37. Wang, S., Ding, Y.: Stability analysis of field oriented doubly-fed induction machine drive based on computer simulation. Electric Machines and Power Systems 21(1), 11–24 (1993)
38. Mei, F., Pal, B.: Modal analysis of grid-connected doubly-fed induction generators. IEEE Transactions on Energy Conversion 22(3) (September 2007)
39. Wu, F., Zhang, X.P., Godfrey, K., Ju, P.: Small signal stability analysis and optimal control of a wind turbine with doubly fed induction generator. IET Proceedings of Generation Transmission and Distribution 1(5), 751–760 (2007)
40. Singh, B., Singh, S.N.: Reactive capability limitation of doubly-fed induction generators. Electric Power Components & Systems 37(4), 427–440 (2009)
41. MATLAB/SIMULINK™, `http://www.mathworks.com`

Appendix

Table A: Parameters of simulated DFIG

Parameters	Values
Rated power	1.5 MW
Stator voltage	575 V
R_s	0.0071 p.u.
R_r (referred to stator)	0.005 p.u.
L_s	0.171p.u.
L_r (referred to stator)	0.156 p.u.
L_m	2.9 p.u.
Number of pole pairs	3
Inertia constant (H)	5.04

Intelligent Controller Design for a Remote Wind-Diesel Power System: Design and Dynamic Performance Analysis

Hee-Sang Ko, Kwang Y. Lee, and Ho-Chan Kim

Abstract. This chapter presents design of intelligent controllers for a wind-diesel power system equipped with a wind turbine driving an induction generator. The goal for the design is to maintain a good power quality under varying wind and load conditions. On the other hand, the controller has to show acceptable closed-loop performance including stability, robustness, optimal energy, and steady-state and transient performance at a permissible level of control effort. Moreover, such a controller has to be highly adaptive to various operating conditions and independent of model parameters that might be uncertain. Toward these goals the concepts of fuzzy-robust controller and fuzzy-neural hybrid controller are applied to design integrated non-linear controllers to provide control input for excitation system and governor system simultaneously.

Index Terms: Wind power generation, diesel driven generators, frequency and voltage control, fuzzy logic control, linear quadratic regulator, sliding mode control, neural networks.

Hee-Sang Ko
Product Development Team/Wind Turbine Division,
Samsung Heavy Industries Co., Korea
e-mail: heesang2008.ko@samsung.com

Kwang Y. Lee
Dept. of Electrical and Computer Engineering, Baylor University,
Waco, TX 76798-7356, USA
Kwang Y. Lee
e-mail: Kwang_Y_Lee@baylor.edu

Ho-Chan Kim
Dept. of Electrical Engineering, Jeju National University, Jeju, Korea
e-mail: hckim@jejunu.ac.kr

L. Wang et al. (Eds): *Wind Power Systems*, Green Energy and Technology, pp. 297–335.
springerlink.com

1 Introduction

Power systems utilizing renewable energy resources such as wind, solar, and micro-hydro require control methods to maintain stability in spite of the real time variation of input energy and load and at the same time, maximize the use of the renewable energy resources. Since the early eighties, the wind-diesel autonomous power system (WDAPS) has been accepted and widely used for power generating systems in remote areas. In such cases, the WDAPS serves an entire isolated load or microgrid, and is responsible for maintaining frequency and voltage stability within the microgrid. The main focus in WDAPS design is to secure both fuel-saving of diesel generator unit and providing reliable power supply to the load [1].

The random power variations of WDAPS can cause relatively large frequency and voltage fluctuations. In a large grid, these fluctuations can have little effect on the overall quality of the delivered power. However, with weak autonomous networks or microgrids, these power fluctuations can have a marked effect [2,3]; hence, the control of the voltage and frequency of a wind-diesel system is considered more challenging than in large grids. Two nonlinearities should be considered in a WDAPS: mechanical torque of the wind turbine and magnetic saturation in synchronous and/or induction machine. The operating points keep changing because of the fluctuating natural wind source. Therefore, a model-based linear controller may not be optimal for such an unpredictable system. Hence, several intelligent control techniques are proposed using a fuzzy-robust and fuzzy-neural hybrid control algorithms.

The system under study consists of a horizontal axis, 3-bladed, stall regulated wind turbine (WT), which drives an induction generator (IG) [4,5]. The IG is connected to an AC bus-bar in parallel with a diesel-generator set consisting of a turbocharged diesel engine (DE) driving a synchronous generator (SG). The two generators together serve the 40 kW electrical load, and the control of voltage and frequency is maintained by the diesel-generator set. A hybrid wind-diesel power system includes battery storage or superconducting magnetic energy storage (SMES) and a dumpload.

In this chapter, several intelligent control techniques are proposed using fuzzy and neural networks: fuzzy linear quadratic regulator (Fuzzy-LQR) [6], fuzzy-sliding mode controller (Fuzzy-SMC) [7], and a fuzzy-neural hybrid controller [8]. The fuzzy-LQR (or fuzzy-SMC) controller is developed based on the Takagi-Sugeno (TS) fuzzy model, the possibility auto-regression model (PARM), and the LQR (or SMC) controller. The TS fuzzy model provides a simple and straightforward way to decompose the nonlinear model into a group of local tasks, which tend to be easier to handle. In the end, the TS fuzzy model provides the mechanism to blend these local tasks (or sub-systems) together to deliver the overall model, where the PARM is used for finding optimal sub-systems. The PARM can be used to make the fuzzy model to be adaptive with observed data. On the other hand, the fuzzy-neural hybrid controller is developed based on the inverse dynamic neural model (IDNM) and the fuzzy logic controller. An inverse input-output relationship of the WDAPS is identified by an IDNM, where the desired output of the IDNM is the control input for the system. Fuzzy logic controller provides a flexible controller covering a wide range of energy/voltage compensation. A neural network inverse model (NNIM) is designed to provide a compensating control for the system.

In the simulation study, the integrated non-linear controller is tested under varying load condition and fluctuating wind speed. The controllers are compared with the conventional proportional-integral (PI) controller and shown to be more effective against disturbances caused by the wind and the load variations; thus, providing better power qualities in a given site.

2 System Description

2.1 System Configuration

The WDAPS consists of a wind turbine having an induction generator (IG), a diesel engine (DE), a synchronous generator (SG), a battery bank connected with a three-phase thyristor-bridge controlled converter, battery storage system (BSS) or superconducting magnetic energy storage (SMES), a dumpload, and the system load. A three-phase dumpload is used with each phase consisting of seven transistor-controlled resistor banks. When wind-generated power is sufficient to serve the load, the DE is disconnected from the SG by electromagnetic clutch, and the synchronous generator acts as a synchronous condenser.

The main purpose of the dumpload, SMES and BSS is to regulate the system frequency. The SG (with/without diesel) is used for reactive power control, which is achieved by the excitation system used for voltage regulation. The SG also contributes in compensating for the lack of reactive power in the induction generator. A current source converter, which is simpler than the voltage source converter, is used for the battery storage system because the charging current can be critical to the battery life. A smooth charging can be achieved using a large inductor in the DC bus minimizing the current fluctuation. The SMES is a control unit for a synchronous machine [9,28]. When there is a sudden rise in the load, the stored energy is immediately released to the power system. As the governor starts working to set the power system to a new operating condition, the SMES unit is charged back to its initial value of current. In the case of sudden fall in the load, the SMES immediately gets charged towards its full value, thus absorbing some portion of the excess energy in the system, and as the system returns to its steady state, the excess energy absorbed is released and the SMES current attains its normal value.

Fig. 1 shows the overall configuration of the WDAPS: C_a is the capacitor bank, Q_d is the fuel flow rate at the governor chamber valve, E_{fd} is the excitation field voltage, f is the frequency, V_b is the bus voltage, L_{filt} and C_{filt} are respectively the

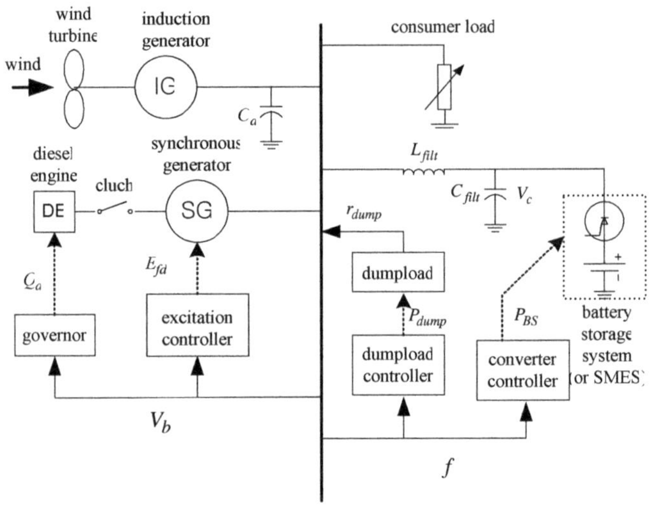

Fig. 1. The overall control configuration of WDAPS

filter inductance and capacitance in the AC side, V_c is the AC side voltage of the-converter, P_{BS} is the battery power, P_{dump} is the dumpload power, and r_{dump} is the dumpload resistance.

2.2 Components Models

The models for the generators are based on the standard Park's transformation [16] that transforms all stator variables to a rotor reference frame described by a direct and quadrature (*d-q*) axis. The set of SG equations are based on the *d-q* axis in accordance with [16,17].

The nonlinear mathematical model of the WDAPS is summarized in this section. The following considerations are taken into account to identify component models: the electrical system is assumed as a perfectly balanced three-phase system with pure sinusoidal voltage and frequency. High frequency transients in stator variables are neglected, which indicates that the stator voltage and currents are allowed to change instantly, because for dynamic study the transient period is focused instead of sub-transient period. Damper-winding models are ignored because their effect appears mainly in a grid-connected system or a system with several synchronous generators running in parallel. Different component models are of equal level of complexity.

2.2.1 Wind-Diesel Mechanical and Electrical Model

The modeling of the SG and the IG generator is based on [17,20].

Diesel-synchronous generator (salient pole):

$$
\begin{aligned}
\dot{Q}_f &= \frac{1}{\tau_c}\left(-Q_f + Q_d(t-\tau_d)\right) \\
\dot{\theta}_{cl} &= \omega_d - \omega_s \\
\dot{\omega}_d &= \frac{1}{J_d}\left(k_c k_v Q_f - (D_d + D_{cl})\omega_d + D_{cl}\omega_s - k_v p_o - C_{cl}\theta_{cl}\right) \\
\dot{\psi}_f &= \frac{1}{\tau'_{do}}\left(-\psi_f + L_{md} I_{sd}\right) + E_{fd} \\
\dot{\omega}_s &= \frac{1}{J_s}\left(C_{cl}\theta_{cl} + D_{cl}\omega_d - T_s - (D_{cl} + D_s)\omega_s\right)
\end{aligned}
\tag{1}
$$

$$
\begin{bmatrix} R_s & -\omega_s L'_d & 1 & 0 \\ \omega_s L_q & R_s & 0 & 1 \\ R_1 & -\omega_s L_1 & -1 & 0 \\ \omega_s L_1 & R_1 & 0 & -1 \end{bmatrix}
\begin{bmatrix} I_{sq} \\ I_{sd} \\ V_{sq} \\ V_{sd} \end{bmatrix}
+
\begin{bmatrix} 0 & 0 \\ 0 & 0 \\ 1 & 0 \\ 0 & 1 \end{bmatrix}
\begin{bmatrix} V_{bq} \\ V_{bd} \end{bmatrix}
=
\begin{bmatrix} \omega_s \frac{L_{md}}{L_f}\psi_f \\ 0 \\ 0 \\ 0 \end{bmatrix}
\tag{2}
$$

where $T_s = -\frac{L_{md}}{L_f}\psi_f I_{sq} - (L'_d - L_q) I_{sq} I_{sd}$ and

ω_s : bus frequency (or angular speed of SG)

E_{fd}, ψ_f : filed voltage and filed flux linkage of SG

V_{sq}, V_{sd} : stator terminal voltage components of SG

R_1, L_1 : resistance and reactance between SG and bus

R_s : stator resistance of SG

L_q, L_d, L_f : *q*-, *d*-axis, and field inductance of SG

I_{sq}, I_{sd} : current component of SG into the bus

L_{md}, L'_d : *d*-axis field mutual inductance and transient inductance.

J_s, D_s : inertia and frictional damping of SG

T_s : air gap torque of SG

τ'_{do} : transient open circuit time constant

Q_f : fuel flow rate into the combustion chamber

Q_d : fuel flow rate at the governor chamber valve

p_o : zero torque pressure when running idle

τ_d : time delay of combustion

k_v : stroke volume of the engine

k_c : a constant describing efficiency of combustion

θ_{cl} : torsional angle between the engine and the generator shaft

Induction generator (squirrel-cage rotor):

$$\begin{aligned}\dot{\psi}_{rq} &= \frac{1}{\tau_o^{'}}(-\psi_{rq} + L_m I_{aq}) + \omega_b(\omega_s - \omega_a)\psi_{rd} \\ \dot{\psi}_{rd} &= \frac{1}{\tau_o^{'}}(-\psi_{rd} + L_m I_{ad}) - \omega_b(\omega_s - \omega_a)\psi_{rq}\end{aligned} \tag{3}$$

$$\begin{bmatrix} R_a & -\omega_s L_s^{'} & 1 & 0 \\ \omega_s L_s^{'} & R_a & 0 & 1 \\ r_2 & -\omega_s L_2 & (\omega_s^2 C_a L_2 - 1) & \omega_s R_2 C_a \\ \omega_s L_2 & R_2 & -\omega_s R_2 C_a & (\omega_s^2 C_a L_2 - 1) \end{bmatrix} \begin{bmatrix} I_{aq} \\ I_{ad} \\ V_{aq} \\ V_{ad} \end{bmatrix} + \begin{bmatrix} 0 & 0 \\ 0 & 0 \\ 1 & 0 \\ 0 & 1 \end{bmatrix} \begin{bmatrix} V_{bq} \\ V_{bd} \end{bmatrix} = \begin{bmatrix} \omega_s \frac{L_m}{L_r}\psi_{rd} \\ -\omega_s \frac{L_m}{L_r}\psi_{rq} \\ 0 \\ 0 \end{bmatrix} \tag{4}$$

where

C_a, ω_a : capacitor bank and angular speed of wind turbine

ψ_{rq}, ψ_{rd} : rotor flux linkage components of SG

$I_{aq}, I_{ad}, V_{aq}, V_{ad}$: stator terminal current and voltage of IG

$R_a, L_s^{'}, L_r$: rotor resistance and inductance of SG

R_2, L_2 : resistance and reactance between IG and bus

$\tau_o^{'}$: transient open circuit time constant

L_m : mutual inductance of SG

Drive train model (shaft between rotor turbine and the IG):

$$\begin{aligned}\dot{\theta}_c &= \omega_t - \omega_a \\ \dot{\omega}_t &= \frac{1}{J_t}\left(\frac{P_w}{\omega_t} - C_c\theta_c - (D_t + D_c)\omega_t + D_c\omega_a\right) \\ \dot{\omega}_a &= \frac{1}{J_a}(C_c\theta_c + D_c\omega_t - (D_a + D_c)\omega_a - T_a)\end{aligned} \tag{5}$$

where

ω_t : bus frequency (or angular speed of IG)

T_a : air gap torque of IG

Current balance form for electrical model:

$$I_{sq} + I_{iq} - I_{lq} - I_{ac,q} = 0, \qquad I_{sd} + I_{id} - I_{ld} - I_{ac,d} = 0 \tag{6}$$

where current component of the load (I_{lq}, I_{ld}) and current component of IG into the bus (I_{iq}, I_{id}) are given

$$I_{lq} = \left(\frac{R_3}{R_3^2 + X_3^2} + \frac{1}{r_{dump}} \right) V_{bq} + \frac{X_3}{R_3^2 + X_3^2} V_{bd},$$

$$I_{ld} = -\frac{X_3}{R_3^2 + X_3^2} V_{bq} + \left(\frac{R_3}{R_3^2 + X_3^2} + \frac{1}{r_{dump}} \right) V_{bd},$$

where R_3, X_3 is equivalent load resistance and reactance, and

$$I_{iq} = I_{aq} + \omega_s C_a V_{ad}, \qquad I_{id} = I_{ad} - \omega_s C_a V_{aq},$$

$$I_{ac,q} = \frac{-\omega_s C_{filt}}{1 - \omega_s^2 C_{filt} L_{filt}} V_{cd} + \frac{1}{1 - \omega_s^2 C_{filt} L_{filt}} I_{cq},$$

$$I_{ac,d} = \frac{\omega_s C_{filt}}{1 - \omega_s^2 C_{filt} L_{filt}} V_{cq} + \frac{1}{1 - \omega_s^2 C_{filt} L_{filt}} I_{cd}$$

where

$I_{cq}, I_{cd}, V_{cq}, V_{cd}$: AC side current and voltage of the converter

$I_{ac,q}, I_{ac,d}$: AC side current before the filters.

2.2.2 Battery Storage System (BSS)

The model of the three-phase thyristor current converter and the battery storage system (Fig. 2) emphasizes the input-output relationship between voltage and current and the connection to the overall electrical system in the rectifier operation. The DC-side of the converter is connected to the battery bank. Thyristor is assumed ideal but with constant loss and harmonic current ratio between AC- and DC-current is constant as 0.955 with a large value of L_b [23].

The ideal no-load maximum DC voltage of the six-pulse converter V_{co} is

$$V_{co} = \frac{3\sqrt{2}}{\pi} V_c \tag{7}$$

and the terminal voltage of the equivalent battery is

$$V_{bt} = V_{co}\cos(\alpha_r) \tag{8}$$

where

V_c : AC side line-to-line voltage $\left(V_c = \sqrt{V_{cd}^2 + V_{cq}^2}\right)$

V_{co} : communicating voltage drop of the six-pulse converter

α_r, I_{bes} : firing angle and DC current flowing into battery

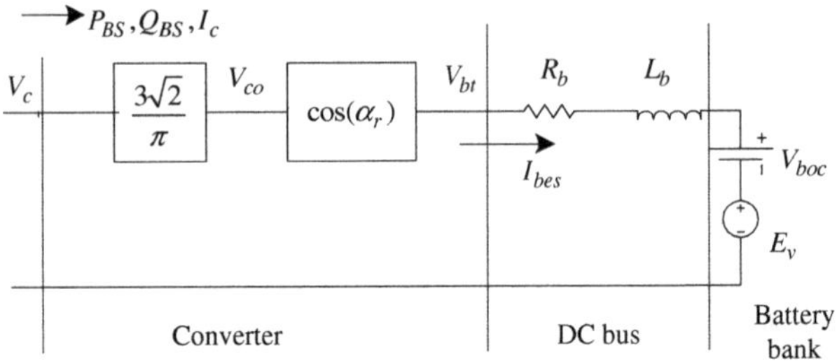

Fig. 2. Equivalent circuit of the battery storage system in the rectifier operation

Here, the firing angle is controlled by the internal PI controller. In Fig. 3, the control scheme of the battery storage system is depicted, where u_c is the output of the PI converter controller, K_{PI} is the proportional gain and T_{PI} is the integral gain. To represent the charging speed of the BSS, the forcing voltage ($E_v = K_{v,ref} I_{bes}$) is made dependent to the DC current (I_{bes}) in DC bus where $K_{v,ref}$ is the forcing gain. If I_{bes} is large, E_v is increased that increases the voltage of the battery bank (i.e., $V^{new}_{bos} = V_{bos} + E_v$). Hence, the sudden change of the battery bank can be prevented.

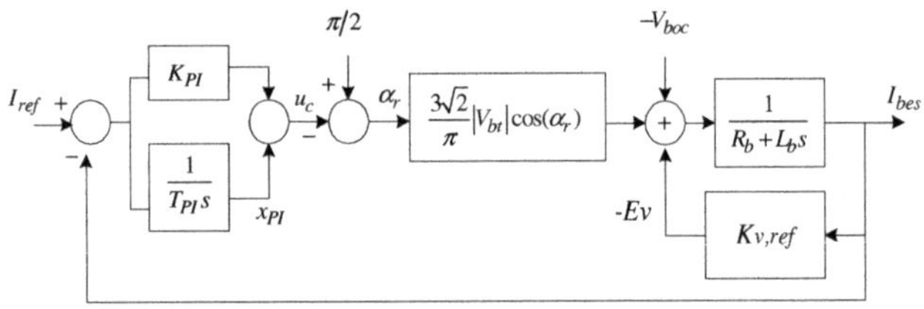

Fig. 3. The control scheme of the converter in the rectifier operation

Since the current setpoint (I_{ref}) for the converter is the control input, P_{BS} is approximated as $K_{ref} I_{ref}$, where K_{ref} is the proportional gain. Such approximation is possible due to the fast response of the converter. Here, K_{ref} is chosen as 10 for I_{ref}, and the forcing gain $K_{v,ref}$ is chosen as 10% of the V_{boc}. The DC current is

$$\dot{I}_{bes} = \frac{1}{L_b}[V_{bt} - R_b I_{bes} - V_{boc} - E_v]$$
$$\dot{x}_{PI} = \frac{1}{T_{PI}}(I_{ref} - I_{bes}) \qquad (9)$$
$$u_c = x_{PI} + K_{PI}(I_{ref} - I_{bes})$$

where

R_b, L_b : total resistance (battery internal + DC line) and DC line inductance
$V_{boc}, \alpha_r (= \pi/2 - u_c)$: open circuit DC voltage and firing angle.

By connecting the battery storage to the AC bus, the charging and discharging power P_{BS}, Q_{BS} can be described as

$$P_{BS} = V_{cd} I_{cd} + V_{cq} I_{cq}, \quad Q_{BS} = V_{cd} I_{cq} - V_{cq} I_{cd} \qquad (10)$$

where the power factor on the AC side of the converter is $\cos(\varphi)=0.955\cos(\alpha)$ and $P_{BS}=V_{bt}I_{bes}+P_{c,loss}$, $Q_{BS}=P_{BS}\tan(\varphi)$. Therefore, the equivalent converter currents I_{cd} and I_{cq} can be obtained. Here, $P_{c,loss}$ is the constant power loss in the thyristor. The currents for current balance form can be obtained as follows:

$$I_{cq} = \frac{V_{cq} P_{BS} + V_{cd} Q_{BS}}{V_c^2}, \quad I_{cd} = \frac{V_{cd} P_{BS} - V_{cq} Q_{BS}}{V_c^2} \qquad (11)$$

2.2.3 Dumpload Model

Fig. 4 is the three-phase dumpload, where each phase consists of 7 transistor-controlled resistor banks with binary resistor sizing in order to minimize quantum effects and provide more-or-less linear resolution. Fig. 5 shows how the transistors are switched to meet the required power. For example, based on the rated AC line voltage of 230*V* and per-phase resistance of R (=120Ω), if the required dumpload power from the dumpload controller is 880*W*, then Step-2 is identified, and only switch S_2 is turned ON.

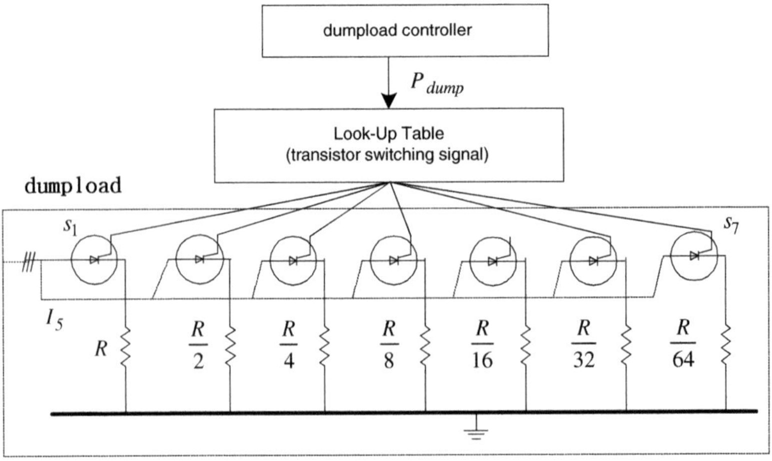

Fig. 4. The structure of the dumpload with binary resistor sizing

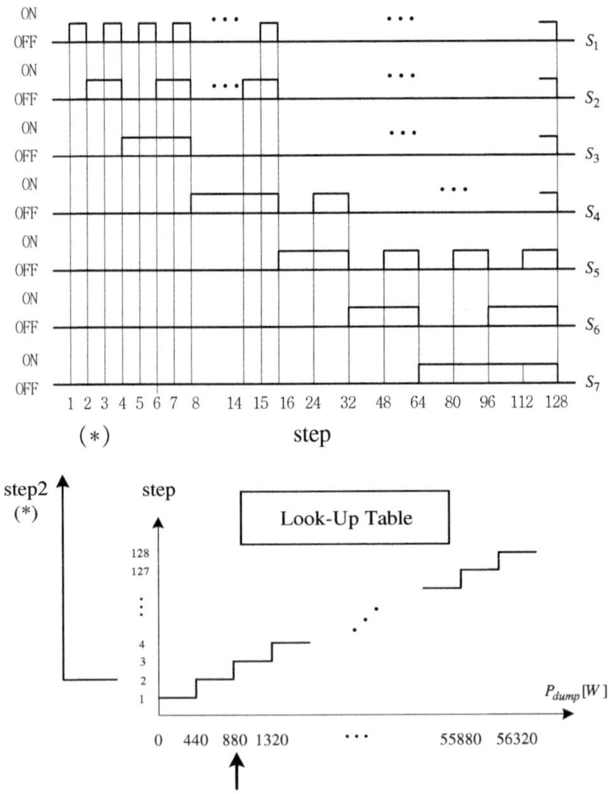

Fig. 5. Transistor switching signal

2.2.4 Per-Unit Base Values

The per-unit system is used in modeling the wind-diesel mechanical and electrical system, where the base values are given below:

Table 1. Per-unit base values

Terms	Symbols	Parameters
Angular speed/Frequency	ω_b	2π50 rad/sec
Power	$S_b=P_b=Q_b$	55000 VA
Line AC voltage	V_b	230 V(rms)
DC voltage	V_{dc}	230 V
AC current	$I_b = S_b/\sqrt{3}V_b$	138 A
DC current	$I_{dc} = S_b/V_{dc}$	239 A
Resistance	$R_{base} = V_b^2/S_b$	0.96 Ω
Inductance	$L_{base} = R_{base}/\omega_b$	3.06 mH
Capacitance	$C_{base} = 1/(R_{base}\omega_b)$	3.31 mF
Torque	$T_b = S_b/\omega_b$	175.1 Nm
Moment of inertia	$J_{base} = S_b/\omega_b^2$	0.557 kgm^2/s
Torsional stiffness	$C_{T,base} = T_{base}/rad$	175.1 Nm/rad
Torsional damping	$D_{T,base} = T_{base}/\omega_b$	0.557 Nms/rad

3 The Intelligent Control Schemes

3.1 Fuzzy-Robust Controller Design

Fig. 6 depicts the input and output relationship of the WDAPS from the control point of view. The control inputs are the excitation field voltage (u_1) of the SG, the fuel flow rate at the governor chamber valve (u_2), the battery power (u_3), and the dumpload power (u_4). The measurements are the voltage amplitude (y_1) and the frequency (y_2) of the AC bus. The wind speed (v_1) and the load (v_2) are considered to be disturbances. From the control point of view, this is a coupled 4×2 multi-input-multi-output nonlinear system, since every input controls more than one output and every output is controlled by more than one input.

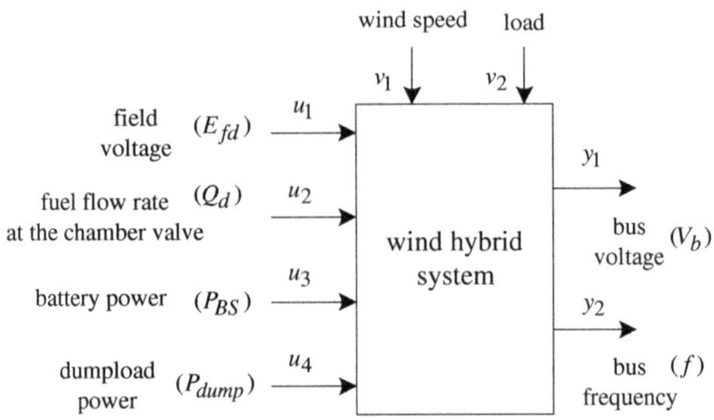

Fig. 6. The control structure of the WDAPS

3.1.1 Reduced-Order Model

Since the nonlinear model presented in Section 2.2.1 is too complex for controller design, there is a need for a reduced-order model derived based on practical reasons; i.e., what is measurable and what can be manipulated. With the considerations in components models of Section 1.2, the reduced-order model assumes that the dynamic response of the converter is much faster than the desired bandwidth of the controlled system. This implies that the differential equations of the converter can be neglected. Also, there is no elasticity in the drive train. Finally, electrical dynamics of the induction generator is not explicitly modeled. Then, the reduced-order model can be represented by the field flux linkage and the angular speed of the SG:

$$\begin{aligned}\dot{\psi}_f &= \frac{1}{\tau_{do}^{'}}\left(-\frac{L_f}{\omega_s L_{md}}(V_{sq} + R_a I_{sq} - \omega_s L_d^{'} I_{sd}) + L_{md} I_{sd}\right) + E_{fd} \\ \dot{\omega}_s &= \frac{1}{J_s}(-D_s\omega_s - T_s - k_v p_o + k_v k_c Q_d)\end{aligned} \tag{12}$$

The control inputs are the field voltage (E_{fd}), the fuel flow rate at the chamber valve (Q_d), the battery power (P_{BS}), and the dumpload power (P_{dump}). The outputs are the voltage (V_b) and the frequency ($f = \omega_s$).

In Eq. 12, the air gap torque of the synchronous generator T_s can be represented as

$$T_s = \frac{P_s}{\omega_s} = \frac{P_{BS} + P_{dump} + P_{load} - P_{ind}}{\omega_s} \tag{13}$$

where P_s, P_{load}, P_{ind}, and P_{BS} are the power of the synchronous generator, the load, the induction generator, and the battery storage system, respectively, P_{dump} is the dumpload power, and ω_s is the angular speed, which is proportional to frequency f.

Applying Eq. 13 into Eq. 12, the reduced-order model becomes

$$\begin{aligned}\dot{\psi}_f &= \frac{1}{\tau'_{do}}\left(-\frac{L_f}{\omega_s L_{md}}(V_{sq} + R_a I_{sq} - \omega_s L'_d I_{sd}) + L_{md} I_{sd}\right) + E_{fd} \\ \dot{\omega}_s &= \frac{1}{J_s}\left(-D_s\omega_s + \frac{P_{ind} - P_{load}}{\omega_s} - \frac{(P_{BS} + P_{dump})}{\omega_s} - k_v p_o + k_v k_c Q_d\right)\end{aligned} \tag{14}$$

At a local operating point, flux linkage ψ_f in Eq. 14 can be represented in terms of the bus voltage and the frequency based on the assumption that the rate of change of voltage is a linear combination of rate of change of rotor flux and angular speed of the SG:

$$\dot{V}_b = \eta_1\dot{\psi}_f + \eta_2\dot{\omega}_s \tag{15}$$

where $\eta_1 = \dfrac{\partial V_b}{\partial \psi_f}$ and $\eta_2 = \dfrac{\partial V_b}{\partial \omega_s}$. Here, η_1 and η_2 are approximated as 1 per-unit. Hence, from Eq. 14 and Eq. 15 the final reduced-order model is derived in the state-space form as

$$\begin{aligned}\dot{x}(t) &= Ax(t) + Bu(t) \\ y(t) &= Cx(t)\end{aligned} \tag{16}$$

where

$$x(t) = [V_b \quad \omega_s]^T, \quad u(t) = \left[E_{fd} \quad Q_d \quad P_{BS} \quad P_{dump}\right]^T$$

$$A = \begin{bmatrix}1 & 1\\ 0 & 1\end{bmatrix}\begin{bmatrix} -\dfrac{L_f V_{sq}}{\tau'_{do} L_{md}\omega_s} & \dfrac{L_f}{\tau'_{do}\omega_s L_{md}}\left(L_d I_{sd} - \dfrac{R_a I_{sq}}{\omega_s}\right) \\ \dfrac{P_{ind} - P_{load}}{J_s V_b \omega_s} - \dfrac{k_v p_0}{J_s} & -\dfrac{D_s}{J_s}\end{bmatrix},$$

$$B = \begin{bmatrix} 1 & \dfrac{k_v k_c}{J_s} & -\dfrac{1}{J_s\omega_s} & -\dfrac{1}{J_s\omega_s} \\ 0 & \dfrac{k_v k_c}{J_s} & -\dfrac{1}{J_s\omega_s} & -\dfrac{1}{J_s\omega_s}\end{bmatrix}, \qquad C = \begin{bmatrix}1 & 0\\ 0 & 1\end{bmatrix}$$

Note that the reduced-order model Eq. 16 is in the linear form for the system matrices A, B and C. However, matrices A and B are not fixed, but changes as functions of state variables, thus making the model nonlinear. Therefore, the fuzzy model-based controller can be effective, taking into account model imperfections and uncertainties. Also, it should be advised that the reduced-order model is only for the purpose of designing controllers; not for overall simulation study. The proposed controller is designed in the following sub-sections.

3.1.2 Fuzzy Model

The Takagi-Sugeno fuzzy model represents a nonlinear system by partitioning the system into sub-systems and then combining them with linguistic rules. Three linear sub-systems are considered for the nonlinear state-space models Eq. 16 as

$$\begin{aligned} \dot{x}(t) &= A_i x(t) + B_i u(t) \\ y(t) &= C_i x(t), \quad i = 1,2,3 \end{aligned} \tag{17}$$

where $A_i \in \Re^{n\times n}$, $B_i \in \Re^{n\times m}$, and $C_i \in \Re^{p\times n}$. Here, n, m and p are the number of states, inputs and outputs, respectively. It can be seen from the reduced-order models that n=p=2 and m=4. The sub-systems are obtained by partitioning the state-space into overlapping ranges of low, medium, and high states. For each subspace, different model (i=1, 2, 3) is applied. The degree of membership functions for states V_b and ω_s are depicted in Fig. 7.

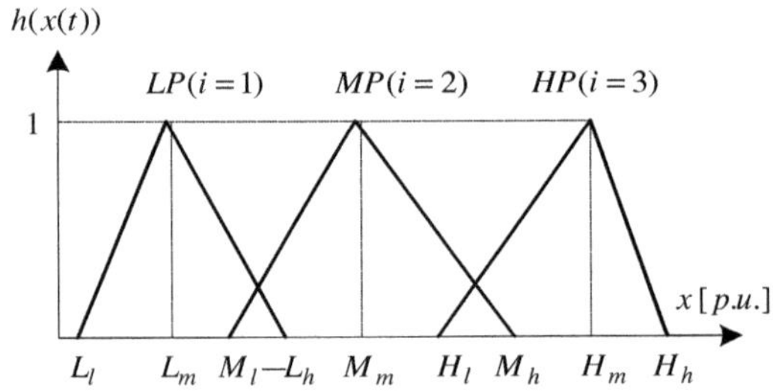

Fig. 7. The membership function for states

Here, LP(i=1), MP(i=2), and HP(i=3) stand for possible low, most possible, and possible high membership functions, respectively. Each membership function also represents model uncertainty for each sub-system. The implicit rule is to apply corresponding sub-systems according to the degree of belongings to the subspaces measured by the membership functions. Therefore, even if the sub-systems are linear models, the composite system represents a nonlinear system. Membership functions can be optimized by the observed data.

Three controllers are designed for the three linear sub-systems, and then the total control output is obtained by defuzzification. Hence, the fuzzy-robust controller output is

$$u_{FR}(t) = \frac{\sum_{i=1}^{3} h_i(x(t))u_i(t)}{\sum_{i=1}^{3} h_i(x(t))} \tag{18}$$

where $u_{FR}(t)$ is the fuzzy-robust controller output, $u_i(t)$ is the controller output for each linear sub-system, and $h_i(x(t))$ is the corresponding membership value.

3.1.3 Possibility Auto-regression Model

Now, there is a question that how the coordinates L_j, M_j, H_j (j=l,m,h.) can be optimally obtained. The possibility auto-regression model (PARM) [18] is introduced for optimal coordinate values of L_j, M_j, and H_j for states V_b and ω_s. The PARM represents the possible distribution of observed data by three auto-regression models in the form of

$$\hat{x}(k)=\overline{A}_1 x(k-1)+\overline{A}_2 x(k-2)+\cdots+\overline{A}_n x(k-n) \tag{19}$$

where $\hat{x}$ is a fuzzy set over an interval $[\hat{x}^L \quad \hat{x}^M \quad \hat{x}^H]$, and $\hat{x}^L,\hat{x}^M$, and $\hat{x}^H$ stand for *possible low*, *the most possible*, and *possible high* models, respectively. Here, n is the total number of observed data, and $\overline{A}_l$ is a fuzzy set defined for the interval of $[a_l^-, a_l, a_l^+]$ and $l=1,2,\cdots,n$.

An auto-regression (AR) model for each of the state variables (V_b and ω_s) is given as

$$x(k)=a_1 x(k-1)+a_2 x(k-2)+\cdots+a_n x(k-n) \tag{20}$$

The coefficients $a_1,\cdots,a_n$ can be found by the least squares minimization [19]. Once the coefficients are obtained, *the most possible* model $\hat{x}^M$ is defined from Eq. 20 as

$$\hat{x}^M(k)=a_1 x(k-1)+a_2 x(k-2)+\cdots+a_n x(k-n). \tag{21}$$

On the other hand, *possible high* and *possible low* models are, respectively, defined as

$$\begin{aligned}\hat{x}^H(k)&=a_1^+ x(k-1)+a_2^+ x(k-2)+\cdots+a_n^+ x(k-n)\\ \hat{x}^L(k)&=a_1^- x(k-1)+a_2^- x(k-2)+\cdots+a_n^- x(k-n)\end{aligned} \tag{22}$$

where

$$\hat{x}^H(k-l+1)\ge\hat{x}^M(k-l+1)\ge\hat{x}^L(k-l+1),\quad l=1,2,\cdots,n \tag{23}$$

The coefficients a_l^+ and a_l^- for the *possible high* and *possible low* models are obtained by minimizing the area enclosed by the two models as following:
Minimize:

$$\sum_{l=1}^{n}\{\hat{x}^H(k-l)-\hat{x}^L(k-l)\} \tag{24}$$

subject to

$$\hat{x}^H(k-l)\ge\hat{x}^M(k-l)\ge\hat{x}^L(k-l),\quad l=1,2,\ldots,n.$$

The Barns algorithm is used to solve Eq. 24 [20]. Only in the beginning of the procedure, a PI controller is used to collect data to model the PARM. Once the PARM is modeled in the form of Eq. 20 and Eq. 21, the membership functions, *LP*, *BP*, *HP*, can be optimally defined by the mean, maximum, and minimum values of the output of $\hat{x}^L$, $\hat{x}^M$, and $\hat{x}^H$ over the observed period. The PARM can be updated with newly observed data, making the controller adaptive.

3.1.4 Non-zero Final-State Compensation

Conventional state-feedback control methods are for zero final states. However, the final states may not be zero but constants such as in the system under study. Therefore, it is necessary to consider such non-zero final state, leading to the, so called, tracking problem [19].

Here, an additional state is introduced to address the tracking problem as

$$\dot{x}_r(t) = r(t) - y(t) \tag{25}$$

where $x_r(t) \in \Re^p$ is the additional state vector and the signal $r(t)$ satisfies

$$\dot{r}(t) = \gamma(r_{ref} - r(t)) \tag{26}$$

with $\gamma \in \Re^{p \times p}$, a positive definite design matrix, and a constant reference signal $r_{ref}(=1)$.

Eq. 25 utilizes the integral action in Eq. 26 that makes steady-state error zero. Therefore, whenever the final state $x(\infty)$ is constant, the signal $r(t)$ makes the state $x_r(t)$ to be zero. Hence, the non-zero final state problem can be solved.

Including the additional state, the states can be defined as

$$\tilde{x}(t) = \begin{bmatrix} x_r(t)^T & x(t)^T \end{bmatrix}^T \tag{27}$$

where $\tilde{x}(t) \in \Re^{p+n}$ and the associated system and input matrices for the augmented system are represented as

$$\dot{\tilde{x}}(t) = \tilde{A}\tilde{x}(t) + \tilde{B}u(t) \tag{28}$$

where $\tilde{A} \in \Re^{(p+n)\times(p+n)}$, $\tilde{B} \in \Re^{(p+n)\times m}$ and with matrices A, B and C of ith subsystem as

$$\tilde{A} = \begin{bmatrix} 0 & -C \\ 0 & A \end{bmatrix}, \quad \tilde{B} = \begin{bmatrix} 0 \\ B \end{bmatrix}$$

The proposed fuzzy-robust controller can then be derived from the augmented matrices in Eq. 28. The signal $r(t)$ will be added in the final control structure.

The overall fuzzy-robust control scheme is given in Fig. 8. Here, $u_F(t)$ is the final control input in the form

$$u_F(t) = r(t) + u_{FR}(t) \tag{29}$$

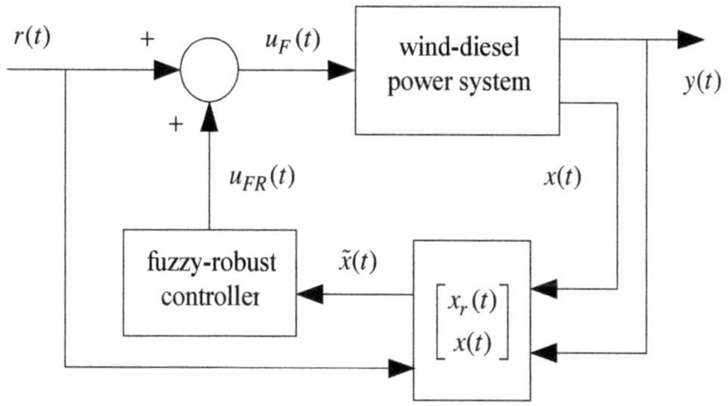

Fig. 8. The overall fuzzy-robust control scheme

3.1.5 Fuzzy-LQR controller

The LQR is designed for each linear sub-system by minimizing the quadratic performance index [19]. The object of the LQR design is to determine the optimal control law u which can transfer the system from its initial state to the final state such that a given performance index is minimized. The performance index is given in the quadratic form

$$J = \frac{1}{2}\int_0^\infty (\tilde{x}(t)^T Q\tilde{x}(t) + u(t)^T Ru(t))dt \tag{30}$$

where Q is a positive-semidefinite, real, symmetric matrix and R is a positive-definite, real, symmetric matrix.

To design the LQR controller, the first step is to select the weighting matrices Q and R. The value R weighs inputs more than the states while the value of Q weighs the state more than the inputs. Then, the feedback gain K can be computed and the closed-loop system responses can be found by simulation. This method has an advantage of allowing all control loops in a multi-loop system to be closed simultaneously, while guaranteeing closed-loop stability. The LQR controller is given by

$$u(t) = -K\,\tilde{x}(t) \tag{31}$$

where K is the constant feedback gain obtained from the solution of the continuous algebraic Ricatti equation:

$$\begin{aligned} K &= R^{-1}\tilde{B}^T P \\ \tilde{A}^T P + P\tilde{A} &+ Q - P\tilde{B}R^{-1}\tilde{B}^T P = 0 \end{aligned} \tag{32}$$

3.1.6 Fuzzy-SMC Controller

To design a controller for each sub-system, the sliding mode control (SMC) is applied that provides robustness against disturbances and uncertainties [19]. The regular state-space model based sliding mode controller has no direct effect on the dynamics of the sliding motion by the hyperplane matrix S. The regular state-space model of Eq. 28 is first obtained, which is expressed as

$$\dot{\tilde{x}}(t) = \hat{A}\tilde{x}(t) + \hat{B}u(t) \tag{33}$$

where $\hat{A} = T_r\tilde{A}T_r^T$, $\hat{B} = T_r\tilde{B}$ and the orthogonal matrix $T_r \in \Re^{(p+n)\times(p+n)}$.

The sliding mode controller is a multivariable controller designed by minimizing the following quadratic performance index [17]:

$$J = \frac{1}{2}\int_0^\infty \tilde{x}(t)^T Q_t \tilde{x}(t)dt \tag{34}$$

where Q_t is a symmetric-positive definite matrix.

The linear controller for each linear sub-system Eq. 17 can be designed as

$$u(t) = -(S\hat{B})^{-1}(S\hat{A} - \xi S)\tilde{x}(t) \tag{35}$$

where S is the hyperplane system matrix and where $\xi \in \Re^{m\times m}$ is a stable design matrix. The detailed design procedure is given in [7].

3.2 Fuzzy-Neural Hybrid Controller Design

3.2.1 Feedback Controller Based on Fuzzy Logic

Fuzzy control systems are rule-based systems in which a set of fuzzy rules represents a control decision mechanism to adjust the effects of certain system conditions. Fuzzy controller is based on the linguistic relationships or rules that define the control laws of a process between input and output [10,11]. This feature draws attention toward a fuzzy controller due to its nonlinear characteristics and there is no need for an accurate system modeling. The fuzzy controller consists of rule base, which represents a fuzzy logic quantification of the expert's linguistic description of how to achieve good control, fuzzification of actual input values, fuzzy inference, and defuzzification of fuzzy output. When the expert's linguistic description is not available, fuzzy controller still can be designed by using the measurement of real-time input/output data [12,13].

A total of 121 rules are used for the power system under study. The general form of the fuzzy rule is given in the *if-then* form as follows:

$$\text{if } x(k) \text{ is } A \text{ and } \Delta x(k) \text{ is } B, \text{ then } y(k) \text{ is } C, \tag{36}$$

where

$x(k)$, $\Delta x(k)$: input signals,

$y(k)$: controller output,

A, B, C : linguistic variables.

The linguistic values extracted from the experimental knowledge are NH (negative high), NL (negative large), NB (negative big), NM (negative medium), NS (negative small), ZE (zero), PS (positive small), PM (positive medium), PB (positive big), PL (positive large), and PH (positive high).

In the power system under study, generator power deviation (ΔP) is chosen for the input of a fuzzy controller. The linguistic descriptions provide experimental expressions of the expert for a control decision-making process and each linguistic variable is represented as triangular membership functions shown in Figs. 9 and 10. In the fuzzy controller, the input normalization factors are chosen to represent the proper membership quantifications of linguistic values. In addition, normalization factors can be used to yield the desired response of the fuzzy controller: g_1, g_2 for a normalization factor for input of fuzzy controller and g_0 for a denormalization factor for output of fuzzy controller.

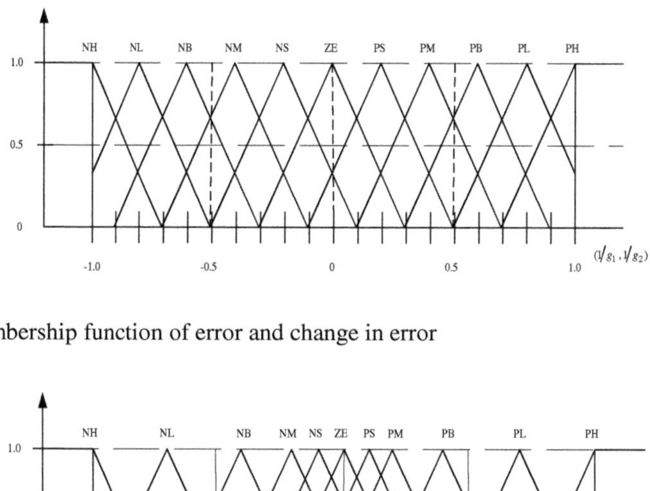

Fig. 9. Membership function of error and change in error

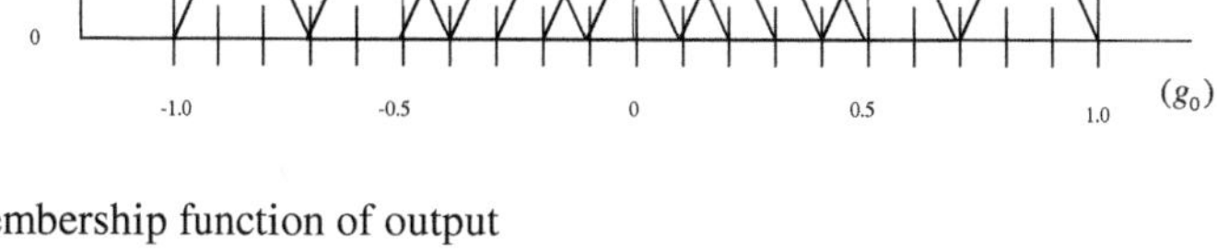

Fig. 10. Membership function of output

In Figs. 9 and 10, the membership functions are overlapped with each other to smooth a fuzzy system output, and a fuzzy controller is designed to regulate a system smoothly when an error and a change in error are near zero. The rules are established to control transient stability problem for all possible cases. Table 2 and Table 3 show the inference rule table for two input fuzzy variables in negative and positive changes in error, respectively.

Table 2. Inference rule table for negative change in error

		Change in error					
		-1	-0.8	-0.6	-0.4	-0.2	0
error	-1	-1	-1	-1	-1	-1	-1
	-0.8	-1	-1	-1	-1	-1	-0.7
	-0.6	-1	-1	-1	-1	-0.7	-0.4
	-0.4	-1	-1	-1	-0.7	-0.4	-0.2
	-0.2	-1	-1	-0.7	-0.4	-0.2	-0.1
	0	-1	-0.7	-0.4	-0.2	-0.1	0
	0.2	-0.7	-0.4	-0.2	-0.1	0	0.1
	0.4	-0.4	-0.2	-0.1	0	0.1	0.2
	0.6	-0.2	-0.1	0	0.1	0.2	0.4
	0.8	-0.1	0	0.1	0.2	0.4	0.7
	1	0	0.1	0.2	0.4	0.7	1

Table 3. Inference rule table for positive change in error

		Change in error				
		0.2	0.4	0.6	0.8	1
error	-1	-1	-1	-1	-1	-1
	-0.8	-0.4	-0.2	-0.1	0	-0.7
	-0.6	-0.2	-0.1	0	0.1	-0.4
	-0.4	-0.1	0	0.1	0.2	-0.2
	-0.2	0	0.1	0.2	0.4	-0.1
	0	0.1	0.2	0.4	0.7	0
	0.2	0.2	0.4	0.7	1	0.1
	0.4	0.4	0.7	1	1	0.2
	0.6	0.7	1	1	1	0.4
	0.8	1	1	1	1	0.7
	1	1	1	1	1	1

It is required to find the fuzzy region for the output for each rule. The centroid or the center of gravity defuzzification method [11] is used which calculates the most typical crisp value of the fuzzy set and the *then* part "y is C" in Eq. 36 can be expressed by Eq. 37.

$$y = \frac{\sum_i \mu_A(y_i) \times y_i}{\sum_i \mu_A(y_i)} \tag{37}$$

where μ_A is a degree of membership function.

3.2.2 Feedforward Compensator Based on Neural Network Inverse Model

A neural network can model an input/output relationship of a dynamic system. A direct or forward model is a mapping that maps a system input to a system output. An inverse model, on the other hand, is an inverse mapping that maps a system output to a system input. In particular, if one set the output to be the reference, then the inverse model could give a desired input for the output to follow the reference or setpoint. The concept of inverse model was used in designing feedforward controls for dynamic systems [14,15]. Park, Choi and Lee [14] and Harnold, et al. [15] approached the problem from the viewpoint of discrete-time model of the nonlinear system, thus avoiding the issues of the invertiblity of nonlinear model.

A two layer neural network is applied to obtain a dynamic feedforward compensator [25]. In general, the output of a system can be described with a function or a mapping of the plant input-output history [25,26]. For a single-input single-output (SISO) discrete-time system, the mapping can be written in the form of a nonlinear function as follows:

$$y(k+1) = f(y(k), y(k-1), ..., y(k-n), u(k), u(k-1), ..., u(k-m)). \tag{38}$$

Solving for the control, Eq. 38 can be represented as following:

$$\begin{aligned} u(k) = g(&y(k+1), y(k), y(k-1), y(k-2), ..., y(k-n), \\ &u(k-1), u(k-2), u(k-3), ..., u(k-m)), \end{aligned} \tag{39}$$

which is a nonlinear inverse mapping of Eq. 38. The objective of the control problem is to find a control sequence, which will drive a system to an arbitrary reference trajectory. This can be achieved by replacing $y(k+1)$ in Eq. 39 with reference output y_{ref}, or the temporary target $y_r(k+1)$, evaluated by

$$y_r(k+1) = y(k) + \alpha(y_{ref} - y(k)), \tag{40}$$

where α is the target ratio constant $(0 < \alpha \le 1)$. The value of α describes the rate with which the present output $y(k)$ approaches the reference output value, and thus has a positive value between 0 and 1 [12,13]. In Fig. 11, the training mode is introduced, where Δ denotes the vector of delay sequence data defined in Eq. 39. Fig. 12 shows the neural network inverse model (NNIM) in training mode. All activation functions in hidden layer are *tanh*(*x*) (described as f_j in Fig. 12) and a linear activation function is used in the output layer (depicted as F_i in Fig. 12).

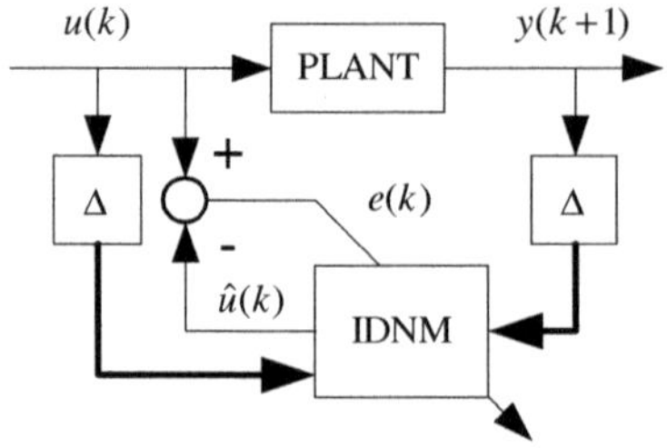

Fig. 11. Training mode of neural network inverse model

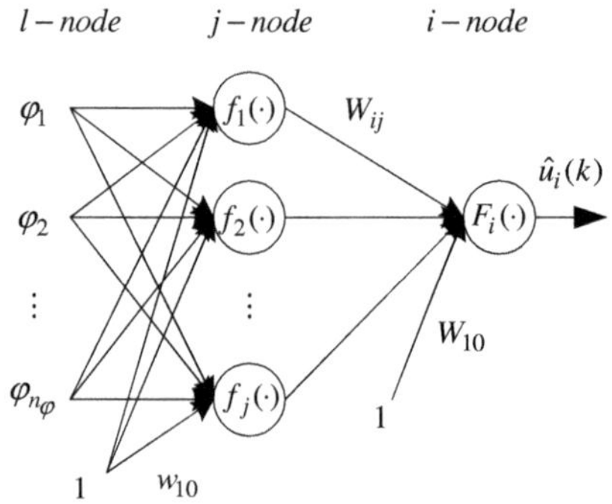

Fig. 12. Neural network inverse model (NNIM)

The output of the NNIM can be represented as

$$\hat{u}_i(k) = F_i\left[\sum_{j=1}^{n_h} W_{ij} f_j\left(\sum_{l=1}^{n_\varphi} w_{jl}\overline{\varphi} + w_{j0}\right) + W_{i0}\right], \tag{41}$$

where

$$\begin{aligned}\overline{\varphi} &= [\varphi_1, \varphi_2, \varphi_3, \ldots, \varphi_{n_\varphi}]^T \\ &= [y(k+1), y(k), \ldots, y(k-n), u(k-1), \ldots, u(k-m)]^T\end{aligned}$$

w_{jl} : weight between input and hidden layers,

n_h, n_φ : number of hidden neurons and external input,

W_{ij} : weight between hidden and output layers.

The above neural network inverse model is trained based on the input-output data as described in Fig. 11. To train the neural network inverse model, Levenberg-Marquardt method is applied which is fast and robust [25,26,27]. The trained NNIM is used as a feedforward compensator.

The total control scheme is indicated in Fig. 13. In the fuzzy controller, the input normalization factors are chosen to represent the proper membership quantifications of linguistic values. In addition, the output normalization factors can be used to yield the desired response of the fuzzy controller. The symbol Δ denotes the vector of delay sequence data. The total control input is $u(k) = u_{fb}(k) + u_{ff}(k)$. The feedback control $u_{fb}(k)$ is the output of the fuzzy controller and the output of the feedforward controller, $u_{ff}(k)$, can be represented as following:

$$u_{ff}(k) = g(y_r(k+1), y_r(k), y_r(k-1), \dots, y_r(k-n), \\ u_{fb}(k-1), u_{fb}(k-2), \dots, u_{fb}(k-m)). \tag{42}$$

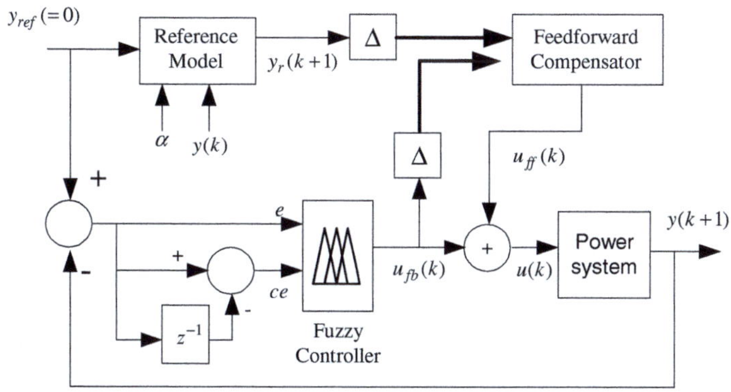

Fig. 13. The fuzzy-neural hybrid control

In Fig. 13, once a signal of a feedforward compensator is given into the control system, the fuzzy controller provides a signal that minimizes the error between the system output and its setpoint. This control scheme can be a soft way of generating a control signal to minimize the tracking error and improve a system performance in the sense that compensating signal is given in advance [27]. This implies the improvment of existing PID-type controller, which is the main purpose of a feedforward controller in a hybrid control scheme.

4 Evaluation by Simulation

The system under study consists of a horizontal axis, 3-bladed, stall regulated wind turbine with a rotor of 16.6 m diameter, that drives an induction generator (IG) rated at 55 kW. The IG is connected to an AC bus in parallel with a diesel-synchronous generator unit that consists of a 50 kW turbocharged diesel engine (DE) driving a 55 kVA brushless synchronous generator (SG). Nominal system frequency is 50 Hz, and the rated line AC voltage is 230 V [21]. The battery storage is connected to the AC bus through a thyristor-bridge controlled current source converter rated at 55 kW. A load is rated at 40 kW. The inertia of the IG is 1.40 kgm^2, and the inertia of the SG is 1.11 kgm^2. The three-phase dumpload is used where each phase consists of 7 transistor-controlled resistor banks with binary resistor sizing in order to minimize quantum effects and provide near linear resolution.

4.1 Fuzzy-Robust Controller

This section describes a simulation performance that tests the proposed controller. The augmented system state $\tilde{x}(t)$ is defined as

$$\tilde{x}(t) = \begin{bmatrix} x_{r1}(t) & x_{r2}(t) & x_1(t) & x_2(t) \end{bmatrix}^T \tag{43}$$

where x_1 and x_2 stand for voltage and frequency, respectively.

For the wind-diesel system, the wind-battery storage system, and the wind-dumpload system, the controller design parameters for the PI controllers of the governor, excitation system, converter, and dumpload are set with the proportional gain 30 and the integral gain 90. The time step size for overall simulation is 1ms. With the PI controller, the possible ranges of the states are obtained by the PARM as shown in Table 4. Three linear models are obtained from Eq. 17 applying L=0.5 and H=1.5 for both V_b and f. In Table 4, 2nd-order AR model is chosen for *the most possible* model. Each possible model is constructed by taking minimum, mean, and maximum value of the PARM.

Table 4. Ranges for possible low, the most possible, and possible high models [p.u.]

		Wind-diesel		Wind-battery storage	
		Frequency	Voltage	Frequency	Voltage
Possible low	min	0.9712	0.9620	0.9183	0.7307
	mean	0.9871	0.9836	0.9483	0.9612
	max	1.0025	1.0018	0.9918	1.0689
Most possible	min	0.9842	0.9780	0.9683	0.7625
	mean	1	1	1	1
	max	1.0159	1.0186	1.0459	1.1154
Possible high	min	1.0002	0.9974	1.0447	0.8103
	mean	1.0166	1.0199	1.0789	1.0660
	max	1.0324	1.0387	1.1284	1.1853

		Wind-dumpload	
		Frequency	Voltage
Possible low	min	0.9639	0.9050
	mean	0.9864	0.9872
	max	1.0078	1.0293
Most possible	min	0.9769	0.9191
	mean	1	1
	max	1.0215	1.0453
Possible high	min	0.9966	0.9402
	mean	1.0199	1.0236
	max	1.0420	1.0693

4.1.1 Fuzzy-LQR Controller

For the LQR controller design parameters, the diagonal terms of Q are Q_{11}=Q_{33}=3000, Q_{22}=Q_{44}=1000, and the off-diagonal terms are zero. Also, the diagonal terms of R are R_{11}=0.5, R_{22}=1, and the off-diagonal terms are zero.

Case 1: Wind turbine + diesel-generator + dumpload

This case is to show how the dumpload contributes the better frequency control when there exists excess power in the network. Fig. 14 shows the wind speed. While the wind turbine is running in parallel with the DG, the load is decreased from 38 kW to 24 kW at 15 sec. In the following figures, the proposed fuzzy-LQR controller is referred to as FZLQR for comparison with the PI controller. Fig. 15 shows the comparison of the active power of the SG, IG, load, and dumpload. Fig. 16 shows the comparison of the system performance of the frequency and the voltage, respectively, between the PI controller and the FZLQR. Figs. 17 and 18 show the results of the wind-diesel system without the dumpload. When the PI controller is tuned, proportional gain (P) is first tuned, and the tuning of the integral gain (I) is followed. Finally, the fine-tuning is done for both P and I until the best performance is achieved.

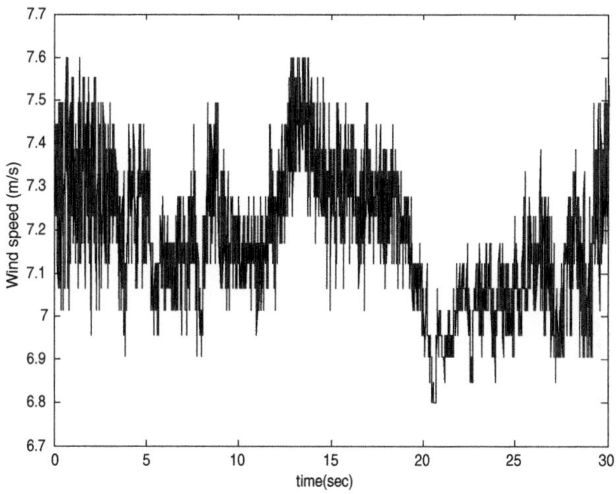

Fig. 14. Wind speed (m/s)

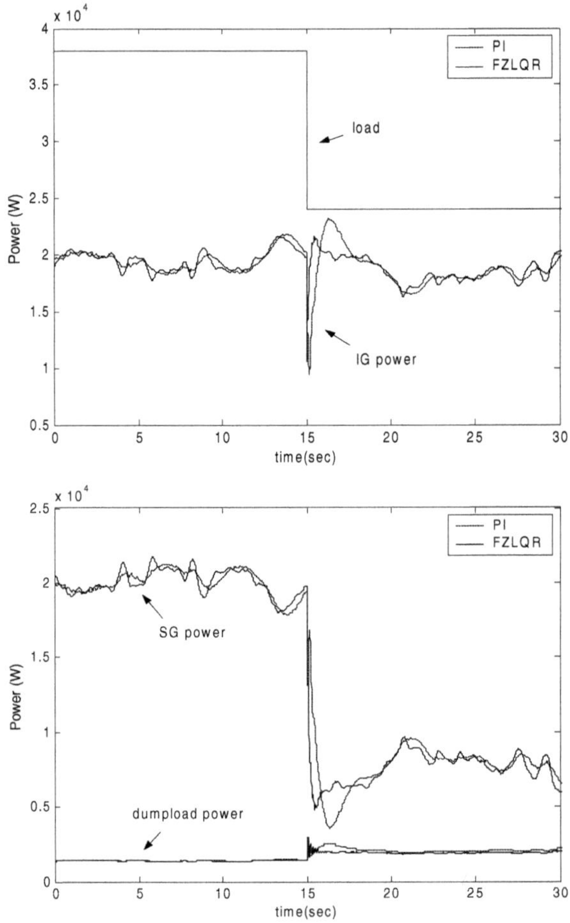

Fig. 15. Power of the load, IG, SG, and dumpload

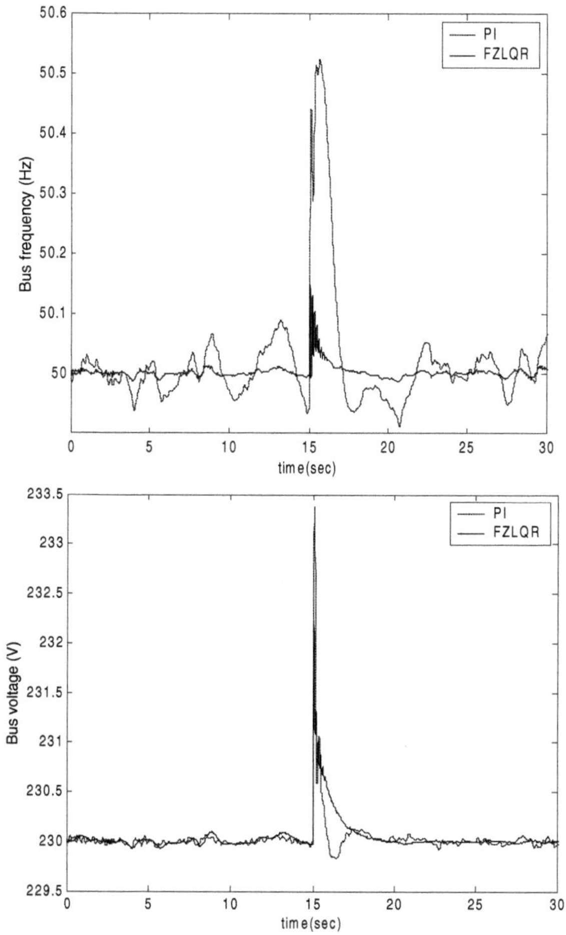

Fig. 16. Bus frequency and bus voltage

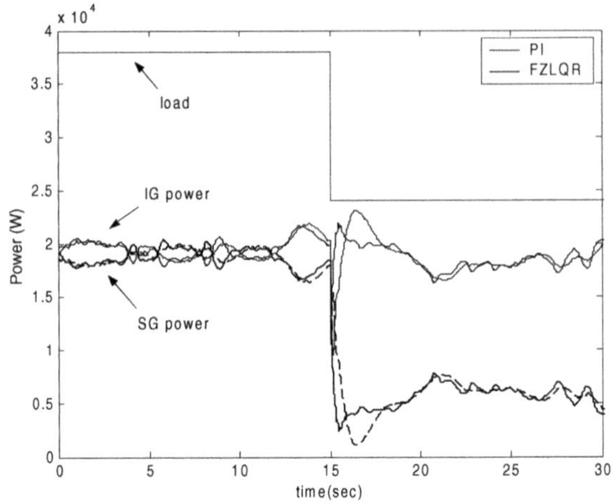

Fig. 17. Power of the load, IG, and SG without the dumpload

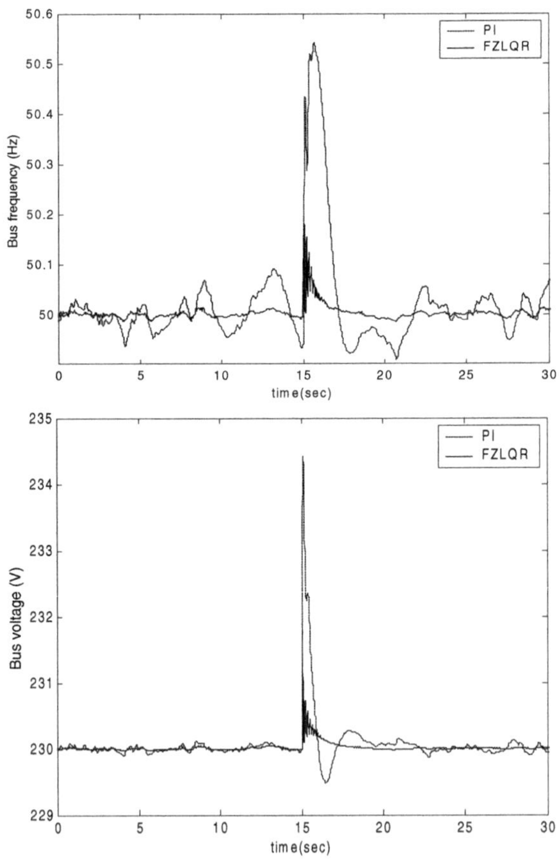

Fig. 18. Bus frequency and bus voltage without the dumpload

Case 2: Wind turbine + battery storage system + dumpload

This case is to examine how the battery storage system (BSS) and the dumpload work for better frequency control, and how they contribute the voltage control with the excitation system in the SG. To represent the charging speed of the BSS, the forcing voltage (E_v) is applied as described in Section 2.2.2. The simulation procedure is as follows: Fig. 19 shows the wind speed. While the DG is shutdown, the load is changed from 38 kW to 20 kW at 5 sec. Figs. 20 and 21 are the comparisons of the system outputs between the PI controller and the FZLQR, where the charging speed during transient period is about 600 watt from 5sec. to 7 sec. for the steady state period, and the charging speed is about 10 watt per second. Fig. 20 shows the comparison of the active power of the IG, load, and dumpload including the power drop in SG. Fig. 21 shows the comparison of the responses of the frequency and the voltage.

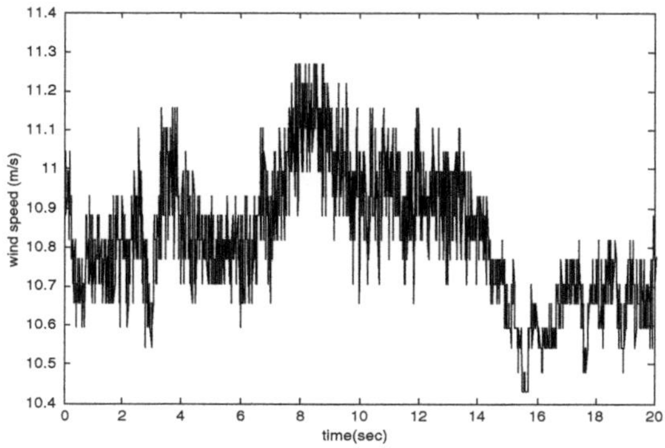

Fig. 19. Wind speed in the charge operation

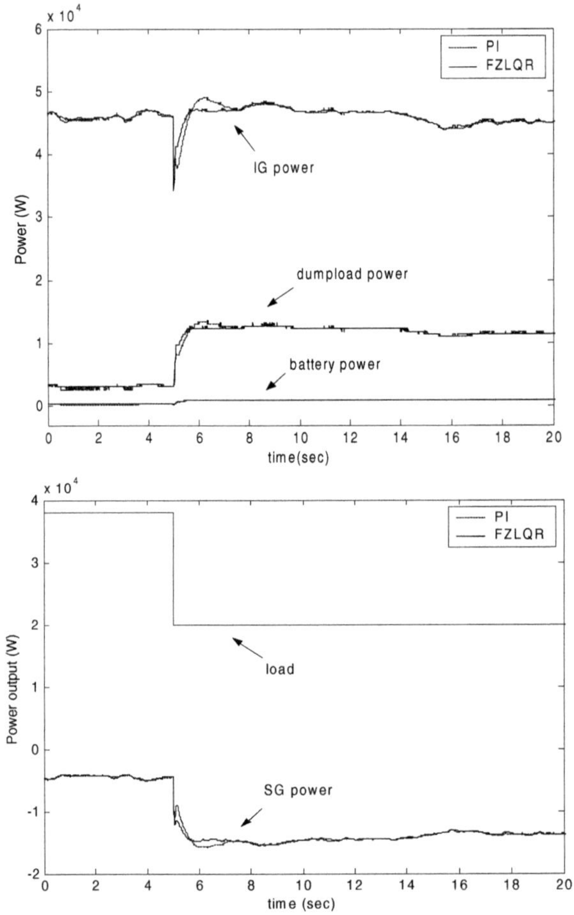

Fig. 20. Power of the IG, dumpload, battery storage, load, and SG

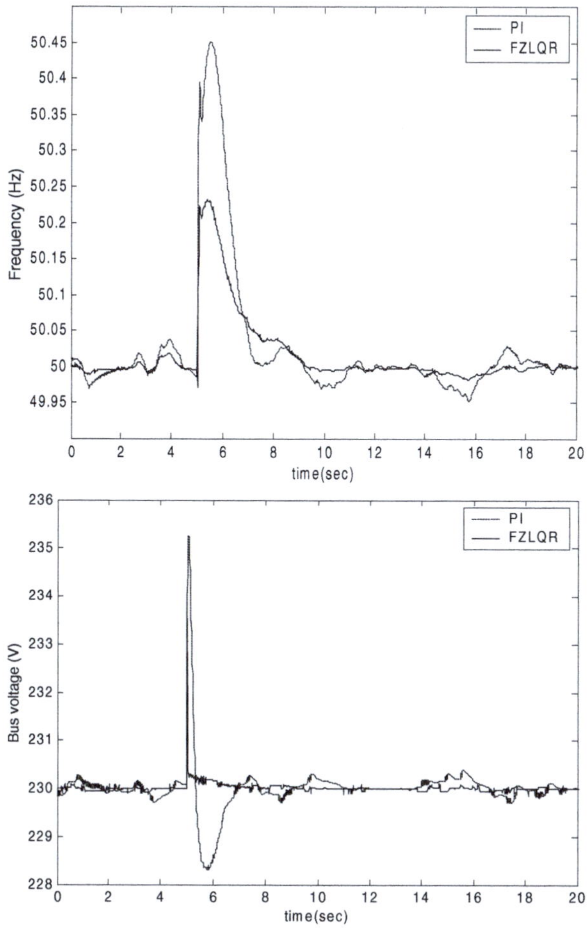

Fig. 21. Bus frequency and bus voltage

From the simulation study, the proposed fuzzy-LQR controller achieves the smoother and tighter power quality control in terms of the voltage and frequency with respect to the nominal condition (i.e., 230 V and 50 Hz). Also, it can be recognized that it is important to choose the correct control mechanism and to design proper controller for the battery performance.

4.1.2 Fuzzy-SMC Controller

<u>*Case 3: Wind turbine + diesel-generator + dumpload*</u>

For fuzzy-SMC controller design parameters, the diagonal matrix Q is with Q_{11}=Q_{33}=1000 and Q_{22}=Q_{44}=2000, and the diagonal matrix ξ is with ξ_{11}=50 and ξ_{22}=80. The rest of the terms are set to zero. The tuned PI controller gains for the governor and excitation system are P_{gov}=20, I_{gov}=60, and P_{efd}=30, I_{efd}=90.

Wind speed is shown in Fig. 22. For the simulation task, a step load change is applied at 5 seconds from the initial loading of 35*kW* to 27*kW*. In the following figures, the proposed fuzzy-SMC controller is referred to as SMLQR for comparison with the PI controller. Fig. 23 shows the power in the IG, the load, and the dump load. In this case, when the load decreases, the dumpload dissipates the excess power to the network.

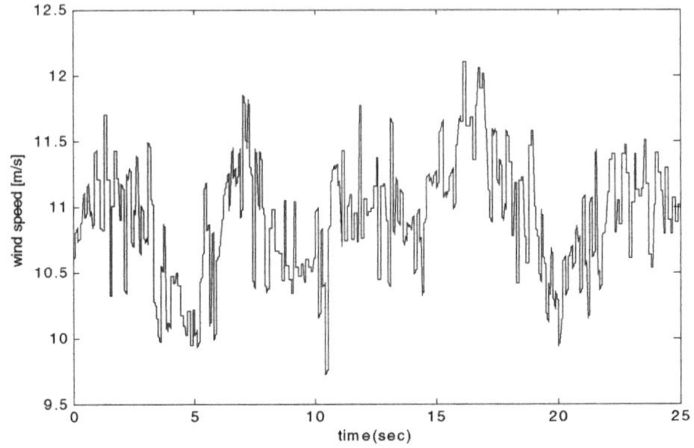

Fig. 22. Wind speed

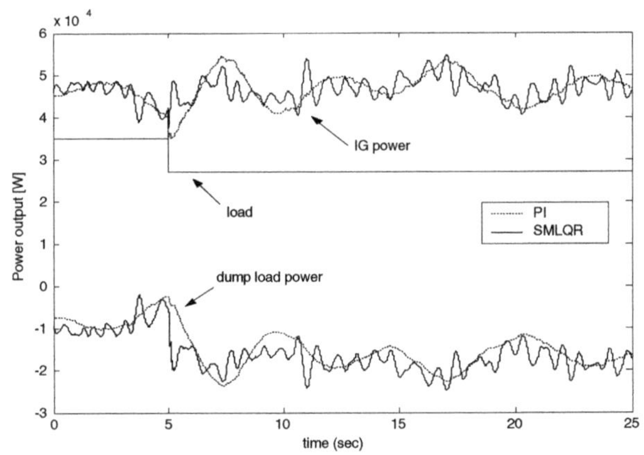

Fig. 23. Power of the IG, load, and dumpload

The proposed fuzzy-SMC control scheme improves the bus frequency and bus voltage compared to the PI controller as shown in Fig. 24.

In this system, the SG is used as a synchronous condenser. By controlling the field excitation, the SG can be made to either generate or absorb reactive power to maintain its terminal voltage. Fig. 25 shows the reactive power from the SG. In SMLQR, the improvement of frequency and voltage in RMS error relative to the PI controller is 51.922% and 52.511% in per unit, respectively.

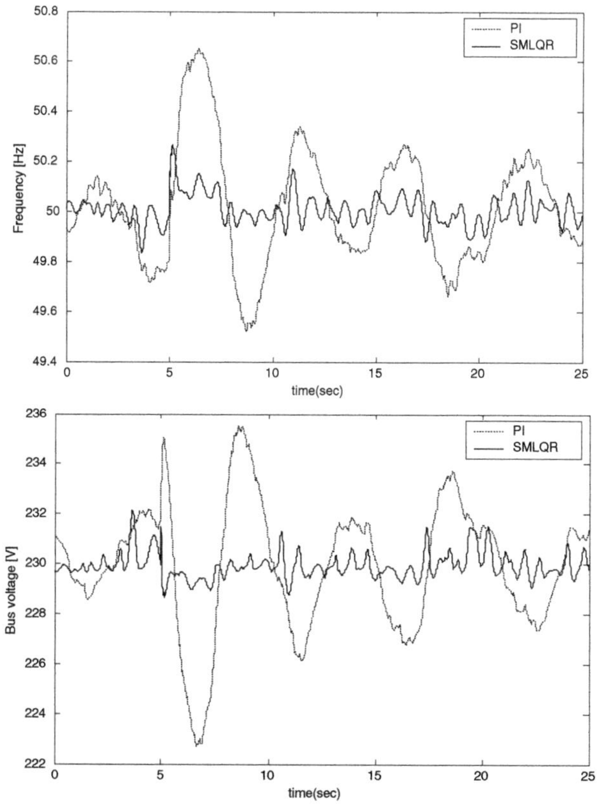

Fig. 24. Bus frequency and bus voltage

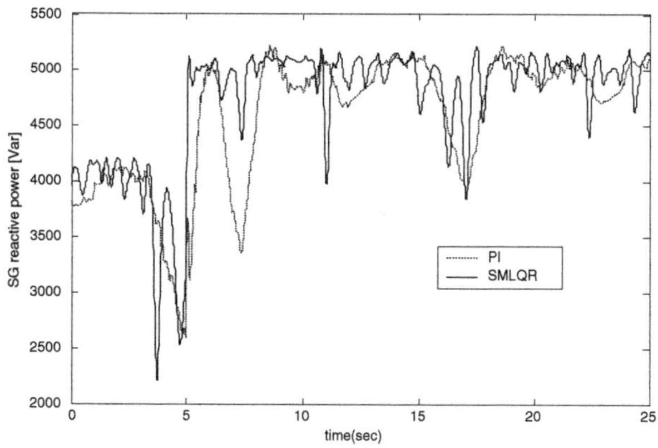

Fig. 25. Reactive power output from the SG

The fuzzy-SMC controller achieves better performance compared to the PI controller. The maximum voltage and frequency deviations are less than 1%. However, the voltage performance of the PI controller shows slow damping. Such poor performance is caused by the neglect of the interaction of variables between the PI controller loops [22]. Clearly, a control method is required that handles a multi-input-multi-output system. In the proposed controller, all performances are smooth and damped. Therefore, the fuzzy-SMC controller provides more effective mechanism for multi-input-multi-output nonlinear system.

4.2 Fuzzy-Neural Hybrid Controller

Pitch control has the potential for producing the highest level of interaction because of the presence of both diesel and wind turbine control loops [8]. When wind power rises above the power set point and SMES unit is fully charged, the pitch control system begins to operate to maintain an average power equal to the set point. The pitch control system consists of a power measurement transducer, a manual power set point control, a proportional plus integral (PI) feedback function, and hydraulic actuator, which varies the pitch of the blades. Variable pitch turbines operate efficiently over a wider range of wind speeds than fixed-pitch machines. In this simulation, turbine-blade pitch control based on fuzzy-neural hybrid control is studied.

First, a fuzzy controller is designed for a feedback controller and a neural network inverse model is obtained for a feedforward compensator. The target ratio constant α is 0.1 and the normalization factors g_1, g_2, g_0 are 5, 50, and 5, respectively, which are determined by trial and error. Levenberg-Marquardt method is applied to train a neural network inverse model. The sampling time is 0.01 sec for the control action. Training is carried out by applying varying white noise signals. Firstly, before training, fuzzy control is implemented with the plant. Secondly, white noise signal is inserted to the fuzzy controller and data set is obtained, with the noise signal as input and the plant output as output. Then, the neural network inverse model (NNIM) is trained by setting the noise signal as output and the plant output as input to the NNIM.

The proposed fuzzy-neural hybrid controller is tested in a WDAPS. Two cases are considered: first, the sudden step load increase of 0.01 [p.u.] and SMES is in discharging mode (inverter mode). Second, the SMES fully discharged and there is a sudden step load increase. In this case, SMES is in recharging mode (rectifier mode).

Case 4: A sudden step load increase

A load is suddenly increased by 0.01 [p.u.]. The SMES releases the charged current (2 [p.u.]). The governor and pitch mechanism start operating for charging current of SMES and damping of WDAPS. In the following figures, the proposed fuzzy-neural hybrid controller is referred to as FNHC in comparison with the PI and FC, where FC stands for fuzzy logic feedback controller. Fig. 26 shows improvement in the system frequency oscillations and power deviations.

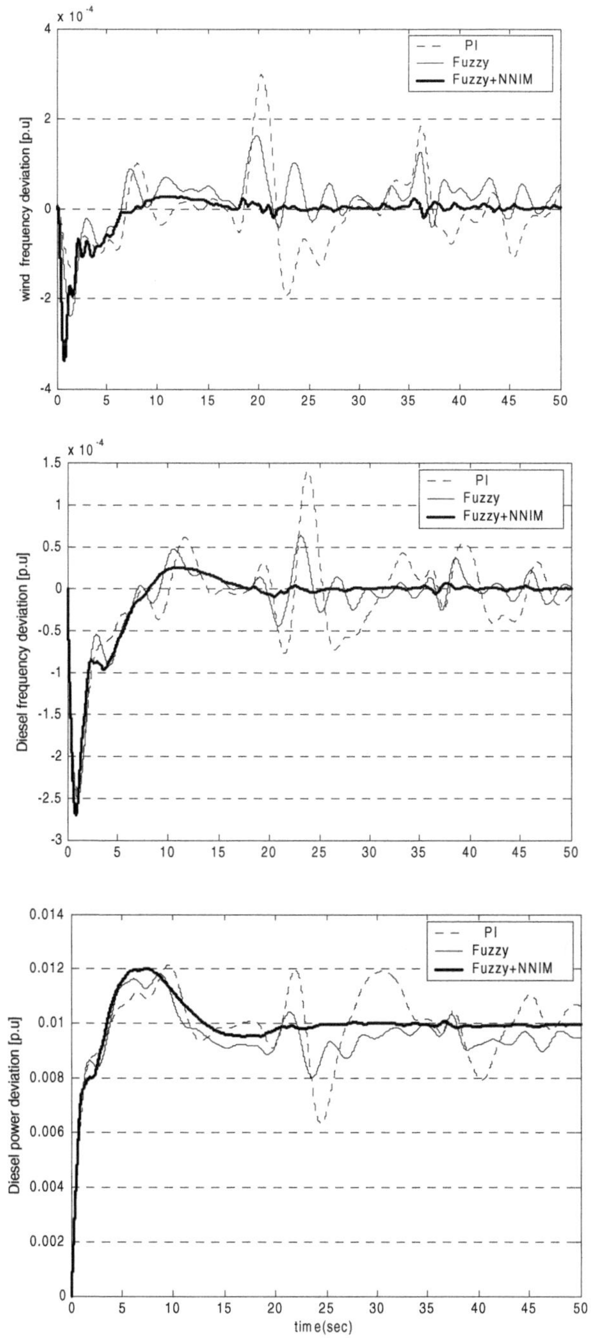

Fig. 26. Comparison of system response among PI, FC, and FNHC

Case 5: Sudden step load increase with fully discharged SMES

In this case, the SMES is fully discharged (0 [p.u.]). Then, the SMES is recharged to set point (2 [p.u.]). The wind power generation from the wind turbine is assumed not sufficient. Fig. 27 also shows that the FNHC performance is much better than the PI and the FC.

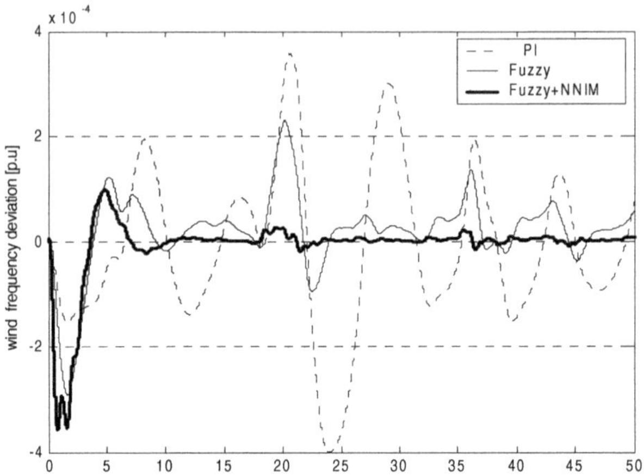

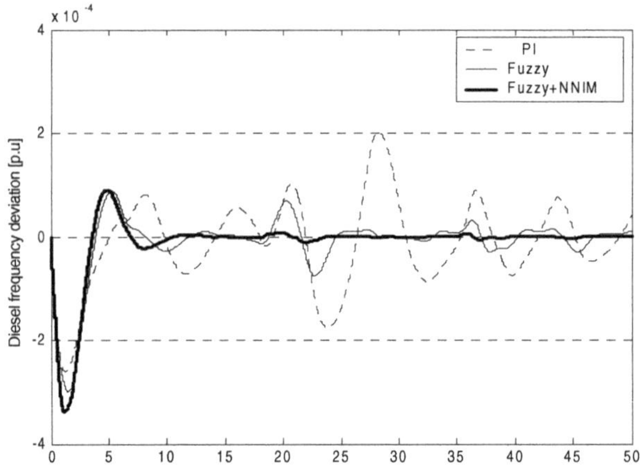

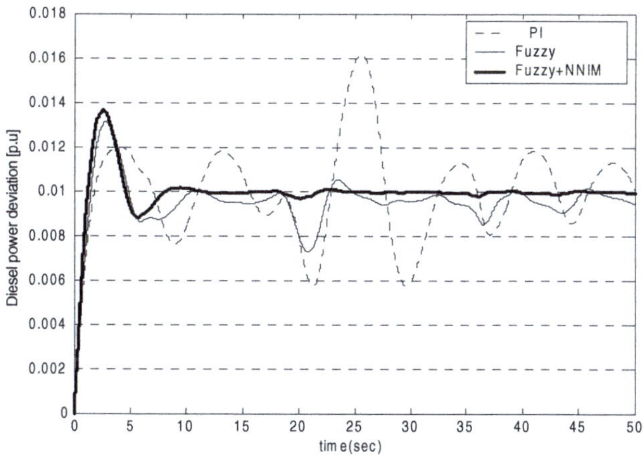

Fig. 27. Comparison of system response among PI, FC, and FNHC

5 Conclusion

Intelligent controllers are designed to control wind-diesel autonomous power system (WDAPS). Specifically, a new type of the fuzzy-robust controller and the fuzzy-neural hybrid controller are presented for the study of the power quality control. In the fuzzy-robust controller, the derived simulation model including the reduced-order model can be applied for different wind-turbine hybrid power system configurations for the study of a power quality control. The choice of control techniques is motivated from the fact that the system is nonlinear and multivariable. Even though the reduced-order model is used to design controllers, model imperfections and uncertainties can be compensated by the fuzzy model. On the other hand, the main idea of fuzzy-neural hybrid control is that the dynamic feedforward control can be used for improving the reference tracking while feedback is used for stabilizing the system and for suppressing disturbances. Feedforward controller is a neural network inverse model (NNIM) and feedback controller is a fuzzy controller.

The proposed intelligent controllers were tested in a WDAPS and compared with the conventional PI controller. The intelligent control schemes provide more effective control for the system to achieve better power quality, which is demonstrated by smooth transition of voltage and frequency. Thus, the usefulness of intelligent controllers is demonstrated.

References

1. Karaki, S.H., Chedid, R.B., Ramadan, R.: Probabilistic production costing of diesel-wind energy conversion systems. IEEE Trans on Energy Conversion 15, 284–289 (2000)

2. Pandiaraj, K., Taylor, P., Jenkins, N.: Distributed load control autonomous renewable energy systems. IEEE Trans on Energy Conversion 16, 14–19 (2001)
3. Stavrakakis, G.S., Kariniotakis, G.N.: A general simulation algorithm for the accurate assessment of isolated diesel –wind turbines systems interaction: part 1: A general multimachine power system model. IEEE Trans on Energy Conversion 10, 577–583 (1995)
4. Uhlen, K., Foss, B.A., Gjosaeter, O.B.: Robust control and analysis of a wind-diesel hybrid power plant. IEEE Trans on Energy Conversion 9, 701–708 (1994)
5. Chedid, R.B., Karaki, S.H., Chadi, E.C.: Adaptive fuzzy control for wind-diesel weak power systems. IEEE Trans on Energy Conversion 15, 71–78 (2000)
6. Ko, H.S., Jatskevich, J.: Power quality control of wind-hybrid power generation system using fuzzy-LQR controller. IEEE Trans on Energy Conversion 22, 516–527 (2007)
7. Ko, H.S., Kang, M.J., Boo, C.J., Jwa, C.K., Kang, S.S., Kim, H.C.: Power quality control of hybrid wind power generation system using fuzzy-robust controller. In: Ishikawa, M., Doya, K., Miyamoto, H., Yamakawa, T. (eds.) ICONIP 2007, Part II. LNCS, vol. 4985, pp. 468–477. Springer, Heidelberg (2008)
8. Ko, H.S., Lee, K.Y., Kang, M.J., Kim, H.C.: Power quality control of an autonomous wind-diesel power system based on hybrid intelligent controller. Neural Networks 21, 1439–1446 (2008)
9. Ise, T., Kita, M., Taguchi, A.: A hybrid energy storage with a SMES and secondary battery. IEEE Trans. on Applied Superconducting 15, 1915–1918 (2005)
10. Passino, K.M.: Fuzzy control: theory and applications. Addison-Wesley Publishing, Reading (1997)
11. Yen, J., Langari, R.: Fuzzy logic: intelligence, control, and information. Prentice-Hall, Englewood Cliffs (1999)
12. Park, Y.M., Moon, U.C., Lee, K.Y.: A self-organizing fuzzy logic controller for dynamic systems using fuzzy auto-regressive moving average (FARMA) model. IEEE Trans on Fuzzy Systems 3, 75–82 (1995)
13. Park, Y.M., Moon, U.C., Lee, K.Y.: A self-organizing power system stabilizer using fuzzy auto-regressive moving average (FARMA) model. IEEE Trans on Energy Conversion 11, 442–448 (1995)
14. Park, Y.M., Choi, M.S., Lee, K.Y.: An optimal tracking neuro-controller for nonlinear dynamic systems. IEEE Trans on Neural Networks 7, 1099–1110 (1996)
15. Harnold, C.L.M., Lee, K.Y., Lee, J.H., Park, Y.M.: Free model based adaptive inverse control for dynamic systems. In: Proc. the 37th IEEE Conf on Decision and Control, Tampa, Florida, pp. 507–512 (1998)
16. Krause, P.C., Wasynczuk, O., Sudhoff, S.D.: Analysis of electrical machinery. McGraw-Hill, New York (1986)
17. International Electrotechical Commision, Publication 34-10, Rotating electrical machines, Part 10: Conventions for description of synchronous machines, Geneve (1975)
18. Niimura, T., Ko, H.S., Ozawa, K.: A day-ahead electricity market price prediction based on fuzzy regression model in deregulated environment. In: IEEE International Joint Conf. on Neural Networks, vol. 2, pp. 1362–1366 (2002)
19. Ogata, K.: Modern control engineering. Prentice-Hall, New Jersy (1986)
20. Barnes, E.R.: Affine transform method. Mathematical Programming 36, 174–182 (1986)
21. Uhlen, K., Foss, B.A., Gjosaeter, O.B.: Robust control and analysis of a wind-diesel hybrid power plant. IEEE Trans on Energy Conversion 9, 701–708 (1994)

22. Garduno-Ramirez, R., Lee, K.Y.: Power plant coordinated-control with wide-range control-loop interaction compensation. In: Proc. the 15th IFAC World Congress, Barcelona, Spain, (CD) paper #2407 (2002)
23. Kassakian, J.G., Schlecht, M.F., Verghese, G.C.: Principles of power electronics. Addison-Wesley Publishing, New York (1992)
24. Utkin, V.I., Guldner, J., Shi, J.: Sliding modes in electromechanical systems. Taylor and Francis, Philadelphia (1999)
25. Haykin, S.: Neural networks: A comprehensive foundation. Prentice Hall, New Jersey (1998)
26. Ng, G.W.: Application of neural networks to adaptive control of nonlinear systems. John Wiley and Sons Inc., Chichester (1997)
27. Madsen, P.P.: Neural network for optimization of existing control systems. In: Proc. IEEE International Joint Conf. on Neural Networks, pp. 1496–1501 (1995)
28. Tripathy, S.C., Kalantar, M., Balasubramanian, R.: Dynamics and stability of wind and diesel turbine generator with superconducting magnetic energy storage unit on an isolated power system. IEEE Trans on Energy Conversion 6, 579–585 (1991)

Adaptive Fuzzy Control for Variable Speed Wind Systems with Synchronous Generator and Full Scale Converter

V. Calderaro, C. Cecati, A. Piccolo, and P. Siano

Abstract. Control systems for variable-speed wind turbines (WTs) are continuously evolving toward innovative and more efficient solutions. Among the various techniques, fuzzy logic is gaining reputation due to its simplicity and effectiveness. In this chapter, after a review of fuzzy logic based control applied to wind energy conversion systems, a sensorless peak power tracking control for maximum wind energy extraction and a voltage control allowing compensation of voltage variations at the WT connection point are proposed. Both the controllers are based on fuzzy logic.

Before that, a data-driven design methodology is introduced, in order to generate the "best" Takagi–Sugeno–Kang fuzzy model, for the maximum power exploitation from a variable-speed wind turbine. The performance of the variable speed wind systems employing a synchronous generator and a full scale converter endowed with the proposed fuzzy controllers are tested under some common operating conditions.

1 Introduction

It is well known by scientists and practitioners that conventional analytical methods can be adopted for solving many problems in power system and converters

V. Calderaro · A. Piccolo · P. Siano
Department of Information and Electrical Engineering,
University of Salerno, via Ponte don Melillo, 84084 Fisciano, SA, Italy
e-mail: vcalderaro@unisa.it, apiccolo@unisa.it, psiano@unisa.it

C. Cecati
Department of Electrical and Information Engineering,
University of L'Aquila, Loc. Monteluco di Roio, 67100 L'Aquila, Italy
e-mail: carlo.cecati@univaq.it

L. Wang et al. (Eds): *Wind Power Systems*, Green Energy and Technology, pp. 337–366.
springerlink.com

planning, operation and control, but their actual formulation and practical results often suffer from restrictive assumptions. Optimal control theory, for instance, can be implemented aiming at performance enhancement of both power systems and converters, but the actual achievement of adaptation and robustness is often problematic and not guaranteed at all.

Wind Energy Conversion Systems (WECSs) are commonly described by mathematical models, but, very often, difficulties arise when looking for an accurate model of the system under study, particularly if it is affected by model uncertainties, non linearities and parameter variations. As a result, Proportional–Integral–Derivative (PID) regulators, tuned using trial-and-test methods, are commonly employed with the drawback that the desired accuracy is satisfied only within a short interval close to the desired operating point.

In those cases, fuzzy logic control (FLC) represents a very interesting choice as it overcomes the lack of system information, allowing significant improvements of system performance over PID controllers.

The application of fuzzy theory to power systems was successful experimented in power converters as well as in wind turbines (WTs) control (El-Hawary 1998), which are becoming very diffused as the electrical generation from wind is very efficient and economically attractive.

Early wind generation systems, supplying either an utility grid or isolated loads, used variable pitch/constant speed WT, coupled with a squirrel cage induction generator or a wound-field synchronous generator. Nowadays, variable speed WTs with electronic control are common in wind farms, and at many different power levels (from few kW up to 10 MW) as well as technical solutions (axial flux machines, doubly-fed generators and so forth) (Simoes and Farret 2007).

In case of low wind speed, the rotor is controlled by varying the generator reaction torque in response to measured rotor speed and/or generated power. In case of high wind speed, control variables are regulated in order to maintain the power at the highest values and without risks for the system (Galdi et al. 2009).

Power limitation can be achieved using either a passive or an active regulation. In the first case, rotor blades are designed such as to stalling close to the rated speed; in the second case the blade pitch is continuously regulated such as to obtain the rated power. Pitch adjustment is made in response to measurement of the rotor speed and/or the generated power.

Control systems are designed such as to alleviate transients through the WT, regulating and smoothing the generated power, ensuring the appropriate dynamics and maximizing the output power. Voltage and frequency regulation can be implemented, too (Leithead and Connor 2000).

A common method for defining control strategies for a variable speed WT is to specify the rotor speed as a function of the wind speed. However, rotor speed, torque and dynamics vary with wind speed, which is estimated from measurements made on the WT itself. Unfortunately, there are many difficulties with this approach because the aerodynamics is non-linear and non-uniquely related with the wind speed. For this reason and considering that several kinds of disturbances, including grid parameters and atmospheric conditions can also exist, fuzzy logic represents an effective method for controlling WTs, assuring fast convergence and insensibility to parameter variations even in presence of noisy and inaccurate signals (Simoes et al. 1997a, Calderaro et al. 2008; Galdi et al. 2008).

Several studies implementing different regulation strategies have been dedicated to fuzzy control of WTs with the main objective of maximizing their output power. In most of them, fuzzy control was applied to WTs with induction generators. In (Hilloowala and Sharaf 1996) a rule-based FLC regulating the output power of a Pulse Width Modulation (PWM) inverter applied to a stand-alone wind generator, is presented. The controller uses as inputs two real-time measurements: the error between the rectifier output power and its reference and the range of change of the error. The output is the control signal determining the power transferred to the local load. In Simoes et al. (1997b) the authors use three distinct FLCs to maximize the output power, enhancing performance through the optimal speed/power characteristic. The first controller searches the best generator speed until the system settles down at the maximum output power condition; the second controller performs a reduction of the core losses, increasing WECS efficiency; the third one provides a robust speed control against wind vortex and turbine oscillatory torques. Differently from the previous FLCs (Chen et al. 2000) the inputs are the variations of the output power and the actual speed of the generator, both obtained by measurements of the electrical variables at the WT terminal.

Previous works show that there are two main approaches for implementing fuzzy control in WT design: the first one is based on the knowledge of the optimal speed/power characteristic, the second one uses the output power, a feedback signal and the actual speed, obtained by means of real measurements.

These approaches suggested many studies. For instance, in (Adzic et al 2008; Kaur et al. 2008) the first control philosophy was implemented in order to maximize output power. In (El Mokadem et al. 2009) a fuzzy logic supervisor was proposed for ensuring a regular primary reserve, even when the generator works below the rated power and avoiding the wind speed measurement and the need of precise WT model. For such a purpose, a power reference was determined to maintain an energy reserve for a large wind power range. Such a reserve can be achieved by actions on the torque of the electrical generator and on the pitch angle.

A unified approach consisting of two alternative control schemes is proposed in (Mirecki et al. 2007). If the WT characteristic is a priori known, it is used for the optimal control of power, torque or speed; if the characteristic is unknown, a fuzzy logic based algorithm is implemented. More in detail, it is based on behavioral rules linked to power and speed variations and optimizes the output power.

There are several papers, such as (Senjyu et al. 2007), employing a fuzzy control of the pitch angle for electrical outputs regulation; WT is controlled by a fuzzy reasoning with three inputs: average wind speed, variance and absolute average of frequency deviation. The paper (Chen and Hsu 2008) presents a unified voltage and pitch angle controller for a WT, aiming at reaching voltage control and stabilization of generator speed and system frequency. A FLC works when wind energy conversion system is subjected to a major disturbance such as grid disconnection. In this case a supplementary pitch angle controller, regulating the output power, is designed using fuzzy logic inference rules.

Other authors address their attention on power converters applying fuzzy control strategies. For instance, (Cecati et al. 2003, 2005) presents the implementation of a FLC to Active Rectifiers, (Mattavelli et al. 1997; Gupta et al. 1997) implements a fuzzy system for dc/dc control and (Jasinski et al. 2002; Saetieo and Torrey 1998) to PWM rectifiers.

In this chapter, a fuzzy logic based control for variable speed WTs is proposed, in order to implement both a sensorless peak power tracking control for maximum wind energy extraction and a voltage control allowing a compensation of the voltage variations at the Point of Common Coupling (PCC).

The major improvements with respect to previous works (Calderaro et al. 2008; Galdi et al. 2008, 2009) are the opportunity to directly generate the duty cycle command to drive the dc/dc boost converter and the capability to compensate the voltage variations at the PCC by controlling the reactive power generated/absorbed by the dc/ac converter. Moreover, a FLC is also implemented to control the voltage applied to the capacitor before the inverter.

In the following sections, after an overview of the basic concepts on the variable-speed WT control, a description of the system is introduced. Hence, a data-driven designing methodology for fuzzy controllers is proposed. The methodology generates the "best" Takagi–Sugeno–Kang (TSK) fuzzy model, for the estimation of the maximum power obtainable from a variable-speed WT. The proposed method combines genetic algorithms (GAs) and recursive least-squares (RLS) estimation for the model parameter adaptation and a fuzzy clustering for partitioning the input–output space. Then, inverter FLCs for controlling the voltages at the PCC and on the capacitor before the inverter are introduced and designed.

The performances of the variable-speed wind system employing a synchronous generator and full scale converter and endowed with the proposed FLCs are evaluated through some case studies.

2 Control of Variable Speed Wind Turbines for Maximum Power Exploitation

Modern WECSs are capable to operate in a wide spread of wind speeds and weather conditions.

As pointed out in (Johnson et al. 2006; Boukhezzar et al. 2006), variable-speed WTs operate within a boundary delimited by the three main operational regions shown in Fig. 1. A stopped turbine or a turbine that is just starting up is considered to be operating in region 1. Here, current control strategies are not critical and a wind speed monitoring determines whether it lies within the specifications for turbine operation: when this condition is satisfied, the system executes the routines necessary to start up the turbine.

In Region 2 (yaw drive), a control of generator torque and blade pitch angle is implemented, aiming at capturing as much wind energy as possible. In region 3, the wind speed is above the rated value and the turbine must limit the captured wind power avoiding failures and maintaining the maximum achievable energy production. Usually, this is done pitching its blades, in order to shed additional power. Yaw control, generator torque and blade pitch strategies can be successful used (Johnson et al. 2006) in order to shed surplus power and limit the captured energy.

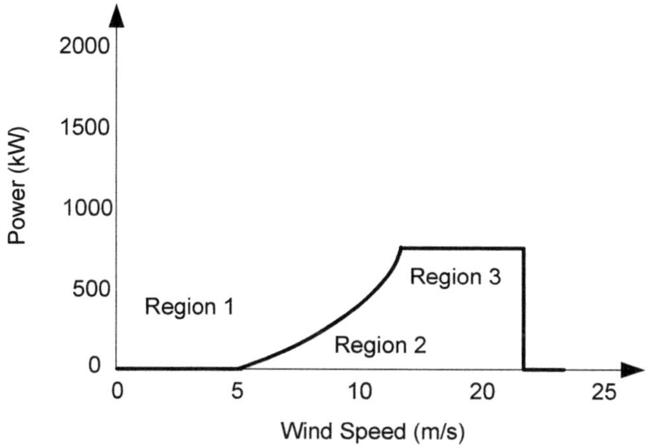

Fig. 1. Example of steady-state power curve

The most interesting region is “2”, where the primary goal is to maximize the captured energy using a variable-speed WT. The power and the torque produced by a WT depends on the available wind power, the power curve of the machine and the machine capability to react to wind variations:

$$P_\omega = \frac{1}{2}\rho C_P(\lambda, \beta) A V_\omega^3 \tag{1}$$

$$T_\omega = \frac{P_\omega}{\omega_r} = \frac{1}{2}\rho C_T(\lambda, \beta) r_m A V_\omega^2 \tag{2}$$

where:

P_ω is the rotor mechanical power (W);

T_ω is the turbine torque (N·m);

V_ω is the wind speed at the center of the rotor (m/s);

$A = \pi r_m^2$ is the wind rotor swept area (m^2);

ρ is the air density (kg/m^3);

$\omega_r = \dfrac{\lambda V_\omega}{r_m}$ is the rotor angular velocity (rad/s);

r_m is the turbine radius (m);

C_P is the rotor power coefficient, i.e. the percentage of the kinetic energy of the incident air mass converted into mechanical energy by the rotor (the maximum value for Betz's limit 59.3%) (Johnson et al. 2006);

C_T is the torque coefficient;

C_P and C_T are nonlinear functions of the tip speed ratio and the pitch angle, correlated by the following relationships: $C_P(\lambda, \beta) = \lambda C_T(\lambda, \beta)$;

β is the pitch angle of rotor blades (deg), it is constant for fixed pitch WTs;

$\lambda = \dfrac{\omega_r r_m}{V_\omega}$ is the tip speed ratio, i.e. the ratio between the blade tip speed and the wind speed upstreaming the rotor.

The maximum power coefficient $C_{P\_\max}$ corresponds to the optimal tip speed ratio λ_{opt}. Clearly, the turbine speed should be changed according to the wind speed such as to maintain the optimum tip speed ratio.

The maximum aerodynamic torque of the WT is given by:

$$T_{opt} = \frac{\rho C_{P\_\max}(\lambda_{opt}, \beta)\pi r_m^5 \omega_{opt}^2}{2\lambda_{opt}^3} = K\omega_{opt}^2 \tag{3}$$

where: $K = \dfrac{\rho C_{P\_\max}(\lambda_{opt}, \beta)\pi r_m^5}{2\lambda_{opt}^3}$.

The main problem using such a kind of control is that the blade aerodynamics can change significantly over the time and consequently there is no accurate way for obtaining K.

2.1 Voltage Control Capabilities Requirements for WECSs

To ensure power systems security and stability, in many European countries, system operators are setting new requirements for WECSs such as:

- operations during a grid fault;
- operations within a certain frequency range: 47–52 Hz;
- active power control during frequency variations, limiting the power increases up to a certain rate (power ramp rate control);
- to supply or to consume reactive power depending on power system requirements (reactive power control), or to apply voltage control by adjusting the reactive power, based on grid measurements (voltage control).

Large wind farms in remote areas or off-shore, are connected with transmission system. Since each voltage node is a local quantity, a voltage control at these far places can be difficult. Therefore, WTs should have intrinsic voltage control capabilities. The latter are expected to become more and more important regarding to grid connection requirements and the turbines market potential (Knight and Peters, 2005). In particular, voltage and reactive power control at the WT connection point are used in order to keep the voltage within the required limits avoiding voltage stability problems. The fundamental requirement is that the steady grid state voltage variations must be maintained within a certain range (e.g. ±10%) even after the connection of a WT. In order to achieve this goal, the WT terminal voltage is generally measured and fed into a voltage controller, which computes the amount of reactive power to be generated or consumed. When the measured voltage is below the set point, reactive power generation is increased; when it is higher, reactive power generation is decreased.

Notice that, while constant speed WTs with squirrel cage induction generators always consume reactive power (the value of which depends on the terminal voltage, the active power generation and the rotor speed), variable speed WTs, equipped with a doubly fed induction generator or with a direct-drive synchronous generator are able to control the terminal voltage.

Variable-speed WTs can optimize both the produced active power and the reactive power (generated or consumed) independently and at every speed (Kana et al. 2001; Freris 1990; Valtchev et al. 2000; Muljadi and Butterfield 2001).

3 System Configuration

Figure 2 shows a typical synchronous generator system with uncontrolled diode rectifier bridge, boost dc/dc converter and three-phase IGBT inverter (Yamamura et al. 1999; Song et al. 2003; Knight and Peters 2005; Tafticht et al. 2006;

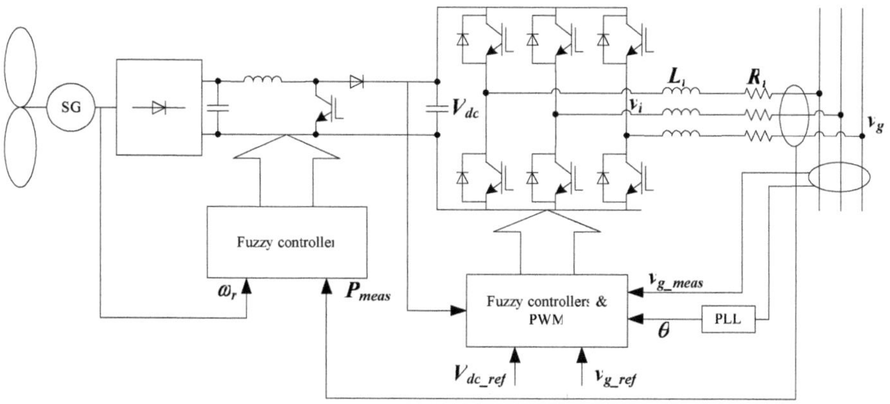

Fig. 2. Schematic diagram of wind energy conversion system connected to the main grid

Chinchilla et al. 2006): the boost converter is controlled for achieving maximum power extraction from the generator, the inverter is controlled for delivering high-quality power to the grid.

Recently, back-to-back converters, i.e. active rectifiers coupled with inverters, are increasing their diffusion, due to their better performance in terms of wind power extraction capability and power quality. This topology, not considered in this paper, avoids the dc/dc converter, increasing reliability (Portillo et al. 2006, Bueno et al. 2008).

This section presents the structure of the considered system, consisting of the models of the distribution network, the synchronous generator, the WT, the dc/dc converter with control for maximum power tracking and the inverter with voltage regulation capability. Each part will be discussed separately in the following sub-sections.

3.1 Wind System Description

The IGBT-based PWM inverter operates at phase-to-phase voltage of 575V, 60Hz. The uncontrolled rectifier output is a dc voltage proportional to the wind speed. The current controlled dc/dc step-up converter ensures the maximum power production, the inverter allows reactive power control and keeps the dc-link voltage on the capacitor before the inverter to a constant value (1100 V). The inverter is interconnected to the grid by means of a low pass filter, thus reducing .current and voltage harmonics to satisfy EMI regulations (THD < 6%).

3.2 Synchronous Generator Model

The WT uses a synchronous machine in which the mechanical subsystem is described by:

$$\Delta\omega(t) = \frac{1}{2H}\int_0^t (T_m - T_e)dt - K_d\Delta\omega(t) \tag{4}$$

$$\omega(t) = \Delta\omega(t) + \omega_0 \tag{5}$$

where $\Delta\omega$ is the speed deviation from rated value, H is the inertia constant, T_m and T_e are the mechanical and electromagnetic torque respectively, K_d is the damping factor representing the effect of damper windings, $\omega(t)$ is the rotor speed and ω_0 the speed of operation.

The electrical part is described by a sixth-order state-space model taking into account the dynamics of the stator, field (f) and damper windings (k):

$$\begin{cases} V_d = R_S i_d + \frac{d}{dt}\varphi_d - \omega_R \varphi_q \\ V_q = R_S i_q + \frac{d}{dt}\varphi_q + \omega_R \varphi_d \\ V'_{fd} = R'_{fd} i'_{fd} + \frac{d}{dt}\varphi'_{fd} \\ V'_{kd} = R'_{kd} i'_{kd} + \frac{d}{dt}\varphi'_{kd} \\ V'_{kq1} = R'_{fq1} i'_{fq1} + \frac{d}{dt}\varphi'_{fq1} \\ V'_{kq2} = R'_{fq2} i'_{fq2} + \frac{d}{dt}\varphi'_{fq2} \end{cases} \quad . \tag{6}$$

Subscripts R, S represent the rotor and the stator quantities, respectively. All rotor parameters are defined in a stator frame of reference. The equivalent circuit of the model is represented in a rotor reference frame (d-q frame).
The flux equations are:

$$\begin{cases} \varphi_d = L_d i_d + L_m (i'_{fd} + i'_{kd}) \\ \varphi_q = L_q i_q + L_{mq} i_{kq} \\ \varphi'_{fd} = L'_{fd} i'_{fd} + L_{md} (i_d + i'_{kd}) \\ \varphi'_{kd} = L'_{kd} i'_{kd} + L_{md} (i_d + i'_{fd}) \\ \varphi'_{kq1} = L'_{kq1} i'_{kq1} + L_{mq} i_q \\ \varphi'_{kq2} = L'_{kq2} i'_{kq2} + L_{mq} i_q \end{cases} \tag{7}$$

where the subscript m represents the magnetizing inductance.

3.3 Fuzzy Control of the Boost Converter for Maximum Wind Power Exploitation

The maximum power tracking control is carried out by the dc/dc boost converter. A TSK (Takagi 1985) adaptive fuzzy peak power tracking controller has been adopted with the aim of extracting the maximum amount of wind energy. The controller, shown in Fig. 3, has two inputs: the measured rotor speed and the active power generated by the WT and one output: the reference duty cycle used to drive the boost converter. By acquiring and processing the inputs at each sample instant, it estimates the duty cycle corresponding to the maximum power that may be generated by the WT.

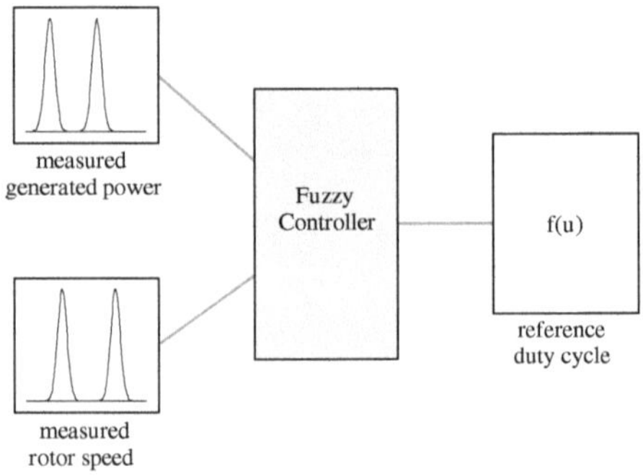

Fig. 3. Schematic diagram of the fuzzy controller for maximum power tracking

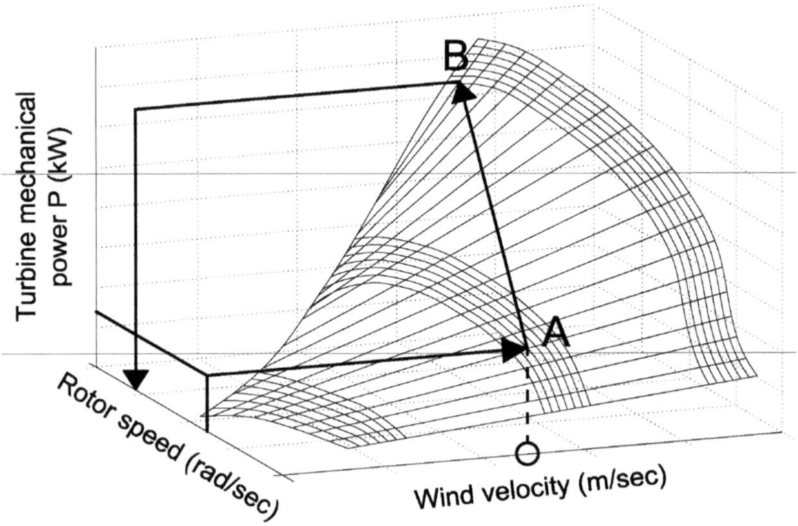

Fig. 4. Turbine power curves

The WT power curves shown in Fig. 4, illustrate the proposed adaptive fuzzy control. Starting from the point A, the controller computes the optimum operating point B according to the measured rotor speed ω_A and the measured turbine power P_A. Hence, the generator speed is controlled in order to reach the speed ω_B, allowing the extraction of the maximum power P_B from the WT, without using any wind velocity measurement.

3.4 Inverter Fuzzy Control for Voltage Regulation

Once the maximum wind power has been extracted, it is applied to the inverter, which allows exchanging real and reactive power with the grid. Vector control has been adopted using a rotating *d-q* reference frame; system phase angle is accomplished through a Phase Locked Loop (PLL).

It is worth noting that the control carried out in this stage does not work on an isolated grid due to the absence of a voltage reference and due to the practical impossibility of a satisfactory balancing of the load demand.

As well known, the use of the vector control approach guarantees a decoupling between active and reactive power. Applying Park's transformation to the three phase voltages at PPC and considering the filter R_t-L_t, follows (Borghetti et al. 2003):

$$\begin{cases} v_{gd}(t) = v_{id}(t) - \left(R_t i_{gd}(t) + L_t \dfrac{di_{gd}(t)}{dt} - \omega_n L_t i_{gq}(t) \right) \\ v_{gq}(t) = v_{iq}(t) - \left(R_t i_{gq}(t) + L_t \dfrac{di_{gq}(t)}{dt} + \omega_n L_t i_{gd}(t) \right) \end{cases} \tag{8}$$

where ω_n is the rated angular frequency.

Notice that in the reference frame synchronized with the grid voltages holds $v_{gq}(t)=0$.

Defining:

$$\begin{cases} v'_{id}(s) = v_{id} - v_{gd} + \omega_n L_t i_{gq} \\ v'_{iq}(s) = v_{iq} - \omega_n L_t i_{gd} \end{cases} \tag{9}$$

and using (8), the system can be described by:

$$\begin{cases} v'_{id} = R_t i_{gd} + L_t \dfrac{di_{gd}}{dt} \\ v'_{iq} = R_t i_{gq} + L_t \dfrac{di_{gq}}{dt} \end{cases} \tag{10}$$

The references to i_{gd} and i_{gq} are obtained by means of two FLCs. The current i_{gd_ref} is the output of the first FLC having the error between the voltage applied to the capacitor before the inverter and its reference and the integral of error itself as inputs. The second FLC has i_{gq_ref} as output and the error between the voltage at the PCC and its reference value and the integral of error itself as inputs. The control system consists also of two PI regulators, guaranteeing stability and zero steady state error for the controlled currents (i_{gd} and i_{gq}).

Since the outputs of the control system are PWM signals, it is necessary to solve (9) with respect to v_{id} and v_{iq}, thus obtaining:

$$\begin{cases} v_{id\_ref} = v'_{id\_ref} + v_{gd} - \omega_n L_t i_{gq} \\ v_{iq\_ref} = v'_{iq\_ref} + \omega_n L_t i_{gd} \end{cases} \tag{11}$$

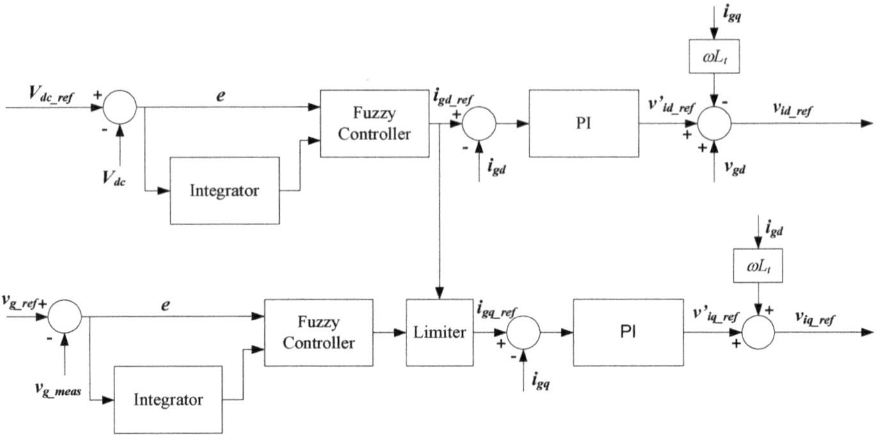

Fig. 5. Block diagram of the ac/ac control

and then to apply the relationship between the Voltage Source Converter (VSC) output voltages and the PWM modulating signals (Mohan, 1995). Fig. 5 shows the schematic diagram of the proposed controller.

In this scheme, V_{dc} and V_{dc_ref} are the measured dc voltage applied to the capacitor before the inverter and its reference value, v_{g_meas} and v_{g_ref} are the measured grid voltage at the PCC and its reference value (1 p.u.).

Considering that, for a particular combination of the active and the reactive power, (12) cannot be complied, the original value of i_{gd_ref} is kept as far as possible, while i_{gq_ref} is modified appropriately. As a result, it keeps its original sign and only its modulus is reduced according to:

$$\left| i_{gq\_ref} \right| = \sqrt{I^2_{g\_peak} - i^2_{gd\_ref}} \tag{12}$$

The application of this criterion implies that high priority is conferred to the active power reference. In fact, since active power depends on i_{gd_ref}, when modifying just i_{gq_ref} only the reactive power changes.

The implementation of the FLCs for the dc/ac converter requires an adequate knowledge base and the ability to transform the latter in a set of fuzzy rules. The knowledge base has been coded in a set of rules consisting of linguistic statements linking a finite number of conditions with a finite number of conclusions (Balazinski et al. 1995; Mohamed et al. 2008; Azli et al. 2005). Such a knowledge can be collected and delivered by human experts and expressed by a finite number ($r = 1, 2, \ldots, n$) of heuristic Multiple Input Single Output MISO fuzzy rules, written in the form:

$$\begin{aligned} &R^{(r)}{}_{\mathrm{MISO}} : \mathrm{IF}\ (x\ is\ A_i^{(r)})\ \mathrm{AND}\ (y\ is\ B_i^{(r)}) \ldots \\ &\mathrm{AND}\ (z\ is\ C_i^{(r)})\ \mathrm{THEN}\ (u\ is\ U_j^{(r)})\text{''} \end{aligned} \tag{13}$$

where $A_i^{(r)}$, $B_i^{(r)}$, ..., $C_i^{(r)}$ are the values of linguistic variables (conditions) $x, y, ...z$, defined in the universes of discourse: $X, Y, ..., Z$, respectively, and $U_j^{(r)}$ is the value of independent linguistic variable u in the universe of discourse U.

Among all the parameters associated with a FLC, membership functions (MFs) have a dominant effect in changing its performance (Mendel and Mouzouris 1997; Green and Sasiadek 2006).

The type of MFs is frequently chosen to fit an expected input data distribution or clusters and can influence both the tracking accuracy and the execution time. Triangular, trapezoidal, and Gaussian membership functions are the common choice even if any convex shape can be adopted. Even though most researchers are inclined to design the input/output fuzzy membership sets using equal span mathematical functions, these do not always guarantee the best solution.

In the proposed approach, the selection of the best membership functions has been performed on the basis of a prior knowledge and on experimentation with the system and its dynamics. In particular, triangular and Gaussian membership functions have been compared. Moreover, in order to design a FLC, shrinking span MFs have been chosen: this guarantees smoother results with less oscillations, large and fast control actions when the system state is far from the set point, and moderate and slow adjustments when it is near to the set point. Thus, when the system is closer to its set point, the fuzzy MFs, for those specific linguistic terms, have narrower spans.

The fuzzy sets of the inputs (variable error, integral of error) and of the output assume the following names: "NVB"= negative-very-big, "NB"= negative-big, "NM" = negative-medium, "NS" = negative-small, "ZE" = zero, and so forth. Triangular shapes have been chosen for input and output membership functions as they give the best results in this case. With regards to the selection of the number of fuzzy rules, implementing as many rules as possible guarantees completeness and ensure appropriate control resolution for accuracy. Nevertheless, since the type and number of MFs influences the size of fuzzy approximation error, a high number of rules may produce an overparameterized system with reduced generalization capability, degraded approximation accuracy, and increased execution time.

In general, the 'best' number of fuzzy rules depends upon the number of input variable MFs, controller and system performance, execution time, type of MF, ease of construction, and adaptability. The number and type of control rules have been obtained by carrying out a sensitivity analysis by varying the number and type of rules. A tuning process, starting from a set of initial insight and practical considerations and progressively modifying the number and type of rules allowed reaching a suitable level of performance.

A Mamdani-based system architecture has been realized using max–min composition techniques and centre of gravity methods in the inference engine and defuzzification, respectively. The FLCs have variables constructed with nine triangular MFs and 64 rules.

Inference rules logic for both the FLCs, can be derived by the control surfaces and are as the following ones: "if error is NVB and integral error is NB than the output is NVB"; "if error is ZE and integral error is PVB than the output is PM". The schematic diagram of the FLC for voltage regulation at the PCC is shown in Fig. 6 while the control surfaces of both FLCs are shown in Fig. 7.

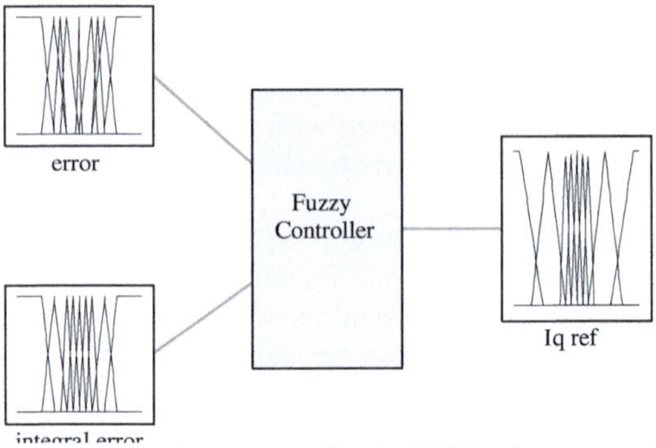

Fig. 6. Schematic diagram of the fuzzy controller for PCC voltage control

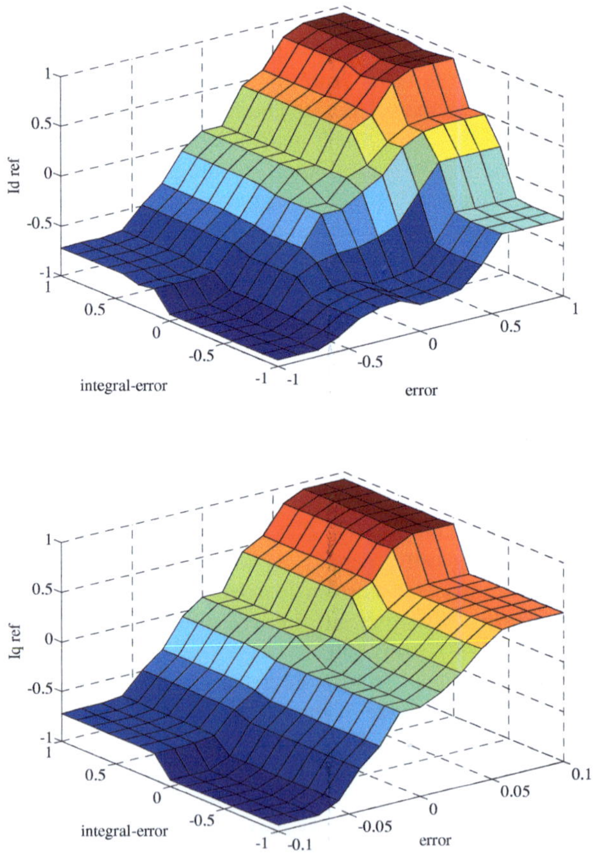

Fig. 7. Fuzzy control surfaces

4 Maximum Power Point Tracking Fuzzy Controller

This section describes the data-driven identification procedure used to identify the structure and parameters of the best TSK FLC. The procedure minimizes, on a typical wind pattern, an objective function based on the Mean Squared Error (MSE) between the duty cycle corresponding to the maximum power estimated by the FLC and the duty cycle corresponding to the maximum power that the WT can supply.

The method for the generation of the FLC is based on a Genetic Algorithms GA, fuzzy clustering (Bezdek 1981) and Recursive Least Square (RLS) procedure (Anstrom and Wittenmark 1989). The GA has a chromosome (representing an individual in a GA population) of two elements: (N_r, r), where N_r is the number of clusters and r is the spread of the membership functions.

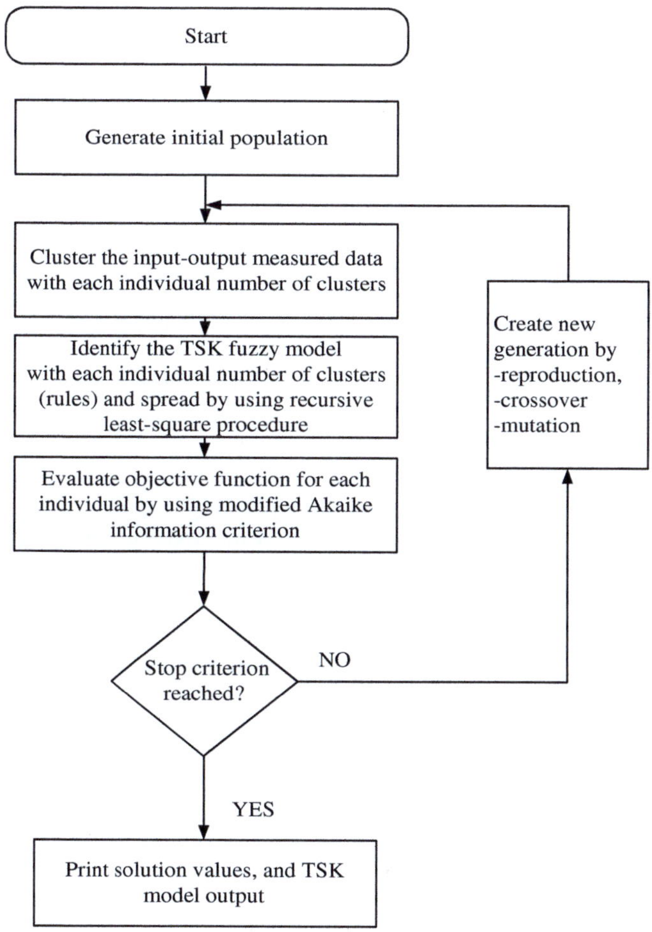

Fig. 8. Flow chart of the FLC identification algorithm

The algorithm flow chart is shown in Fig. 8: for each possible chromosome, the corresponding FLC is identified within two steps:

- a fuzzy clustering technique is applied, choosing a number of clusters equal to N_r;
- assuming the centres furnished by the previous step, the number of rules equal to N_r and the spreads of the memberships functions equal to *r*, the model parameters are identified by a RLS procedure.

Once the FLC is identified, its fitness function (i.e. the function to be minimized by the GA) is evaluated and the GA stops when the prefixed stop criterion is reached, as described in the following.

For each couple of the rotor speed and the generated power, the proposed method requires the knowledge of the duty cycle corresponding to the maximum power extractable from the WT. The generation of power curves similar to the ones shown in Fig. 4 are consequently required.

The power curves are generated considering a range of wind speeds by the following procedure (Hui 2005):

1. for each wind speed value $V_\omega(j)$, in the considered range the rotor speed is regulated to a constant value $\omega_r(i)$ by varying the dc-dc converter duty cycle;
2. the corresponding turbine power $P_m(i,j)$ is measured;
3. the rotor speed is updated to the next constant $\omega_r(i+1)$ by varying the dc-dc converter duty cycle;

The previous steps are repeated until the data of most operation points have been collected and the power curves are generated.

By using the power curves a data set of samples can be obtained. Each sample consists of two inputs (measured rotor speed and generated power) and one output (dc-dc converter duty cycle allowing maximum extractable power from the WT for the corresponding inputs).

In order to perform a partitioning of the input-output space, various approaches can be used. Among them, pattern-recognition methods of fuzzy clustering, such as fuzzy c-means (FCM) (Bezdek 1981), are suitable tools for the partitioning process.

Only for clarity, the FCM algorithm is here applied to a set of unlabeled patterns $X = (X_1, X_2, \dots, X_N)$, $X_i \epsilon R^s$ where N is the number of patterns and *S* is the dimension of pattern vectors.

The prototypes are selected to minimize the following objective function:

$$F_m(\mathbf{U},\mathbf{W}) = \sum_{j=1}^{C}\sum_{i=1}^{N}(\mu_{ij})^m d_{ij}^2 \tag{14}$$

subject to the following constraints on $\mathbf{U}$:

$$\begin{cases} \mu_{ij} \in [0,1] & i=1,\dots N \qquad j=1,\dots C \\ \sum_{j=1}^{C} \mu_{ij} = 1 & i=1,\dots N \\ 0 < \sum_{i=1}^{N} \mu_{ij} < N & j=1,\dots C \end{cases} \tag{15}$$

and where $d_{ij} = \left\| \mathbf{x_i} - \mathbf{w_j^{(t)}} \right\|$.

Where:

$\mathbf{U}$ is the membership function matrix;

$\mathbf{W}$ is cluster centre matrix;

d_{ij} is the Euclidean distance;

μ_{ij} is the value of the membership function of the *i*th pattern belonging to the *j*th cluster;

m is a constant to control fuzziness or amount of clusters overlapping;

C is the number of clusters;

N is the number of patterns in $\mathbf{X}$;

$\mathbf{w_j^{(t)}}$ is the cluster centre of the *j*th cluster for the *t*th iteration.

The objective function measures the quality of partitioning that divides a dataset into C clusters by comparing the distance from pattern $\mathbf{x_i}$ to the current candidate cluster centre $\mathbf{w_i}$ with the distance from pattern $\mathbf{x_i}$ to other candidate cluster centres.

Equation (14) describes a constrained optimization problem, which can be converted to an unconstrained optimization problem by using the Lagrangian multiplier technique.

To minimize the objective function under fuzzy constraints, given a fixed number of clusters C, m and a small positive constant, ε, the FCM algorithm, starts with a set of initial cluster centres, or arbitrary membership values, then it generates randomly a fuzzy *c*-partition and set iteration number t=0.

A two-step iterative process works as follows: given the membership values $\mu_{ij}^{(t)}$, the cluster centre matrix $\mathbf{W}$ is calculated by:

$$w_j^{(t)} = \frac{\sum_{i=1}^{N} (\mu_{ij}^{(t-1)})^m x_i}{\sum_{i=1}^{N} (\mu_{ij}^{(t-1)})^m} \qquad j=1,\dots C \tag{16}$$

Given the new cluster centres $\mathbf{W}^{(t)}$, the membership values $\mu_{ij}^{(t)}$ are updated by:

$$\mu_{ij} = \frac{1}{\sum_{l=1}^{C}\left(\frac{d_{ij}}{d_{il}}\right)^{\frac{2}{(m-1)}}} \qquad \begin{matrix} i = 1,\ldots N \\ j = 1,\ldots C \end{matrix} \tag{17}$$

where if $d_{ij} = 0$ then $\mu_{ij} = 1$ and $\mu_{ij} = 0$ for $l \neq j$. The process stops when $\left\|\mathbf{U}^{(\mathbf{t})} - \mathbf{U}^{(\mathbf{t-1})}\right\| \leq \varepsilon$, or a predefined number of iterations is reached. With this approach, clustering becomes the basis of the fuzzy model identification algorithm, inducing the rule-base structure of the fuzzy model.

Using such a method, data samples are organized in clusters, each of which is associated to a centre by using the FCM algorithm; in this manner the TSK model is based on a set of fuzzy IF-THEN rules, extracted by using the FCM clustering technique.

The set of IF-THEN rules have the following form (Yen and Wang 1998):

$$R_h : \text{IF } x_1 \text{ is } A_h^1 \text{ and} \ldots \text{and } x_p \text{ is } A_h^p \text{ THEN } y \text{ is } f_h(x), \quad h = 1,\ldots,R \tag{18}$$

where:

$$f_h(x) = a_{0h} + a_{1h}x_1 + a_{2h}x_2 + \cdots + a_{ph}x_p \tag{19}$$

in which $x_{1,\ldots,p}$ are the input variables, y is the output variable, $A_h^{1,\ldots,p}$ are the fuzzy sets, $f_h(x)$ is as a linear function of the inputs and R is the number of rules.

The *h-th* fuzzy rule of the collection describes the local behaviour associated to the fuzzy input region characterized by the antecedent of the fuzzy rule.

For each input $\tilde{x}$, the inferred value of the TSK model, is calculated as:

$$\tilde{y} = \frac{\sum_{h=1}^{R} A_h(\tilde{x}) * f_h(\tilde{x})}{\sum_{h=1}^{R} A_h(\tilde{x})} = \frac{\sum_{h=1}^{R} \tau_h * f_h(\tilde{x})}{\sum_{h=1}^{R} \tau_h} \tag{20}$$

where the degree of firing of each rule τ_h, for the current input $\tilde{x}$, is determined by the Gaussian law, which ensures the greatest possible generalization:

$$\tau_h = e^{-\alpha\left\|\tilde{x} - x_h^*\right\|^2}, \qquad h = 1,2,\ldots,R; \tag{21}$$

where x_h^* is the centre of a rule, $\alpha = 4/r^2$ and r is a positive constant defining the zone of influence of the rule.

Since the consequent of each rule is linear, its parameters, which minimize the overall error between the FLC and the system being modeled, can be determined recursively by using a least-squares procedure. The FLC has to be identified by the following two sub-tasks:

1. identification of the antecedent part of the model, by determining the centres (x_h^*, $h=1,2,\ldots,R;$) and spreads (r) of the membership functions;
2. identification of the parameters (a_{ih}, $i=0,1,\ldots p;$ $h=1,2,\ldots,R;$) of the consequent part by least squares technique.

The GA fitness function to be minimized is based on the modified Akaike Information Criterion (AIC) in which a penalty for the over fitting is introduced so that the complexity of the fuzzy model is determined by the number of fuzzy rules in the model and not only by the number of antecedent and consequent parameters of the rules.

In order to balance the reduction of the fitting error and the increasing model complexity, the modified AIC, tends to minimize both the output error and the order of the model. It can be defined as in (Yen and Wang 1998):

$$AIC = N_c log(MSE) + 2m_p \tag{22}$$

where N_c is the number of data samples, m_p is the number of parameters in the model and the MSE exhibited by the identified model, when used for estimation of an input-output association for a given data set. Based on the considerations, the single parameter m_p in the above statistical information criteria is replaced and a "complexity function" defined as:

$$s(m_a, m_c, m_r) = m_a + m_c + cm_r \tag{23}$$

where:

- m_a is the number of antecedent parameters, m_c is the number of consequent parameters,
- m_r is the number of fuzzy rules constituting the model,
- c is a constant which allows the user incorporating heuristics regarding the relative importance for reducing the number of fuzzy rules.

Therefore the GA fitness function has been defined as follows:

$$f(N,r) = MSE + \frac{2(m_a + m_c + cm_r)}{N_c} \tag{24}$$

where the MSE is written as:

$$MSE = \frac{\sum_{k=1}^{N_C}\left(Dutycycle_k^{estimated} - Dutycycle_k^{P\max}\right)^2}{N_c} \tag{25}$$

where $Dutycycle_k^{P\max}$ and $Dutycycle_k^{estimated}$ are the duty cycles corresponding to the maximum extractable power obtained from the power curves and the FLC, for each time step k, respectively.

The GA randomly generates the initial population of individuals and, at each generation, creates a new set of improved individuals by selecting individuals according to their fitness. The selection mechanism uses a normalized geometric ranking scheme. After the new population is selected, each genetic operator is applied to selected individuals for a discrete number of times. These are: simple crossover and binary mutation; an elitism mechanism, ensuring the best member of the population is not lost, is adopted, too. The iteration process continues until the stop criteria are reached: then the process stops either when the improvement of the best objective function value is below a threshold for a given number of generations or if the total number of generations is higher than a maximum number.

5 Case Study

5.1 Network and WTs Description

A 2 MW wind turbine, connected to a 25 kV distribution system exporting power to a 120 kV grid through a 30 km, 25 kV feeder is considered (Mathworks 2007). WT parameters are summarized in Table 1; using a per unit (p.u.) representation. The base for power and voltage are 2/0.9 MVA and 730 V, respectively.

Table 1. Parameters of the wind turbine and of the synchronous generator system

Rated wind turbine mechanical output	2 MW
Generator type	Three-phase synchronous generator
Generator rated power	2/0.9 MVA
Generator rated phase to phase voltage (rms)	730 V
Stator resistance	0.006 p.u
Leakage inductance	0.18 p.u.
Synchronous reactance (d-axis)	1.305 p.u
Synchronous reactance (q-axis)	0.474 p.u.
Transient reactance (d)	0.296 p.u.
Sub-transient reactance (d-axis)	0.252 p.u.
Sub-transient reactance (q-axis)	0.243 p.u.
Base frequency	60 Hz;
Number of poles	2
Wind turbine inertia constant	4.32 sec
Friction coefficient	0.01 p.u.

5.2 Maximum Power Point Tracking TSK Fuzzy Control Performances

Three different wind data sets acquired by the Wind Engineering Research Field Laboratory (http://www.winddata.com/) were used: wind speed time histories consist of 50.000 observations (data samples) within a 2.000 seconds interval, with a sampling rate fixed to 25 Hz.

The WT power curves were generated by means of the previously described procedure and by using the Matlab/SimPowerSystems models of the wind system. The FLC for the dc/dc converter was identified by using the data set "1", obtained considering the wind profile shown in Fig. 9, as training set.

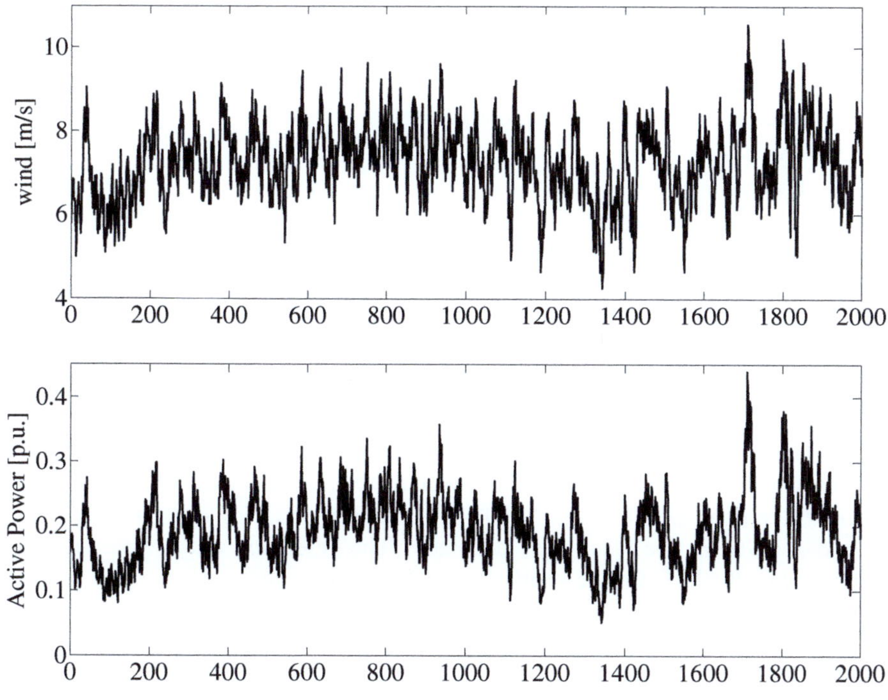

Fig. 9. Wind pattern and active power generated for the training set

In this application the Euclidean norm was used for clustering, the fuzzyness parameter, m, was set equal to 2.0, the value of ε was set to 10^{-5} and the maximum number of iterations was set to 100.

For tuning the GA, a population size of 10 and a maximum number of 100 generation was set. The simulation experience and carried out sensitivity analysis evidenced that these values, for the problem under study, guarantee the convergence of the algorithm to a satisfactory solution (Galdi et al. 2009).

For mapping, each term is set on the domain of the corresponding linguistic variable; the Gaussian MFs were used.

The learning procedure identified an optimal FLC characterized by 2 rules and 2 MFs with spreads equal to 0.1134. This choice guaranties a performance, expressed in terms of the MSE, equal to 1.02×10^{-5}, a percentage Maximum Absolute Error (MAE), defined as the maximum percentage absolute error over all the data set, equal to 6.07%, and a Maximum Percentage Error (MPE), defined as the MPE on maximum power estimate, equal to 0.08. The identified model has the structure and the parameters shown in table 2.

Table 2. FLC model structure

Number of rules	2
Number of membership functions	2
Membership functions	Gaussian
Spreads of the membership functions	0.1134
And method	product
Implication method	min
Aggregation method	max
Defuzzification method	weighted average

In order to evaluate the generalization capability and the accuracy of the FLC, two wind speed profiles, different from the first one were used. By comparison between the duty cycle estimated by the FLC and the effective one (Table 3) for maximum extractable power for the training and the two validation sets, a great accuracy of the FLC can be evidenced with a MSE less than 8×10^{-5}, a MPE less than 0.6% and a MAE less than 21%. The implemented FLC exhibits generalization capability as well as great accuracy. The reference duty cycle and that generated by the FLC for the validation set 1 are shown in fig. 10.

Table 3. FLC performances

	MPE%	MAE %	MSE [MW^2]
Data set 1 (training set)	0.08	6.07	1.02×10^{-5}
Data set 2 (validation set 1)	-0.55	5.58	2.63×10^{-5}
Data set 3 (validation set 2)	-0.36	21.13	7.75×10^{-5}

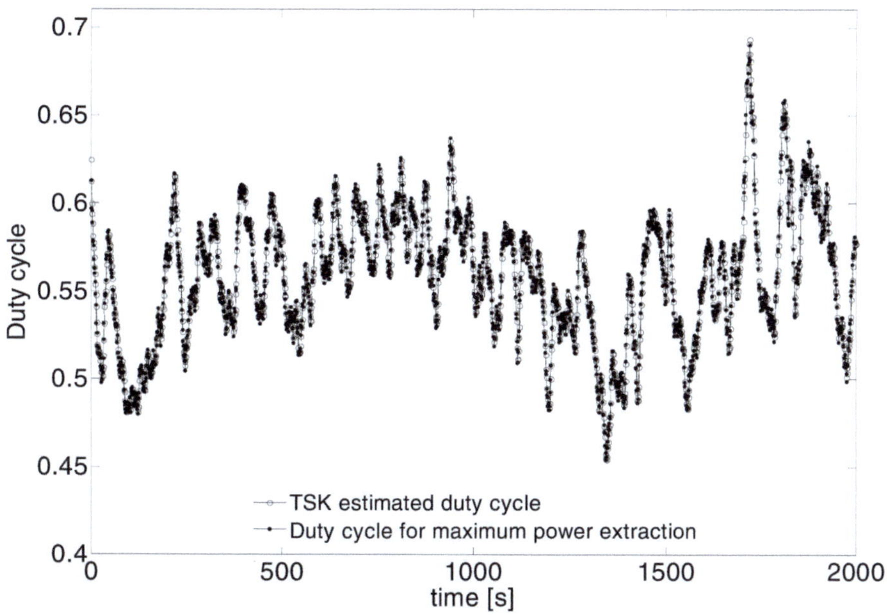

Fig. 10. Duty cycle considering the validation set 1

5.3 Network Voltage Variation and Voltage Control Capability of the System

This section reports the evaluation of the inverter performance. Voltage variations are imposed to the 120 kV grid as described in Fig. 11.

The wind data set acquired by the Wind Engineering Research Field Laboratory (http://www.winddata.com/) were considered: the wind speed time history consists of 17500 observations within a 700 seconds interval, being the sampling frequency equal to 25 Hz. The wind profile and the generated active power are shown in Fig. 12.

Simulation results, shown in Figs. 12 to 15 show that the inverter works correctly, varying its outputs in good agreement with the inputs and the reference values. As shown in Fig. 13, the FLC for the Point of Common Coupling voltage control tries to maintain the ac voltage at the PCC as close as possible to its rated value (1 p.u.) and compensates the high voltage grid voltage variations by modifying the amount of generated/absorbed reactive power. When the measured voltage is below the set point, the reactive power generation increases; when it is higher, the reactive power generation decreases. The proposed FLC compensates, within a certain boundary, the voltage variations at the PCC. These limits depend on the generator rating and type, as well as power converter rating.

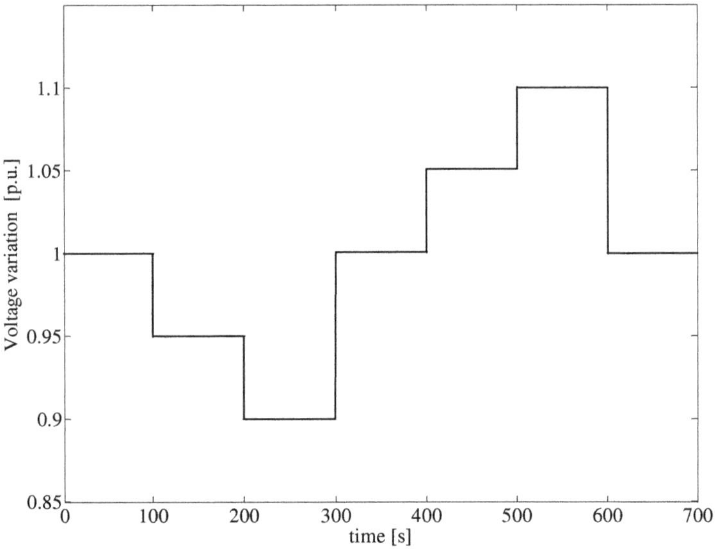

Fig. 11. Voltage variation are imposed to the 120 kV grid

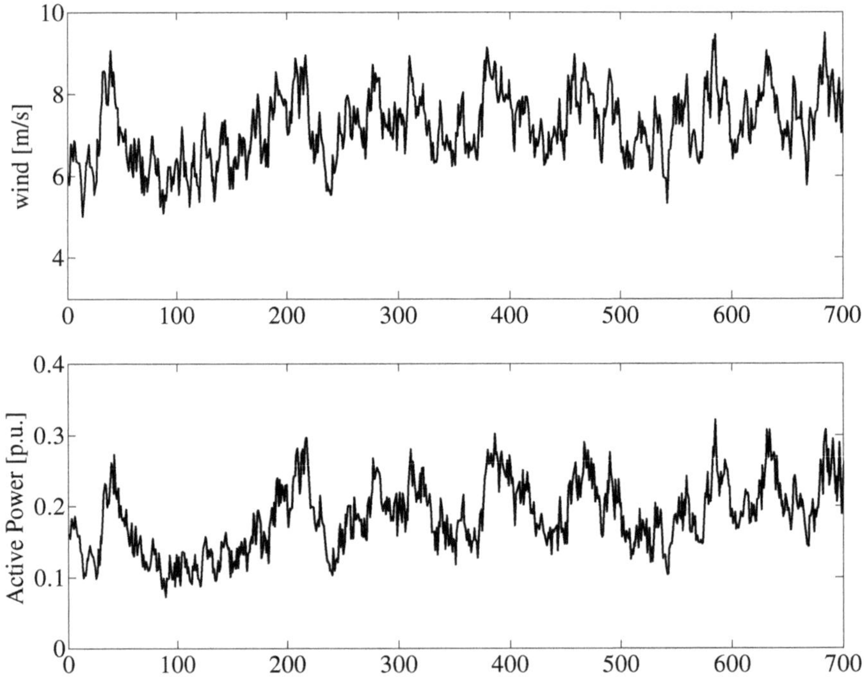

Fig. 12. Wind profile and the generated active power

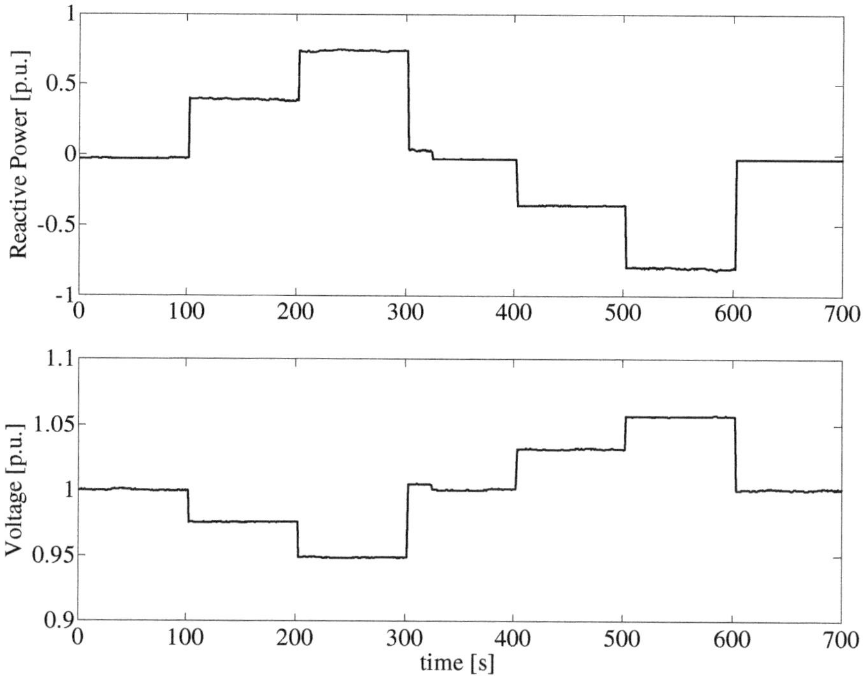

Fig. 13. Reactive power and voltage at the PCC

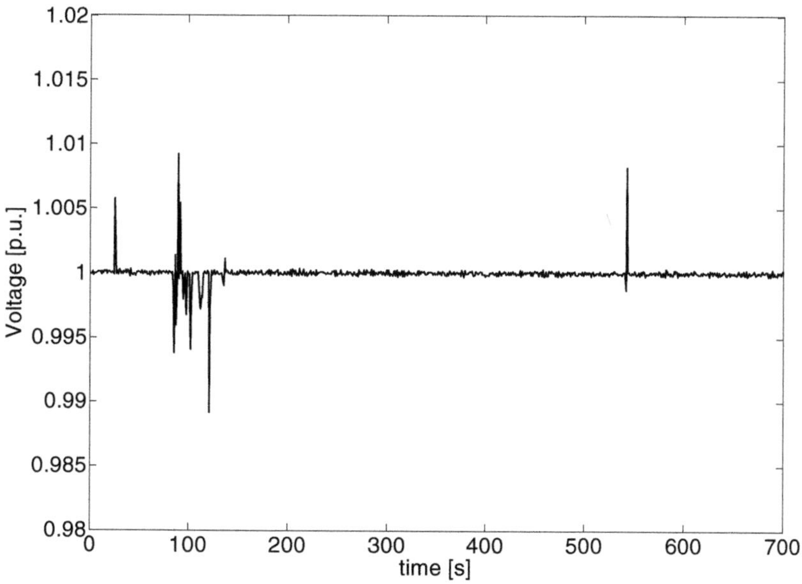

Fig. 14. Voltage on the capacitor before the inverter

The FLC controlling the dc/dc voltage regulates the dc voltage with error below 10 V (1% of the nominal value). The voltage variations around the 100 s instant, (Fig. 14), are caused by low wind speed value (around 6 m/s). As shown in Fig. 15, both the references values supplied by the inverter FLCs are tracked with high accuracy.

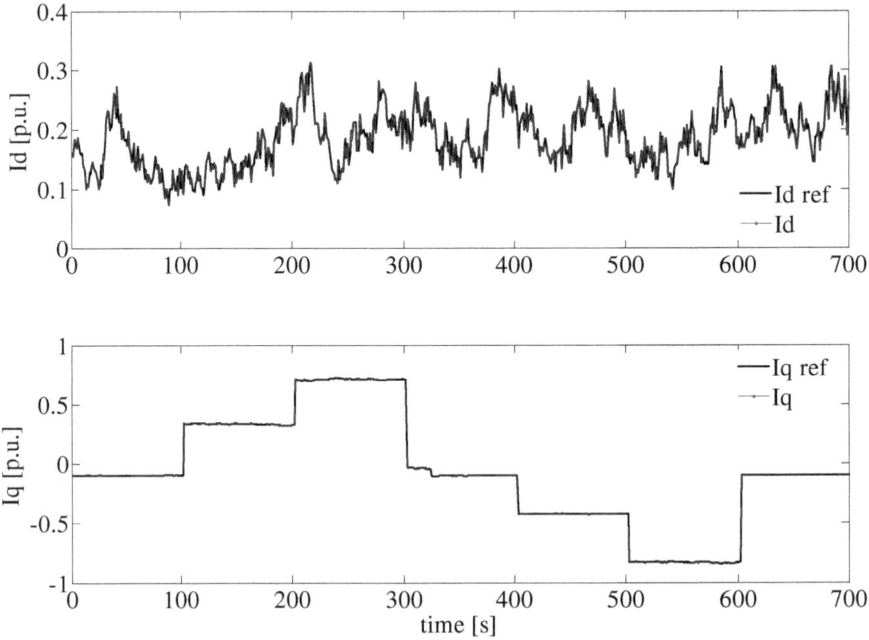

Fig. 15. Measured i_{gd} and i_{gq} and their reference values obtained by the FLCs

6 Implementation Issues and Future Work

The design of digitally controlled power converters is affected by several problems such as software portability/re-usability and very tight real-time constraints imposed in digital power converter design.

If the microprocessor capabilities are not sufficient, multiple microprocessors can be employed. In such cases, a microcontroller provides to signal acquisition, conditioning and part of the computational tasks, a more powerful microprocessor or DSP executes the main control loop. This solution is expensive and technically challenging, moreover it is not successful in case of high computational complexity.

Nowadays, Field Programmable Gate Array circuits (FPGA) are becoming more and more diffused. In fact, they allow the design of custom integrated circuits implementing complex functions and calculations in hardware, thus satisfying the high throughput demand (Tzou and Hsu 1997; Cecati et al. 2004). Additionally, they have excellent code portability as both the VHDL, the main hardware description language, and its development tools are almost device independent. One field which can obtain significant advantages by the use of FPGA is multilevel converters for wind systems, allowing to deal with high voltage and power. For instance, (Cecati et al. 2009), implements a FLC on a FPGA.

An experimental study with a real variable speed WT and synchronous generator system of 20kW is in progress (Calderaro et al. 2009). The WT is intended to be controlled by FLCs programmed on a FPGA and can be monitored by a Supervisory Control And Data Acquisition (SCADA) system aided by wireless communication.

7 Discussion

Even if the fundamental requirement for WTs is that the balance within a power system between power supply and power demand is guaranteed, many recent guidelines impose further restrictive constraints about the disconnection of wind power plants. For instance, in Italy, the recent upgrading of standard CEI 11-32, requires a system for automatic disconnection of WTs from the grid. This requirement is due to the need of modulation of energy production in case of abnormal network state. Obviously, if this requirement is justified for the distribution system operator (that must ensure proper system operation) it clashes with the needs of the independent power producer that under favorable conditions should limit output power.

In the presented chapter, the capabilities of variable speed WTs, equipped with a voltage FLC, to compensate the grid voltage drop and to keep the voltage at its reference value have been evaluated.

Dynamic simulations evidenced that variable speed WT, endowed with the proposed FLC for PCC voltage regulation, is capable of smoothing, within certain limits, terminal voltage. Unfortunately, in many cases this implies higher costs related to power electronic converters with a rating higher than that for operating at unity power factor.

The fuzzy control system has been completed with a FLC that assures maximum power extraction from the WT. With regards to the maximum power tracking FLC, simulations have demonstrated that it allows maximum wind power exploitation with a high degree of accuracy. It is characterized by a low memory occupancy, fault tolerance and learning capability. Moreover it overcomes some disvantages of classical control methods and well adapts to be implemented on FPGA. Its on site adaptivity and learning ability can be significant for WT manufacturers and can represent an alternative to the lookup table approach, often used for maximum power tracking. FLCs are active simultaneously and are suitable to be applied to other speed variable WTs with different WEACS.

8 Conclusions

In this chapter the use of fuzzy logic is discussed and a novel fuzzy logic approach for variable speed WTs control has been proposed. The fuzzy logic control consists of three FLCs. The first one is a TSK adaptive fuzzy peak power tracking controller which aims at maximum wind power exploitation by determining the reference duty cycle used to drive the dc-dc boost converter. The other two FLCs allow the exchanging of real and reactive power with the grid, meeting the reference power furnished by the dc/dc boost converter and also regulating the voltage at the PCC. The benefits deriving from the implementation of the proposed FLCs have been confirmed by simulated case studies.

References

Adzic, E., Ivanovic, Z., Adzic, M., Katic, V.: Optimum fuzzy logic control of induction generator in wind turbine application. In: Proc. of 6th International Symposium on Intelligent Systems and Informatics, Subotica, pp. 1–5 (2008)

Anstrom, K.J., Wittenmark, B.: Adaptive Control. Addison-Wesley, Reading (1989)

Azli, N.A., Wong, S.N.: Development of a DSP-based Fuzzy PI Controller for an Online Optimal PWM Control Scheme for a Multilevel Inverter. In: Proc. of International Conference on Power Electronics and Drives Systems, vol. 2, pp. 1457–1461 (2005)

Balazinski, M., Czogala, E., Gravelle, S.: Automatic tool selection using a fuzzy decision support system. In: Proc. of IEEE International Conference on Fuzzy Systems, vol. 2, pp. 615–620 (1995)

Bezdek, J.C.: Pattern Recognition with Fuzzy Objective Function Algorithms. Plenum Press, New York (1981)

Borghetti, A., Caldon, R., Guerrieri, S., Rossetto, F.: Dispersed generators interfaced with distribution systems: dynamic response to faults and perturbations. In: Proc. of IEEE Power Tech Conference Proceedings, vol. 2 (2003)

Boukhezzar, B., Siguerdidjane, H., Hand, M.: Nonlinear control of variable speed wind turbines for load reduction and power optimisation. In: Proc. of 44th AIAA aerospace sciences meeting and exhibit, Nevada (2006)

Bueno, E.J., CÓbreces, S., RodrÍguez, F.J., HernÁndez, A., Espinosa, F.: Design of a Back-to-Back NPC Converter Interface for Wind Turbines With Squirrel-Cage Induction Generator. IEEE Trans. On Energy Conversion 23(3), 932–945 (2008)

Calderaro, V., Galdi, V., Piccolo, A., Siano, P.: A fuzzy controller for maximum energy extraction from variable speed wind power generation systems. Electric Power Systems Research 78, 1109–1118 (2008)

Calderaro, V., Galdi, V., Piccolo, A., Siano, P.: Grid Impact And Integration Of Micro Wind Generation. In: Proc. of 20th Int. Conf. on Electr. Distr. CIRED, vol. 1 (2009)

Cecati, C., Dell'Aquila, A., Liserre, M., Ometto, A.: A fuzzy-logic-based controller for active rectifier. IEEE Trans. on Industry Applications 39(1), 105–112 (2003)

Cecati, C., Dell'Aquila, A., Lecci, A., Liserre, M., Monopoli, V.G.: FPGA-based multilevel modulations for H-bridge-based converters. In: Proc. of IEEE International Symposium on Industrial Electronics, vol. 2, pp. 95–962 (2004)

Cecati, C., Dell'Aquila, A., Lecci, A.: Implementation Issues of a fuzzy-logic-based three-phase active rectifier employing only voltage sensors. IEEE Trans. on Industrial Electronics 52(2), 378–385 (2005)

Cecati, C., Piccolo, A., Siano, P.: Multilevel Inverters and Fuzzy Logic for Fuel Cells Power Conditioning and Control. IEEE Trans. on Industry Applications (2009) (submitted)

Chen, W.L., Hsu, Y.Y.: Unified voltage and pitch angle controller for wind-driven induction generator system. IEEE Trans. on Aerospace and Electronic Systems 44(3), 913–926 (2008)

Chen, Z., Gomez, S.A., McCormick, M.: A fuzzy logic controlled power electronic system for variable speed wind energy conversion systems. In: Proc. of Eighth International Conference on Power Electronics and Variable Speed Drives, IET, London, pp. 114–119 (2000)

Chinchilla, M., Arnaltes, S., Burgos, J.C.: Control of permanent magnet generators applied to variable-speed wind-energy systems connected to the grid. IEEE Trans. on Energy Conversion 21(1), 130–135 (2006)

El-Hawary, M.E.: Fuzzy theory in electric power systems. In: El-Hawary, M.E. (ed.) Electric power applications of fuzzy systems, pp. 7–11. IEEE Press, New York (1998)

El Mokadem, M., Courtecuisse, V., Saudemont, C., Robyns, B., Deuse, J.: Fuzzy logic supervisor-based primary frequency control experiments of a variable-speed wind generator. IEEE Trans. on Power Systems 24(1), 407–417 (2009)

Freris, L.L.: Wind Energy Conversion Systems, pp. 182–184. Prentice-Hall, Englewood Cliffs (1990)

Galdi, V., Piccolo, A., Siano, P.: Designing an adaptive fuzzy controller for maximum wind energy extraction. IEEE Trans. on Energy Conversion 28, 559–569 (2008)

Galdi, V., Piccolo, A., Siano, P.: Exploiting maximum energy from variable speed wind power generation systems by using an adaptive Takagi-Sugeno-Kang fuzzy model. Energy Conversion And Management 50, 413–421 (2009)

Green, A., Sasiadek, J.Z.: Heuristic Design of a Fuzzy Controller for a Flexible Robot. IEEE Trans. on Control Systems Technology 14(2), 293–300 (2006)

Gupta, T., Bourdreaux, R.R., Nelms, R.M., Hung, J.Y.: Implementation of a fuzzy controller for DC-DC converters using an inexpensive 8-b microcontroller. IEEE Trans. on Industrial Electronics 44(5), 661–669 (1997)

Jasinski, M., Liserre, M., Blaabjerg, F., Cichowlas, M.: Fuzzy logic current controller for PWM rectifiers. In: Proc. of 28th Annual Conference of the IEEE Industrial Electronics Society, pp. 1300–1305 (2002)

Johnson, K.E., Pao, L.Y., Balas, M.J., Fingersh, L.J.: Control of variable-speed wind turbines: standard and adaptive techniques for maximizing energy capture. IEEE Control Systems Magazine 26(3), 70–81 (2006)

Kana, C.L., Thamodharan, M., Wolf, A.: System management of a wind energy converter. IEEE Trans. on Power Electronics 16(3), 375–381 (2001)

Kaur, K., Chowdhury, S., Chowdhury, S.P., Mohanty, K.B., Domijan, A.: Fuzzy logic based control of variable speed induction machine wind generation system. In: Proc. of Power and Energy Society General Meeting, pp. 1–11. IEEE, Pittsburgh (2008)

Knight, A.M., Peters, G.E.: Simple wind energy controller for an expanded operating range. IEEE Trans. on Energy Conversion 20(2), 459–466 (2005)

Hilloowala, R.M., Sharaf, A.M.: A rule-based fuzzy logic controller for a PWM inverter in a stand alone wind energy conversion scheme. IEEE Trans. on Industry Applications 32(1), 57–65 (1996)

Li, H., Shi, K.L., McLaren, P.G.: Neural-network-based sensorless maximum wind energy capture with compensated power coefficient. IEEE Trans. on Industry Applications, vol 41(6), 1548–1556 (2005)

Leithead, W.E., Connor, B.: Control of variable speed wind turbines: design task. International Journal of Control 73(13), 1189–1212 (2000)

Mattavelli, P., Rossetto, L., Spiazzi, G., Tenti, P.: General-Purpose fuzzy controller for DC-DC converters. IEEE Trans. on Power Electronics 12(1), 79–86 (1997)

Mendel, J.M., Mouzouris, G.C.: Designing Fuzzy Logic Systems. IEEE Trans. on Circuits and Systems 44(11), 885–895 (1997)

Mirecki, A., Roboam, X., Richardeau, F.: Architecture Complexity and Energy Efficiency of Small Wind Turbines. IEEE Trans. on Industrial Electronics 54(1), 660–670 (2007)

Mohamed, S.N.F., Azli, N.A., Salam, Z., Ayob, S.M.: Fuzzy Sugeno-type fuzzy logic controller (SFLC) for a Modular Structured Multilevel Inverter (MSMI). In: Proc. of Power and Energy Conference, pp. 599–603 (2008)

Mohan, N., Undeland, T., Robbins, W.P.: Power electronics: converters, applications and design, pp. 200–225. John Wiley & Sons, USA (1995)

Muljadi, E., Butterfield, C.P.: Pitch-controlled variable-speed wind turbine generation. IEEE Trans. on Industry Applications 37(1), 240–246 (2001)

Portillo, R.C., Prats, M.M., Leon, J.I., Sanchez, J.A., Carrasco, J.M., Galvan, E., Franquelo, L.G.: Modeling Strategy for Back-to-Back Three-Level Converters Applied to High-Power Wind Turbines. IEEE Trans. on Industrial Electronics 53(5), 1483–1491 (2006)

Saetieo, S., Torrey, D.A.: Fuzzy logic control of a space vector PWM current regulator for three-phase power converters. IEEE Trans. on Power Electronics 13(3), 419–426 (1998)

Senjyu, T., Kaneko, T., Yona, A., Urasaki, N., Funabashi, T., Yamada, F.: Output power control for large wind power penetration in small power system. In: Proc. of Power Engineering Society General Meeting, pp. 1–7. IEEE, Tampa (2007)

Simoes, M.G., Farret, F.A.: Alternative Energy Systems: Design and Analysis with Induction Generators. CRC Press Taylor & Francis Group (2007)

Simoes, M.G., Bose, B.K., Spiegel, R.J.: Fuzzy logic based intelligent control of a variable speed cage machine wind generation system. IEEE Trans. on Power Electronics 12(1), 87–95 (1997a)

Simoes, M.G., Bose, B.K., Spiegel, R.J.: Design and performance evaluation of a fuzzy-logic-based variable-speed wind generation system. IEEE Trans. on Industry Applications 33(4), 956–965 (1997b)

Song, S.H., Kang, S., Hahm, N.K.: Implementation and control of grid connected AC–DC–AC power converter for variable speed wind energy conversion system. In: Proc. of Appl. Power Electron. Conf. Expo., vol. 1, pp. 154–158 (2003)

Tafticht, T., Agbossou, K., Cheriti, A., Doumbia, M.L.: Output power maximization of a permanent magnet synchronous generator based standalone wind turbine. In: Proc. of IEEE ISIE, Montreal, pp. 2412–2416 (2006)

Takagi, T., Sugeno, M.: Fuzzy identification of systems and its applications to modeling and control. IEEE Trans. on Systems, Man and Cybernetics 15, 116–132 (1985)

The MathWorks, SimPowerSystems For Use with Simulink, User's Guide Version 4 (2007)

Tzou, Y., Hsu, H.: FPGA Realization of Space-Vector PWM Control IC for Three-Phase PWM Inverters. IEEE Trans. on Power Electronics 12(6), 953–963 (1997)

Valtchev, V., Bossche, A., Ghijselen, J., Melkebeek, J.: Autonomous renewable energy conversion system. Renewable Energy 19(1), 259–275 (2000)

Yen, J., Wang, L.: Application of statistical information criteria for optimal fuzzy model construction. IEEE Trans. on Fuzzy Systems 6(3), 362–372 (1998)

Yamamura, N., Ishida, M., Hori, T.: A simple wind power generating system with permanent magnet type synchronous generator. In: Proc. of IEEE Int. Conf. Power Electron. Drive Syst., vol. 2, pp. 849–854 (1999)

http://www.winddata.com/

Application of TS-Fuzzy Controller for Active Power and DC Capacitor Voltage Control in DFIG-Based Wind Energy Conversion Systems

S. Mishra, Y. Mishra, Fangxing Li, and Z.Y. Dong

Abstract. This chapter focuses on the implementation of the TS (Tagaki-Sugino) fuzzy controller for the Doubly Fed Induction Generator (DFIG) based wind generator. The conventional PI control loops for mantaining desired active power and DC capacitor voltage is compared with the TS fuzzy controllers. . DFIG system is represented by a third-order model where electromagnetic transients of the stator are neglected. The effectiveness of the TS-fuzzy controller on the rotor speed oscillations and the DC capacitor voltage variations of the DFIG damping controller on converter ratings is also investigated. The results from the time domain simulations are presented to elucidate the effectiveness of the TS-fuzzy controller over the conventional PI controller in the DFIG system. The proposed TS-fuzzy controller can improve the fault ride through capability of DFIG compared to the conventional PI controller.

S. Mishra
The Department of Electrical Engineering at IIT Delhi, India
e-mail: sukumar@ee.iitd.ac.in

Y. Mishra
The School of ITEE, The University of Queensland,
Australia and presently also a visiting scholar at
The University of Tennessee, Knoxville, TN, USA
e-mail: ymishra@itee.uq.edu.au; ymishra@utk.edu

Fangxing Li
The Department of Electrical Engineering, The University of Tennessee,
Knoxville, TN, USA
e-mail: fli6@utk.edu

Z.Y. Dong
The Department of Electrical Engineering, Hong Kong Polytechnic University, Hong Kong
e-mail: zydong@ieee.org

L. Wang et al. (Eds): *Wind Power Systems*, Green Energy and Technology, pp. 367–382.
springerlink.com

1 Introduction

Recently there has been a growing amount of interest in wind energy conversion systems (WECS). Among various other techniques of wind power generation, the doubly fed induction generator (DFIG) has been popular because of its higher energy transfer capability, low investment and flexible control [1]. DFIG is different from the conventional induction generator in a way that it employs a series voltage-source converter to feed the wound rotor. The feedback converters consist of Rotor side converter (RSC) and Grid side converter (GSC). The control capabilities of these converters give DFIG an additional advantage of flexible control and stability over other induction generators.

The dynamic behavior of DFIG has been investigated by several authors in the past. A third order model for transient stability using PSS/E has been reported in [2]. Furthermore, the detailed model of the grid connected DFIG has been presented in [3] whereas the modal analysis has been discussed in [4]. The change in modal properties for different operating conditions and system parameters is discussed in [4]. However, the detailed model for the converters and the controllers was either neglected or overly simplified. The performance of decoupled control of active and reactive power of DFIG is presented in [5]. The control methods for DFIG to make it work like a synchronous generator and the fault ride through behavior have been reported in [6] and [7] respectively.

The DFIG control strategy is based on conventional Proportional Integral (PI) technique which is well accepted in the industry. The decoupled control of DFIG has following controllers namely P_{ref} , V_{sref} , V_{dcref} and q_{cref}. These controllers are required to maintain maximum power tracking, stator terminal voltage, DC voltage level and reactive power level at GSC respectively. However, the intelligent controllers like fuzzy and neural network controllers, capturing the system operators' experience, outperform the conventional PI controllers and have been reported in the past [8-16]. The TS-fuzzy logic control has been successfully applied for UPFC in [17] for a multi machine power system.

The fuzzy logic approach provides the design of a non-linear, model free controller and hence, can be used for the coordinated control of RSC and GSC in the DFIG system. The Mamdani type fuzzy logic controller may not be able to provide superior control over a wide range of operation [18]. Instead, a Takagi-Sugeno (TS) type fuzzy controller can provide a wide range of control gain variation by utilizing both linear and non-linear rules in the consequent expression of the fuzzy rule base [18]. As new methods have been outlined for the design of TS fuzzy controllers, the purpose of this chapter is to highlight the application of TS fuzzy controllers to provide regulation of the active power output and DC capacitor voltage of the DFIG. The simulation results presented highlight the effectiveness of the TS-fuzzy controller in damping rotor speed oscillations and in controlling the DC voltage variations.

According to the present grid code, the wind farm should be able to ride through any fault in the system. Hence fault ride through capability is required by the system operators as mentioned in [11]. Therefore, the contributions of this chapter are: (i) to study the effectiveness of the TS-fuzzy controller on the variation of the DC voltage across capacitor and rotor speed oscillations (ii) the efficacy of the TS-fuzzy controller in improving the fault ride through capability of the system. This chapter is structured as follows: Section II presents the modeling of the DFIG system. The detailed control methodology is discussed in Section III. Section IV describes the TS-fuzzy controller and its application to the DFIG. Section V discusses simulation and results followed by conclusions in Section VI.

2 Modeling of DFIG

The grid connected single machine infinite bus system is as shown in Fig. 1. The stator and rotor voltages of the doubly excited DFIG are supplied by the grid and the power converters respectively.

Simulation of the realistic response of the DFIG system requires the modeling of the controllers in addition to the main electrical and mechanical components. The components considered include, (i) turbine, (ii) drive train, (iii) generator and (iv) the converter system.

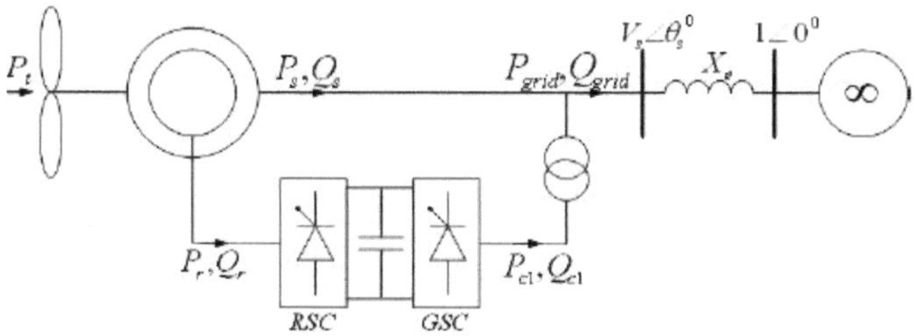

Fig. 1. DFIG system

2.1 *Turbine*

The turbine in DFIG system is the combination of blades and hub. Its function is to convert the kinetic energy of the wind into the mechanical energy, which is available for the generator. In general the detailed models of the turbine are used for the purpose of design and mechanical testing only.

The stability studies done in this chapter do not require detailed modeling of the wind turbine blades and hence it is neglected in this chapter. Inputs to the wind turbine are the wind speed, pitch angle and the rotor speed and the output from the wind turbine is the mechanical torque.

2.2 *Drive Train*

In stability studies, when the response of a system subjected to any disturbance is analyzed, the drive train system should be modeled as a series of rigid disks connected via mass-less shafts. The two-mass drive train model is considered for the stability studies of DFIG system and the dynamics can be expressed by the differential equations below [4],

$$2H_t \frac{d\omega_t}{dt} = T_m - T_{sh} \tag{1}$$

$$2H_g \frac{d\omega_r}{dt} = T_{sh} - T_e \tag{2}$$

$$\frac{d\theta_{tw}}{dt} = (\omega_t - \omega_r)\omega_B \tag{3}$$

$$T_{sh} = K\theta_{tw} + D\frac{d\theta_{tw}}{dt} \tag{4}$$

where H_t and H_g [s] are the turbine and generator inertia, ω_t and ω_r [p.u] are the turbine and DFIG rotor speed, and T_{sh} is the shaft torque, T_m is the mechanical torque and T_e is the electrical torque. θ_{tw} [rad] is the shaft twist angle, K[p.u./rad] the shaft stiffness, and D[p.u. s/rad] the damping coefficient.

2.3 Generator

The most common way of representing DFIG for the purpose of simulation and control is in terms of direct and quadrature axes (*dq* axes) quantities, which form a reference frame that rotate synchronously with the stator flux vector [3].

$$\frac{dE_q'}{dx} = -s\omega_s E_d' + \omega_s \frac{L_m}{L_{rr}} v_{dr} - \frac{1}{T_0'}[E_q' - (X_s - X_s')i_{qs}] \tag{5}$$

$$\frac{dE_d'}{dx} = s\omega_s E_q' - \omega_s \frac{L_m}{L_{rr}} v_{qr} - \frac{1}{T_0'}[E_d' + (X_s - X_s')i_{qs}] \tag{6}$$

Whereas, the parameters are defined as: $X_s = \omega_s L_{ss} = x_s + X_m$, $X_s' = \omega_s(L_{ss} - \frac{L_m^2}{L_{rr}})$ and $T_0' = \frac{L_{rr}}{R_r}$. The algebraic equations can be expressed as

$$P_s = -E_d^{'} i_{ds} - E_q^{'} i_{qs} \tag{7}$$

$$Q_s = E_d^{'} i_{qs} - E_q^{'} i_{ds} \tag{8}$$

$$E_d^{'} = -r_s i_{ds} + X_s^{'} i_{qs} + v_{ds} \tag{9}$$

$$E_q^{'} = -r_s i_{qs} - X_s^{'} i_{ds} + v_{qs} \tag{10}$$

where s is the rotor slip; P_s is the output active power of the stator of the DFIG; L_{ss} is the stator self-inductance; L_{rr} is the rotor self-inductance; L_m is the mutual inductance; ω_s is the synchronous angle speed; X_s is the stator reactance; x_s is the stator leakage reactance; x_r is the rotor leakage reactance; $X_s^{'}$ is the stator transient reactance; $E_d^{'}$ and $E_q^{'}$ are the d and q axis voltages behind the transient reactance, respectively; $T_0^{'}$ is the rotor circuit time constant; i_{ds} and i_{qs} are the d and q axis stator currents, respectively; v_{ds} and v_{qs} are the d and q axis stator terminal voltages, respectively; v_{dr} and v_{qr} are the d and q axis rotor voltages, respectively; Q_s is the reactive power of the stator of the DFIG. The voltage equations and the flux linkage equations of the DFIG are based on the motor convention.

2.4 Converter Model

The converter model in DFIG system comprises of two pulse width modulation invertors connected back to back via a dc link. The rotor side converter (RSC) is a controlled voltage source as since it injects an AC voltage at slip frequency to the rotor. The grid side converter (GSC) acts as a controlled voltage source and maintains the dc link voltage constant.

The power balance equation for the converter model can be written as:

$$P_r = P_{gc} + P_{dc} \tag{11}$$

where P_r, P_{gc}, P_{dc} are the active power at RSC, GSC and DC link respectively, which can be expressed as,

$$P_r = v_{dr} i_{dr} + v_{qr} i_{qr} \tag{12}$$

$$P_{gc} = v_{dgc} i_{dgc} + v_{qgc} i_{qgc} \tag{13}$$

$$P_{dc} = v_{dc} i_{dc} = -C v_{dc} \frac{dv_{dc}}{dt} \tag{14}$$

3 Controllers for DFIG

This section describes the controllers used for the DFIG system. As mentioned above, there are two back to back converters hence we need to control these two converter sides. Primarily, these controllers are known as RSC and GSC controllers. For controlling the aerodynamic power beyond certain point, pitch angle controller is used. This section also introduces a new auxiliary control signal which is added to the active power control loop to enhance the damping. This is known as the damping controller.

3.1 Rotor Side Converter (RSC) Controller

The RSC is used to control the wind turbine output power and the voltage measured at the grid terminals.

The power is controlled such that it follows a pre-defined power-speed characteristics, named tracking characteristic. This characteristic is illustrated by the ABCD curve in Fig. 2 superimposed to the mechanical power characteristics of the turbine obtained at different wind speeds. The speed of the turbine ωr is measured and the corresponding mechanical power of the tracking characteristic is used as the reference power for the power control loop. The tracking characteristic is defined by four points: A, B, C and D. From zero speed to speed of point A the reference power is zero. Between point A and point B the tracking characteristic is a straight line. Between point B and point C the tracking characteristic is the locus of the maximum power of the turbine (maxima of the turbine power *versus* turbine speed curves). The tracking characteristic is a straight line from point C and point D. The power at point D is one per unit (1 p.u.). Beyond point D the reference power is a constant equal to 1 p.u. The power control loop is illustrated in Fig. 3. For RSC, the d-axis of the rotating reference frame used for d-q transformation is aligned with air-gap flux. The actual electrical output power, measured at the grid terminals of the wind turbine, is added to the total power losses (mechanical and electrical) and is compared with the reference power obtained from the tracking characteristic. A Proportional-Integral (PI) regulator is used to reduce the power error to zero. The output of this regulator is the reference rotor current I_{qr_ref}, that must be injected in the rotor by the RSC. This is the current component that produces the electromagnetic torque T_{em}. The actual I_{qr} component is compared to I_{qr_ref} and the error is reduced to zero by a current regulator (PI). The output of this current controller is the voltage V_{qr} generated by the RSC. The voltage at grid terminals is controlled by the reactive power generated or absorbed by the RSC. The reactive power is exchanged with the grid, through the generator. In the exchange process, generator absorbs reactive power to supply its mutual and leakage reactance. The excess of reactive power is sent to the grid or to RSC. The control loop is shown in Fig. 4. The wind turbine control implements the V-I characteristic illustrated in Fig. 5. As long as the reactive current stays within the maximum current values ($-I_{max}$, I_{max}) imposed by the converter rating, the voltage is regulated at the reference voltage V_{ref}.

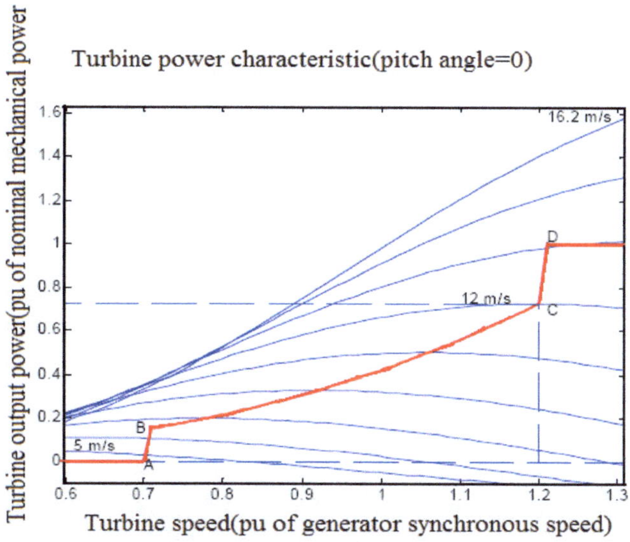

Fig. 2. Turbine characteristics and tracking characteristic

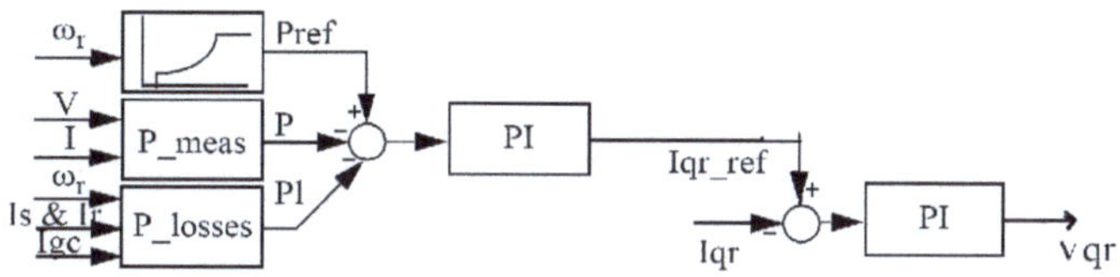

Fig. 3. RSC active power controller

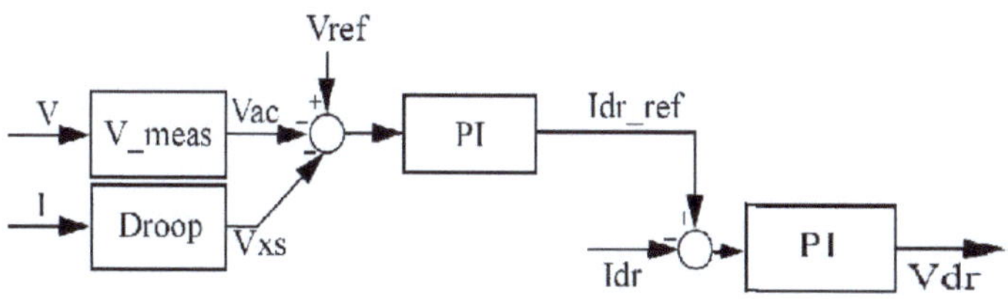

Fig. 4. RSC grid voltage controller

3.2 Grid Side Converter (GSC) Controller

The GSC is used to regulate the voltage of the DC capacitor. The control schematic is illustrated in Fig. 5. The d-axis of the rotating reference frame used for *d-q* transformation is aligned with the positive sequence of grid voltage. This controller consists of: (i) a measurement system measuring the *d* and *q* components of AC currents to be controlled as well as the DC voltage (V_{dc}); (ii) an outer regulation loop consisting of a DC voltage regulator. The output of the DC voltage regulator is the reference current I_{dgc_ref} for the current regulator (I_{dgc} is the current in phase with grid voltage which controls active powerflow); (iii) an inner current regulation loop consisting of a current regulator. The current regulator controls the magnitude and phase of the voltage generated by converter (V_{gc}) as shown in Fig. 6.

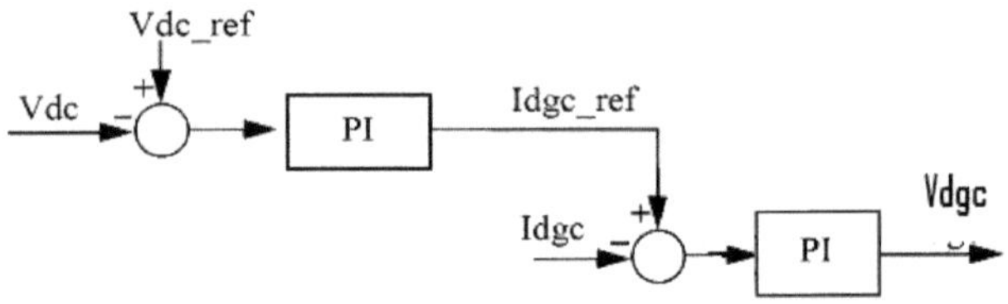

Fig. 5. Grid side converter control (DC capacitor voltage control)

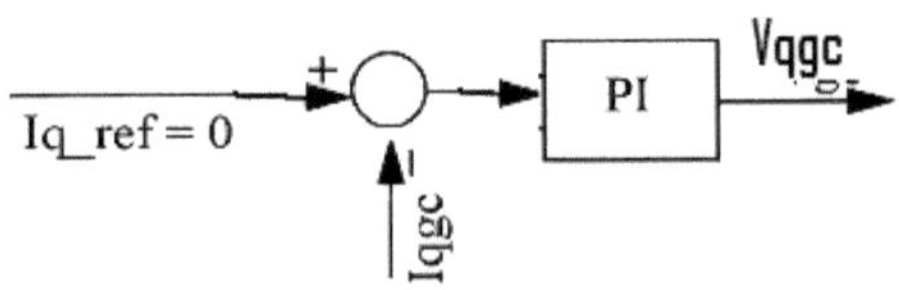

Fig. 6. Grid side converter control (Reactive power control)

3.3 *Pitch Angle Controller*

The pitch angle is kept constant at zero degree until the speed reaches point D speed of the tracking characteristic. Beyond point D the pitch angle is proportional to the speed deviation from point D speed. The construction of the pitch angle controller is shown in Fig. 7.

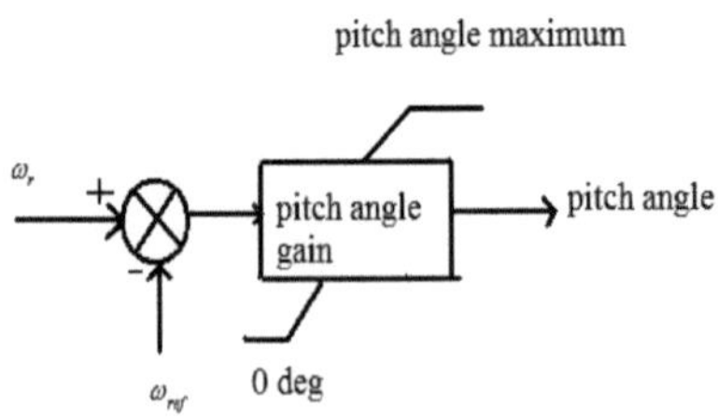

Fig. 7. Pitch angle controller

4 Design of TS Fuzzy Controller for DFIG

The fuzzy controllers are conventional non-linear controllers and can produce satisfactory results when constructed properly using the experience of the system operator. The design of fuzzy logic controller consists of (i) determining the inputs, (ii) setting up the rules and (iii) the design method for converting the rules into a crisp output signal, known as defuzzification. First of all, the input signal, in this case is the error and rate of change of error signal, is measured and depending on the crisp value of the signal, it can be expressed in terms of the degree of membership of the fuzzy sets. The shape of the fuzzy sets can be determined by the expert knowledge of the system. The next step is to construct the fuzzy rules, again based on the expert knowledge of the control problem, to accommodate all the possible combinations of memberships.

The TS-fuzzy controller differs from the Mamdani-fuzzy in its rule consequent. The linguistic rule consequent is made variable by means of its parameters. As the rule consequent is variable, the TS fuzzy control scheme can produce an infinite number of gain variation characteristics. In essence, the TS fuzzy controller is capable of offering more and better solutions to a wide variety of non-linear control problems. The quadrature-current component of the RSC, i_{qr-ref}, and the direct-current component of the GSC, $i_{dgc-ref}$, are controlled by active power deviation and DC voltage deviation respectively as shown in Fig. 8 and Fig. 9 respectively.

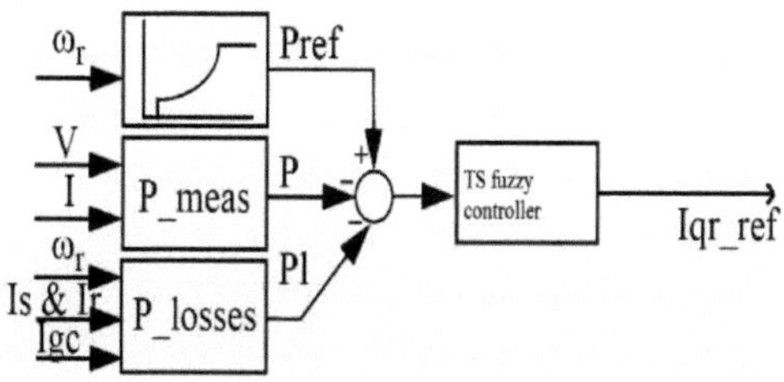

Fig. 8. Rotor side controller with TS fuzzy controller

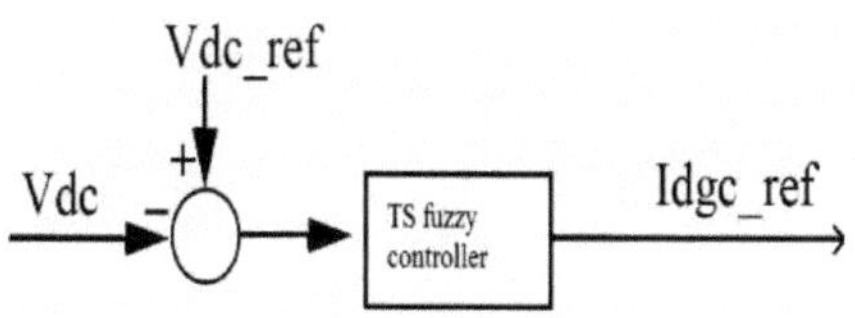

Fig. 9. Grid side converter DC voltage controller with TS fuzzy controller

The active power and DC voltage deviations are fuzzified using two input fuzzy sets P (positive) and N (negative). The membership function used for the positive set is defined by (15) and can be represented as shown in Fig. 10.

$$\mu_P(x_i) = \begin{cases} 0 & x_i < -L \\ \frac{x_i+L}{2L} & -L \leq x_i \leq L \\ 1 & x_i > L \end{cases} \tag{15}$$

Where, $x_i(k)$ denotes the input to the fuzzy controller at the kth sampling instant given by

$$x_1(k) = e(k) = P_{ref} - P \quad \text{or} \quad V_{DC-ref} - V_{DC} \tag{16}$$

$$x_2(k) = \int e(k) \tag{17}$$

The membership functions for x_1 and x_2 is shown in Fig.11. The values of L_1 and L_2 are chosen on the basis of the maximum value of real power or DC voltage error and the integral of the error. The maximum value of errors and its integral is determined observing these variations by running the programs once with the PI controllers.

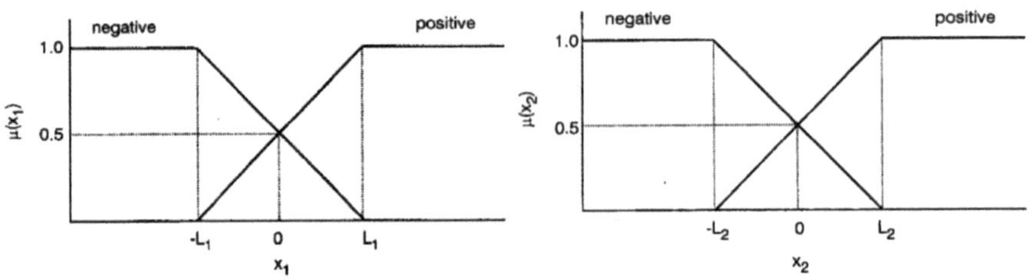

Fig. 10. Membership functions

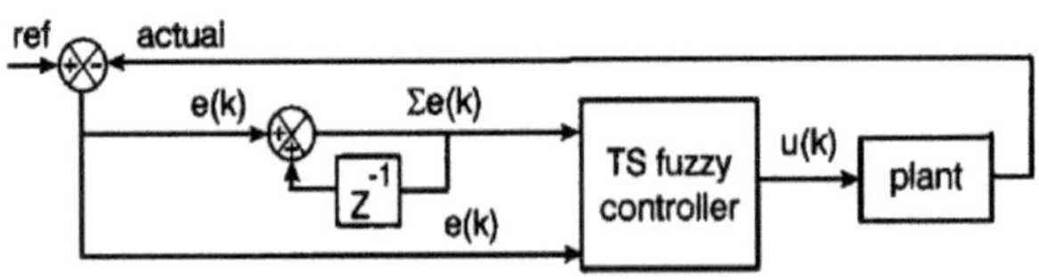

Fig. 11. TS Fuzzy control scheme with error and integral of error

The TS fuzzy controller uses the four simplified rules as

Rule 1:

If $x_1(k)$ is ***P*** and $x_2(k)$ is ***P*** then $u_1(k) = K_1(a_1x_1(k) + a_2x_2(k))$ (18)

Rule 2:

If $x_1(k)$ is ***P*** and $x_2(k)$ is ***N*** then $u_2(k) = K_2(u_1(k))$ (19)

Rule 3:

If $x_1(k)$ is ***N*** and $x_2(k)$ is ***P*** then $u_3(k) = K_3(u_1(k))$ (20)

Rule 4:

If $x_1(k)$ is ***N*** and $x_2(k)$ is ***N*** then $u_4(k) = K_4(u_1(k))$ (21)

In the above rule base u_1, u_2, u_3, and u_4 represent the consequent of the TS fuzzy controller. The output of the TS fuzzy controller is defined as follows:

$$u(k) = \frac{\sum_{j=1}^{4} \mu_j u_j(k)}{\sum_{j=1}^{4} \mu_j} \quad (22)$$

5 Simulation Results and Discussion

The TS fuzzy controller was implemented in DFIG power control and DC voltage control, in MATLAB/SimPower Systems environment, while the parameters of the wind turbine (WT) with DFIG are given in the Appendix. The parameters of all the other controllers are taken from the MATLAB DFIG system model and are modified to improve the response of rotor speed and DC oscillations. Using hit and trial method, the parameters of active power and DC voltage control loops are adjusted to achieve the lowest possible peaks for rotor and DC voltage oscillations. Then, the tuned TS-fuzzy controllers are compared with these PI parameters of the DFIG model. The Single Machine Infinite Bus (SMIB) system shown in Fig.12 is taken for the case study and the simulations are performed to verify the effectiveness of the TS fuzzy and the damping controller in improving the transient stability, the system damping and the fault ride-through capability of the WT with DFIG.

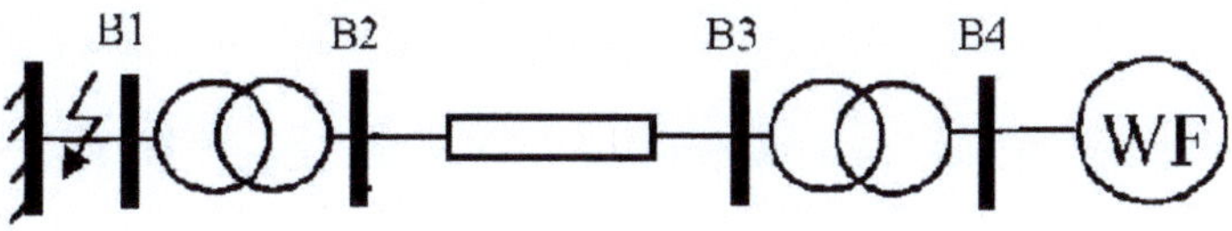

Fig. 12. SMIB systemï

5.1 Parameters of the Wind Systems

The parameters of turbine, generator, converter, controller, and TS fuzzy controller coefficients are given as follows:

Turbine data

Nominal wind turbine mechanical output power in MW =9
Pitch angle controller gain =500
Maximum rate of change of pitch angle (deg/sec) =2
Inertia constant of turbine in seconds =2

Generator data

Nominal power in MVA = 10
Nominal voltage (L-L) in volts = 575
Stator resistance in p.u. =0.00706
Stator inductance in p.u. =0.171
Rotor resistance in p.u. =0.005
Rotor inductance in p.u. =0.156
Magnetizing inductance =2.9
Inertia constant in seconds =0.4
Friction factor or damping factor in p.u. =0.01
Pair of poles (P) =3

Converter data

Converter maximum power in p.u. =0.5
Grid-side coupling inductor r =0.0015 and x=0.15, both in p.u.
Nominal DC voltage in volts =1200
DC capacitor value in mF = 60

Controller data

Grid voltage regulator gains K_P =1.25 and K_I =300
Droop X_s in p.u. = 0.02
Power regulator gains K_P=2 and K_I =10
DC bus voltage regulator gains K_P= 0.002 and K_I = 0.05
Grid side converter current regulator gains K_P=1 and K_I =100
Rotor side converter current regulator gains K_P=0.3 and K_I= 8
Damping controller proportional gain K_P =12

TS fuzzy controller coefficients

Power controller K_1 =2.5, K_2 =2.1, K_3 =1.0, and K_4 =0.5
DC voltage controller K_1 =1.0, K_2 =0.5, K_3 =5.0, and K_4 =5.0

5.2 Effect of TS Fuzzy Controller at Wind Speed 10m/s

The damping of the wind turbine with DFIG using TS Fuzzy controller and PI controller in its power control loop and DC voltage control loop is compared under 3-phase bus fault at Bus B1, which is cleared after 120 ms. The wind speed is kept constant 10 m/s. The improvement in the dynamic response of rotor speed oscillations with TS fuzzy compared to the PI controller is shown in Fig. 13. The change in the response of the real power with the implementation of the TS-fuzzy controller is hardly noticed as in Fig 14.

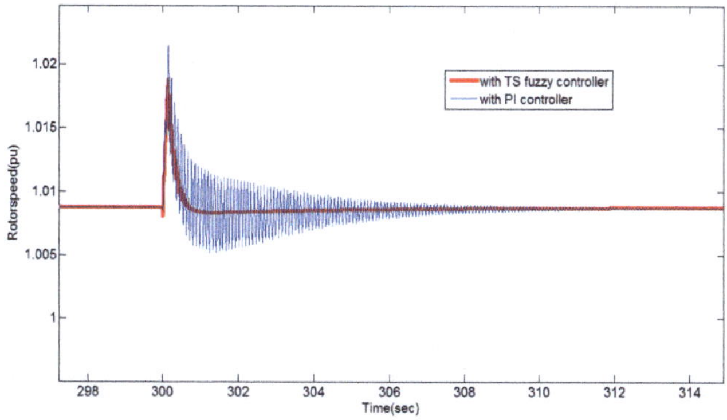

Fig. 13. Variation of rotor speed following a fault

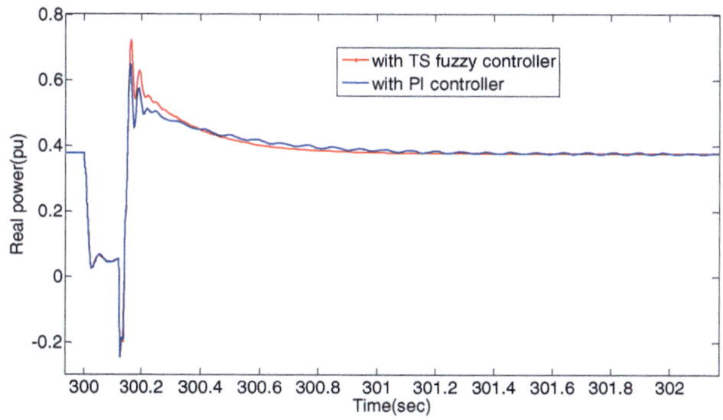

Fig. 14. Variation in electrical power output at 10 m/s for 120ms fault

The oscillations in the DC link capacitor voltage are also compared. The positive peak value of DC link voltage with the PI controller is 1600 V, whereas this is reduced to only 1400 V in the case of TS-fuzzy controller as shown in the Fig. 15. The improvement is beneficial for the operation of converters, since this reduces the stress on the RSC and GSC converters. Moreover, the oscillation/peak in the DC link capacitor voltage beyond the protection limit would trip the convertors. With the implementation of TS-fuzzy controller, the system will not reach the threshold and hence can sustain the fault for longer duration, thereby enhancing the system stability margin.

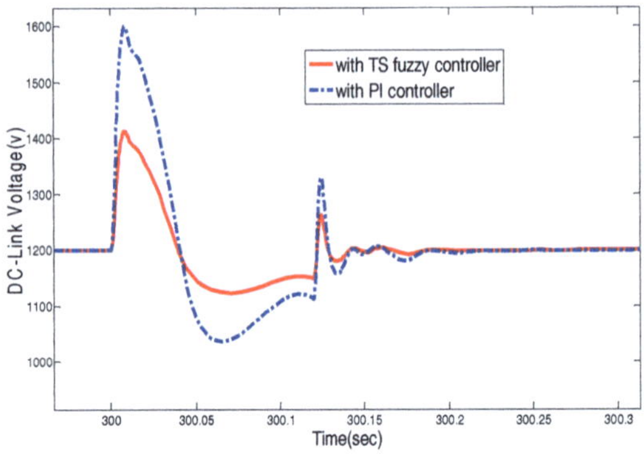

Fig. 15. DC Voltage variation at 10m/s for 120 ms fault

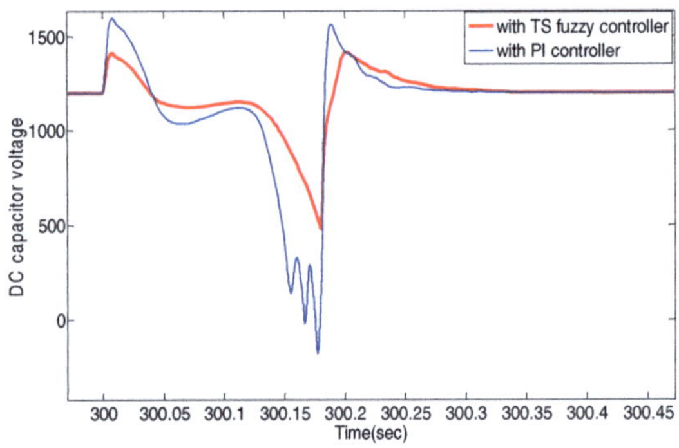

Fig. 16. DC Voltage variation at 10m/s for 180 ms fault

When the fault duration is increased to 180 ms, the DC Link capacitor voltage of DFIG with PI controller is going towards negative (-200 V) which will initiate the trip circuit to trip the DFIG from the grid. Whereas, the TS-fuzzy controller keeps the DC voltage positive thereby prevents the tripping of protection relays. This is shown in Fig. 16. Hence, with the help of TS fuzzy controller in DFIG fault ride through capability is improved.

5.3 Effect of TS Fuzzy Controller at Wind Speed 14m/s

The performance of the TS-fuzzy controller is investigated at the changes wind speed. Fig. 17 shows the comparison of the PI and the TS-fuzzy controller with the wind speed of 14 m/s. TS-fuzzy is has better response by bringing rotor speed to the steady state value quickly.

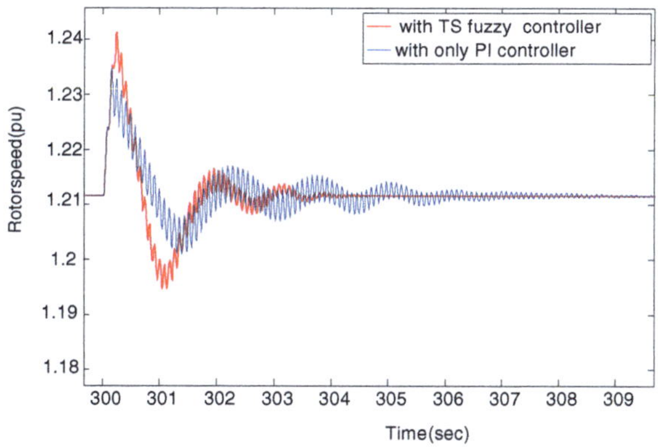

Fig. 17. Rotor speed oscillations followed by a 3 phase fault at 120kv bus.

6 Conclusion

A TS-fuzzy controller is developed for controlling the active power and DC capacitor voltage of the DFIG based WT system. It is observed that the damping of the rotor oscillations are improved with the implementation of TS-fuzzy controller compared to its counterpart PI controller. The positive and negative peak oscillations in DC capacitor voltage, following a 3 phase fault, is reduced to only 1400V and 500V in the case of TS-fuzzy controllers. Instead, these peaks are 1600V and -200V for the conventional PI controller. This reduction in the peak rise in the DC link voltage would not only help in reducing the stress on RSC and GSC convertors but would also help in designing the appropriate protection system for the reliable/secure operation of the DFIG system. =This would, in turn improve the fault ride through capability of DFIG as the system can sustain the fault for longer duration of time compared to PI controllers.

Fuzzy controllers, in contrast to the conventional PI controllers, can take care of the non-linearity in the control law and hence are known to have better performance than PI under variable operating conditions. Moreover, the TS-fuzzy is better than the mamdani type fuzzy controllers in terms of the number of fuzzy sets for the input fuzzification, number of rules used and the number of coefficients to be optimized. Therefore, in this chapter, the TS-fuzzy based controller is proposed for the active power and DC voltage control loops of the DFIG system. Furthermore, the application of these controllers for the multi machine DFIG systems would be tested and hence would be the next part of our research.

References

[1] Eriksen, P.B., Ackermann, T., Abildgaard, H., Smith, P., Winter, W., Rodriguez Garcia, J.M.: System operation with high wind penetration. IEEE Power Energy Mag 3(6), 65–74 (2005)

[2] Lei, Y., Mullane, A., Lightbody, G., Yacamini, R.: Modeling of the wind turbine with a doubly fed induction generator for grid integration studies. IEEE Transactions on Energy Conversion 21(1), 257–264 (2006)
[3] Mei, F., Pal, B.C.: Modeling snd small signal analysis of a grid connected doubly fed induction generator. In: Proc. of IEEE PES General Meeting, San Francisco, pp. 358–367 (2005)
[4] Mei, F., Pal, B.C.: Modal analysis of grid connected doubly fed induction generator. IEEE Transactions on Energy Conversion 22(3), 728–736 (2007)
[5] Yamamoto, M., Motoyoshi, O.: Active and reactive power control for doubly-fed wound rotor induction generator. IEEE Transactions on Power Electronics 6(4), 624–629 (1991)
[6] Hughes, F.M., Anaya-Lara, O., Jenkins, N., Strbac, G.: Control ofdfig-based wind generation for power network support. IEEE Transactions on Power Systems 20(4), 1958–1966 (2005)
[7] Morren, J., de Haan, S.W.H.: Ride through of wind turbines with doubly-fed induction generator during a voltage dip. IEEE Transactions on Energy Conversions 20(2), 435–441 (2005)
[8] Bansal, R.C.: Bibliography on fuzzy sett theory application to power systems (1994-2001). IEEE Trans. Power Syst. 18(4), 1291–1299 (2003)
[9] Tomsovic, K., Chow, M.Y. (eds.): Tutorial on Fuzzy Logic Applications in Power Systems. IEEE-PES Winter Meeting, Singapore (January 2000)
[10] Song, Y.H. (ed.): Modern Optimization Techniques in Power Systems. Kluwer Academic Publishers, Netherlands (1999)
[11] Song, Y.H., Johns, A.T.: Applications of Fuzzy Logic in Power Systems: Part 2. Comparison and Integration with Expert Systems, Neural Networks and Genetic Algorithms. IEE Power Engineering Journal 12(4), 185–190 (1998)
[12] Song, Y.H., Johns, A.T.: Applications of Fuzzy Logic in Power Systems: Part 3. Example Applications. IEE Power Engineering Journal 13(2), 97–103 (1999)
[13] Song, Y.H., Dunn, R.: Fuzzy Logic and Hybrid Systems, in Artificial Intelligence Techniques in Power Systems. In: Wardwick, K., Ekwue, A., Aggarwal, R. (eds.) Artificial Intelligence Techniques in Power Systems, London, UK. IEE Power Engineering Series, vol. 22, pp. 68–86 (1997)
[14] Song, Y.H., Johns, A.T.: Applications of Fuzzy Logic in Power Systems: Part 1. General Introduction to Fuzzy Logic. IEE Power Engineering Journal 11(5), 219–222 (1997)
[15] Madan, S., Bollinger, K.E.: Applications of Artificial Intelligence in Power Systems. Electric Power System Research 41(2), 117–131 (1997)
[16] Laughton, M.A.: Artificial Intelligence Techniques in Power Systems. In: Wardwick, K., Ekwue, A., Aggarwal, R. (eds.) Artificial Intelligence Techniques in Power Systems, London, UK. IEE Power Engineering Series, vol. 22, pp. 1–18 (1997)
[17] Mishra, S., Dash, P.K., Panda, G.: TS-fuzzy controller for UPFC in a multi machine power system. In: IEE proceedings on generation, transmission and distribution, January 2000, vol. 147(1), pp. 15–22 (2000)
[18] Ying, H.: Constructing non-linear variable gain controllers via the Takagi-Sugeno fuzzy control. IEEE Trans. Fuzzy Syst. 6(2), 226–235 (1998)
[19] Hughes, F.M., Lara, O.A., Jenkins, N., Strbac, G.: A power system stabilizer for dfig-based wind generation. IEEE Transactions on Power Systems 21(2), 763–772 (2006)

Fuzzy Logic as a Method to Optimize Wind Systems Interconnected with the Grid

Paulo J. Costa, Adriano S. Carvalho, and António J. Martins

Abstract. The wind systems have become more and more complex, namely due to increase of the turbines and wind farms power, what implies higher significance of either system non-linearities or system multivariable characteristics. It adds as direct consequence from wind systems high power that wind farms are becoming active elements in the wind power system. So the wind system performance and behaviour become critical elements of the grid. The performance of the turbine control is improved keeping on the power coefficient always near to maximum value, optimizing the energy extraction from the wind speed and so increasing the global system efficiency. It adds that the operation within a wind farm context demands for optimization of the whole operation. This chapter analyses wind systems biased by the increasing of wind power penetration into the grid and intends to deal with the requirement of controlling either each wind system or wind farm pointing out a methodology to design fuzzy controllers in order to optimize turbine operation and farm operation taking in accounting its operation as power system element. Some results are carried out to validate main conclusions, namely the need of including in the control process data on very short-term wind condition through adopting new techniques of real-time forecasting.

Keywords: Wind turbine, wind farm, fuzzy controllers, grid integration.

1 Introduction

Nowadays the all societies have a high level of energy demand for support their development putting on the table different energetic issues. Energy efficiency and

Paulo J. Costa
Escola Superior de Tecnologia e Gestão do Instituto de Viana do Castelo
Avenida do Atlântico
4900-348 Viana do Castelo, Portugal
e-mail: pjc@estg.ipvc.pt

Adriano S. Carvalho · António J. Martins
Faculdade de Engenharia da Universidade do Porto
Dr. Roberto Frias
4200-465 Porto Portugal
e-mail: asc@fe.up.pt, ajm@fe.up.pt

L. Wang et al. (Eds): *Wind Power Systems*, Green Energy and Technology, pp. 383–405.
springerlink.com

the use of renewable energy become two important vectors in the modern energy policy. There are three important sources of renewable energy for the planet Earth: solar, gravitational and geothermic energy. Wind energy is one indirect consequence from incident solar energy, promoting the air circulation between hot and cold zones. The kinetic energy transported in the wind can be transformed into mechanical energy by using a wind turbine. The kinetic energy of a stream of air with mass, m, and moving with a velocity, v, is given by

$$E = \frac{1}{2} m v^2 \tag{1}$$

Considering a wind rotor of cross sectional area A exposed to this wind stream the kinetic energy of the air stream available for the turbine can be expressed as

$$E = \frac{1}{2} \rho \upsilon v^2 \tag{2}$$

where ρ is the air density and υ is the volume of air parcel available to the rotor. The air parcel interacting with the rotor per unit time has a cross-sectional area equal to the rotor area (A) and thickness equal to the wind velocity (v). Hence, the energy per unit time, that is power, can be expressed as

$$P = \frac{1}{2} \rho A v^3 \tag{3}$$

Theoretically this equation shows the power available in a wind stream. However, a turbine cannot extract this power completely from the wind. When the wind stream passes the turbine, a part of its kinetic energy is transferred to the rotor and the air leaving the turbine carries the remaining away. The actual power produced by a rotor will be determined by the efficiency of this process. This efficiency is usually related to the power coefficient (C_p). The power coefficient of a turbine depends on many factors such as the profile of the rotor blades, blade arrangement and setting, etc. A designer would try to fix these parameters at their optimum level in such a way that it is possible to extract a maximum value of power in a wide range of wind velocities.

$$P_t = \frac{1}{2} \rho A C_p v^3 \tag{4}$$

The power coefficient is dependent on the ratio between the linear velocity of the blade tip ($R^*\omega_t$) and the wind velocity (v). This ratio, known as the tip-speed ratio, is defined as

$$\lambda = \frac{\omega_t R}{v} \tag{5}$$

where R is the radius of the turbine.

Is the turbine operation is designed to run at constant or variable speed. Usually, the performance of a wind turbine is expressed by means of a non-dimensional

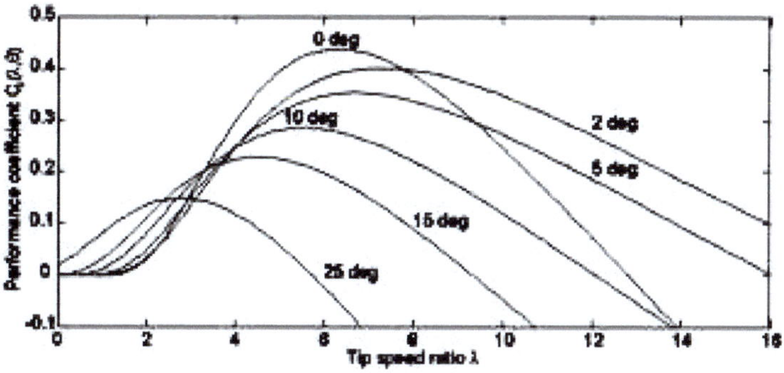

Fig. 1. Power coefficient C_p as a function of tip speed ratio, λ, and pitch angle for a specific blade

performance curve, the C_p–λ curve, from which the performance can be measured regardless of the conditions under which the turbine is operated. The wind turbine power coefficient increases with the tip ratio from zero until its maximum. From then on, it decreases for higher tip ratios [1]. The design of the turbine implies lower values of C_p for high wind speeds, a way of handling higher input power, but it presents similar behavior – low C_p when the wind speeds remain at low range. Fig. 1 depicts the relation between the power coefficient, C_p, and tip speed ratio, λ, showing its characteristics parameterized with different β angles, attack angle of the turbine rotor to air in movement [2].

Despite the great technologic development on the wind systems, the performance of the first conversion is still far from the Betz limit, 59% [2].

1.1 Wind Energy Capture Control

In order to increase the wind energy captured by the turbine, two different control strategies have been designed and implemented: pitch control and stall control. In the first of these strategies an electronic controller checks the turbine's power output at several times per second. When the power output becomes too high, it sends a signal to the blade pitch mechanism, which immediately turns the rotor blades slightly out of the wind, adapting the attack angle. Conversely, the blades are turned back into the wind whenever the wind drops again. The wind turbines with this kind of control mechanism are known as pitch controlled wind turbines. In another control strategy, the rotor blades are bolted onto the hub at a fixed angle. However, the geometry of the rotor blade profile is aerodynamically designed in such a way that it ensures that from the moment the wind speed becomes too high, it is caused turbulence on the side of the rotor blade which is not facing the wind, creating a stall which prevents the lifting force of the rotor blade from acting on the rotor. The wind turbines with this kind of control mechanism are known as stall controlled wind turbines.

Of course, modern turbines adopt the mechanism of pitch control to take advantage on controlling the output power simultaneously operating at variable speed in order to control tip speed ratio and so the power extraction for different wind speeds. Every controller is designed to extract the rated power to output depending on the wind speed.

1.2 Wind Turbines

Different types of wind turbines are available on the market: fixed and variable speed turbines with asynchronous or synchronous electrical generators. The different types of wind turbines have their own advantages and disadvantages. Fixed-speed wind turbines normally cause a voltage drop during start-up [3]. The voltage drop is mainly caused by reactive power consumption during magnetization of the generator. Another problem concerned with fixed speed wind turbines is the flicker produced during normal operation of the wind turbine. The variations in the generated power are mainly caused by the wind turbines themselves through the wind speed variations and the tower shadow effect. These variations can lead to flicker emission. Variable-speed wind turbines can reduce these power variations and eliminate the flicker caused by power pulsation. One small disadvantage of these systems is the harmonic currents produced by the inverter. However, efficient control systems associated with LCL filters produce high quality currents in the grid. The wind generation systems are completed with electrical generators coupled to turbine rotational axis either directly, for synchronous generators, or through a gear box, adapting the rotational speed, for asynchronous generators.

The wind farm is formed by a set of wind turbines with certain field geometry. This field geometry is an important parameter that has to be considered because this can introduce important wind speed attenuation. Studies show that, for turbines that are spaced 8 to 10 rotor diameters, apart in the prevailing downwind direction and 5 to 7 rotor diameters apart in the crosswind direction have typically less 10% of the initial wind velocity [4].

1.3 Grid Interconnection

Wind systems have three different phases: transform kinetic into mechanical energy, convert mechanical into electrical energy by an electrical machine and finally use an electronic converter to supply energy to the grid. So, it is important that the whole system captures the optimum quantity of energy from the wind, transform it and inject it into the grid according to the grid codes imposed by the TSO (Transmission System Operator) [5], [6]. The satisfaction of these conditions have become important objectives in the wind energy industry. During the first years of wind energy development the power of each wind turbine was small and the energy injected into the grid was insignificant compared with other electrical sources, but today they can reach up to 5 MW for only a single electrical machine; so if a wind farm has ten to fifteen wind turbines the installed power will be between 50 MW and 75 MW.

Nowadays the power systems start to face problems of integrating significant levels of wind power as it is decentralized, distributed over large extensions. So, planning, operation and control of power systems with large wind power become an important task. The main power system problems from the wind energy technology come from the lack of control on the active and reactive powers. The active and reactive power control is very important to keep the frequency and voltage stable within limits. Lack of reactive power can lead to voltage problems and no control in the active power can cause frequency deviations.

The integration of high levels of wind power in the power system can lead to some problems. In a rough approach, the integration of wind power in the power system can be analyzed at two levels, [7]:

- The installed wind power is relatively small compared to the conventional power system. In this case, the wind energy counts for a small part of the total energy production in the power system. It can be assumed that the power system has enough spinning reserve of active power and the frequency is kept constant. Therefore only voltage problems are of concerning.
- The installed wind power is comparable to the conventional power stations. The wind power counts for a large part of the total energy production in the power system. The large scale integration can cause power quality or stability problems and in some cases the frequency can be affected by the wind turbines.

In the two cases the power system must supply reliable and high quality power to the loads. To achieve reliability the power system must have reserves and controllers that can deliver the power when it is demanded, task mainly supplied by conventional generators and controllers installed throughout the power system. The voltage and frequency variations are compensated by controllers keeping the power quality within limits.

2 Wind Turbine/Farm Fuzzy Control

To increase the energy extraction requires to design the right controller able of optimizing the power coefficient Cp, especially at low wind speeds. So the control method imposes the turbine rotational speed to achieve always the maximum value of the power coefficient, being possible to be independent from the wind speed measurement and specific system characteristics. In fact, the effective wind speed and its variation on a wind turbine are very complex to be measured and so to adopt them demands for sophisticated techniques as to optimize the power extraction is a core requirement.

In order to show the algorithm validity two different wind turbines are analyzed. The first one considers the characteristics of a commercial wind turbine, with pitch control and delivering mechanical power to a synchronous generator. Fig. 2 illustrates its real operating characteristics, in terms of power coefficient and output power. The second one is also a commercial one, with pitch control and delivering mechanical power to an asynchronous generator. Fig. 3 illustrates its real operating characteristics, in terms of power coefficient and output power.

2.1 Real Performance Analysis of Two Different Wind Turbines

In the first case (Machine 1), the pitch-controlled wind turbine data is recorded and shown in fig. 2. Also, the power coefficient is calculated and displayed. In fig. 3 (Machine 2) the stall-controlled turbine is shown.

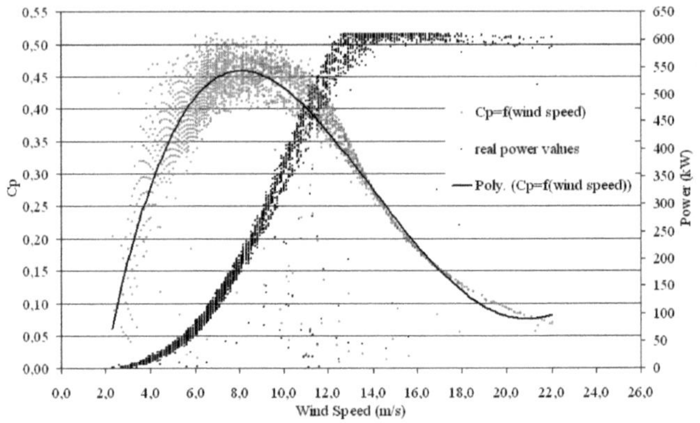

Fig. 2. Machine 1 real behavior: output power and power coefficient

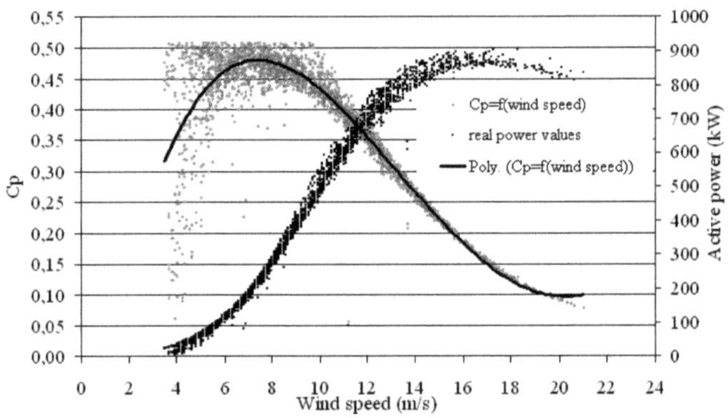

Fig. 3. Machine 2 real behavior: output power and power coefficient

Through the data analysis from the two different farms, it is possible to observe that both machines have many points away from their maximum power coefficient.

So, it can be concluded that during the operation time of these systems the converted power is not maximized. Different methods have been developed in order to maintain the power coefficient always near its maximum value, optimizing the energy extraction from the wind speed and so increasing the global system efficiency. Two different approaches are possible: first one does not apply wind speed information, it has high immunity level to the wind speed information; the second one needs to have data on wind speed to make the best control of the wind system [8], [9], [10], [11] and [12]. In this case the wind speed information is supplied to the control system by measuring the wind speed or by a forecast process. The forecast process involves many different time scales, but in this particular case the most important is the so-called very short-time (in order of a second to a few minutes) [13], [14] and [15].

2.2 Fuzzy Logic Control

Modern systems include large and complex subsystems and sophisticated technical devices that need to be controlled. This requires the construction of mathematical models. However, these models are rather cumbersome and include some vagueness. So it is hard classical mathematics that processes these models [16].

The philosophy of fuzzy logic is supported on a mathematical framework where imprecise conceptual phenomena in modeling and decision making may be precisely and rigorously studied. It lets mathematical models describe rather unmodelled situations and finds solutions of "unsolvable" problems. Fuzzy logic is applied to wind farm control with the goal of pass-through the complex, nonlinearity and uncertainty of these control systems. In [17] some limitations of conventional controllers are presented: nonlinear models are computationally intensive and have complex stability problems, a plant does not have accurate models due to uncertainty and lack of complete knowledge, uncertainty in measurements and multivariables and multiloops systems have complex constraints and dependencies.

Some benefits of fuzzy controllers are also presented: developing a fuzzy controller is cheaper than developing a model-based or other controller to do the same system, they are robust because they can cover a much wider range of operating conditions and they can operate with noise and disturbances of different natures, are customizable; indeed it is easier to understand and modify their rules, and easy to learn how fuzzy controllers operate and how to design and apply them to a real application.

2.3 Fuzzy Control and Adaptive Fuzzy Control

Generally, the operation of fuzzy controllers can be represented in a similar way as a conventional controller. A formulation in time discrete, with k=kT, where T represents the period of operation of the controller, can be:

$$u(k) = f\left[e(k), e(k-1), ..., e(k-n), u(k-1), u(k-2), ...u(k-m)\right] \tag{6}$$

being $u(k)$ the controller output and $e(k)$ the error. The function f, the control law, is described by a set of rules.

2.3.1 Fuzzy Controller

The main topologies of fuzzy logic based controllers are: PI/PID-type, Takagi-Sugeno (TS) and sliding mode type, developed from the general control law. In control processes, the most usual type is the PI/PID with or without adaptive parameters. The fuzzy PID-type controller can be used alone or also with other functions namely in hybrid configurations.

Indeed the simple linguistic model can not contain, in an explicit form, quantitative knowledge about the controlled process. This disadvantage can be eliminated with another fuzzy model type.

The Takagi-Sugeno fuzzy controller is a hybrid model, combining a fuzzy antecedent with a deterministic consequent, thus allowing the integration of qualitative knowledge and quantitative information.

2.3.2 Adaptive Fuzzy Controller

The objective of a controller is that the output of a process remains in a given reference value, despite the disturbances that might occur in system variables, or to make the output follow a particular path.

In many control systems some parameters change over time getting different values or even being not measurable in real-time condition. So, in these systems it is necessary to adapt the controller to the parametric variation and consequently keep on the system as stable. Moreover, its adaptation capacity could allow self-calibration of the controller to the process model. In this context from the beginning the controllers based on fuzzy logic have been studied to provide the ability to be adaptive controllers. In the work of Procyk and Mamdani it appears the first adaptive controller (Ref).

The tuning of a controller based on fuzzy logic presents more difficulties than the tuning of conventional controllers. The reasons for this increased complexity are that the fuzzy controller is extremely flexible, and its performance depends on a wide range of parameters, from the membership functions to the inference mechanism; as it is a nonlinear controller. It there is not a unique method for tuning and adapting fuzzy controllers. Several approaches are possible, from the knowledge base built with experts' information about the controlled process to similarities between fuzzy controllers and conventional PID controllers. The main methods aimed at designing and analyzing the dynamics and the stability of adaptive fuzzy controllers are: the fuzzy self-organising controller, the model reference, and the neuro-fuzzy method.

The method of adapting fuzzy controllers based on the architecture of neural networks uses the structure and mechanisms of learning/adaptation of a neuronal network in association with the process of fuzzy logic reasoning.

Two main approaches in model-reference fuzzy adaptive controllers are common: direct control and indirect control. In direct control, the fuzzy controller is adapted directly from the difference between the model output and the process output. In indirect control, a process parametric model is identified in order to determine the plant error. Since there are no analytical methods to direct adaptive control of nonlinear processes the self-organizing fuzzy controller uses a performance index to directly adapt the low level controller. The self-organizing controller implements two simultaneous tasks: 1- monitors the process and its environment and determines appropriate control actions (long term rulebase) and 2- uses the results of the control actions using a performance index to improve the controller behaviour.

2.4 Stability of Fuzzy Controlled Systems

Controlled systems based on fuzzy logic controllers are supported by several stability analysis methods. The formal approach to the stability analysis of systems with fuzzy controllers and the design of fuzzy controllers based on stability criteria are well studied and documented. Mainly, the stability analysis based on the Lyapunov criteria for TS-type controllers [18], and the design of direct and indirect adaptive fuzzy controllers, with guaranteed stability [19] should be referred.

The need to provide a formalism to obtain stability of the controlled process was felt from the beginning of the application of fuzzy controllers. When the systems have few variables, the graphical evolution analysis of the controlled system can be used to obtain stability information. From the several graphical analysis methods, the two more important are the vector field analysis and the circle criterion, [20].

The energy and controllability concepts have been the base of several studies of stability on dynamic fuzzy systems in association with the fuzzy logic extension principle. The controllability and stability of fuzzy dynamic systems have been studied in several ways: the characterization of the stability concept, assigning to the system degrees of stability such as "very stable", "more or less stable", etc; the introduction of the smooth controllability concept in fuzzy dynamic systems, complementary to the above one; the introduction of stability concepts based on energy variations of the fuzzy system.

This stability analysis shows some important features, such as a physical and intuitive interpretation, being applied to systems described by fuzzy relational equations; it can be applied to adaptive systems.

2.5 Application of Fuzzy Controllers in Wind Systems

2.5.1 Wind Turbine

The control algorithm has the main objective of keeping the tip speed ratio always inside of a small window near to the optimum value, independently from the existence of the pitch controller, because this has a larger time constant. So, the emphasis of the control is on the turbine rotational speed and its time behaviour due to the balance between aerodynamic power, energy from the wind, and electric power generated by the electrical generator, extracted from the turbine shaft. Fig. 4 illustrates the global framing of this controller in the wind system.

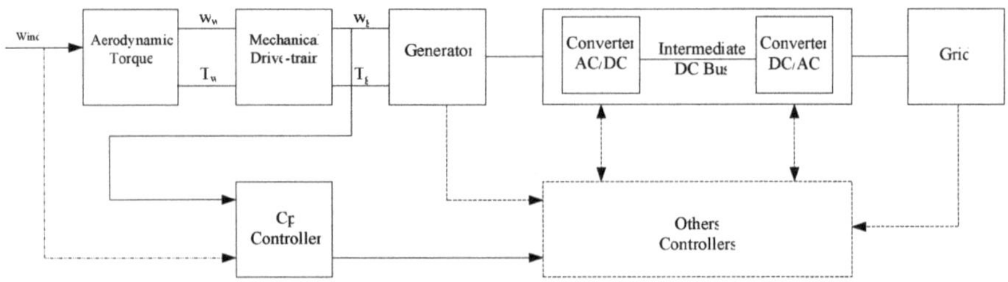

Fig. 4. Integration of power coefficient controller on the global wind control structure

The control algorithm is based in one global controller that receives the all information, prepares the input variables and sends them to its nuclear fuzzy control system. This module processes the behaviour of the input variables and finds the right solution to apply in the wind turbine speed to achieve the main goal - maximize the operation point by achieving a maximum value of the power coefficient and so increasing the power extraction. The process begins with a wind measurement (real or not) or in fault situation [21], [22].

This approach is based on the development of a fuzzy controller for the power coefficient parameter. So, an appropriate structure and rules of inference for this controller allows at optimizing the operation of the wind turbine. The structure used is presented in fig. 5. Linguistic variables are used to translate real values into linguistic values. The possible values of a linguistic variable are not numbers but so called 'linguistic terms'.

The fuzzy controller has three inputs: speed error (error), speed slope (dv) and last torque change (ldt), and one output variable (output), the change of the generator torque reference.

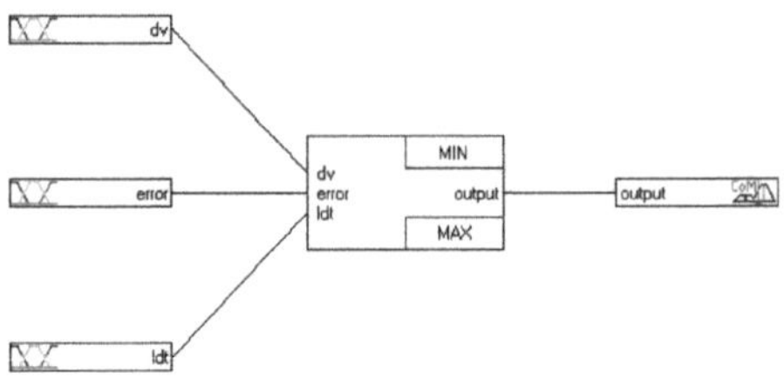

Fig. 5. Input and output variables of the fuzzy logic controller

The rules 'if' part describes the condition for which the rules are designed. The 'then' part describes the response of the fuzzy system to this condition. The degree of support is used to weigh each rule according to its importance.

The resulting control surface using the min-max and centroid operators is shown in fig. 6 and 7. These figures show a 3D surface result of the applied rules, with the two input variables on horizontal axes, while the output is plotted on the vertical axis. In order to have a three dimensional representation, the third input value is considered constant. In the first figure (fig. 6) it is considered that the last torque change has no variation, while in fig. 7 the speed slope is constant.

In this case it is possible to increase the energy extraction using the right controller, especially at low wind speeds. This control method imposes the turbine rotational speed to achieve always the maximum value of the power coefficient, independent from the wind speed measurement and specific systems characteristics.

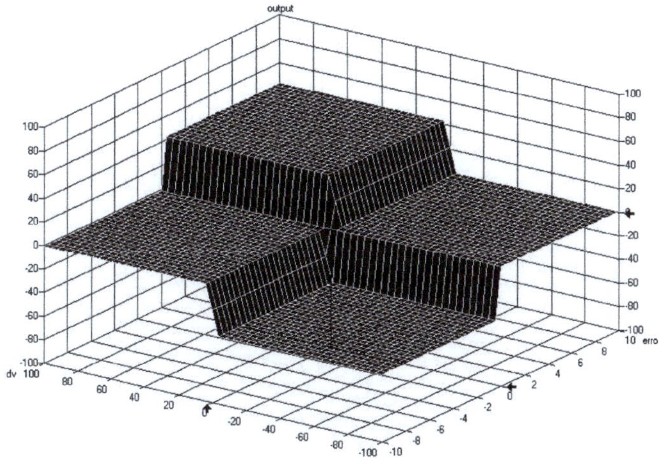

Fig. 6. Fuzzy controller surface (last torque change as a constant input)

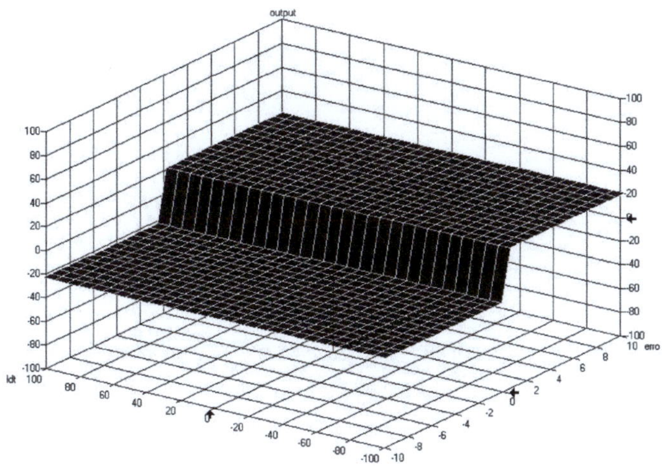

Fig. 7. Fuzzy controller surface (speed slope as a constant input)

The wind speed variation on a wind turbine is very complex and demand sophisticated techniques to optimize the power extraction. In order to show the algorithm validity the behaviour and performance of the two different commercial wind turbines described above were analyzed. The first simulation considers the characteristics of the wind turbine associated to an asynchronous generator, with results shown in fig. 8. In the second case, see fig. 9, the wind turbine associated to a synchronous generator is analyzed.

The performance of the first energy conversion depends strongly on the rotation speed; so it is important to have a control system that follows continuously the

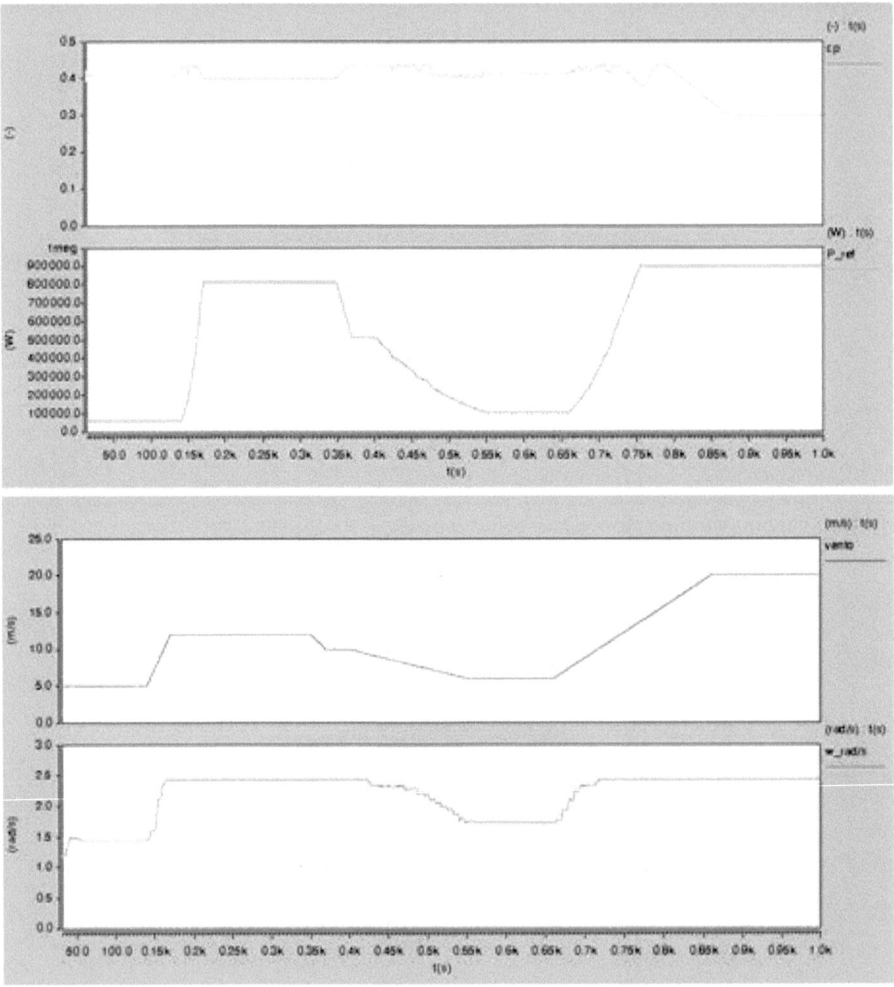

Fig. 8. The operation of the controller applied to a wind turbine MM52/900 for optimize the power coefficient. Traces from top: Cp - Power coefficient, P_ref - Power reference, vento - Wind speed and w_rad/s - wind turbine speed.

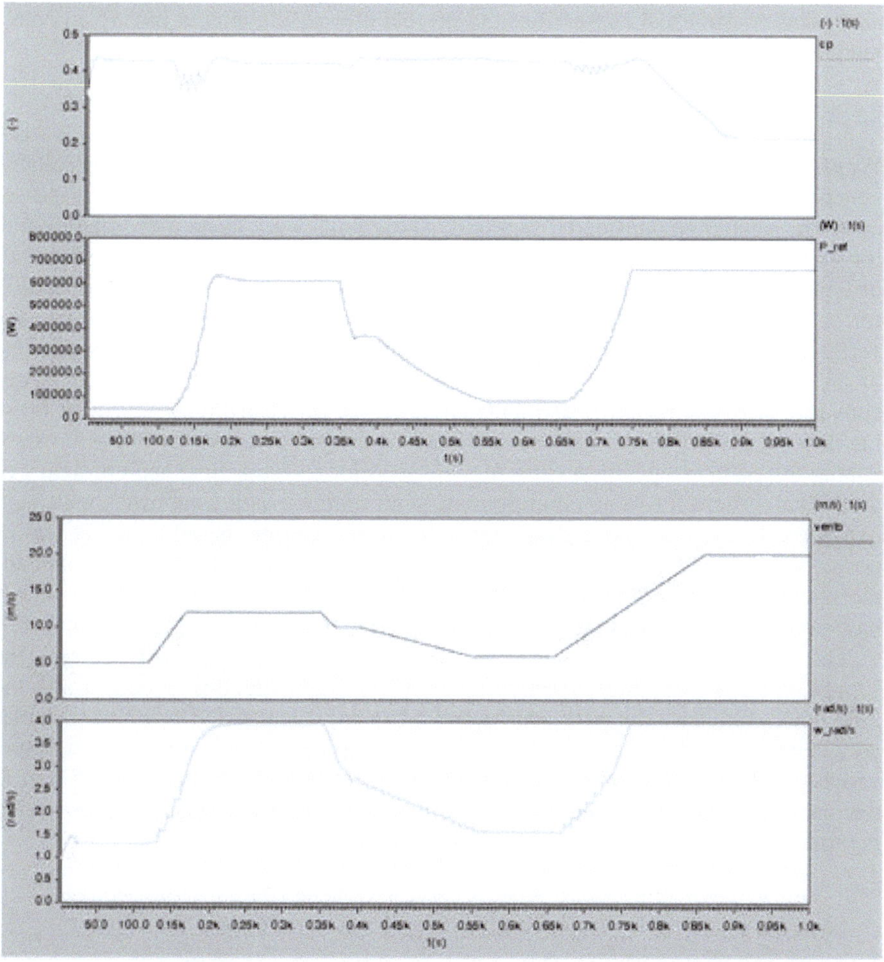

Fig. 9. The operation of the controller applied to a wind turbine E44/600 for optimize the power coefficient. Traces from top: Cp - Power coefficient, P_ref - Power reference, vento - Wind speed and w_rad/s - wind turbine speed.

evolution of the relation between these two variables, aerodynamic torque and the electromagnetic torque. The power coefficient should be kept close to its maximum value; So the relation between rotation speed of the wind turbine and the wind speed must be so constant as possible.

The control algorithm developed has assumed one proposal: all system could be maximized and optimized considering only the speed wind turbine control. The use of fuzzy logic allows solving the problems of non-linear and strong complexity of the power coefficient behaviour. The input of this algorithm is based on the relation between aerodynamic and electromagnetic torques and the effect that it has in the speed rotational wind turbine. So, this control method is not based on analytical or empirical expressions but on real effects allowing at having a control method independent from system characteristics.

2.5.2 Wind Farms

Wind farms are energy generation systems being autonomous and controllable generating units. At current technological status, each set turbine-generator-power converter should be considered as a unit. These units work as an individual system for variable speed systems. Consequently, through the controller and conditions of different wind regimens, it is possible to establish an optimum point of operation for each generation system. The connection to the grid is made by a power converter, independently from the wind system being based on a synchronous machine or an asynchronous one as electric generator. With this configuration the power converter controls the DC bus voltage and then the active and reactive power flowing into the grid. The control characteristics of the power converter allow at establishing the point of operation in rotational speed for each machine and efficiency of the global system for same conditions of wind regimen as it establishes with high dynamics the system load. Indeed power converter operates based on references of power to be generated. In this context it is demanded to establish a controller for the operation of the farm as a whole. Thus, the farm becomes a control system based on a set of the machines that integrate it running as one integrated system.

The controller implements a fuzzy system as shown in fig. 10. The first step is the sum of all the active power produced by each unit. Then with this value and with the total power reference the error and its derivative are calculated. These two variables are the inputs of a fuzzy controller that produce a control output variable. This output variable is then used to determine the output reference for each machine and sent to the machine unit control. Finally each machine unit control adjusts its operation point in order to get the reference imposed by the wind farm controller.

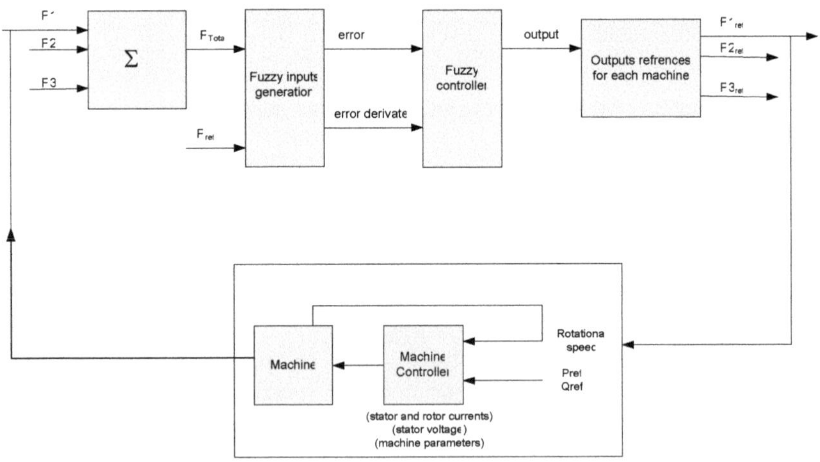

Fig. 10. Simplified structure of the wind farm control system

The approach is based on the development of a fuzzy controller for the entire farm. So, an appropriate structure and rules of inference for this controller allows at optimizing the operation characteristics of the farm. The structure used is presented in fig. 11. Linguistic variables are used to translate real values into linguistic values. The possible values of a linguistic variable are not numbers but so called “linguistic terms”. The controller takes the total error of power to be generated face to the generated power and its derivative. Thus the developed controller operates in the fuzzy surface that fig. 12 presents, with the error of power and its derivative as inputs and the zz axis as the controller output. The rule blocks contain the control strategy of a fuzzy logic system.

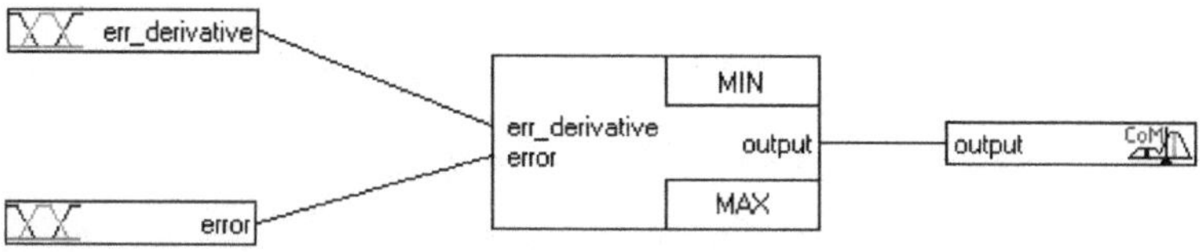

Fig. 11. Input and output variables of the wind farm fuzzy logic controller

The rules 'if' part describes the situation, for which the rules are designed. The 'then' part describes the response of the fuzzy system in this situation. The degree of support is used to weigh each rule according to its importance. For validation analysis it is considered a set of three wind turbines.

The farm controller is constituted by the following nine rules:

If error is ”N” And err_derivative is ”N” Then output is “PB”
If error is ”N” And err_derivative is ”Z” Then output is “PM”
If error is ”N” And err_derivative is ”P” Then output is “PS”
If error is ”Z” And err_derivative is ”N” Then output is “PM”
If error is ”Z” And err_derivative is ”Z” Then output is “Z”
If error is ”Z” And err_derivative is ”P” Then output is “NM”
If error is ”P” And err_derivative is ”N” Then output is “NS”
If error is ”P” And err_derivative is ”Z” Then output is “NM”
If error is ”P” And err_derivative is ”P” Then output is “NB”

with the following significance:

N - Negative	NS – Negative Small	PS – Positive Small
Z – Zero	NM – Negative Medium	PM – Positive Medium
P – Positive	NB – Negative Big	PB – Positive Big

The resulting control surface using the min-max and centroid operators is shown in fig. 12.

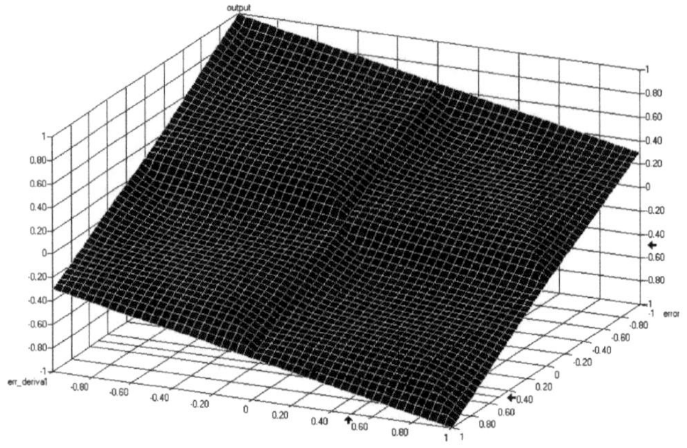

Fig. 12. Fuzzy controller surface

The power variations from wind turbines is very complex and demand more sophisticated techniques to cope with the spatial distribution of wind turbines than a simple scale up from a single wind turbine. In order to show the model validity there are presented three simulations. The farm nominal power is 1500 kW. This power characteristic causes an active power margin of 20%, eventually needed for participating in frequency control or to impose a positive *dP/dt*. In the first simulation from [0, 2] s; [4, 7] s; [11, 12] s the sum of all production is equal or bigger than 1500 kW. From t=7 s to t=10 s one machine stopped due to a mechanical problem (fig. 13). In the second simulation, fig. 14, from t=4 s to t=12 s all machines have conditions to produce nominal power of 600 kW. Finally, fig. 15 shows the simulation conditions of fig. 13 but with a grid voltage drop of 15 %.

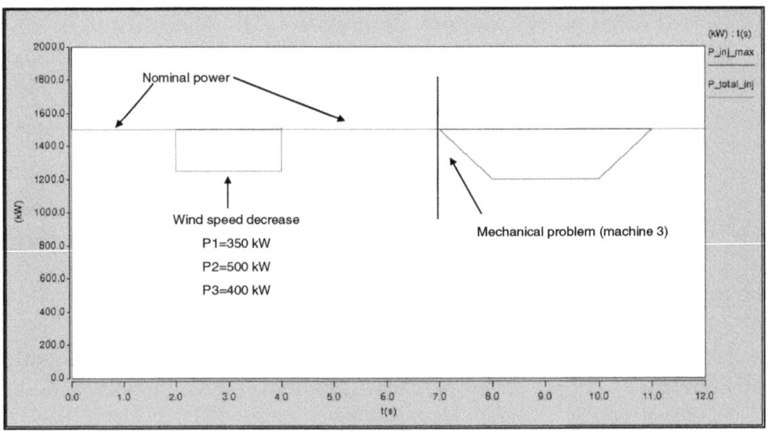

Fig. 13. The operation of the fuzzy controller applied to a wind farm of three machines. Between t=2 s and t=4 s, the wind speed decreased leading to the following generated powers: P1=350 kW; P2= 500 kW; P3= 400 kW. At t=7 s, machine 3 stopped. The new supplied powers are: P1= 600 kW; P2= 600 kW; P3=0 kW.

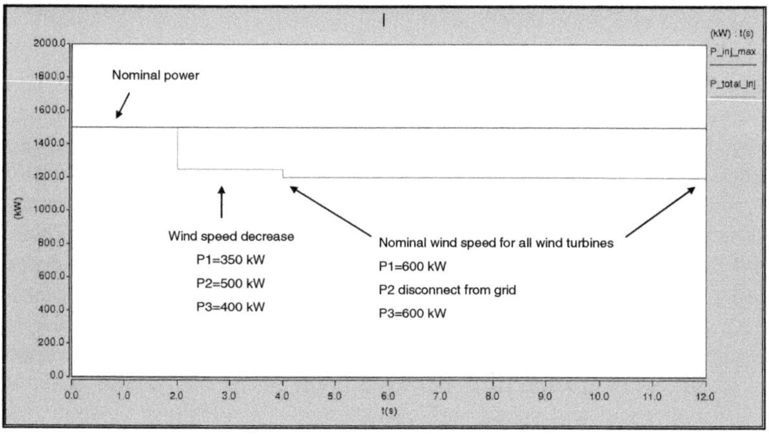

Fig. 14. The operation of a wind farm of three machines without controller. Between t=2 s and t=4 s the wind speed decreased and the new reference powers are: P1= 350 kW; P2= 500 kW; P3= 400 kW. At t=4 s, the wind speed returns to the nominal value, but the farm is unable to reach the rated power.

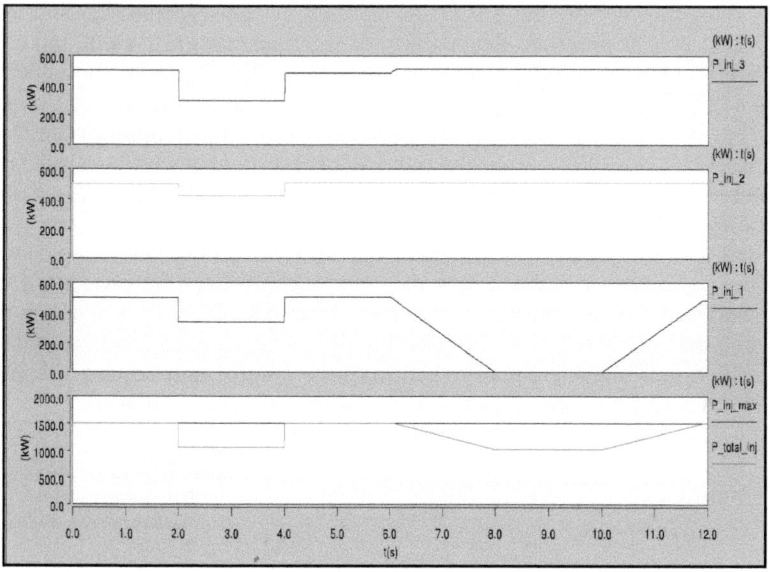

Fig. 15. Operation of the fuzzy controller applied to a wind farm of three machines with a grid voltage drop of 15%. Traces from top: P3, P2, P1, farm nominal power, total injected power. Wind speed conditions: nominal during the intervals [0, 2] s and [4, 12] s; lower during [2, 4] s. Machine conditions: machine 3 disconnect at t=6 s and reconnected at t=10 s.

So, planning operation and control of power systems with large wind power becomes new task as the wind farm becomes a grid element controllable by TSO operator. These results demonstrate the need of a whole wind farm controller. Fuzzy controller takes control on the point of operation for each machine, taking into consideration always the total generated power, with the main goal of obtaining a good quality of the generated power. It is shown that it is possible to have all wind turbines working and injecting power into the grid at the right operating point without needing to disconnect one or more wind turbines for the same power.

Wind farms are energy generation systems being autonomous and controllable generating units. At current technological status, each turbine-generator-power converter set must be considered as a unit. These units work as a single system for variable speed systems. Variable speed wind systems demand for an appropriate control of operating point at different wind regimens as well as they have to operate at different rotational speed and, consequently, at different system efficiency. Control characteristics of each machine are based on associated power converter that implements control on power flow to the grid.

In this work it is discussed the design of the controller for the wind farm. However, the electric system presents stability problems, what implies that the injection of wind energy can not be made without respecting some rules. So, it is necessary to develop a supervision controller for the whole wind farm that not only manages the operation point of each machine but also takes into account quality rules imposed by the TSO (transmission systems operators). This controller can also help to improve the stability of the transmission system and can provide information on wind power variations with the objective of reducing the degree of randomness that wind energy shows. The approach is based on the development of a fuzzy controller for the entire farm. So, an appropriate structure and inference rules allow at optimizing the operation characteristics of the farm. The first step is the sum of all the active powers produced each unit and with the TSO power indication establishes a total power reference, fig. 16. Then with this value and with the total power reference the error and its derivative are calculated. These two variables are the inputs of a fuzzy controller that produce a control output variable. This output variable is then divided for n machines, and is used to determine the output reference considering the power of each machine and sent to the machine unit control. Finally each machine control unit adjusts its operation point in order to get the reference imposed by the wind farm controller.

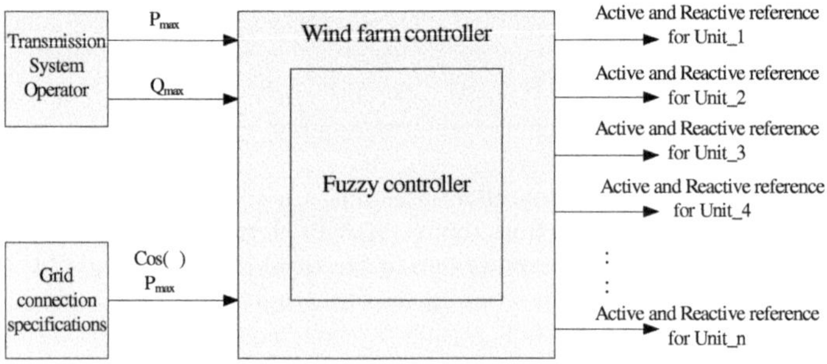

Fig. 16. Structure of active control system for a wind farm

The performance of the new controller can be analyzed through the results presented in fig. 17. In this figure, from top to bottom, it implicitly appears represented the allowed maximum power in the grid connection, 1500 kW, and the power reference received from the TSO of 1200 kW. This controller must adjust the references of each machine in order to respect this limit and simultaneously to store some capacity of energy. So, continuing from top to bottom, the curves of power generated without control (blue) and the curves of power with the controller (green) for Machine 3, Machine 2 and Machine 1 are presented respectively.

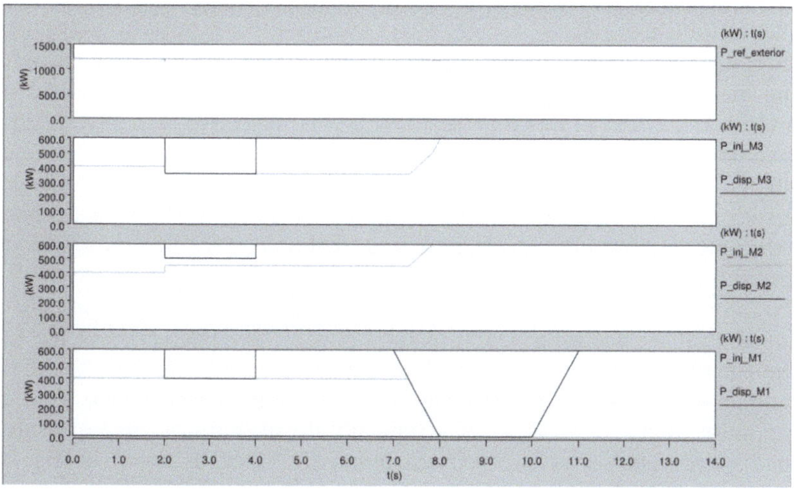

Fig. 17. Performance of a wind farm controller face to an external imposition of active power reference injection into the grid

The results obtained with this controller show that the wind farm production is optimized through non linear supervision, considering internal and external constraints. It is also possible to guarantee a well-defined output power even with a small bandwidth variation in the wind speed input. These controller characteristics are essential to build a supervision controller to integrate wind energy into power systems.

The last functionality incorporated in this controller is the capability of predicting the wind power generation in order to improve the technical and economical integration of wind energy into the electricity supply system. These controllers are being applied either to a whole wind farm or to a set of wind farms helping to control the wind energy of a region.

2.6 *Forecasting Analysis*

Electric systems present stability problems when they have to integrate a large volume of wind generation. Also, the effect of the unpredictability of wind generation output causes some problems on the dispatch process. So, it becomes necessary to improve the wind power prediction systems.

Basically, there are two main approaches to wind power prediction: statistical methods and physical ones. Usually the time horizon covered by the wind forecasts is given according to scheduling schemes of the conventional power plants and conditions on electricity markets which are typically of the order of one to two days ahead. The time horizon required for wind forecasting can be divided in different time scales: very-short time for previous in the next hour; short-term for a day-ahead; medium term extending the range for 3 to 5 days; and long-term, 5-7 days in advance for maintenance planning of large power plant components, wind turbines or transmission.

Persecuting this goal the authors present some research results on wind forecasting methods for very short-term window trying to find the standard of the wind speed behaviour in this time scale, [23]. The objective is to have in advance an accurate behaviour for the next seconds to 2 hours and to incorporate these data into the farm controller, allowing at incorporating into the controller some derivative behavior. This purpose is achieved, firstly by doing digital signal processing with the real data from the recent past, and then processing this information by a numerical method, [24].

The information of two or more seconds ahead is very important for the wind turbine controller. The information of the optimal rotational speed of the turbine can be calculated by the forecasting method and so the reference for the right operation point is obtained in advance. The first step is to measure the wind speed, getting a time series of wind speed – typically the all turbines have measurement of wind speed; The series is processed by an IIR filter, as shown in fig. 18. In order to smooth it for the dynamics required by the controller.

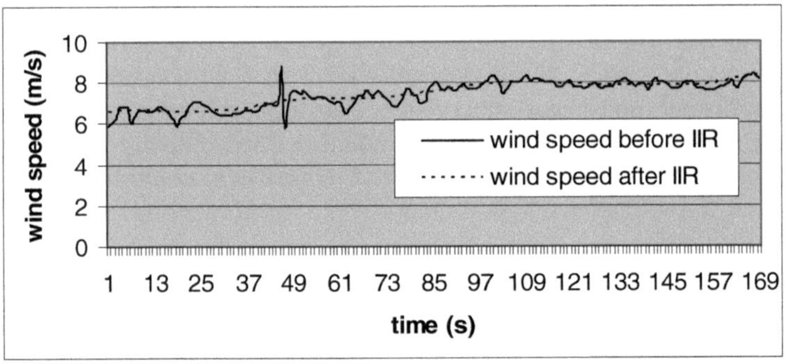

Fig. 18. Wind speed at the input and output of the IIR filter

The results show that it is possible to eliminate the all disturbances introduced by the sampling process and obtain a clean signal that maintains the main characteristics of the real wind speed within the wanted dynamics.

The second step is to apply the filtered wind speed in a forecasting method. Among different approaches, a method based on cubic splines was used, [24].

Fig. 19 shows the end sliding window and the prediction of some points ahead. It is possible to see a small difference between the real value and its prediction for few points ahead; if the number of points increases the difference become more significant. At this point it is necessary to discard the first point of the sliding window and integrate the real point of the first wind speed predicted.

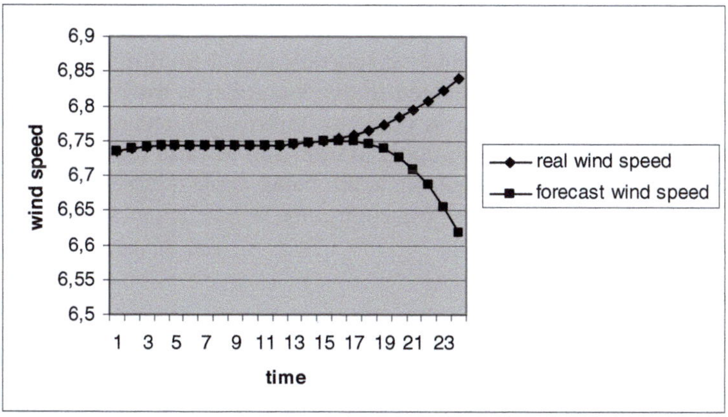

Fig. 19. Real and forecast wind speed for a window time variation

The developed approach within this time scale is better than the physical method, based on a mathematical interpolation method. The results obtained with this approach are good ones, presenting low RMS errors for the time scale from seconds to one to two hours. So, it is possible to change the approach for the turbine control from a mechanical view to an electrical perspective improving the turbine performance as well as solving some problems related to mechanical stress for a short time as the rotational speed becomes more stable. It means that to get a good wind speed prevision for very-short time becomes significant data to establish an appropriate mechanical operating point as well as its dynamics of change.

3 Conclusions

The spatial distribution of the wind turbines gives some benefits in reducing the power variations from the wind energy. But, some power systems start to face problems to integrate an increasing level of wind power. So, planning, operation and control of power systems with large wind power become a new and important task.

Nowadays the electric system demands for a whole wind farm controller such that must be developed using some kind of artificial intelligence namely fuzzy logic. In a wind farm, the fuzzy controller/supervisor takes control on the point of operation for each machine, taking always the total farm generated power into consideration, with the main aim of obtaining a high quality of the generated power. So, it is possible to have the all wind turbines working and producing energy in the right operation point without the necessity of disconnecting one or more wind turbines of the wind farm.

The results obtained with the fuzzy controller show that the wind farm production is optimized through nonlinear supervision, considering internal and external constraints. It is also demonstrated a well-defined output power, even if some wind speed dynamics occurs at the input. These controller characteristics are very important and essential to build a supervision controller to integrate wind energy in power systems, easing the planner to schedule power for different wind farms.

The last functionality incorporated in this controller is the capacity of predicting the wind power production in order to improve the technical integration of wind energy into the electricity supply system and to raise the economical value for central dispatches. These controllers are being applied either to a whole wind farm or to a set of wind farms helping control the wind energy of a region.

References

1. Manwell, J.F., Mcgowan, J.G., Rogers, A.L.: Wind Energy Explained – Theory, Design and Application. Wiley, England (2002)
2. Dobesch, H., Kury, G.: Basic Meteorological Concepts and Recommendations for the exploitation of wind energy in the atmospheric boundary layer, Vienna, Austria (2001)
3. Gasch, R., Twele, J.: Wind Power Plants - Fundamentals, Design, Construction and Operation. James & James Science Publishers Ltd., London (2004)
4. Spera, D.: Wind Turbine Technology - Fundamental Concepts of Wind Turbine Engineering. Asme Press, New York (1994)
5. Heier, S.: Grid Integration of Wind Energy Conversion Systems. John Wiley, Chichester (1998)
6. Rosas, P.: Dynamic Influences of Wind Power on the Power System. PhD thesis. Technical University of Denmark (2003)
7. Akhmatov, V.: Analysis of Dynamic Behaviour of Electrical Power Systems with Large Amount of Wind Power. PhD thesis, Lyngby, Denmark (2003)
8. Hansen, A., Bindner, H., Rebsdorf, A.: Improving Transition between Power Optimization and Power Limitation of Variable Speed/Variable Pitch Wind Turbines. In: Proceedings of European Wind Energy Conference and Exhibition, Nice, pp. 889–892 (1999)
9. Simões, M.G., Bose, B.K., Spiegel, R.J.: Design and Performance Evaluation of a Fuzzy-Logic-Based Variable-Speed Wind Generation System. IEEE Transactions on Industry Applications 33(4), 956–963 (1997)
10. Ekelund, T.: Modelling and Linear Quadratic Optimal Control of Wind Turbines. PhD thesis. Göteborg, Chalmers University of Technology, Sweden (1997)
11. Vihriala, H.: Control of Variable Speed Wind Turbines. PhD thesis. Tampere University of Technology, Tampere, Finland (2002)
12. Tanaka, T., Oumiya, T.: Output Control by Hill-Climbing Method for a Small Wind Power Generating System. Renewable Energy 12(4), 387–400 (1997)
13. Chapman, C.: The Analysis of Time Series. Chapman & Hall/CRC, New York (1995)

14. Costa, A., Crespo, A., Navarro, J., Madsen, H., Feitosa, E.: A Review on the Young History of the Wind Power Short-Term Prediction. Renewable and Sustainable Energy Reviews 12, 1725–1744 (2008)
15. Lange, M., Focken, U.: Physical Approach to Short-Term Wind Power Prediction. Springer, Berlin (2005)
16. Ross, T.J.: Fuzzy Logic with Engineering Applications. John Wiley, Chichester (2004)
17. Reznik, L.: Fuzzy Controllers. Newnes, Oxford (1997)
18. Tanaka, K., Wang, H.O.: Fuzzy Control Systems Design and Analysis: A Linear Matrix Inequality Approach. John Wiley, Chichester (2001)
19. Wang, L.-X.: Adaptive Fuzzy Systems and Control. Design and Stability Analysis. Prentice-Hall, New Jersey (1994)
20. Aracil, J., Gordillo, F.: Stability Issues in Fuzzy Control. Physica-Verlag, Heidelberg (2000)
21. Costa, P., Martins, A., Carvalho, A.: Optimization of Energy Generation in Wind farm Through Fuzzy Control. In: European Wind Energy Conference, EWEC 2004, London, United Kingdom (2004)
22. Costa, P., Martins, A., Carvalho, A.: Wind Energy Extraction and Conversion: Optimization through Variable Speed Generators and Non Linear Fuzzy Control. In: European Wind Energy Conference, EWEC 2006, Athens, Greece (2006)
23. Carvalho, A., Costa, P., Martins, A.: Increasing Power Wind Generation Through Optimization of the Dynamics of Control System Based on Accurate Forecasting of the Very Short Term Wind. In: German Wind Energy Conference, DEWEK 2008, Bremen, Germany (2008)
24. Hayes, M.: Digital Signal Processing. McGraw-Hill, New York (1999)

Intelligent Power System Frequency Regulations Concerning the Integration of Wind Power Units

H. Bevrani, F. Daneshfar, and R.P. Daneshmand

As the use of wind power turbines increases worldwide, there is a rising interest on their impacts on power system operation and control. Frequency regulation in interconnected networks is one of the main challenges posed by wind turbines in modern power systems. The wind power fluctuation negatively contributes to the power imbalance and frequency deviation. Significant interconnection frequency deviations can cause under/over frequency relaying and disconnect some loads and generations. Under unfavorable conditions, this may result in a cascading failure and system collapse.

This chapter presents an overview of the key issues on frequency regulation concerning the integration of wind power units into the power systems. Following a brief survey on the recent developments, the impact of power fluctuation produced by wind units on system frequency performance is presented. An updated frequency response model is introduced, and the inertia contribution of wind turbine in the overall system inertia is properly considered. The need for the revising of frequency performance standards is emphasized, and an intelligent agent based load frequency control (LFC), using multi-agent reinforcement learning (MARL) is proposed. Finally, nonlinear time-domain simulations on a 39-bus test power system are used to demonstrate the capability of the proposed control structure, and to analyze the system frequency performance in the presence of high wind power penetration and associated issues.

1 Introduction

The increasing need for electrical energy in the twenty-first century, as well as limited fossil fuel reserves, very high transportation and fuel cost and the increasing concerns with environmental issues for the reduction of carbon dioxide (CO2) and

H. Bevrani · F. Daneshfar · R.P. Daneshmand
Department of Electrical and Computer Engineering, University of Kurdistan, Sanandaj, PO Box 416, Iran

L. Wang et al. (Eds): *Wind Power Systems*, Green Energy and Technology, pp. 407–437.
springerlink.com

other greenhouse gasses [40], causes fast development in the area of renewable energy sources (RESs).

RESs are derived from natural sources such as the sun, wind, hydro-power, biomass, geothermal, oceans, and fuel cells that replenished themselves over a relatively short period of time. The most kind of the RES used is wind energy which is clean and available in nature. The wind turbine generators have attracted a lot of attention. Nowadays, the electric power industry has become more complicated than ever. It is necessary to interconnect more distributed generations in power systems because of environmental concerns. Primary concern includes global environmental and energy depletion issues. In addition to the energy security and deregulation matters which affect the so-called business chances. Therefore, the distributed generations, especially wind power generations, attract attentions in many countries.

Since, the primary energy source (wind) cannot be stored and is uncontrollable, the controllability and availability of wind power significantly differs from conventional power generation. In most power systems the output power of wind turbine generators varies with wind speed fluctuation, this fluctuation results into frequency variation [15, 19, 30]. Some reports are recently addressed the power system frequency control issue, in the presence of wind turbines [10, 13, 23, 25, 26, 27, 29, 32, 33, 37]. In this direction, the load-frequency control (LFC) is one of important control problems in concerning the integration of wind power turbine in a multi-area power system [2, 8, 18, 20, 21, 31].

Using conventional linear control methodologies for the LFC design in a modern power system is not more efficient, because they are only suitable for a specific operating point in a traditional structure. If the dynamic/structure of system varies; they may not perform as expected. Most of conventional control strategies provide model based controllers that are highly dependent to the specific models, and are not useable for large-scale power systems concerning the integration of RES units with nonlinearities, undefined parameters and uncertain models. If the dimensions of the power system increase, then these control design may become more different as the number of the state variables also increases, significantly.

Therefore, design of intelligent controllers that are more adaptive and flexible than conventional controllers is become an appealing approach. Intelligent control has been already used for the frequency regulation issue in the power systems [14, 24, 44]; however there are just few reports on the frequency control design in the presence of RES units [11].

One of the adaptive and nonlinear intelligent control techniques that can be effectively applicable in the frequency control design is reinforcement learning (RL). Some efforts are addressed in [3, 4, 5, 7, 16, 17]. RL based controllers learn and are adjusted to keep the area control error small enough in each sampling time of a LFC cycle. Since, these controllers are based on learning methods; they are independent of environment conditions and can learn a wide range of operating conditions. The RL based frequency control design is a model-free design and can easily scalable for large scale systems and suitable for frequency variation caused by wind turbine fluctuation.

In the continuation of the authors' recent works [9, 11], the present chapter addresses the LFC design using an agent based reinforcement learning for a large interconnected power system concerning the integration of wind power units. In this chapter, a multi-agent RL based control structure is proposed. Each control area includes an agent that communicates with each other to control the frequency among whole interconnected system. Each agent (controller agent) provides an appropriate control action according to the area control error *(ACE)* signal, using reinforcement learning. In a multi-area power system, the learning process is considered as a multi-agent RL process and agents of all areas learn together (not individually).

The above technique has been applied to the LFC problem in a network with the same topology as IEEE 10 generators 39-bus test system integrated with wind power units, as a case study. The rest of chapter is organized as follows: In Section 2.2, the impact of wind power generation on power system frequency and the structure of a network which the above architecture is implemented for, are discussed. A generalized frequency response model is introduced, and the need for revising of frequency performance standards is emphasized.

In Section 2.3, a brief introduction to single-agent and multi-agent based RL is presented. Section 2.4 and 2.5 address the proposed intelligent frequency control technique using agent based RL. It is explained that how the designed controllers for the test system can work. Simulation results are provided in Section 2.6 and the chapter is concluded in Section 2.7.

2 Impacts of Wind Power Generation on the Power System Frequency and Frequency Regulation

Dynamic behavior of a power system in the presence of wind power generation systems might be different from conventional power plants. The power outputs of such sources are dependent on weather conditions, seasons, and geographical location. When wind power is a part of the power system, additional imbalance is created when the actual wind power deviates from its forecast due to wind velocity variations. So, scheduling conventional generator units to follow load (based on the forecasts) may also be affected by wind power output [11]. Furthermore, the effect of wind farms on the dynamic behavior of power system may cause a different system frequency response to a disturbance event (such as load disturbance).

Since, the system inertia determines the sensitivity of overall system frequency; it plays an important role in this consideration [11]. The lower system inertia, lead to faster changes in the system frequency following a load generation variation. The addition of synchronous generation to a power system intrinsically increases the system inertial response. This intrinsic increase does not necessarily with the addition of wind turbine generators (WTGs) due to their differing electromechanical characteristic [28]. So, the impact of wind farms on power system inertia is a key factor in investigating the power system frequency behavior in the presence of high penetration wind power generation.

Depending on the technology is used in installed wind turbine generators; proportion of wind power generation on system inertia could be different. Thus, in order to analyze the impact of WTGs on system inertia (and consequently on system frequency) it is important to consider the wind generation technologies.

2.1 Wind Generation Technologies

Generally, there are two basic categories of wind turbine generators; fixed speed and variable speed WTGs. Here, a brief description is given for these wind turbine technologies.

Fixed Speed WTGs

Fixed Speed WTGs generally use squirrel-cage induction generators. Stator winding of the generator is directly connected to the grid. The rotor of the turbine blades is coupled to the rotor of the generator through a gearbox. So, the turbine could utilize the kinetic energy stored in the turbine blades and contributes to the system frequency stability by providing spinning inertia [41].

When the system frequency deviates from its nominal value, the relationship between the system frequency and electromagnetic torque of induction generator will determine the inertial response. Strong coupling between the squirrel-cage induction generator stator and power system and low nominal slip (about one to two percent [34]) of them cause any deviation in the system speed lead to a change in rotational speed. Such linking between rotor speed and system frequency gives rise to an inertial response from the fixed speed WTGs during a system frequency falling.

Generally, the fixed speed turbines have a simple construction. However, as they cannot track wind speed fluctuations the energy capture is not as efficient as in variable speed systems. As a wind rotor presents maximum power coefficient at its design tip speed ratio (TSR), for constant speed operation this maximum power coefficient can be reached to its designed wind speed [28, 36, 41]. Fig. 1 shows a general configuration for a fixed speed wind turbine generator.

Variable Speed WTGs

Variable speed WTGs could be equipped with either synchronous generators or doubly-fed induction generators (DFIG).

In the case of synchronous generators, the wind turbine is allowed to spin at whatever speed that is needed to reached maximum power. So, the electrical output frequency would vary due to instantaneous variation in the wind velocity. As a result, the generator cannot be directly connected to the grid due to its poor power quality. In the variable speed application these generators are decoupled from the grid through the use of a back-to -back *ac*/*dc*/*ac* converter attached to the stator of the synchronous generator.

First, the variable frequency *ac* output of generator is rectified into *dc* using high power switching transistors; then, the *dc* converts back to *ac* at grid

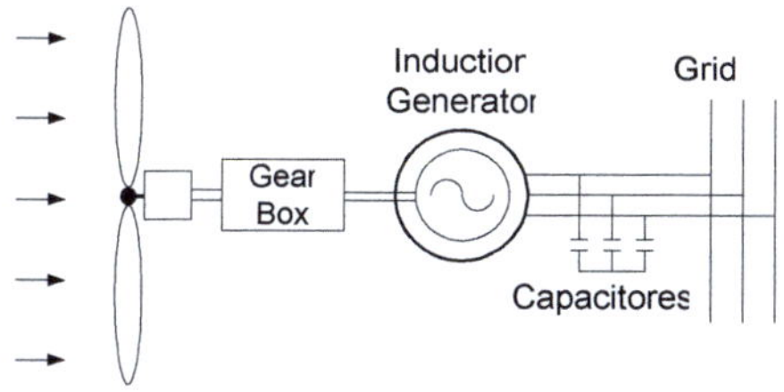

Fig. 1. Fixed speed wind turbine generator

frequency through an inverter before feeding to the grid. This design causes the stator be isolated from power system and do not touch frequency variations. Therefore, during a frequency event wind turbine generator has no inertial response and its power output does not change.

Another and more common variable speed wind turbine technology is the DFIG which uses wound-rotor induction machines. In this type, stator winding is directly connected to the grid. However, the rotor winding is connected to the power system by employing a back-to-back *ac/dc/ac* converter which varies the electrical frequency as acceptable by the grid. Thus, the electrical frequency will be different from the mechanical frequency which allows for variable speed application. This converter allows for the control of active and reactive power using constant power factor or constant voltage [36, 41].

When a frequency event occurs, the inertial response provided by the DFIG depends on the relationship between electromagnetic torque of the machine and power system frequency. This relationship depends on the control structure used in the converter and their parameters [35].

Variable speed WTGs present higher energy capturing, lower mechanical stress, more constant output power, and reduced noise compared with fixed speed machines [1]. This type of wind power technology is more common than the other. From this type, DFIGs are more popular and comprise the predominant portion of the installed wind energy. Fig. 2 shows simple configurations for the variable speed wind turbine generators.

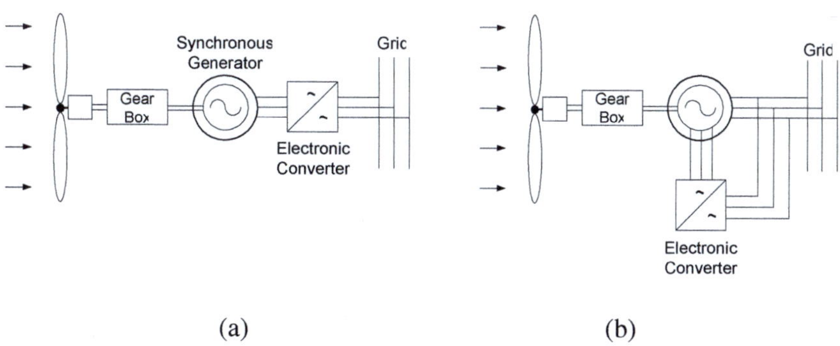

Fig. 2. Variable speed wind turbine generators; (a) with synchronous generator, (b) with doubly fed induction generator (DFIG)

2.2. System Response Analysis

To investigate the impact of wind power generation on the power system (particularly on system frequency), a simulation study is provided in the Simpower environment of MATLAB software. Here, the power system response for a test system is shown under normal and alert conditions in the presence of WTGs.

Test System

As mentioned, the wind power generation could affect the dynamic behavior of the power system. The frequency response characteristic of a power system with a high penetration of wind power may be different from that of the conventional system [11]. This section provides a simulation study on the impacts of wind power unit, particularly DFIG wind power generation on the power system frequency. For this purpose, a network with the same topology as the well-known IEEE 10 generators 39-bus test system is considered as a case study.

This test system is widely used as a standard system for testing of new power system analysis and control synthesis methodologies. A single-line diagram of the system is given in Fig. 3. It represents a greatly reduced model of the power system in New England. It can be used to study both static and dynamic problems in the power system. This system has 10 generators, 19 loads, 34 transmission lines, and 12 transformers. The simulation parameters for the generators, loads, lines, and transformers of the test system are given in Appendix.

The 39 buses system is organized into 3 areas. Total system installed capacity are 841.2 MW of conventional generation and 45.34 MW of wind power generation. There are 198.96 MW of conventional generation, 22.67 MW of wind power generation and 265.25 MW load in Area 1. In Area 2, there are 232.83 MW of conventional generation, and 232.83 MW load. In Area 3, there are 160.05 MW of conventional generation, 22.67 MW of wind power generation and 124.78 MW of load.

All power plants in the power system are equipped with speed governor and power system stabilizer (PSS). However, only one generator in each area is responsible for the LFC task using a well-tuned proportional integral (PI) controller; G1 in Area 1, G9 in Area 2, and G4 in Area 3.

For the sake of simulation, random variations of wind velocity have been considered. Dynamics of WTGs including the pitch angle control of the blades are also considered. The start up and rated wind velocity for the wind farms are specified as about 8.16 (m/s) and 14 (m/s), respectively. Furthermore, the pitch angle controls for the wind blades are activated only beyond the rated wind velocity. The pitch angles are fixed to zero degree at the lower wind velocity below the rated one.

Impact of WTGs on System Frequency in a Normal Operating Condition

When wind power plants are introduced into the power system, an imbalance is created when the actual wind output deviates from its forecast [11]. This power imbalance may lead to frequency deviations from nominal value (60 Hz in the present example). Fig. 4 illustrates the frequency deviation due to wind power fluctuations.

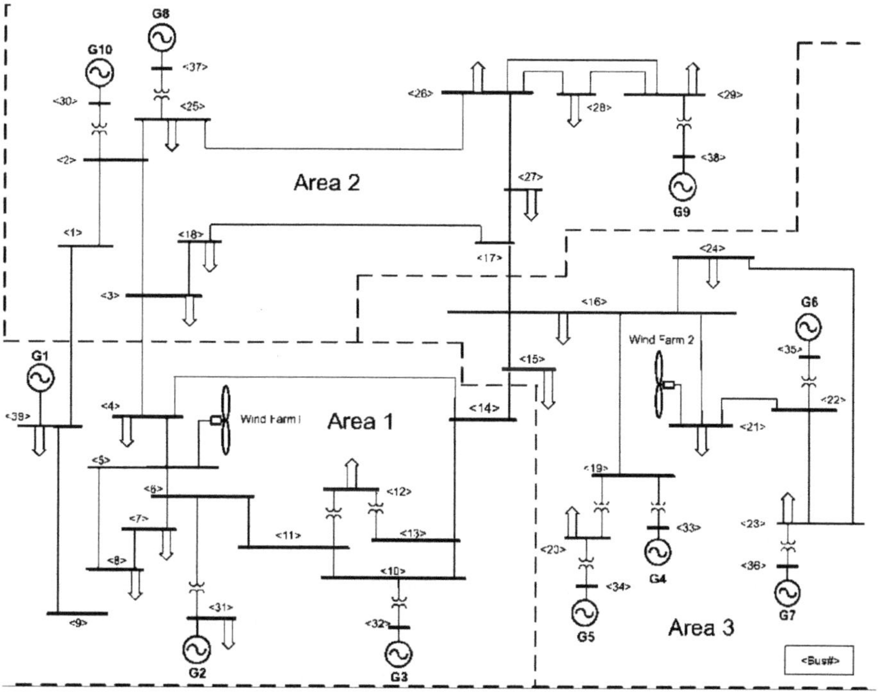

Fig. 3. Single-line diagram of 39-bus test system

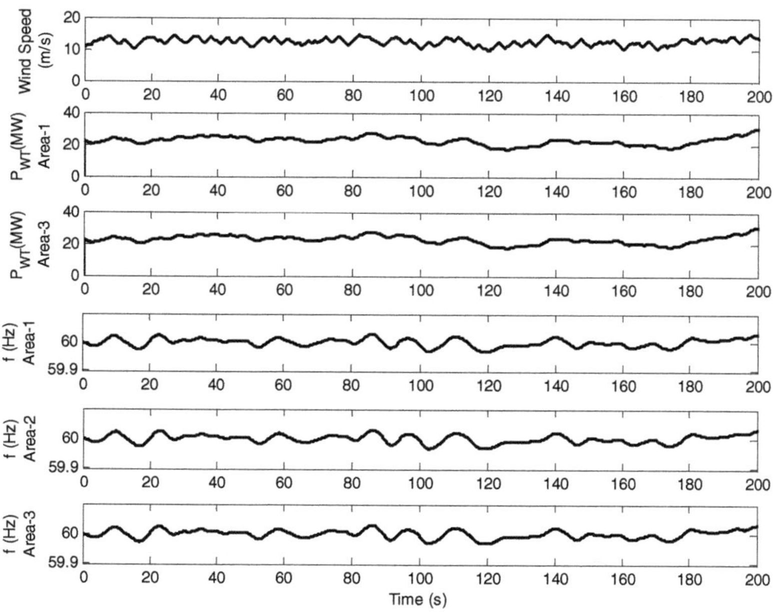

Fig. 4. System response to wind power fluctuations

From the power quality point of view, frequency deviations should be limited in a specified standard band. As shown in the simulation results, in the presence of wind power generation, the frequency regulation performance is significantly decreased.

Simulation results show that the generator units equipped with conventional PI controllers (G1, G9 and G4) are unable to provide a desirable frequency regulation following fluctuation caused by wind power variations.

Impacts of WTGs on the System Frequency under Load Disturbance Condition

Beside the normal condition, impact of high penetration wind power generation on system frequency at abnormal condition is also noticeable. As mentioned, adding of wind generators to a power system leads to increase in total system inertia. The most pronounce effect of high values of inertia is to reduce the initial rate of frequency decline, and to delay and reduce in the maximum deviation [6].

To investigate above issue, system response following a step load disturbance is investigated. A step load increase is considered at 10 s in each area as follows. 3.8 percent of area load at bus 8 in Area 1, 4.3 percent of area load at bus 3 in Area 2, and 6.4 percent of area load at bus 16 in Area 3 have been changed. System response is shown in Fig. 5. For having a clear comparison, a zoomed view of Fig. 5 around 10s is redrawn in Fig. 6.

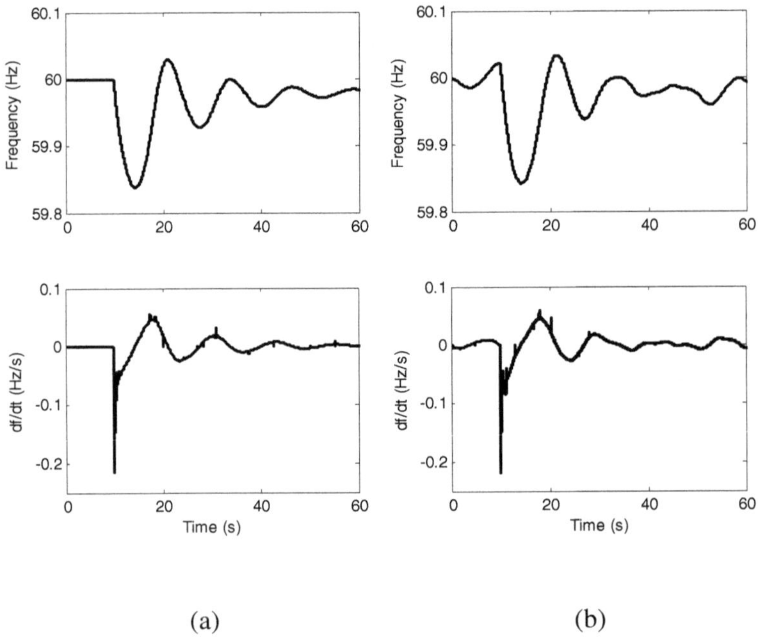

Fig. 5. Overall system frequency and rate of frequency change to a simultaneous step load changes in three areas; (a) without WTGs, (b) with WTGs

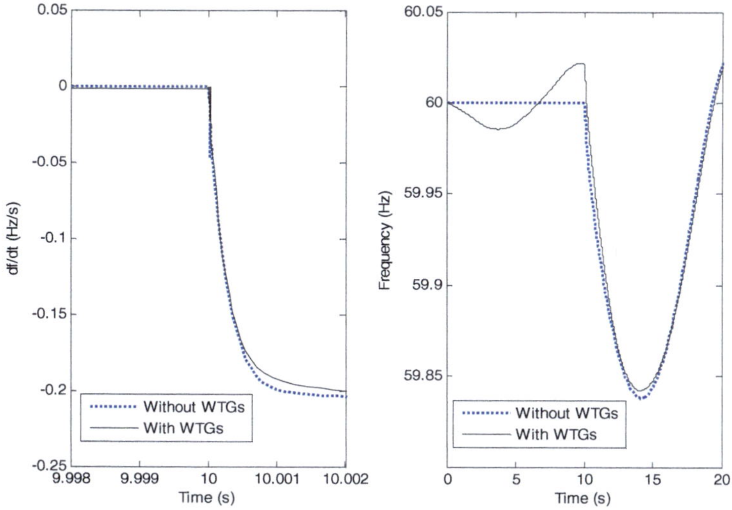

(a) (b)

Fig. 6. Frequency deviation following step load disturbances at 10 sec; (a) rate of frequency change, (b) system frequency

As shown in Fig. 6a, the initial rate of frequency change following a load disturbance, has reduced in the presence of WTGs; because the wind power penetration in the power system can increase the overall system inertia [11]. However, as DFIGs provide small inertial response, this reduction is not noticeable.

Maximum frequency deviation range following a step load disturbance is shown in Fig. 6. It could be seen that in the presence of WTGs, the frequency drop has reduced. It is justifiable by considering the larger inertia due to the addition of large wind farms to the system. The higher inertia value results in a lower drop in frequency. Since the response is slower for the higher inertia, the governor has more time to response and therefore limits the maximum frequency deviation to smaller value [6].

The Impact of High Penetration Wind Generation on the Required LFC Reserve

When WTGs are introduced to the power system, as they generate a part of power system loads, much portion of conventional nominal power can be available for using in supplementary control. However, as the variable wind farms power output may or may not be available during peak demand and abnormal periods, due to unpredictable nature of wind; it might be that these resources cannot contribute to the overall system frequency regulation and reliability. On the other hand, the additional power variation from WTGs results in frequency deviation. It seems that for a large wind power penetration, this deviation will be so larger and as a result, the conventional LFC reserve may be insufficient to maintain frequency within the bounds for service quality.

It was found that wind power, does not impose major extra variations on the system until a substantial penetration is reached [11]. Large geographical spreading of wind power will reduce variability, increases predictability, and decrease the occasions with near zero or peak output [22]. It is investigated in [11] that the power fluctuation from geographically dispersed wind farms will be uncorrelated with each other, hence smoothing the sum power and not imposing any significant requirement for additional frequency regulation reserve, and required extra balancing is small.

The fluctuation of the aggregated wind power output in a short term (e.g., tens of seconds) for a larger number of wind turbines are much smoothed. It is investigated that the wind turbines aggregation has positive effects on the regulation requirement. Relative regulation requirement decreases whenever larger aggregations are considered [22].

To show the impact of large aggregation of wind turbines on wind power outputs smoothing and required supplementary reserve, a simulation study was applied on the test case system with considering high penetration wind power in Area 1 and Area 3. The effect of wind power on frequency, following the step load disturbance can be investigated in Fig. 7. To show the impact of wind farm size on power smoothing two conditions are considered: occurrence of the load disturbance in with considering average wind power output (neglecting wind power fluctuations), and with actual wind farms output.

It can be also seen in Fig. 7, that because of smoothing effect on wind power output fluctuations due to large penetration of wind farms, wind power fluctuations does not impose too much variations on the response of generators equipped with LFC, respect to the considering of average wind power output.

2.3 Generalized Area Frequency Response Model

To analyze the additional variation caused by WTGs, the total effect is important, and every change in wind power output does not need to be matched one for one by a change in another generating unit moving in the opposite direction. Instantaneous fluctuations in load and WTG power output might amplify each other, be completely unrelated to each other, or they may cancel each other out [11]. However, the slow WTG power fluctuation dynamics and total average power variation negatively contribute to the power imbalance and frequency deviation, which should be taken into account in the well-known LFC control scheme. This power fluctuation must be included in the conventional LFC structure.

A generalized LFC model in the presence of WTGs is shown in Fig. 8. Here, to cover the variety of generation types in the control area, different values for turbine-governor parameters and the generator regulation parameters are considered. Fig. 8 shows the block diagram of typical control area with n generator units. The shown blocks and parameters are defined as follows: Δf is frequency deviation, ΔP_m is mechanical power, T_{ij} is synchronizing coefficient, ΔP_C is supplementary control action, ΔP_L is load disturbance, H_{Sys} is equivalent inertia constant, D_{Sys} is equivalent damping coefficient, B is frequency bias, R_i is drooping characteristic, ΔP_P is primary control, α_i is participation factors, ΔP_{WTG} is WTG power fluctuation, $K(s)$ is LFC controller, and $M_i(s)$ is governor-turbine model.

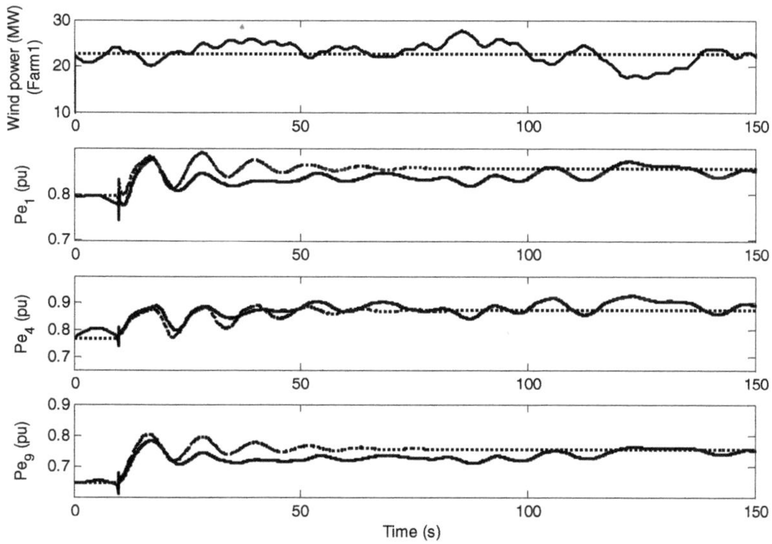

Fig. 7. Impact of wind power fluctuation on required supplementary LFC reserve; actual wind power (solid), average wind power (dotted)

Following a load disturbance within the control area, the frequency of the area experiences a transient change and the feedback mechanism generates appropriate rise or lower signal to the participating generator units according to their participation factors α_i to make generation follow the load. In the steady state, the generation is matched with the load, driving the tie-line power and frequency deviations to zero. As there are many conventional generators in each area, the control signal has to be distributed among them in proportion to their participation.

As shown in Fig. 8, the frequency performance of a control area is represented approximately by a lumped load generation model using equivalent frequency, inertia and damping factors [9]. Because of range of use and specific dynamic characteristics such as a considerable amount of kinetic energy, the wind units are more important than the other renewable energy resources. The equivalent system inertia can be defined as:

$$H_{sys} = H_C + H_W = \sum_{i=1}^{N1} H_{Ci} + \sum_{i=1}^{N2} H_{Wi} \quad (1)$$

The H_C and H_W are the total inertia constants due to conventional and wind turbine generators, respectively. The inertia constant for wind power is time dependant. The typical inertia constant for the wind turbines is about two to six seconds [11].

In Fig. 8, the filtered total effect of power fluctuation ΔP_{WTG} is considered. For a large WTGs penetration, the resulting *ACE* signal must reflect the total WTGs power generation changes which is usually smoothed compared to variations from the individual wind turbine units.

$$ACE = B\Delta f + \sum (P_{Con,act} - P_{Con,sched}) + \sum (P_{WTG,act} - P_{WTG,estim}) \tag{2}$$

where $P_{Con,act}$, $P_{Con,sched}$, $P_{WTG,act}$, and $P_{WTG,estim}$ are actual conventional power, scheduled conventional power, actual WTGs power, and estimated WTGs power, respectively.

In typical LFC implementations, the system frequency gradient and *ACE* signal must be filtered to remove noise effects before use. The *ACE* signal then is often applied to a proportional integral (PI) control block [9, 10]. Control dead band and ramping rate are different for various systems [11]. The control can send higher/lower pulses to generating plants if its *ACE* signal exceeds a standard limits.

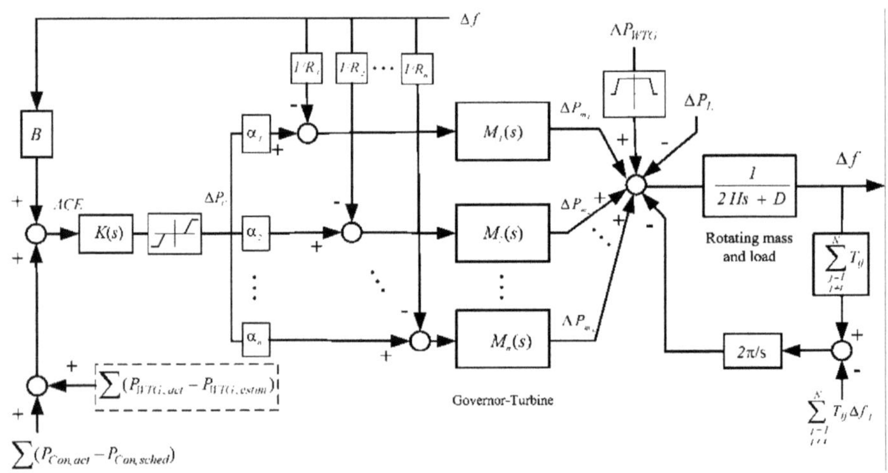

Fig. 8. LFC model with considering WTG power fluctuation

2.4 The Need for Revising of Frequency Performance Standards

Power system frequency control is an issue that may evolve into new guidelines. The existing frequency operating standards need to change to allow for the introduction of renewable power generation, and allow for modern distributed generator technologies.

It is investigated that the slow component of renewable power fluctuation negatively affects the performance standards such as policy P1 of UCTE (Union for the Coordination of Transmission of Electricity) performance standard, or the control performance standards CPS1 and CPS2 introduced by NERC (North American Electric Reliability Council) [11].

The standards redesign must be done in both normal and abnormal conditions, and should take account of operational experience on the initial frequency control schemes and again used measurement signals including tie-line power, frequency, and rate of frequency change settings. The new set of frequency performance standards are under development in many countries. The new standards introduce the update high and low trigger, abnormal, and relay limits applied on the interconnection frequency excursions.

The revised standards may bring an element of a more centralized frequency control through a better coordination among control areas, delegating more authority to the control areas performing frequency monitoring functions, and perhaps creating distributed or inter-area control centers to decentralized frequency control through the creation of corresponding ancillary service markets.

As mentioned, the rate of frequency change (df/dt) following a disturbance is proportional to the power imbalance, and it also depends on the equivalent system inertia [11]. Since large wind farms can considerably increase the overall system inertia, the df/dt will be significantly changed. From an operational point of view, a larger variable renewable power in the power system causes a smaller frequency rate change following a sudden loss of generation or load disturbance. This issue is important for those networks that use the protective df/dt relays to re-evaluate their tuning strategies.

The performance standards revision has already commenced in many countries [11]. In Australia, the Australian Electricity Market Commission (AEMC) is proposing revised technical rules for generator connection, including wind generators. As well as meeting technical standards, generators are required to provide information on energy production via the system operator's SCADA system. National Electricity Market Management Company (NEMMCO) sets out functional requirement for an Australian Wind Energy Forecasting System (AWEFS) for wind farms in market regions. In the USA, NERC is working to revise the conventional control performance standards. The existing market rules and priority rules for the transport of RES electricity is also under re-examination by UCTE in Europe.

3 A Background on Agent Based RL

The present section presents a brief background on the single-agent and multi-agent RL (MARL) [38]. First, the single-agent RL is defined and its solution is described. Then, the multi-agent task is defined. The discussion is restricted to finite state and action spaces, as the major part of MARL results are given for finite spaces [12].

3.1 Single-Agent RL

Reinforcement learning (RL) is learning what to do, how to map situations to actions, so as to maximize a numerical reward signal [16]. In fact the learner will discover which action should be taken by interacting with the system and trying the different actions which may lead to the highest reward. The RL will evaluate the actions taken and gives the learner a feedback of how good the action taken was and whether it should repeat this action in the same situation or not. In another word, the RL methods learn to solve a problem by interacting with a system. The learner is called the agent and the system it interacts with, is known as the environment. During the learning process, the agent interacts with the environment and takes an action a_t from a set of actions, at time t. These actions will affect the system and will take it to a new state x_{t+1}. Therefore, the agent is provided with the corresponding reward signal (r_{t+1}). This agent-environment interaction is repeated until the desired goal is achieved. In this text what is meant by the state is the required information for making a decision, therefore what we would like, ideally, is a state signal that summarizes past perceptions in a way that all relevant information is retained. A state signal that succeeds in retaining all relevant information is said to be Markov, or to have the Markov property [16] and a reinforcement learning task that satisfies this property is called a finite Markov Decision Process (MDP). If an environment has the Markov property, then its dynamics enable us to predict the next state and expected next reward given the current state and action.

In the present work, it is assumed that the environment has the Markov property; therefore a MDP problem is solved. In each MDP the objective is to maximize sum of returned rewards over time, then the expected sum of discounted rewards defined by the following equation:

$$R = \sum_{k=0}^{\infty} \gamma^k r_{t+1} \tag{3}$$

where $0 < \gamma < 1$ is a discount factor, which gives the most importance to the recent rewards.

Another term is value function that is defined as the expected return or reward (E) when starting at state x_t while following policy $\pi(x,a)$ (see Eq. 4). Policy is the way the agent maps the states to the actions [16].

$$V^{\pi}(x) = E_{\pi}\left\{\sum_{k=0}^{\infty} \gamma^k r_{t+k+1} | x_t = x\right\} \tag{4}$$

The optimal policy is one that maximizes the value function. Therefore, once the optimal state value is derived, the optimal policy can be found as follows:

$$V^*(x) = \max_{\pi} V^{\pi}(x) \text{ }, \forall x \in X \tag{5}$$

In most RL methods, instead of calculating the state value, another term known as the action value is calculated (6), which is defined as the expected discounted reward while starting at state x_t and taking action a_t.

$$Q^{\pi}(x,a) = E_{\pi}\left\{\sum_{k=0}^{\infty} \gamma^{k} r_{t+k+1} | x_t = x, a_t = a\right\} \tag{6}$$

Bellman's equation [16], as shown below, is used to find the optimal action value. In general, an optimal policy is one that maximizes the Q-function defined in the following relation:

$$Q^{*}(x,a) = \max_{\pi} E_{\pi}\left\{r_{t+1} + \gamma \max_{\acute{a}} Q^{*}(x_{t+1}, \acute{a}) \mid x_t = x, a_t = a\right\} \tag{7}$$

Different RL methods have been proposed to solve the above equations. In some algorithms the agent will first approximate the model of the system in order to calculate the Q-function. The method used in this chapter is of temporal difference type which learns the model of the system under control. The only available information is the reward achieved by each taken action and the next state. The algorithm, called Q-learning, will approximate the Q-function and by the computed function the optimal policy, which maximizes this function, is derived.

3.2 Multi-agent RL

A multi-agent system [43] can be defined as a group of autonomous, interacting entities (agents) [44] sharing a common environment, which they perceive with sensors and upon which they act with actuators [42]. Multi-agent systems can be used in a wide variety of domains including robotic teams, distributed control, resource management, collaborative decision support systems. Well-understood algorithms with good convergence and consistency properties are available for solving the single-agent RL task, both when the agent knows the dynamics of the environment and the reward function (the task model), and when it does not. However, the scalability of algorithms to realistic problem sizes is problematic in single-agent RL, and is one of the great reasons to use multi-agent reinforcement learning (MARL) [12]. In addition to scalability and benefits owing to the distributed nature of the multi-agent solution, such as parallel computation, multiple RL agents may utilize new benefits from sharing experience, e.g., by communication, teaching, or imitation [12]. These properties make RL attractive for multi-agent learning. However, several new challenges arise for RL in multi-agent systems. In multi-agent systems other adapting agents make the environment no longer stationary, violating the Markov property that traditional single agent behavior learning relies on, this nonstationarity properties decrease the convergence properties of most single-agent RL algorithms [46]. Another problem is the difficulty of defining a good learning goal for the multiple RL agents [12]. Only then it will be able to coordinate its behavior with other agents. These challenges make the MARL design and learning difficult in large-scale applications, therefore it uses a special learning algorithm as below. Using this learning algorithm, the violating the Markov property problem causes from multi-agent structure and other problems will be solved.

Learning Algorithm

The generalization of the Markov decision process to the multi-agent case is as follows:

Suppose a tuple $< X, A_1, \ldots, A_n, p, r_1, \ldots, r_n >$ where n is the number of agents, X is the discrete set of environment states, A_i , $i = 1, \ldots, n$ are the discrete sets of actions available to the agents, yielding the joint action set $A = A_1 \times \cdots \times A_n$, $p: X \times A \times X \rightarrow [0,1]$ is the state transition probability function, and $r_i : X \times A \times X \rightarrow R, i = 1, \ldots, n$ are the reward functions of the agents.

In the multi-agent case, the state transitions are the result of the joint action of all the agents, $a_k = [a_{1,k}^T, \ldots, a_{n,k}^T], a_k \in A, a_{i,k} \in A_i$ (T denotes vector transpose). As a result, the rewards $r_{i,k+1}$ and the returns $R_{i,k}$ also depend on the joint action. The policies $h_i : X \times A_i \rightarrow [0,1]$ form together the joint policy h. The Q-function of each agent depends on the joint action and is conditioned on the joint policy, $Q_i^h : X \times A \rightarrow R$ [33].

4 The Proposed Intelligent Control Framework

In practice, the LFC system is traditionally using a proportional-integral (PI) type controller as shown in Fig. 8. In this section, an intelligent control design algorithm for such a controller using MARL technique is presented. The design objective is to regulate the frequency in power system concerning the integration of wind power units with various load disturbances.

Fig. 9 shows the proposed model for area *i*, including an intelligent controller. The controller is responsible to produce an appropriate control action (ΔP_{ci}) according to the *ACE* signal using RL.

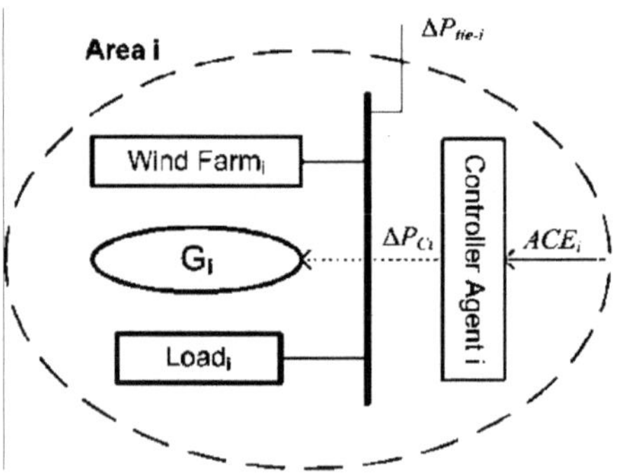

Fig. 9. The proposed model for area *i*

4.1 Controller Agent

The intelligent controller (controller agent) system functions as follows. At each instant (on a discrete time scale k), $k = 1,2,\cdots$ the controller agent observes the current state of the system, x_k, and takes an action, a_k. The state vector consists of some quantities, which are normally available to the controller agent. Here, the average of ACE signal over the time interval $k-1$ to k as the state vector at the instant k is used. For the algorithm presented in this paper, it is assumed that the set of all possible states X, is finite. Therefore the values of various quantities that constitute the state information should be quantized.

The possible actions of the controller agent are the various values of ΔP_c, that can be demanded in the generation level within an LFC interval. ΔP_c is also discretised to some finite number of levels. Now, since both X and A are finite sets, a model for this dynamic system can be specified through a set of probabilities.

Here, an RL algorithm is used for estimating Q^* and the optimal policy. It is similar to the introduced algorithm in [4]. Suppose we have a sequence of samples (x_k, x_{k+1}, a_k, r), $k = 1,2,\cdots$. Each sample is such that x_{k+1} is the (random) state that resulted when action a_k is performed in state x_k and $r_k = g(x_k, x_{k+1}, a_k)$ is the consequent immediate reinforcement. Such a sequence of samples can be obtained either through a simulation model of the system or observing the actual system in operation. This sequence of samples (called training set) can be used to estimate Q^*. Using a specific algorithm. Suppose Q^k is the estimate of Q^* at kth iteration. Let the next sample be (x_k, x_{k+1}, a_k, r), then we obtain Q^{k+1} as:

$$Q^{k+1}(x_k, a_k) = Q^k(x_k, a_k) + \alpha\left[g(x_k, x_{k+1}, a_k) + \gamma \max_{\acute{a}\in A} Q^k(x_{k+1}, \acute{a}) - Q^k(x_k, a_k)\right] \tag{8}$$

where $0 < \alpha < 1$ is a constant called the step size of learning algorithm.

At each time step (as determined by the sampling time for the LFC action) the state input vector, x, to the LFC is determined, then an action in that state is selected and applied to the model, the model is integrated for a time interval equal to the sampling time of LFC to obtain the state vector $\acute{x}$ at the next time step.

Here, the exploration policy for choosing actions in different states is used. It is based on a Learning automata algorithm called pursuit algorithm [39]. This is a stochastic policy where, for each state x, actions are chosen based on a probability distribution over the action space. Let P_x^k denote the probability distribution over the action set for state vector x at the kth iteration of learning. That is, $P_x^k(a)$ is the probability of choosing action a in state x at iteration k. A uniform probability distribution is considered at $k = 0$, that is

$$P_x^0(a) = \frac{1}{|A|} \quad \forall a \in A \ \forall x \in X \tag{9}$$

At the kth iteration, let the state x_k be equal to x. An action a_k, at random based on $p_x^k(.)$ is chosen. That is, $Prob(a_k = a) = p_x^k\ (a)$. Using the performed simulation model, the system is gone to the next state x_{k+1} by applying action a in the state x and is integrated for the next time interval. Then, Q^k is updated to Q^{k+1} using (8) and the probabilities is updated as follows, too.

$$P_x^{k+1}(a_g) = P_x^k(a_g) + \beta\left(1 - P_x^k(a_g)\right)$$

$$P_x^{k+1}(a) = P_x^k(a)(1-\beta) \quad \forall a \in A, a \neq a_g \tag{10}$$

$$P_y^{k+1}(a) = P_y^k(a) \quad \forall a \ \in A, \forall y \in X, y \neq x$$

where $0 < \beta < 1$ is a constant. Thus, at iteration k the probability of choosing the greedy action a_g in state x is slightly increased and the probabilities of choosing all other actions in state x are proportionally decreased.

Learning the Controller

In the present algorithm, the aim is to achieve the well-known LFC objective and to keep the *ACE* within a small band around zero. This choice is motivated by the fact that all the existing LFC implementations use this as main control objective and hence, it will be possible for us to compare the proposed RL approach with the designed linear PI based LFC approaches. As mentioned above, in this formulation, each state vector consists of the average value of *ACE* as state variable. The control action of the LFC is to change the generation set point, ΔP_c. According to the RL algorithms application, usually a finite number of states are assumed. In this direction, state variable and action variable should be discretised to finite levels, too.

The next step is to choose an immediate reinforcement function by defining the function g. The reward matrix initially is full of zero, at each time step we get the average value of *ACE* signal, then according to its discritised values, determine the state of the system, whenever the state is desirable (i.e. $|ACE|$ is less than ε) then reward function $g\ (x_k\ , x_{k+1}, a_k\ )$ is assigned at zero value. When it is undesirable (i.e. $|ACE| > \ \varepsilon$), then $g(x_k\ , x_{k+1}, a_k)$ is assigned a value -$|ACE|$ (we penalized all actions which cause to go to an undesirable state with a negative value).

5 Application to the 3-Control Area Test System

To illustrate the effectiveness of the proposed control strategy, and to compare the results with a conventional PI control design, the described 39-bus system (Section 2.2) is considered as a test case study. A block schematic diagram of the model used for simulation studies is shown in Fig. 10 and the proposed multi-agent structure is shown in Fig. 11.

Here, the purpose is essentially to clearly show the various steps in implementation and illustrate the method. After design choices are made, the controller is trained by running the simulation in the learning mode as explained in Section 2.4. After completing the learning phase, the control actions at various states have converged to their optimal values.

The simulation is run as follows: At each LFC instant k, controller agents of each area, average all corresponding ACE signal instances gained every 0.1 seconds. Three average values of ACE signal instances (each related to one area) form the current state vector, x_k, that is obtained according to the quantized states. When all area's state vectors are ready, then the controller agents choose the action signal a_k that consists of three ΔP_c values for three areas (action signal is gained according to the quantized actions and the exploration policy mentioned above) to change the set points of the governors using the values given by a_k.

In the performed simulation studies, the input variable is obtained as follows. As the LFC decision cycle time chosen, three values of ACE are calculated over a decision cycle. The averages of these values for three areas are the state ble $(x^1_{avg1}, x^1_{avg2}, x^1_{avg3})$. Since, we use the multi-agent reinforcement learning process and agents of all areas are learning together, the state vector is also consisted of all state vectors of three areas, the action vector is consisted of all action vectors of three areas as shown in term $< (X_1, X_2, X_3), (A_1, A_2, A_3), p, (r_1, r_2, r_3) >$ or $< X, A, p, r >$.

Here $X_i = x^1_{avgi}$ is the discrete set of each area states, X is the joint state, A_i is the discrete set of each area actions available to the area i, and A is the joint action. In each instant time after averaging of ACE_i for each area (over three instances), depending on the current joint state (X_1, X_2, X_3), the joint action $(\Delta P_{c1}, \Delta P_{c2}, \Delta P_{c3})$ is chosen according to the exploration policy. Consequently, the reward r is also depends on the joint action which whenever the next state (X) is desirable (i.e. all $|ACE_i|$ are less than ε), then reward function r is assigned a zero value. When the next state is undesirable (i.e. $\exists\ ACE_i \mid |ACE_i| > \varepsilon$) then r is assigned average value of -$|ACE_i|$. In this algorithm, since all agents learn together, parallel computation causes to speed up the learning process. Also this reinforcement learning algorithm is more scalable than single-agent RL.

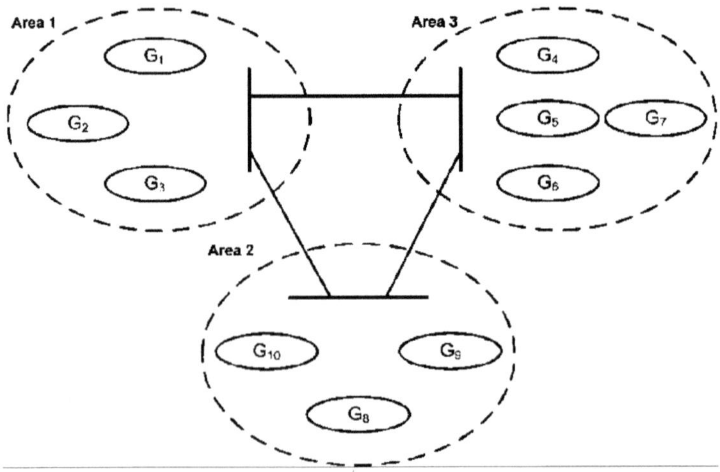

Fig. 10. 3-control area power system

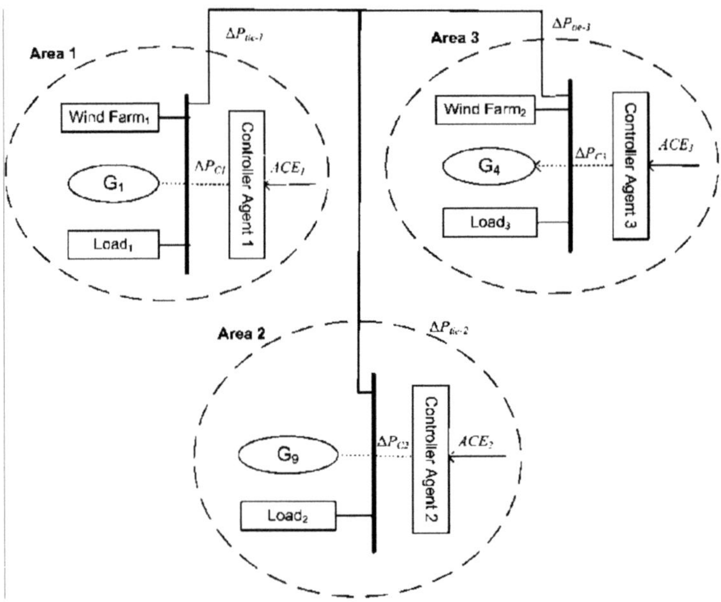

Fig. 11. The proposed multi-agent structure for 3-control area power system

6 Simulation Results

To demonstrate the effectiveness of the proposed control design, some simulations were carried out. In these simulations, the proposed controllers were applied to the model described in Section 2.2. In this section, the performance of the closed-loop system using the well-tuned conventional PI controllers is compared to the designed MARL controllers for the various possible load disturbances.

As a serious test scenario, the following load disturbances (step increase in demand) are applied to three areas: In Area 1, 3.8% of total area load at bus 8, 4.3% of total area load at bus 3 in Area 2, and 6.4% of total area load at bus 16 in Area 3 have been simultaneously increased in a step form. The frequency deviation (Δf), and area control error (ACE) signals of the closed-loop system are shown in Fig. 12, Fig. 13, and Fig. 14.

As shown in the simulation results, using the proposed method, the area control error and frequency deviation of all areas are properly driven close to zero. Furthermore, regarding that the proposed algorithm is an adaptive algorithm and it is based on the learning methods - in each state it finds the local optimum solution to gain the system objectives (ACE signal near zero) - therefore the intelligent controllers provide smoother control action signals and areas frequency deviation is less than the frequency deviation in the system with PI controllers.

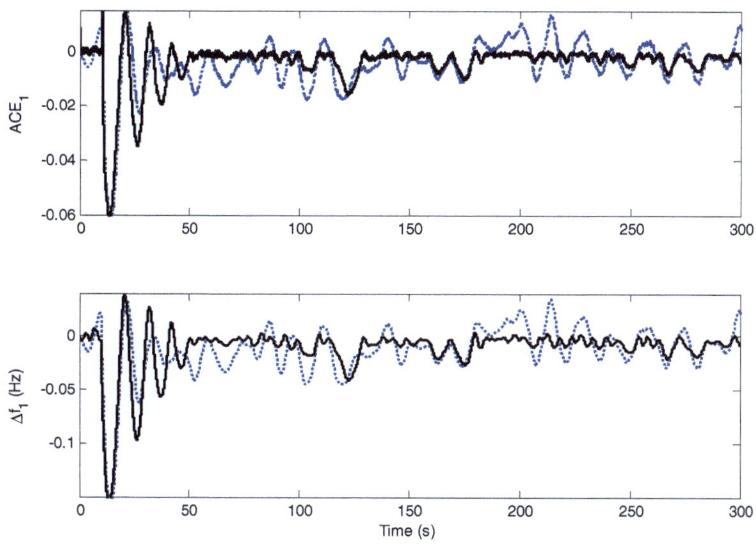

Fig. 12. Area-1 responses; proposed intelligent method (solid), linear PI control (dotted)

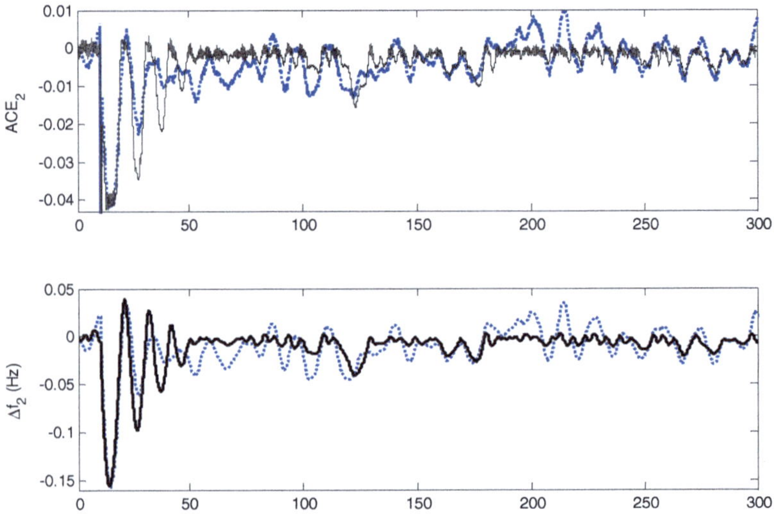

Fig. 13. Area-2 responses; proposed intelligent method (solid), linear PI control (dotted)

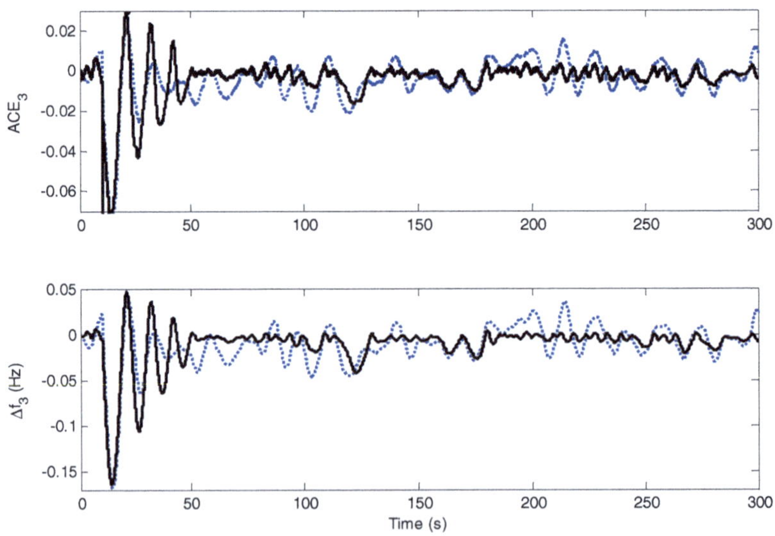

Fig. 14. Area-3 responses; proposed intelligent method (solid), linear PI control (dotted)

7 Summary

This chapter presents an overview of the key issues concerning the integration of WTGs into the power system frequency regulation, that are of most interest today. The most important issues with the recent achievements in this literature are briefly reviewed. The impact of WTGs on frequency control problem is described. An updated LFC model is introduced. Power system frequency response in the presence of WTGs and associated issues is analyzed, and the need for the revising of frequency performance standards is emphasized.

A new method for frequency regulation concerning the integration of wind power units, using MARL has been proposed. The proposed method was applied to a network with the same topology, known as New England 10-generators 39-bus system. The results show that the new algorithm performs well, in comparison of the performance of a PI control design. Two important features of new approach, model independence and flexibility in specifying the control objective; make it very attractive for application in power system operation and control. However, the scalability of MARL to realistic problem sizes is one of the great reasons to use it. In addition to scalability and benefits owing to the distributed nature of the multi-agent solution, such as parallel computation, multiple RL agents may utilize new benefits from sharing experience, e.g., by communication, teaching, or imitation.

References

1. Abbey, C., et al.: Transient Modeling and Comparison of Wind Generator Topologies. In: IPST 2005, p. IPST05 - 131 (2005) (in Canada)
2. Abe, K., Ohba, S., Iwamoto, S.: New load frequency control method suitable for large penetration of wind power generations. In: Power Engineering Society General Meeting (2006)
3. Ahamed, T.P.I.: A neural network based automatic generation controller design through reinforcement learning. International Journal of Emerging Electric Power Systems 6 (2006)
4. Ahamed, T.P.I., Rao, P.S.N., Sastry, P.S.: A reinforcement learning approach to automatic generation control. Electric Power Systems Research 63, 9–26 (2002)
5. Ahamed, T.P.I., Rao, P.S.N., Sastry, P.S.: Reinforcement learning controllers for automatic generation control in power systems having reheat units with GRC and deadband. International journal of power and energy systems 26, 137–146 (2006)
6. Anderson, P.M.: Power System Protection. IEEE/Wiley, New York (1999)
7. Atic, N., Feliachi, A., Rerkpreedapong, D.: CPS1 and CPS2 compliant wedge-shaped model predictive load frequency control. In: Power engineering society general meeting, vol. 1, pp. 855–860. IEEE, Los Alamitos (2004)
8. Banakar, H., Luo, C., Teck Ooi, B.: Impacts of wind power minute-to-minute variations on power system operation. IEEE Transactions on Power Systems 23, 150–160 (2008)

9. Bevrani, H., Hiyama, T.: On load-frequency regulation with time delays: design and real-time implementation. IEEE Transaction on Energy Conversion 24, 292–300 (2009)
10. Bevrani, H., Hiyama, T.: Robust load-frequency regulation: a real-time laboratory experiment. Optimal Control Appl. Methods 28, 419–433 (2007)
11. Bevrani, H.: Robust power system frequency control, 1st edn. Springer, Heidelberg (2009)
12. Busoniu, L., Babuska, R., De Schutter, B.: A comprehensive survey of multi-agent reinforcement learning. IEEE Transaction on Syst., Man., Cyber., Part C: Applications and Reviews 38, 156–172 (2008)
13. Chedid, R.B., Karaki, S.H., El-Chamali, C.: Adaptive fuzzy control for wind-diesel weak power systems. IEEE Transactions on Energy Conversion 15, 71–78 (2000)
14. Demiroren, A., Zeynelgil, H.L., Sengor, N.S.: Automatic generation control for power system with SMES by using neural network controller. Electr. Power Comp System 31, 1–25 (2003)
15. Dokopoulos, P.S., Saramourtsis, A.C., Bakirtzis, A.G.: Prediction and evaluation of the performance of wind-diesel energy systems. IEEE Transactions on Energy Conversion 11, 385–393 (1996)
16. Eftekharnejad, S., Feliachi, A.: Stability enhancement through reinforcement learning: load frequency control case study. Bulk Power System Dynamics and Control VII, 1–8 (2007)
17. Ernst, D., Glavic, M., Wehenkel, L.: Power system stability control: reinforcement learning framework. IEEE Transaction on Power System 19, 427–436 (2004)
18. Gjengedal, T.: System control of large scale wind power by use of automatic generation control (AGC). In: CIGRE/PES, 15–21 (2003)
19. Hall, D.J., Colclaser, R.G.: Transient modeling and simulation of a tubular solid oxide fuel cell. IEEE Transactions on Energy Conversion 14, 749–753 (1999)
20. Hiyama, T., Zuo, D., Funabashi, T.: Multi-agent based automatic generation control of isolated stand alone power system. In: International conference on power system technology (2002)
21. Hiyama, T., Zuo, D., Funabashi, T.: Multi-agent based control and operation of distribution system with dispersed power sources. In: Transmission and Distribution Conference and Exhibition, Asia Pacific. IEEE/PES (2002)
22. Holttinen, H.: Impact of hourly wind power variation on the system operation in the Nordic Countries. Wind Energy 8(2), 197–218 (2005)
23. Horiuchi, N., Kawahito, T., Sizuki, T.: Power control of induction generator by V/F control for wind energy conversion system. Transactions of IEEJ 118-B, 1170–1176 (1998)
24. Karnavas, Y.L., Papadopoulos, D.P.: AGC for autonomous power system using combined intelligent techniques. Electric power systems research 62, 225–239 (2002)
25. Kodama, N., Matsuzaka, T., Inomata, N.: The power variation control of a wind generator by using probabilistic optimal control. Transactions of IEEJ 121-B, 22–30 (2001)
26. Lalor, G., Mullane, A., O'Malley, M.: Frequency control and wind turbine technologies. IEEE Transaction on Power System 20, 1905–1913 (2005)
27. Lalor, G., Ritchie, J., Rourke, S., Flynn, D., O'Malley, M.J.: Dynamic frequency control with increasing wind generation. In: Power Engineering Society General Meeting (2004)
28. Lalor, G., et al.: Frequency control and wind turbine technologies. IEEE Transaction on Power System 20, 1905–1913 (2005)

29. Lindgren, E., Söder, L.: Minimizing regulation costs in multi-area systems with uncertain wind power forecasts. Wind Energy-Wiley Inter-science 11, 97–108 (2007)
30. Lukas, M.D., Lee, K.Y., Ghezel-Ayagh, H.: Development of a stack simulation model for control study on direct reforming molten carbonate fuel cell power plant. IEEE Transactions on Energy Conversion 14, 1651–1657 (1999)
31. Luo Far, C., Banakar, H.G., Pin-Kwan Keung, H., Ooi, B.T.: Estimation of Wind Penetration as Limited by Frequency Deviation. In: Power Engineering Society General Meeting (2006)
32. Morren, J., de Haan, S.W.H., Kling, W.L., Ferreira, J.A.: Primary power/frequency control with wind turbines and fuel cells. In: Power Engineering Society General Meeting (2006)
33. Morren, J., de Haan, S.W.H., Kling, W.L., Ferreira, J.A.: Wind turbines emulating inertia and supporting primary frequency control. IEEE Transactions on Power System 21, 433–434 (2006)
34. Mullane, A.P.: Advanced control of wind energy conversion systems. Ph.D. dissertation, Nat. University of Ireland, Univ. College Cork, Cork, Ireland (2004)
35. Mullane, A., O'Malley, M.: The inertial-response of induction-machine based wind-turbines. IEEE Transaction on Power System 20, 1496–1503 (2005)
36. Sathyajith, M.: Wind Energy Fundamentals, Resource Analysis and Economics, 1st edn., pp. 112–114. Springer, Heidelberg (2006)
37. Senjyu, T., Hayashi, D., Urasaki, N., Funabashi, T.: Oscillation frequency control based on $H\infty$ controller for a small power system using renewable energy facilities in isolated island. In: Power Engineering Society General Meeting (2006)
38. Sutton, R.S., Barto, A.G.: Reinforcement learning: an introduction. MIT Press, Cambridge (1998)
39. Thathachar, M.A.L., Harita, B.R.: An estimator algorithm for learning automata with changing number of actions. International Journal of General Systems 14, 169–184 (1988)
40. The United Nations Framework Convention on Climate Change, The Kyoto Protocol (1997), http://unfccc.int/resource/docs/convkp/kpeng.pdf (accessed June 28, 2008)
41. Vittal, E., et al.: Varying Penetration Ratios of Wind Turbine Technologies for Voltage and Frequency Stability. In: Power and Engineering Society General Meeting, vol. 20, pp. 1–6 (2008)
42. Vlassis, N.: A concise introduction to multi-agent systems and distributed AI. Fac. Sci. Univ. Amsterdam, Amsterdam, The Netherlands, Tech. Rep. (2003)
43. Weiss, G. (ed.): Multi-agent systems: a modern approach to distributed artificial intelligence. MIT Press, Cambridge (1999)
44. Wooldridge, M., Weiss, G. (eds.): Intelligent agents, in multi-agent systems, pp. 3–51. MIT Press, Cambridge (1999)
45. Du, X., Li, P.: Fuzzy logic control optimal realization using GA for multi-area AGC systems. International Journal of Information Technology 12, 63–72 (2006)
46. Gu, Y.: Multi-agent reinforcement learning for multi-robot systems: a survey. Technical Report, CSM-404 (2004)

Appendix

39-Bus Test System Parameters

Generators

All generators are conventional thermal type and are treated as the sixth order machine model. Parameters for two axis model of the synchronous machine are shown in Table 1. For each generator, all values are given in per unit on its base values.

PSS

The power system stabilizer (PSS) is modeled as shown in Fig. 15. All generators are equipped with identical PSSs. PSS parameters are shown in Table 2.

Table 1. Generator parameters

Unit no.	H [s]	x_d'' [pu]	x_d' [pu]	x_d [pu]	x_q'' [pu]	x_q' [pu]	x_q [pu]	x_l [pu]	R_a [pu]	T_{do}'' [s]	T_{do}' [s]	T_{qo}'' [s]	T_{qo}' [s]	T_d'' [s]	T_d' [s]	T_q'' [s]	T_q' [s]	P_n [MW]	V_n [kv]
1	4.768	0.134	0.174	1.220	0.134	0.250	1.160	0.078	0.004	0.033	8.970	0.070	0.500	0.023	1.280	0.023	0.640	120	15.5
2	4.768	0.134	0.174	1.220	0.134	0.250	1.160	0.078	0.004	0.033	8.970	0.070	0.500	0.023	1.280	0.023	0.640	120	15.5
3	6.1867	0.130	0.185	1.050	0.130	0.850	1.240	0.070	0.0031	0.038	6.100	0.070	0.500	---	---	0.023	0.640	75	13.8
4	6.1867	0.130	0.185	1.050	0.130	0.360	0.980	0.070	0.0031	0.038	6.100	0.070	0.500	---	---	0.023	0.640	73.2	13.8
5	6.1867	0.130	0.185	1.050	0.130	0.850	1.240	0.070	0.0031	0.038	6.100	0.070	0.500	---	0.882	0.023	0.640	64	13.8
6	6.1867	0.130	0.185	1.050	0.130	0.360	0.980	0.070	0.0031	0.038	6.100	0.070	0.500	---	---	0.023	0.640	75	13.8
7	6.1867	0.130	0.185	1.050	0.130	0.850	1.240	0.070	0.0031	0.038	6.100	0.070	0.500	---	0.882	0.023	0.640	66	13.8
8	6.1867	0.130	0.185	1.050	0.130	0.850	1.240	0.070	0.0031	0.038	6.100	0.070	0.500	---	0.882	0.023	0.640	64	13.8
9	4.768	0.134	0.174	1.220	0.134	0.250	1.160	0.078	0.004	0.033	8.970	0.070	0.500	0.023	1.280	0.023	0.640	120	15.5
10	6.1867	0.130	0.185	1.050	0.130	0.850	1.240	0.070	0.0031	0.038	6.100	0.070	0.500	---	0.882	0.023	0.640	64	13.8

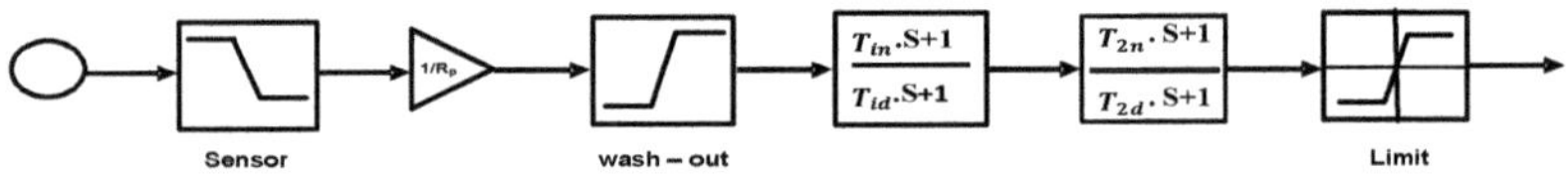

Fig. 15. PSS block diagram

Table 2. PSS parameters

Senor Time constant	K	Wash-out Time constant	T_{1n}	T_{1d}	T_{2n}	T_{1d}	V_{smin}	V_{smax}
0.015	3.125	1	0.060	1	0	0	-0.15	0.15

Governor

The turbine-governors in the test system are modeled as shown in Fig.16, and the related parameters are shown in Table 3.

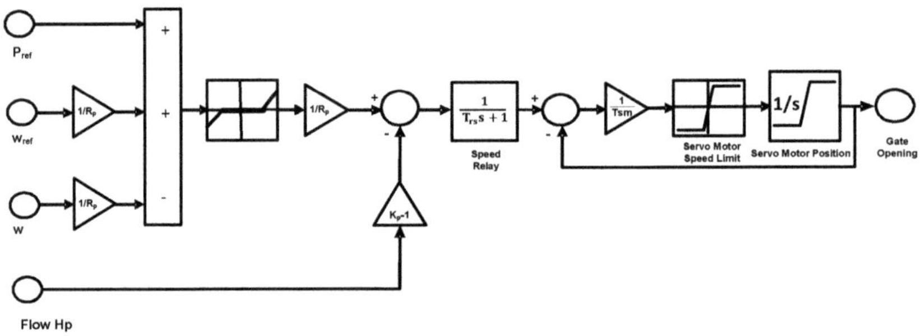

Fig. 16. Turbine-governor block diagram

Table 3. Governor Parameters

k_p	R_p[pu]	DZ[pu]	T_{sr}[s]	T_{sm}[s]	V_{gmin}	V_{gmax}	g_{min}[pu]	g_{max}[pu]
1	0.05	0	0.001	0.15	-0.1	0.1	0	1

Loads

Loads are treated as constant impedance loads and shown in Table 4. The active and reactive powers absorbed by the loads are proportional to the square of the system voltage.

Table 4. Load values

Bus	P[MW]	Q[MVar]
1	0	0
2	0	0
3	21.47	0.24
4	50	18.4
5	0	0
6	0	0
7	23.38	8.4
8	52.2	17.66
9	0	0
10	0	0
11	0	0
12	0.75	8.8
13	0	0
14	0	0
15	32	15.3
16	32.9	3.23
17	0	0
18	10.533	2
19	0	0
20	62.8	10.3
21	27.4	11.5
22	0	0
23	24.75	8.46
24	15.43	9.22
25	11.2	2.36
26	13.9	1.7
27	18.73	5.03
28	20.6	2.76
29	28.35	2.69
31	0.92	0.46
39	138	31.52

Line/Transformers

The line parameters of the test system are shown in the Table 5. For each transformer, all values are given in per unit (on its MVA base), and are given in Table 6.

Table 5. Line parameters

		Line Data			
From bus	To bus	R [ohm/km]	L [H/km]	C [F/km]	Length
1	2	35e-4	1.0902 e-4	18.534 e-8	10
1	39	1e-3	6.6315 e-5	19.894 e-8	10
2	3	3.25e-3	10.0135 e-5	17.056 e-8	4
2	25	28e-3	9.1248 e-5	15.4912 e-8	2.5
3	4	2.6e-3	11.3e-5	11.7456 e-8	5
3	18	2.75 e-3	8.8197 e-5	14.178 e-8	4
4	5	26.667 e-4	11.3176 e-5	11.866 e-8	3
4	14	26.667 e-4	11.406 e-5	12.2197 e-8	3
5	6	40 e-4	13.7934 e-5	23.024 e-8	0.5
5	8	26.667 e-4	9.903 e-5	13.0506 e-8	3
6	7	24 e-4	9.7616 e-5	11.9896 e-8	2.5
6	11	28 e-4	8.7004 e-5	14.7376 e-8	2.5
7	8	26.667 e-4	8.1346 e-5	13.7933 e-8	1.5
8	9	2.5555 e-3	10.6988 e-5	11.2111 e-8	9
9	39	1 e-3	6.6315 e-5	31.831 e-8	10
10	11	40 e-4	11.406 e-5	19.337 e-8	1
10	13	40 e-4	11.406 e-5	19.337 e-8	1
13	14	30 e-4	8.9303 e-5	15.2347 e-8	3
14	15	2.7692 e-3	8.8555 e-5	14.9361 e-8	6.5
15	16	30 e-4	8.3113 e-5	15.1197 e-8	3
16	17	28 e-4	9.4432 e-5	14.2392 e-8	2.5
16	19	3.2 e-3	10.345 e-5	16.1278 e-8	5
16	21	20 e-4	8.9525 e-5	16.897 e-8	4
16	24	20 e-4	10.4333 e-5	12.0253 e-8	1.5
17	18	28 e-4	8.7004 e-5	13.9952 e-8	2.5
17	27	2.6 e-3	9.178 e-5	17.0614 e-8	5
21	22	17.777 e-4	8.2524 e-5	15.1198 e-8	4.5
22	23	20 e-4	8.4883 e-5	16.3223 e-8	3
23	24	2.75 e-3	11.605 e-5	11.9697 e-8	8
25	26	3.5555 e-3	9.5198 e-5	15.12 e-8	9
26	27	3.5 e-3	9.7482 e-5	15.889 e-8	4
26	28	3.5833 e-3	10.4775 e-5	17.2458 e-8	12
26	29	3.3529 e-3	9.7523 e-5	16.0558 e-8	17
28	29	3.5 e-3	10.0135 e-5	16.5122 e-8	4

Table 6. Transformer parameters

From bus	To bus	R [ohm/km]	L [H/km]	C [F/km]	S_n [MVar]	Tap Data Magnitude [pu]	Tap Data Angle [degree]
12	11	0.08e-3	2.175 e-3	0	50	1	0
12	13	0.08e-3	2.175 e-3	0	50	1	0
6	31	0	6.250 e-3	0	250	1	0
10	32	0	5.000 e-3	0	250	1	0
19	33	0.175 e-3	3.550 e-3	0	250	1	0
20	34	0.225 e-3	4.500 e-3	0	250	1	0
22	35	0	3.575 e-3	0	250	1	0
23	36	0.125 e-3	6.800 e-3	0	250	1	0
25	37	0.150 e-3	5.800 e-3	0	250	1	0
2	30	0	4.525 e-3	0	250	1	0
29	38	0.2 e-3	3.900 e-3	0	250	1	0
19	20	0.175 e-3	3.450 e-3	0	250	1	0

Author Index

MIX
Papier aus verantwortungsvollen Quellen
Paper from responsible sources
FSC® C105338

If you have any concerns about our products,
you can contact us on
ProductSafety@springernature.com

In case Publisher is established outside the EU,
the EU authorized representative is:
Springer Nature Customer Service Center GmbH
Europaplatz 3, 69115 Heidelberg, Germany

Printed by Libri Plureos GmbH
in Hamburg, Germany